CW00857815

Geomechanics and Geology

Geological Society books refereeing procedures

The Society makes every effort to ensure that the scientific and production quality of its books matches that of its journals. Since 1997, all book proposals have been refereed by specialist reviewers as well as by the Society's Books Editorial Committee. If the referees identify weaknesses in the proposal, these must be addressed before the proposal is accepted.

Once the book is accepted, the Society Book Editors ensure that the volume editors follow strict guidelines on refereeing and quality control. We insist that individual papers can only be accepted after satisfactory review by two independent referees. The questions on the review forms are similar to those for *Journal of the Geological Society*. The referees' forms and comments must be available to the Society's Book Editors on request.

Although many of the books result from meetings, the editors are expected to commission papers that were not presented at the meeting to ensure that the book provides a balanced coverage of the subject. Being accepted for presentation at the meeting does not guarantee inclusion in the book.

More information about submitting a proposal and producing a book for the Society can be found on its website: www.geolsoc.org.uk.

It is recommended that reference to all or part of this book should be made in one of the following ways:

TURNER, J. P., HEALY, D., HILLIS, R. R. & WELCH, M. J. (eds) 2017. *Geomechanics and Geology*. Geological Society, London, Special Publications, **458**.

GULMAMMADOV, R., COVEY-CRUMP, S. & HUUSE, M. 2017. Geomechanical characterization of mud volcanoes using P-wave velocity datasets. *In*: TURNER, J. P., HEALY, D., HILLIS, R. R. & WELCH, M. J. (eds) *Geomechanics and Geology*. Geological Society, London, Special Publications, **458**, 273–292. First published online May 24, 2017, updated version published online June 9, 2017, https://doi.org/10.1144/SP458.2

GEOLOGICAL SOCIETY SPECIAL PUBLICATION NO. 458

Geomechanics and Geology

EDITED BY

J. P. TURNER
Radioactive Waste Management Ltd, UK

D. HEALY
University of Aberdeen, UK

R. R. HILLIS
Deep Exploration Technologies Co-operative Research Centre, Australia

and

M. J. WELCH
Technical University of Denmark, Denmark

2017
Published by
The Geological Society
London

THE GEOLOGICAL SOCIETY

The Geological Society of London (GSL) was founded in 1807. It is the oldest national geological society in the world and the largest in Europe. It was incorporated under Royal Charter in 1825 and is Registered Charity 210161.

The Society is the UK national learned and professional society for geology with a worldwide Fellowship (FGS) of over 10 000. The Society has the power to confer Chartered status on suitably qualified Fellows, and about 2000 of the Fellowship carry the title (CGeol). Chartered Geologists may also obtain the equivalent European title, European Geologist (EurGeol). One fifth of the Society's fellowship resides outside the UK. To find out more about the Society, log on to www.geolsoc.org.uk.

The Geological Society Publishing House (Bath, UK) produces the Society's international journals and books, and acts as European distributor for selected publications of the American Association of Petroleum Geologists (AAPG), the Indonesian Petroleum Association (IPA), the Geological Society of America (GSA), the Society for Sedimentary Geology (SEPM) and the Geologists' Association (GA). Joint marketing agreements ensure that GSL Fellows may purchase these societies' publications at a discount. The Society's online bookshop (accessible from www.geolsoc.org.uk) offers secure book purchasing with your credit or debit card.

To find out about joining the Society and benefiting from substantial discounts on publications of GSL and other societies worldwide, consult www.geolsoc.org.uk, or contact the Fellowship Department at: The Geological Society, Burlington House, Piccadilly, London W1J 0BG: Tel. +44 (0)20 7434 9944; Fax +44 (0)20 7439 8975; E-mail: enquiries@geolsoc.org.uk.

For information about the Society's meetings, consult *Events* on www.geolsoc.org.uk. To find out more about the Society's Corporate Affiliates Scheme, write to enquiries@geolsoc.org.uk.

Published by The Geological Society from:
The Geological Society Publishing House, Unit 7, Brassmill Enterprise Centre, Brassmill Lane, Bath BA1 3JN, UK

The Lyell Collection: www.lyellcollection.org
Online bookshop: www.geolsoc.org.uk/bookshop
Orders: Tel. +44 (0)1225 445046, Fax +44 (0)1225 442836

British Library Cataloguing in Publication Data

A catalogue record for this book is available from the British Library.
ISBN 978-1-78620-320-5
ISSN 0305-8719

Distributors
For details of international agents and distributors see:
www.geolsoc.org.uk/agentsdistributors

Typeset by Nova Techset Private Limited, Bengaluru & Chennai, India
Printed and bound by CPI Group (UK) Ltd, Croydon CR0 4YY

Contents

Acknowledgements

The conference that led to this volume was jointly organized by the Petroleum Group and Tectonic Studies Group.

The Geological Society thanks the companies below for their generous sponsorship of the conference. The Petroleum Group is thanked for its contribution towards the colour printing in this volume.

Petroleum Group corporate sponsors at the time of the conference:

Sponsors of The Geology of Geomechanics conference:

Geomechanics and geology: introduction

JONATHAN P. TURNER[1]*, DAVE HEALY[2], RICHARD R. HILLIS[3] & MICHAEL J. WELCH[4]

[1]*Radioactive Waste Management Ltd, Building 587, Curie Avenue, Harwell, Oxfordshire OX11 0RH, UK*

[2]*University of Aberdeen, School of Geosciences, King's College, Aberdeen AB24 3UE, UK*

[3]*Deep Exploration Technologies Co-operative Research Centre, PO Box 66, Export Park, Adelaide Airport, SA 5950, Australia*

[4]*Technical University of Denmark, Centre for Oil and Gas, Danish Hydrocarbon Research and Technology Centre, Elektrovej Building 375, 2800 Kongens Lyngby, Denmark*

**Correspondence: jonathan.paul.turner@gmail.com*

Geomechanics investigates the origin, magnitude and deformational consequences of stresses in the crust. Perhaps the earliest description of geology and mechanics was from the sandbox experiments of Willis (1891), and many of the guiding principles were developed by Anderson (1951), Hubbert & Willis (1957), Jaeger & Cook (1979) and Engelder (1992), with input from engineering disciplines (e.g. Griffith 1921). Subsequently, geomechanics has grown such that it now constitutes an important subdiscipline within the geosciences, as witnessed by the increase in SPE papers with 'geomechanics' in their titles (**Addis 2017**). In recent years, awareness of geomechanical processes has been heightened by societal debates on fracking, human-induced seismicity, natural geohazards and safety issues with respect to petroleum exploration drilling, carbon sequestration and radioactive waste disposal.

This volume includes a selection of the papers presented at the October 2015 meeting 'Geomechanics and Geology' held at the Geological Society, sponsored by the Petroleum Group and Tectonic Studies Group. The meeting was convened to explore the common ground linking geomechanics with *inter alia* economic and petroleum geology, structural geology, petrophysics, seismology, geotechnics, reservoir engineering, and production technology. A rich diversity of case studies showcased applications of geomechanics to hydrocarbon exploration and field development, natural and artificial geohazards, reservoir stimulation, contemporary tectonics, and subsurface fluid flow. This introduction selects some of the highlights from the meeting and identifies common themes from papers contained in the present volume and/or presented at the meeting.

What do we understand by geostresses? Couples (2015) observed that concepts of stress are essentially a normalization of forces that work well in homogeneous bodies. But rocks are fundamentally heterogeneous, and stress transmission within them often does not conform to continuum mechanics. A good analogy is photoelastic analysis of beads that show how stress is transmitted in granular materials through load-bearing 'force chains' surrounded by relatively unloaded zones. Couples (2015) suggests crustal stresses are best thought of, alternatively, in terms of elastic energy within rocks.

Stress azimuth can vary from uniformity over very large areas to pronounced swings over distances of a few metres. Inherent heterogeneity of large-scale geosystems is demonstrated by the degree of variation of stress azimuths shown by the *World Stress Map* (Heidbach *et al.* 2016), a compilation of maximum horizontal stress measurements from >6000 wells in >100 basins worldwide (Tingay 2015) and an excellent example of industry–academic collaboration. Tingay concludes that stress measured at any one point is the net result of all forces combining to act on it, from the plate-scale to the local-scale. Main processes controlling horizontal stress are:

- 'far-field' plate tectonic forces generated at forearcs, retroarcs, rifts, ocean ridges, passive margins, cratons, etc.;
- intraplate stress sources: for example, plumes;
- different types of sedimentary basin: for example, compare horizontal stress azimuth in rifts and foredeeps;
- isostasy and topographical body forces, particularly regions of only partially compensated positive and negative 'dynamic topography';

From: Turner, J. P., Healy, D., Hillis, R. R. & Welch, M. J. (eds) 2017. *Geomechanics and Geology*. Geological Society, London, Special Publications, **458**, 1–5.
First published online July 17, 2017, https://doi.org/10.1144/SP458.15

- deglaciation;
- detachment zones: for example, isolation from far-field stresses of supra-detachment sequences in modern deltas;
- geological structures on various scales: for example, stress refraction around major faults and bending stresses within folds;
- mechanical stratigraphy: for example, vertical changes in stress gradient due to changes in elastic properties and focusing of higher magnitude stresses in lenses of stronger rocks within ductile shear zones

Tassone *et al.* (2017) provide an example from SE Australia of contradictory evidence for the state of contemporary stress from 'local' measured stress v. that inferred from plate boundary forces and Recent structures. Neotectonic deformation is dominated by thrust faulting and related folding, consistent with New Zealand plate collision, whilst leak-off tests from the Otway and Gippsland basins indicate strike-slip or normal faulting regimes. They attribute the difference to: (i) depth-controlled differences in mechanical stratigraphy; (ii) compartmentalization of stress according to whether neotectonic stress is accommodated by folding or faulting; and (iii) underestimating horizontal stress magnitude due to assuming that leak-off pressures are accommodated only by tensile failure.

Friction and faulting is investigated by **Fetter *et al.* (2017)** and **Richardson & Seedorff (2017)**. **Fetter *et al.* (2017)** describe Recent large-displacement, low-angle normal faults that offset the seafloor in the highly stretched and thinned crust of the Santos Basin, offshore Rio de Janeiro to investigate the influence of detachment zones on stress orientation. Underlying salt-cored listric faults are shown to have caused local rotation of the stress field such that none of the principal stresses are vertical. The Andersonian model that predicts fault type (i.e. steep normal faults, low-angle thrusts, vertical strike-slip) therefore no longer applies, allowing markedly 'non-Andersonian' fault angles to develop. They compare these structures with similarly active low-angle faults in California and Nevada, USA.

A good example of how mechanical stratigraphy controls stress patterns is provided by *in situ* stress measurements from the coal-bearing Bowen Basin, Queensland, Australia (**Tavener *et al.* 2017**). Regional stress is controlled by interplays between far-field plate boundary processes and more local basin-controlling structures. But at reservoir-scale, the stress state is highly variable laterally and vertically, changing from shallow (<600 m) thrust regime to deeper strike-slip. They attribute this stress complexity to the mechanical stratigraphy, particularly the low Young's modulus and Poisson's ratio of coals relative to their encasing clastics, meaning that coals are most highly stressed in the shallower thrust regime and vice versa at depth in the strike-slip regime. This observation is a powerful tool for predicting how reservoir sequences respond to fracture stimulation – the coals being easier to stimulate in strike-slip settings with fractures better confined to the coals, and vice versa.

Mechanical stratigraphy and the processes controlling it was the subject of a novel study of lava flows by Bubeck (2015). She observed from CT scans that vesicles in lavas become increasingly ellipsoidal towards the bases of the flows, their long axes orientated horizontally. This is attributed to progressive distortion of the vesicles with burial and loading. In the same way as the 'pointy' ends of an egg have relatively higher compressive strength, the lower parts of lava flows containing the most ellipsoidal vesicles are weak under vertical loading but much stronger in horizontal compression. This recognition of significant mechanical stratigraphy within lava flows has important implications for understanding volcano stability.

Several contributions showed how the influence of rock fabric on geomechancial behaviour can lead to phenomena that appear to deviate from well-established norms. Hackston (2015) compared frictional behaviour of mechanically contrasting sandstones using triaxial experimental apparatus. They found that: (i) failure angle in compression was always smaller than in extension, suggesting either stress refraction and/or the influence of microfractures (so-called Griffith cracks); and (ii) deviation of failure angle from classical Mohr–Coulomb theory, suggesting the active role of the intermediate stress σ_2. **Descamps *et al.* (2017)** examined the control that texture and diagenesis exert on geomechanical properties in chalk. They show that clay in argillaceous chalks increases rock strength because it promotes greater compaction and earlier diagenesis.

It is noteworthy that of the 30 papers presented at the meeting, 17 dealt substantially with the role of geofluids in facilitating rock deformation. Like stress, geofluids are a phenomenon that cannot usually be observed in action directly, but it is clear that understanding the impact they have on geomechanical processes is fundamental. We assert that almost no macro-scale brittle deformation in the upper crust takes place in the absence of elevated pore pressures because deviatoric stresses are not high enough to overcome frictional sliding resistance. Mechanisms that generate overpressure include compressional inversion (analogous to liberating porewater by wringing a sponge), exhumation (e.g. tensile failure linked to gas generation at peak burial: **English & Laubach 2017**; **English *et al.* 2017**), deglaciation (due to isothermal decompression), disequilibrium compaction (i.e. rapid burial leading to partial dewatering), metamorphic dehydration reactions and maturation of organic matter (especially volume expansion associated with conversion of kerogens to liquids and, with deeper burial, cracking of liquids to gases).

Several papers presented at the meeting examined relationships between pore pressures and deformation (**Gulmammadov *et al.* 2017**; **Lahann & Swarbrick 2017**; **Roberts *et al.* 2017**; **Sibson 2017**). For example, **Sibson (2017)** observes that megathrust earthquakes appear only to occur where pore pressure/lithostatic pressure $\geq$0.9 and infers that much of the seismogenic crust is critically stressed (i.e. on the verge of failure). This assertion is based mainly on the accumulation of evidence for fluid-driven failure from earthquakes generated by fluid injection down boreholes (e.g. Oklahoma: Keranen *et al.* 2013), and from reservoir-induced seismicity in various fields such as the Groningen gas field in The Netherlands (Grasso 1992).

Collettini (2015) investigated why earthquakes often nucleate in carbonates (e.g. Zagros, Italy, Oklahoma). Many limestones exhibit high permeabilities and it is therefore more difficult to maintain significant pore fluid overpressure. This is important because overpressures promote stable sliding, thus slowly dissipating elastic strains otherwise manifested by seismicity. Experiments reported by Collettini (2015) used a reshearing stage to simulate realistic crustal deformation in which fluid flow is induced under horizontal and vertical loads. Their results demonstrate that the tendency to hydrostatic pore pressures in carbonates leads to more stick–slip behaviour and thus a greater propensity for earthquake activity.

Geological observations and measurements made in the field and the laboratory have profound implications for our understanding of stress systems and their impact on rock deformation (e.g. **Gillespie & Kampfer 2017**). A particularly good example are the horizontal hydrofractures depicted in the cover photograph of this volume. Hydrofractures comprise mineralized fractures that open in response to high pore pressure and/or high differential stress. Zanella (2015) used modelling and worldwide examples to discuss physical conditions for the development of a type of hydrofracture termed 'beef'. By virtue of their horizontal attitude, the presence of beef indicates conditions in which, at least locally, fluid pressures exceed vertical (lithostatic) stress. Given that beef is usually confined to organic-rich shales, an intriguing possibility is that the origin of high pore pressure is the conversion of kerogens to hydrocarbons, leading to the possibility that the presence of beef can be used as a proxy for source rock maturation.

Hydrofracturing is a critical element of the fault-valve model first hypothesized by Sibson *et al.* (1988), also discussed by Myhill (2015). Meredith (2015) used experimental data to address an important implication of fault valving: that veins are critical to the re-sealing of pressure cells, thereby enabling them to build up to the next overpressure cycle. His data suggest that whilst sealing requires only for crack aperture to reduce, and thus occurs fairly rapidly, the process of crystal nucleation and growth on the fracture wall (healing) is slow – a 0.3 μm fracture aperture taking some 100 h to heal.

Application of geomechanics to oil and gas field developments has become increasingly important over the past 40 years, and geomechanics specialists are commonly recruited as permanent members of asset teams in larger development projects. Advances in the characterization and modelling of fractured petroleum reservoirs was a major theme that included case studies from the North Sea (e.g. Freeman 2015; **Wynn *et al.* 2017**) and from reservoir analogues in the Pyrenees (Gutmanis 2015). Another recurring theme of papers from the oil and gas industry was the impact that geomechanics understanding can have on planning wells. Batchelor (2015) examined how the complex relationship between geology and geomechanics presents challenging drilling conditions in the Eocene formations of the UK Central North Sea. The area is characterized by very weak stratigraphy (e.g. sand-in-sand injectites, semi-plastic mudrocks) in which the mud weight required to maintain wellbore stability often exceeds the fracture gradient.

Addis (2017) uses multiple case studies to demonstrate how stress fields may be complex, adjusting to changes in reservoir pressure over time (e.g. Brent Field, North Sea) and varying according to local contrasts in mechanical stratigraphy. For example, the Cusiana Field, Colombia is situated in an active thrust belt in the northernmost Andes and presented significant drilling challenges during development. *In situ* measurements indicated that vertical stress was much higher than would be predicted from Andersonian dynamics. Subsequent modelling revealed a highly compartmentalized stress system in which relatively strong reservoir sandstones acted as 'stress guides', refracting the minimum stress to a horizontal attitude. As a consequence of this greater understanding, the delivery of safer and more stable wells led directly to significant improvements in the performance of the field.

Geomechanics is a rapidly developing field that brings together a broad range of subsurface professionals seeking to use their expertise to solve current challenges in applied and fundamental geoscience. This introduction provides a flavour of the diversity and ingenuity of many of the contributions presented at the Geomechanics and Geology meeting, and hopefully encourages you to delve further into the volume. We hope that the papers herein provide a representative snapshot of the exciting state of geomechanics and establish it firmly as a flourishing subdiscipline of geology that merits broadest exposure across the academic and corporate geosciences.

We are grateful to all the poster presenters and speakers for contributing to a successful meeting. The efforts of authors and reviewers of papers contained herein led directly to this excellent volume and a worthy addition to the Geological Society's unrivalled set of special publications. We thank Tamzin Anderson, Jo Armstrong and Angharad Hills at the Geological Society Publishing House for helping to bring it to fruition. The staff of the conference office in Burlington House assisted us greatly in organizing the meeting itself. Mark Tingay is thanked for making his conference notes available. The meeting was sponsored generously by AGR, Badley Earth Sciences, Tracs, Tectonic Studies Group and The Petroleum Group.

References

ADDIS, M.A. 2017. The geology of geomechanics: petroleum geomechanical engineering in field development planning. *In*: TURNER, J.P., HEALY, D., HILLIS, R.R. & WELCH, M.J. (eds) *Geomechanics and Geology*. Geological Society, London, Special Publications, **458**. First published online June 28, 2017, https://doi.org/10.1144/SP458.7

ANDERSON, E.M. 1951. *The Dynamics of Faulting and Dyke Formation with Applications to Britain*. 2nd edn. Oliver & Boyd, Edinburgh.

BATCHELOR, T. 2015. Case studies of the complex relationship of geology and geomechanics in the Eocen formations of Quad 9 and Quad 15 based on 30 years' experience. Abstract presented at the Geology of Geomechanics Conference, 28–29 October 2015, Geological Society, London.

BUBECK, A. 2015. ¿Como se lava? How representative are 'typical lavas' in volcano stability models? Abstract presented at the Geology of Geomechanics Conference, 28–29 October 2015, Geological Society, London.

COLLETTINI, C. 2015. The role of fluid pressure in frictional stability and earthquake triggering: Insights from rock deformation experiments. Abstract presented at the Geology of Geomechanics Conference, 28–29 October 2015, Geological Society, London.

COUPLES, G. 2015. Some stressful realisations about the concept of stress in geomaterials. Abstract presented at the Geology of Geomechanics Conference, 28–29 October 2015, Geological Society, London.

DESCAMPS, F., FAŸ-GOMORD, O., VANDYCKE, S., SCHROEDER, C., SWENNEN, R. & TSHIBANGU, J.-P. 2017. Relationships between geomechanical properties and lithotypes in NW European chalks. *In*: TURNER, J.P., HEALY, D., HILLIS, R.R. & WELCH, M.J. (eds) *Geomechanics and Geology*. Geological Society, London, Special Publications, **458**. First published online May 25, 2017, https://doi.org/10.1144/SP458.9

ENGELDER, T. 1992. *Stress Regimes in the Lithosphere*. Princeton University Press, Princeton, NJ.

ENGLISH, J.M. & LAUBACH, S.E. 2017. Opening-mode fracture systems: insights from recent fluid inclusion microthermometry studies of crack-seal fracture cements. *In*: TURNER, J.P., HEALY, D., HILLIS, R.R. & WELCH, M.J. (eds) *Geomechanics and Geology*. Geological Society, London, Special Publications, **458**. First published online May 24 2017, https://doi.org/10.1144/SP458.1

ENGLISH, J.M., FINKBEINER, T., ENGLISH, K.L. & YAHIA CHERIF, R. 2017. State of stress in exhumed basins and implications for fluid flow: insights from the Illizi Basin, Algeria. *In*: TURNER, J.P., HEALY, D., HILLIS, R.R. & WELCH, M.J. (eds) *Geomechanics and Geology*. Geological Society, London, Special Publications, **458**. First published online May 30, 2017, https://doi.org/10.1144/SP458.6

FETTER, M., MORAES, A. & MULLER, A. 2017. Active low-angle normal faults in the deep water Santos Basin, offshore Brazil: a geomechanical analogy between salt tectonics and crustal deformation. *In*: TURNER, J.P., HEALY, D., HILLIS, R.R. & WELCH, M.J. (eds) *Geomechanics and Geology*. Geological Society, London, Special Publications, **458**. First published online May 26, 2017, https://doi.org/10.1144/SP458.11

FREEMAN, B. 2015. Predicting sub-seismic fracture density and orientation: A case study from the Gorm Field, Danish North Sea. Abstract presented at the Geology of Geomechanics Conference, 28–29 October 2015, Geological Society, London.

GILLESPIE, P. & KAMPFER, G. 2017. Mechanical constraints on kink band and thrust development in the Appalachian Plateau, USA. *In*: TURNER, J.P., HEALY, D., HILLIS, R.R. & WELCH, M.J. (eds) *Geomechanics and Geology*. Geological Society, London, Special Publications, **458**. First published online June 12, 2017, https://doi.org/10.1144/SP458.12

GRASSO, J.-R. 1992. Mechanics of seismic instabilities induced by the recovery of hydrocarbons. *Pure and Applied Geophysics*, **139**, 507–534.

GRIFFITH, A.A. 1921. The phenomena of rupture and flow in solids. *Philosophical Transactions of the Royal Society*, **A221**, 163–197.

GULMAMMADOV, R., COVEY-CRUMP, S. & HUUSE, M. 2017. Geomechanical characterization of mud volcanoes using P-wave velocity datasets. *In*: TURNER, J.P., HEALY, D., HILLIS, R.R. & WELCH, M.J. (eds) *Geomechanics and Geology*. Geological Society, London, Special Publications, **458**. First published online May 25, 2017, updated version published online June 9, 2017, https://doi.org/10.1144/SP458.2

GUTMANIS, J. 2015. Reservoir characterization by integration of outcrop analog with in situ stress profiling of a fractured carbonate reservoir. Abstract presented at the Geology of Geomechanics Conference, 28–29 October 2015, Geological Society, London.

HACKSTON, A. 2015. Faulting and friction of sandstones. Abstract presented at the Geology of Geomechanics Conference, 28–29 October 2015, Geological Society, London.

HEIDBACH, O., RAJABI, M., REITER, K., ZIEGLER, M. & WSM TEAM 2016. *World Stress Map Database*

Release. GFZ Data Services, https://doi.org/10.5880/WSM.2016.001

Hubbert, M.K. & Willis, D.G. 1957. Mechanics of hydraulic fracturing. *Transactions of Society of Petroleum Engineers of AIME*, **210**, 153–168.

Jaeger, J. & Cook, N.G.W. 1979. *Fundamental of Rock Mechanics*. 3rd edn. Chapman & Hall, London.

Keranen, K.M., Savage, H.M., Abers, G.A. & Cochran, E.S. 2013. Potentially induced earthquakes in Oklahoma, USA: links between wastewater injection and the 2011 Mw 5.7 earthquake sequence. *Geology*, **41**, 699–702.

Lahann, R.W. & Swarbrick, R.E. 2017. An improved procedure for pre-drill calculation of fracture pressure. *In*: Turner, J.P., Healy, D., Hillis, R.R. & Welch, M.J. (eds) *Geomechanics and Geology*. Geological Society, London, Special Publications, **458**. First published online May 30, 2017, https://doi.org/10.1144/SP458.13

Meredith, P.G. 2015. Strength recovery and vein growth during self-sealing of faults in Westerly granite. Abstract presented at the Geology of Geomechanics Conference, 28–29 October 2015, Geological Society, London.

Myhill, D. 2015. Clumped isotope thermometry: A tool to further detail fluid processes in fault zones, a view from the South Pennines Orefield, Peak District, UK. Abstract presented at the Geology of Geomechanics Conference, 28–29 October 2015, Geological Society, London.

Richardson, C.A. & Seedorff, E. 2017. Estimating friction in normal fault systems of the Basin and Range province and examining its geological context. *In*: Turner, J.P., Healy, D., Hillis, R.R. & Welch, M.J. (eds) *Geomechanics and Geology*. Geological Society, London, Special Publications, **458**. First published online May 25, 2017, https://doi.org/10.1144/SP458.8

Roberts, J.J., Wilkinson, M., Naylor, M., Shipton, Z.K., Wood, R.A. & Haszeldine, R.S. 2017. Natural CO_2 sites in Italy show the importance of overburden geopressure, fractures and faults for CO_2 storage performance and risk management. *In*: Turner, J.P., Healy, D., Hillis, R.R. & Welch, M.J. (eds) *Geomechanics and Geology*. Geological Society, London, Special Publications, **458**. First published online June 19, 2017, updated version published online June 23, 2017, https://doi.org/10.1144/SP458.14

Sibson, R.H. 2017. The edge of failure: critical stress overpressure states in different tectonic regimes. *In*: Turner, J.P., Healy, D., Hillis, R.R. & Welch, M.J. (eds) *Geomechanics and Geology*. Geological Society, London, Special Publications, **458**. First published online May 24, 2017, https://doi.org/10.1144/SP458.5

Sibson, R.H., Robert, F. & Poulson, K.H. 1988. High-angle reverse faults, fluid-pressure cycling, and mesothermal gold-quartz deposits. *Geology*, **16**, 551–555.

Tassone, D.R., Holford, S.P., King, R., Tingay, M.R.P. & Hillis, R.R. 2017. Contemporary stress and neotectonics in the Otway Basin, southeastern Australia. *In*: Turner, J.P., Healy, D., Hillis, R.R. & Welch, M.J. (eds) *Geomechanics and Geology*. Geological Society, London, Special Publications, **458**. First published online May 25, 2017, https://doi.org/10.1144/SP458.10

Tavener, E., Flottmann, T. & Brooke-Barnett, S. 2017. *In situ* stress distribution and mechanical stratigraphy in the Bowen and Surat basins, Queensland, Australia. *In*: Turner, J.P., Healy, D., Hillis, R.R. & Welch, M.J. (eds) *Geomechanics and Geology*. Geological Society, London, Special Publications, **458**. First published online May 24, 2017, https://doi.org/10.1144/SP458.4

Tingay, M. 2015. The present-day stress field in sedimentary basins. Abstract presented at the Geology of Geomechanics Conference, 28–29 October 2015, Geological Society, London.

Willis, B. 1891. *The Mechanics of Appalachian Structures*. United States Government Printing Office, Washington, DC.

Wynn, T.J., Kumar, R., Jones, R., Howell, K., Maxwell, D. & Bailey, P. 2017. Chalk reservoir of the Ockley accumulation, North Sea: *in situ* stresses, geology and implications for stimulation. *In*: Turner, J.P., Healy, D., Hillis, R.R. & Welch, M.J. (eds) *Geomechanics and Geology*. Geological Society, London, Special Publications, **458**. First published online May 30, 2017, https://doi.org/10.1144/SP458.3

Zanella, A. 2015. Load transfer, chemical compaction, seepage forces and horizontal hydrofractures within mature source rocks: evidence from theory, physical models and geological examples. Abstract presented at the Geology of Geomechanics Conference, 28–29 October 2015, Geological Society, London.

The geology of geomechanics: petroleum geomechanical engineering in field development planning

M. A. ADDIS

Rockfield Software Ltd, Ethos, Kings Road, Swansea Waterfront SA1 8AS, UK
Tony.Addis@rockfieldglobal.com

Abstract: The application of geomechanics to oil and gas field development leads to significant improvements in the economic performance of the asset. The geomechanical issues that affect field development start at the exploration stage and continue to affect appraisal and development decisions all the way through to field abandonment. Field developments now use improved static reservoir characterization, which includes both the mechanical properties of the field and the initial stress distribution over the field, along with numerical reservoir modelling to assess the dynamic stress evolution that accompanies oil and gas production, or fluid injection, into the reservoirs.

Characterizing large volumes of rock in the subsurface for geomechanical analysis is accompanied by uncertainty resulting from the low core sampling rates of around 1 part per trillion (ppt) for geomechanical properties and due to the remote geophysical and petrophysical techniques used to construct field models. However, some uncertainties also result from theoretical simplifications used to describe the geomechanical behaviour of the geology.

This paper provides a brief overview of geomechanical engineering applied to petroleum field developments. Select case studies are used to highlight how detailed geological knowledge improves the geomechanical characterization and analysis of field developments. The first case study investigates the stress regimes present in active fault systems and re-evaluates the industry's interpretation of Andersonian stress states of faulting. The second case study discusses how the stress magnitudes and, potentially, stress regimes can change as a result of production and pore pressure depletion in an oil or gas field. The last case study addresses the geomechanical characterization of reservoirs, showing how subtle changes in geological processes are manifested in significant variations in strength. The studies presented here illustrate how the timely application of petroleum geomechanical engineering can significantly enhance field development, including drilling performance, infill drilling, completion design, production and recovery.

Geomechanical analysis is based on a comparison of rock mechanical properties and strengths with the stresses acting *in situ*. When the magnitudes of the stresses acting in the rock are lower than the yield strength, the rocks behave elastically and deformations are small. However, if the changes in stress resulting from excavation during drilling, or from pore pressure or thermal variations exceed the compressive yield, peak or pore collapse strengths, the rock will fail through shearing or compaction, leading to non-linear irreversible deformation. If, on the other hand, the stresses become tensile, fracturing of the formation can occur.

A brief timeline of the important developments in geomechanical engineering applied to oil and gas extraction (Fig. 1) shows that these issues have been pursued since the earliest days of widespread hydrocarbon extraction, attracting the attention of some of the most notable figures in the industry.

The geological expression of the geomechanical, or structural, processes that are responsible for the development of many oilfields include folding, faulting, fracturing and diapirism. The stress regimes and strains accompanying these deformations can control the present day initial stresses and textures in many reservoirs. However, oil field developments lead to geomechanical changes, including the following:

(1) rock excavation during drilling results in wellbore stability and sand production problems;
(2) removing pore fluid and pressures from the rock results in compaction and subsidence;
(3) injection into the reservoirs increases the pressure and induces thermal changes, which result in rock fracturing or shearing to enhance the natural geologically controlled permeability. This is accompanied by significant changes in the reservoir and overburden stresses and strains.

In short, field development activities, from drilling to stimulation, perturb the initial geological conditions. It is the job of the petroleum geomechanical engineer to characterize the initial stress regime and the material properties, including the strength of rocks, to predict and plan for any changes in

From: TURNER, J. P., HEALY, D., HILLIS, R. R. & WELCH, M. J. (eds) 2017. *Geomechanics and Geology*. Geological Society, London, Special Publications, **458**, 7–29.
First published online June 28, 2017, https://doi.org/10.1144/SP458.7

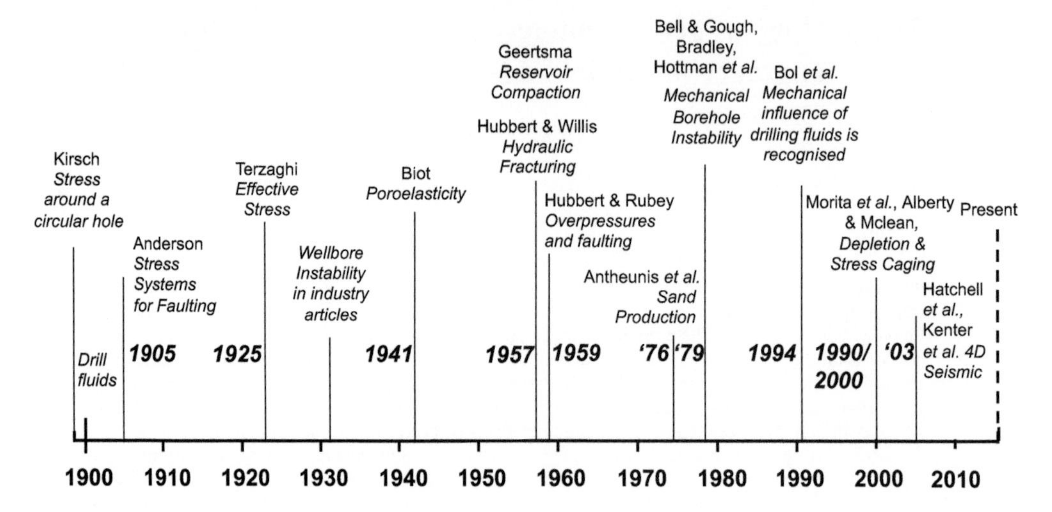

Fig. 1. Timeline of notable developments in the adoption of geomechanics in the oil and gas industry. Modified from SPE Distinguished Lecture by D. Moos (2014).

either of the stress magnitudes or deformations accompanying the field development. The perturbations can be small and linear in nature, going unnoticed in many developments. However, when large changes occur, the response of the formations may become non-linear, or plastic, leading to safety breaches and lost production, costing billions of dollars in design changes in the worst-affected fields. As the industry pushes for increased efficiency, greater safety and improved economics in increasingly challenging fields, all aspects of the development are expected to deliver improved field performance, including the discipline of geomechanics.

The application of geomechanics to oil and gas field development has become increasingly common over the past 40 years, with asset teams in the more challenging and larger field developments adopting geomechanical engineers as a permanent part of the team. The growth of unconventional oil and gas developments has accelerated this trend because the need to hydraulically fracture reservoir formations, or to shear the natural fracture network, to generate artificial reservoir permeability relies on geomechanical knowledge of the mechanical properties and initial stresses, as well as on the evolution of the reservoir stresses during field development.

The continual adoption of geomechanical engineering in the industry is difficult to judge quantitatively. However, a good indicator of the level of integration of petroleum geomechanics in the industry is the number of papers published annually by the Society of Petroleum Engineers (SPE) and stored in the SPE library containing the term 'geomechanics' in the publication title (Fig. 2).

The number of publications increased gradually through the 1990s, with a significant increase around 2000, predating the successful development of shale gas in the USA in 2004 or the increase in US shale oil production after 2010. This measure of industry uptake does not take into account different aspects of geomechanical analysis, such as compaction, hydraulic fracturing and wellbore stability, which have been used by the industry for a much longer period than that indicated in Figure 2. Compaction and subsidence has attracted considerable activity and interest since the late 1950s, with the notable examples of the Wilmington, Ekofisk, Groningen and the Central Luconia fields, demonstrating the extent to which reservoirs are dynamic during the field development. Wellbore stability and sand production have been a focus in the industry since the late 1970s, particularly after 1979 when four seminal wellbore stability papers were published (Bradley 1979*a*, *b*; Bell & Gough 1979; Hottman *et al.* 1979). These presaged a step change in drilling inclined, high-angle, extended and horizontal wells, both onshore in the Austin chalk, the North Slope of Alaska, and offshore in the North Sea and Gulf of Mexico in the 1980s and early 1990s. Figure 2 reflects the uptake of the now-accepted discipline of geomechanics into field development.

Numerous textbooks comprehensively address both the theory and application of geomechanical engineering to field planning (Charlez 1991, 1997; Fjaer *et al.* 2008; Zoback 2008; Aadnoy & Looyeh 2011). This paper, in contrast, presents an individual view and geomechanical insights from significant field developments, with emphasis on the geological

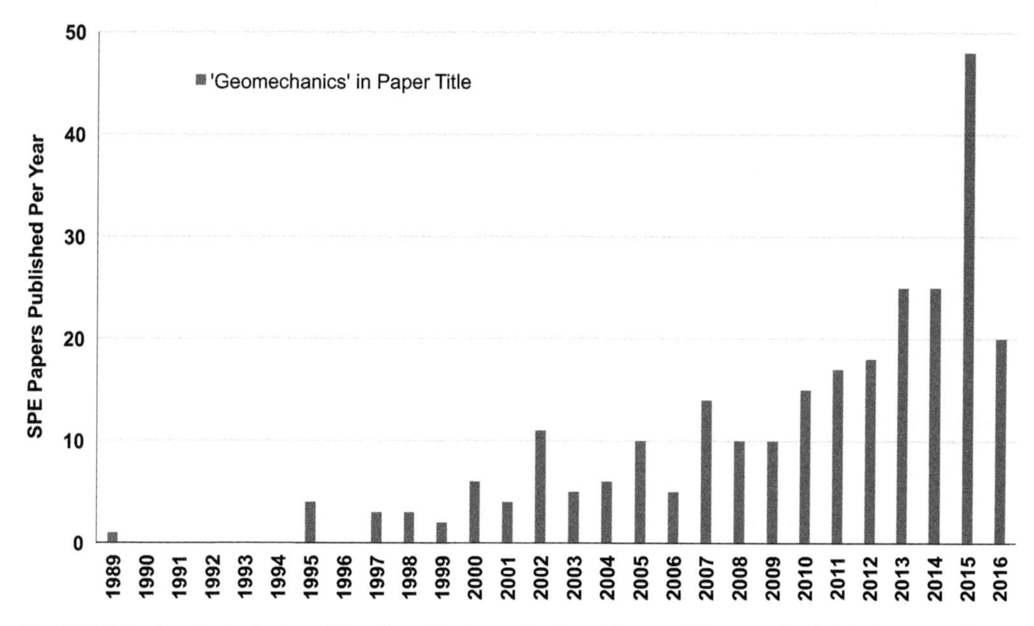

Fig. 2. Publications in the Society of Petroleum Engineers OnePetro library with 'geomechanics' in the paper title (up to the first quarter of 2016).

considerations for stress regime estimation and rock strength determination, along with some insights of their impact on production and development success.

Stress estimation in fault blocks: the limitations of using the Andersonian system

Stress magnitude estimation in the absence of field data

Subsurface stress magnitudes and orientations differ considerably from passive basins and margins to extensional basins and compressional regimes, with wide variations within these structural regimes. The determination of present day subsurface stresses centres around the estimation of the two horizontal total stress magnitudes and directions; the vertical total stress is typically assumed to be equal to the weight of the overburden.

The magnitudes of the two horizontal total stresses (σ_H, σ_h), which result solely from the weight of the overlying column of rock or sediments (σ_v), have been calculated in civil engineering using linear elastic theory since Terzaghi (1943). This approach considers 'passive basin' stresses, expected in flat-lying strata in geologically passive environments, where no horizontal compression or extension occurs and the magnitudes of the two horizontal stresses are equal. The horizontal stress in such a simple case is caused by an element of rock trying to expand laterally against the adjacent elements. The horizontal stress magnitude is therefore directly related to the vertical effective stress (σ'_v) and the Poisson's ratio of the formation. This Eaton approach was formulated for passive basins (Eaton 1969) yet, in the absence of area-specific data or measurements, it is commonly used by the industry as a first approximation regardless of the geological environment.

Geologically, the assumption of a passive basin is inappropriate for the majority of oil and gas accumulations, which are more commonly found 'on structure' and associated with compressional or extensional geological environments. Horizontal stress anisotropy occurs in these structural regimes when a significant difference is observed between the magnitudes of the maximum and minimum total horizontal stresses, σ_H and σ_h, respectively. Two different approaches are used to estimate the magnitudes of these stresses from geological considerations. The first imposes a horizontal strain on the formations, whereas the second considers a constant horizontal stress, or stress gradient, compressing or extending the formations – strain v. stress horizontal boundary conditions. Both methods commonly rely on the assumption of isotropic formation properties.

The strain boundary, or deformation boundary, approach results in different horizontal stress magnitudes in different lithologies, calculated from both the Poisson's ratio and the horizontal stiffness (Young's modulus) of the formations. This method requires estimates of the present day strains applied to the formation in the two horizontal directions, which are either obtained empirically through calibration with any available stress data from the field, or through numerical modelling. The stress boundary method, on the other hand, results in lithology-dependent horizontal stress magnitudes controlled by the contrasts in Poisson's ratio between formations. Two common stress boundaries include a constant stress gradient (or effective stress ratio, σ'_h/σ'_v), which is determined by calibration with any existing stress data from the field, or on the assumption of active faulting.

Andersonian fault systems

The assumption of active faulting relies on estimates of the relative magnitudes of the two horizontal stresses and the vertical stress consistent with the Andersonian stress system for faulting (Anderson 1905, 1951). In this approach, for a given pore pressure, the relative total stress magnitudes required to generate different faulting systems are considered to be:

normal faulting: $\sigma_v \geq \sigma_H \geq \sigma_h$ (where $\sigma_H = \sigma_{//}$)
strike-slip faulting: $\sigma_H \geq \sigma_v \geq \sigma_h$ (where $\sigma_v = \sigma_{//}$)
thrust faulting: $\sigma_H \geq \sigma_h \geq \sigma_v$ (where $\sigma_h = \sigma_{//}$)

where compressive stresses are considered to be positive and $\sigma_{//}$ is the stress oriented parallel to the strike of the fault.

The stress magnitudes associated with these fault systems require that basins are at the limit equilibrium for faulting – that is, on the point of slip, where the stresses required to activate the faults are either the maximum compressional stresses in thrust and strike-slip regimes, or the minimum extensional stresses in normally faulted basins. The following case study illustrates that the measured stresses, particularly the magnitude and orientation of the minimum total stress, are not always consistent with the deformational style of faulting and the assumed stress regime.

Case study 1. Cusiana field, Colombia: identifying the stress regime acting in an active thrust fault environment

The Cusiana field in Colombia is located in the foothills of the Andes at the deformation front, where the foothills reach the Llanos plain, and consists of an anticlinal structure bounded by two thrust faults, the Yopal and Cusiana faults (Fig. 3). The underlying Cusiana fault defining the structure of the field is active, as observed in the ground movements and deformation in buildings prior to the start of the field development. The movement on the main fault blocks consists of thrust deformation with little or no strike-slip motion.

Severe wellbore instability was encountered by BP while drilling the first exploration well (Cusiana 1) and subsequent appraisal wells on the Cusiana structure (Skelton *et al.* 1995). Cusiana 1 was the first well to successfully reach the Mirador reservoir in the region, despite the attempts of several operators. The wellbore instabilities were most common, and challenging, in the overburden that included the Leon and Carbonera formations. The Carbonera Formation was particularly challenging to drill because it consists of four layers of strong sandstone units where mud losses and tight hole were experienced, with hole sections generally in-gauge or slightly under-gauge, separated by more shale-rich units, which experienced severe breakouts in the early exploration and appraisal wells. In one example, a four-arm caliper tool measured breakouts that had extended the wellbore to >1 m (>40 inches) in diameter in one direction, with essentially an in-gauge 31 cm (12¼ inch) borehole diameter in the perpendicular direction (Addis *et al.* 1993; Last *et al.* 1995). These early wellbore stability problems led to significant non-productive time drilling the wells, numerous sidetracks and lengthy drilling programmes.

Cusiana, as an anticlinal structure bound by active thrust faults, was expected to have a typical thrust fault stress system, where the minimum stress is oriented vertically and equal in magnitude to the overburden stress (Anderson 1905, 1951; Hubbert & Willis 1957). Analysis of the wellbore stability problems confirmed that the maximum horizontal stress direction, as determined from breakout orientations, was perpendicular to the strike of the fault, consistent with pure dip-slip (plane strain) thrust deformation. However, the analysis of leak-off test and mud loss data indicated that the minimum stress magnitudes were similar to those expected in passive basins, and considerably smaller in magnitude than the vertical stress. This had significant implications for optimizing the stability of the inclined wells required to develop the field from a number of limited drilling pad locations and any hydraulic fracture stimulation of the reservoir.

Analytical stress analysis

This unexpectedly low magnitude of minimum horizontal stress was explained using a simple isotropic linear elastic analysis of relatively undeformed fault blocks bound by active faults in plane strain conditions (no strike-slip deformation). This approach showed that it is possible for active thrust faults to exhibit a 90° rotation of the minimum stress acting on the fault blocks, from vertical pre-faulting to horizontal post-faulting (Addis *et al.* 1994). This

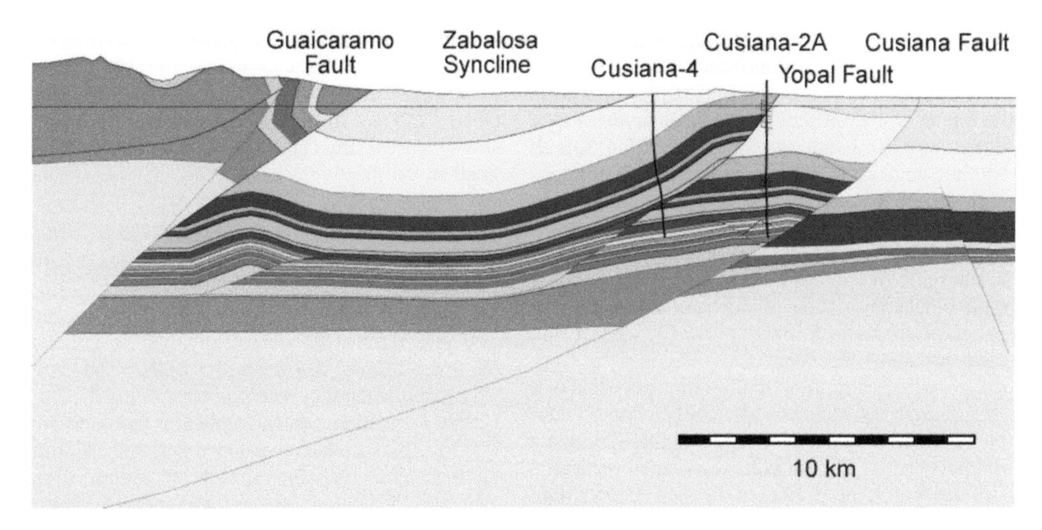

Fig. 3. Cross-section of the Cusiana structure in Colombia showing the bed inclination adjacent to the thrust faults (Willson *et al.* 1999). © 1999, Society of Petroleum Engineers. Reproduced with permission of SPE. Further reproduction prohibited without permission.

rotation of the minimum stress direction resulted from low fault friction, which causes the magnitude of the horizontal total stress acting parallel to the fault ($\sigma_{//}$) to drop below the magnitude of the vertical total stress (σ_v) post-faulting. The stress calculations are shown in Figure 4.

The *y*-axis of Figure 4 shows the magnitude of the horizontal 'plane strain' stress ($\sigma_{//}$) in the fault blocks aligned parallel to the strike of the thrust fault, relative to the magnitude of the vertical stress (σ_v). If this $\sigma_{//}/\sigma_v$ ratio >1, the minimum stress is oriented vertically, as expected for thrust faults.

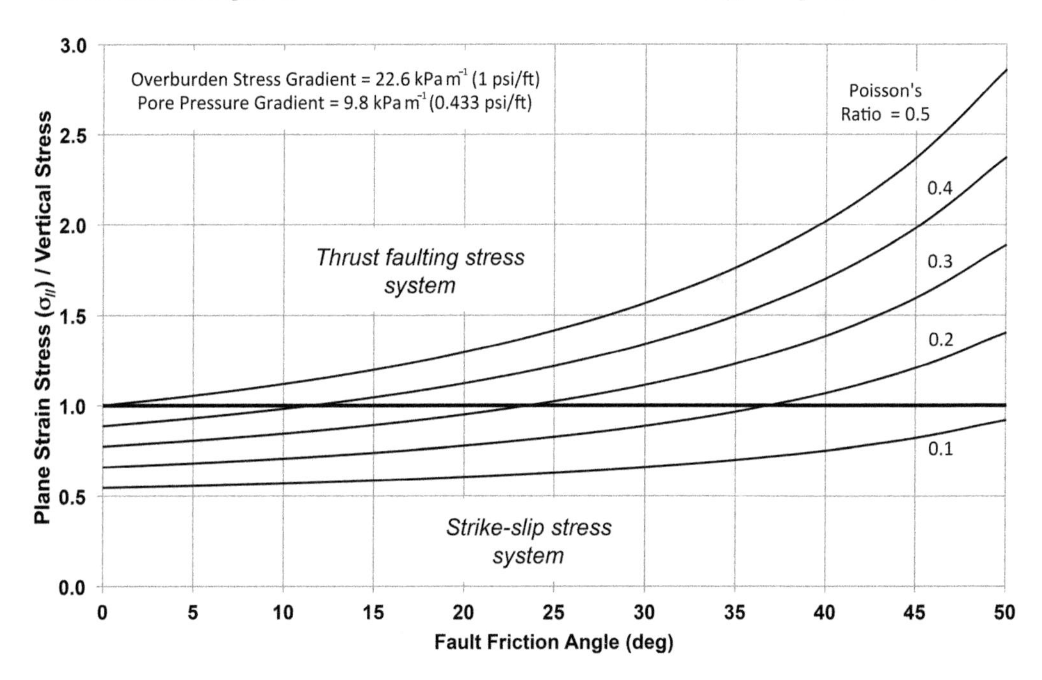

Fig. 4. Variation of the minimum horizontal stress acting in intact fault blocks oriented parallel to a thrust fault, normalized by the vertical stress magnitude, for different friction angles on the thrust fault (Addis *et al.* 1994). © 1994, Society of Petroleum Engineers. Reproduced with permission of SPE. Further reproduction prohibited without permission.

A ratio <1 indicates that the horizontal stress oriented parallel to the strike of the fault is the minimum stress ($\sigma_h = \sigma_{//}$) and a strike-slip stress regime exists, even though the fault deforms as a thrust fault. This ratio is plotted for different fault friction angles on the *x*-axis and for formations with different Poisson's ratios, shown as the contours on the plot.

Figure 4 illustrates that for an active thrust fault at the point of slip, with plane strain deformation, fault blocks consisting of different lithologies, as in the Carbonera Formation in Cusiana, could have lithology-dependent stress regimes: a sandstone with a low Poisson's ratio may have a minimum stress oriented horizontally, while a juxtaposed shale characterized by a higher Poisson's ratio may have a minimum stress oriented vertically. In other words, the stresses in the sand layers within the Carbonera Formation are inconsistent with Anderson's stress regime for the formation of a thrust fault, whereas the stress system in the shale conforms to these pre-faulting stresses.

Numerical stress models

The conclusions of this analytical approach were supported by a finite difference numerical model of the Cusiana structure (Fig. 5). BP's numerical simulation of tectonic deformation in the three Cusiana blocks involved gravitational loading due to the weight of the overburden. Horizontal tectonic movement was then introduced by displacing the Yopal thrust fault overlying the Cusiana structure by 100 m (330 ft) to the SE and displacing the underlying Cusiana thrust fault by 20–25 m (65–82 ft). This ratio of displacements matched the observed fault-throw ratio on the fault surface. The results of this numerical approach illustrate the relatively low horizontal stress magnitude (S_h $_{SW-NE}$) oriented parallel to the strike of the thrust faults (Addis *et al.* 1993; Last *et al.* 1995), in line with the analytical estimates.

While drilling additional wells on the Cusiana structure, these low minimum horizontal stress magnitudes of 14.7–17.0 kPa m^{-1} (0.65–0.75 psi/ft) and vertical overburden stresses of 23.8–24.7 kPa m^{-1} (1.05–1.09 psi/ft) were repeatedly observed across the field (Last *et al.* 1995).

The analytical estimates of horizontal stress magnitudes, illustrated in Figure 4, assume a simple flat-lying fault block geometry. However, in reality, rollover anticlines defining the Cusiana field result in non-horizontal and non-vertical principal stresses. Last & McLean (1996) described the stress rotation based on numerical modelling of the Cusiana structure and the impact on wellbore stability analysis.

This additional complexity does not seem to significantly influence the minimum horizontal stress magnitude in Cusiana: The stresses used to model the stress rotation in Cusiana by Last & McLean (1996) were equivalent to 31.7:24.9:15.8 kPa m^{-1} (1.4:1.1:0.7 psi/ft) for the maximum horizontal stress, vertical stress and minimum horizontal stress gradients, respectively.

Model validation with additional measurements during late field development

More reliable data on the stress magnitudes became available during the completion and stimulation of the Cusiana development wells. The low-permeability Mirador reservoir underlying the Carbonera Formation was hydraulically fractured and the stress measurements obtained during these stimulations again largely support the earlier stress estimates, indicating stress gradients in the sandstone reservoir of $\sigma_v = 23.8$ kPa m^{-1} (1.05 psi/ft), $\sigma_H = 24.9$–28.3 kPa m^{-1} (1.1–1.25 psi/ft) and $\sigma_h = \sigma_{//} = 13.1$–17.6 kPa m^{-1} (0.58–0.78 psi/ft) (Osorio & Lopez 2009).

Microseismic data measured during reservoir stimulation confirmed the development of vertical hydraulic fractures. The microseismic events were aligned with the maximum horizontal stress (NW–SE) direction determined from breakout analysis and were oriented perpendicular to the strike of the fault planes (Osorio *et al.* 2008). This again indicates that the minimum stress magnitude is oriented horizontally and parallel to the strike of the thrust fault ($\sigma_h = \sigma_{//}$) and, while consistent with the analytical and numerical modelling, is apparently at odds with the common interpretation of stress regime associated with the Andersonian thrust deformation of the faults.

Business impact

Despite the very severe drilling problems experienced in the early exploration and appraisal wells on the Cusiana field, the efforts taken to understand the complexity of the stress field allowed the development team to better address the wellbore instability and to improve well design and drilling practices. This led to a significant improvement in performance in this challenging field, with the non-productive time reducing from 47% in 1993 to 27% in 1996 (Last *et al.* 1998).

Stress evolution from numerical sandbox modelling of thrust fault systems

The transformation of the stress state in the compressional thrust fault blocks, pre- and post-faulting (Fig. 4), results from low fault friction, either initial or residual friction. In the latter case, the maximum

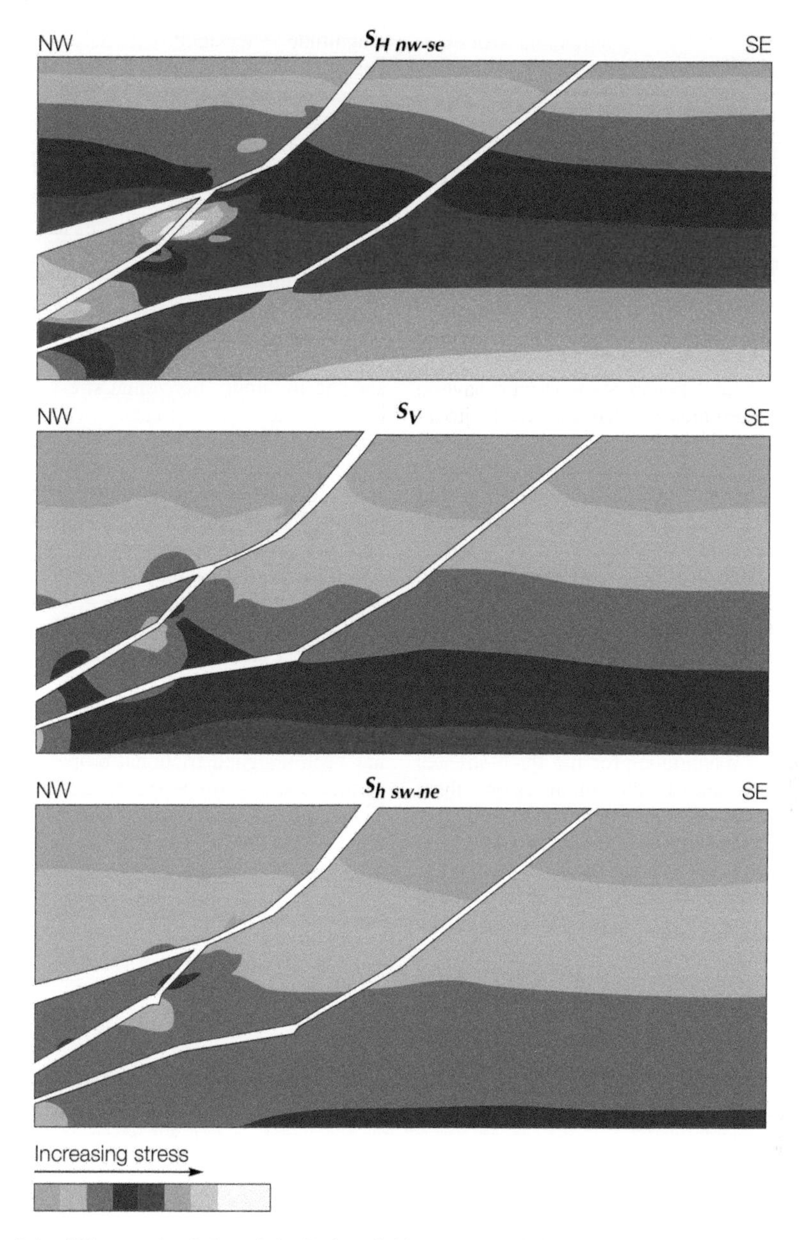

Fig. 5. BP's finite difference simulation of the Cusiana field stresses. Analysis was made on sections oriented NW–SE, parallel to the tectonic compression. The maps show the three principal stress magnitudes for one of the sections. In general, $S_{\mathrm{H\ nw-se}} > S_{\mathrm{v}} > S_{\mathrm{h\ sw-ne}}$. From Addis *et al.* (1993), reproduced with permission of Schlumberger.

compressive stress required to initiate the thrust fault plane exceeds that required to mobilize the fault plane at large strains, the difference between peak strength and residual strength. Using Sibson's characterization of faulting styles (Sibson 1977, 2003), this may be more common at relatively shallow depths in a 'brittle' zone, where strain softening rock behaviour might be expected, rather than at great depths with larger confining pressures, where the fault planes may display perfect plasticity during rock failure.

The results of the finite difference numerical modelling of Cusiana (Fig. 5) only considered the late-stage movements on existing fault planes with an existing structure. However, simulations of sandbox experiments that consider the formation and

Table 1. *Summary of sandstone properties used in the numerical sandbox simulations*

Property	Value		
Young's modulus (Pa)		1.5×10^5	
Poisson's ratio		0.25	
Density (g cm^{-3})		1.7	
Effective plastic strain	0	0.05	0.1
Cohesion (Pa)	140	50	50
Friction angle (°)	36	36 (32)	36 (28)
Dilation angle (°)	30	15	0

development of fault planes show similar results; Crook *et al.* (2006) present such numerical models for extensional faulting.

A large strain finite element numerical modelling code was used to simulate a compressional sandbox experiment. The material being compressed laterally was sandstone, with a characteristically low Poisson's ratio and the elastic and Mohr–Coulomb properties shown in Table 1. The sandstone properties assigned to the numerical sandstone are representative of high porosity sands.

The compressional numerical experiments considered both high and low fault frictions (Fig. 6). The total stress magnitudes for the three stresses (σ_v, σ_H, σ_h) are shown by the colour scales, where red indicates low compressive stresses and blue shows high compressive stresses. The scale for the top row of figures differs from the scale for the figures in the middle and bottom rows by an order of magnitude. The sandbox models simulated in this numerical modelling consist of a uniform formation compressed by the left-hand boundary, with a static right-hand boundary.

The top panels (Fig. 6) show the calculated initial stress conditions due to gravitational loading, prior to compression, where the two horizontal stresses have similar magnitudes and distributions, equivalent to passive basin conditions. The observed differences at the boundaries reflect the friction assigned to the sides of the sandbox.

The panels in the middle row show the vertical stress, compressional maximum horizontal stress and the minimum horizontal stress (perpendicular to the plane strain boundary, $\sigma_h = \sigma_{//}$) after the sandbox experiment had undergone compressional strain. This middle row of simulations uses the sand properties shown in Table 1, which are elasto-plastic, and the friction angle remains constant at 36°, independent of the amount of strain or slip developed on the failure planes. From the analytical solutions presented in Figure 4, these simulations with a sand Poisson's ratio of 0.25 would result in the vertical stress as the minimum principal total stress ($\sigma_3 = \sigma_v$).

The bottom panels again present the three stresses, but for conditions where the sandstone has been assigned frictional properties that reduce as the plastic strain on the failure planes develops. This approach reproduces a fault plane mobilizing residual friction as the fault slips. The properties assigned to this post-peak residual friction are shown in Table 1 in parentheses for the friction

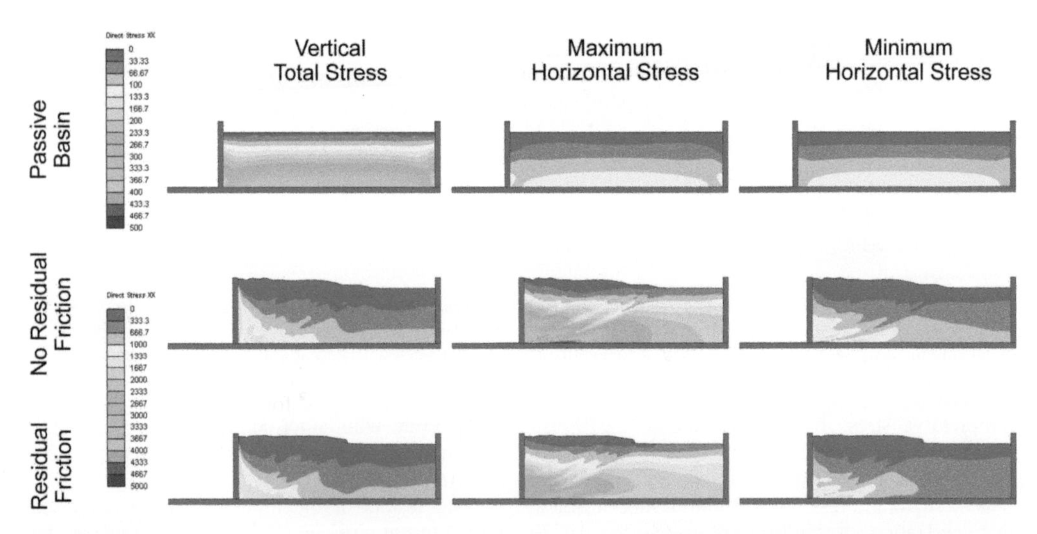

Fig. 6. Numerical simulations of a sandbox experiment illustrating the stress development post-failure. Top row: passive basin conditions, gravitational loading with no lateral compression. Middle row: gravitational loading with lateral compression – sandstone with constant friction angle. Bottom row: Gravitational loading with lateral compression – sandstone with residual friction angle.

angle, where the friction angle drops from 36 to 28° as the failure plane develops more plastic strain. Using the analytical solution in Figure 4, this lower fault residual friction should result in the minimum principal total stress becoming oriented horizontally ($\sigma_3 = \sigma_h = \sigma_{//}$).

These numerical sandbox simulations show that the maximum horizontal stress develops as expected, accompanying the lateral compression. The stress contours outline the developed shear failure bands. The vertical stress is bound by the free surface in the *y*-axis direction, but shows variations resulting from the development of the shear failure bands.

In the middle row of Figure 6, the minimum horizontal stress ($\sigma_h = \sigma_{//}$) is of a similar magnitude to the vertical stress. In the relatively undeformed fault blocks, next to the right-hand boundary of the figures, the minimum horizontal stress slightly exceeds the magnitude of the vertical stress, in line with the Andersonian description and the analytical estimates in Figure 4. This is consistent with the stress system responsible for the formation of the fault planes.

The results from the last suite of simulations are shown in the bottom row, where the friction angle of the sands was reduced as the failure planes developed. These simulations show that once residual friction is developed and the fault plane friction is sufficiently reduced, the stresses generated in the relatively undeformed fault block on the right-hand boundary of the simulations has the minimum horizontal stress oriented parallel to the strike of the fault with a magnitude lower than the vertical stress ($\sigma_3 = \sigma_h = \sigma_{//}$).

These simulations again support the field data as well as the analytical and numerical models used to explain the anomalously low magnitudes of horizontal stress acting in the Cusiana field. The earlier discussion of the stresses present in the Cusiana active thrust fault is also evident in other compressional thrust faulting regions, such as the Canadian Rockies (Woodland & Bell 1989) and the highlands of Papua New Guinea (Hennig *et al.* 2002), which also indicate minimum horizontal stress magnitudes significantly lower than the vertical stress magnitude ($\sigma_v > \sigma_h = \sigma_{//}$). The common interpretation and application of the Andersonian stress system should therefore be treated with caution in similar structural regimes.

Stress regimes in active normal faulting systems

This discussion has focused on thrust fault environments, but many fields are developed in normally faulted regimes. The analytical approach described in Figure 4 was applied to normal faulting and a similar picture emerged: stress rotations in the undeformed fault block can also accompany normal fault development.

In Figure 7, the *y*-axis again plots the ratio of the horizontal stress acting parallel to the strike of the normal fault, which is normally assumed to be the intermediate stress, relative to the magnitude of the vertical stress ($\sigma_{//}/\sigma_v$). The horizontal stresses are always less than the vertical stress in this normal faulting extensional environment and the values on the *y*-axis are always <1. The relative magnitude of this 'fault-parallel' horizontal stress ($\sigma_{//}$) is again plotted for different values of fault plane friction and formation Poisson's ratios.

This plot also shows the magnitude of the extensional horizontal stress acting perpendicular to the strike of the fault plane relative to the magnitude of the vertical stress; the fault mobilization curve. This is the horizontal total stress required to mobilize or slip normal faults for given fault friction angles.

Figure 7 shows that in fault blocks defined by normal faults with no strike-slip component, but with high fault friction angles, isotropic formations with high Poisson's ratios are likely to have minimum horizontal stresses acting perpendicular to the strike of the fault, in line with expectations of the pre-fault stress state. By contrast, in formations with low Poisson's ratios and in the presence of faults with low friction, in the area below the fault mobilization curve, the minimum horizontal stress magnitude may be re-oriented parallel to the normal fault ($\sigma_3 = \sigma_h = \sigma_{//}$), which again contrasts with the Andersonian fault system required to form the faults.

A note on stress polygons

The stress polygon (Anderson 1951) and its modification (Zoback 2008) is commonly used in the petroleum industry for the stress estimation of faulted environments. It can be used to illustrate this rotation of the minimum horizontal total stress and the non-coincidence of the principal stress and deformational axes when low friction or weakening on the fault occurs during slip or for formations with contrasting Poisson's ratios.

Figure 8a shows the standard stress polygon for the Andersonian stresses required to generate the main faulting types for a fault friction of 30°, while Figure 8b illustrates the analytical estimates for formations with low Poisson's ratios (=0.1) for the stress states presented in Figures 4 and 7 for thrust and normal faulting, respectively. The stress polygons show for plane strain conditions that thrust fault deformations can result in stress magnitudes normally attributed to strike-slip faulting and that for thrust and normal fault deformations

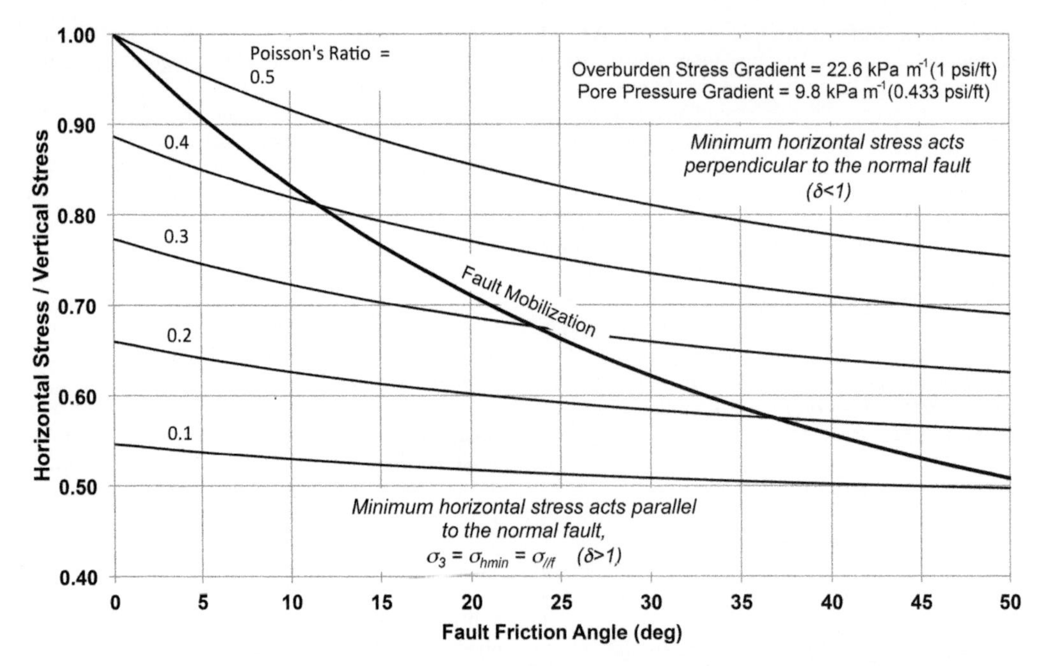

Fig. 7. Variation of the minimum horizontal stress magnitude acting in intact fault blocks oriented parallel to a normal fault, normalized by the vertical stress magnitude, for different friction angles on the normal fault (Addis *et al.* 1994). © 1994, Society of Petroleum Engineers. Reproduced with permission of SPE. Further reproduction prohibited without permission.

the minimum stress may become oriented parallel to the fault plane ($\sigma_3 = \sigma_h = \sigma_{//}$). Figure 8c, calculated using a Poisson's ratio of 0.35, demonstrates that a continuum of stress states may not occur between faulting styles for the plane strain conditions presented in Figures 4 and 7. Figure 8 illustrates that a range of different properties and sizes of the stress polygons exists depending on the values of the fault friction angle and Poisson's ratio for a specified vertical total stress and pore pressure.

Discussion of Andersonian systems and post-faulting stress regimes

Anderson (1905, 1951) described the relative stress magnitudes required to generate normal, strike-slip and thrust faults, specifically up to the point when the faults form, not post-faulting. Anderson used the phrase, 'Suppose now that the stresses are so great as to lead to actual fracture'. That describes the stress regime acting up to the initiation and genesis of the fault, but does not address the stress regimes that could develop after the faults have formed, stating that, 'The effect of all faulting is to relieve the stress and bring conditions nearer to . . . the standard state', i.e. closer to isotropic conditions. The assumption of a minimum stress equal to the vertical stress in a thrust fault regime, or perpendicular to normal faulting, may therefore be incorrect under certain conditions.

The decrease in stress after fault formation can be explained by the well-documented large post-peak 'brittle' or weakening response that accompanies the formation of a well-defined discontinuity (Bishop 1974; Sibson 1977; Mandl 2000). This 'brittle' or weakening response as the fault develops leads to low residual friction at large strains due to a number of possible contributing mechanisms, including: grain disaggregation, a large clay content, or low effective normal stresses resulting from high 'undrained' pore pressures on the fault plane, accompanying the shear stress build-up and deformation. This is akin to the undrained mechanical response described by Skempton's 'A' pore pressure coefficient (Skempton 1954; Yassir 1989).

This section has considered only simple compressional or extensional events. However, geological history is commonly complex, involving multiple loading events that affect the current day stresses. Stress inversion events are well documented in the Tertiary of the North Sea (Biddle & Rudolph 1988), the NW Shelf of Australia (Bailey *et al.* 2016) and globally (Cooper & Williams 1989; Buchanan & Buchanan 1995). Stress inversion also plays a part in the development of the

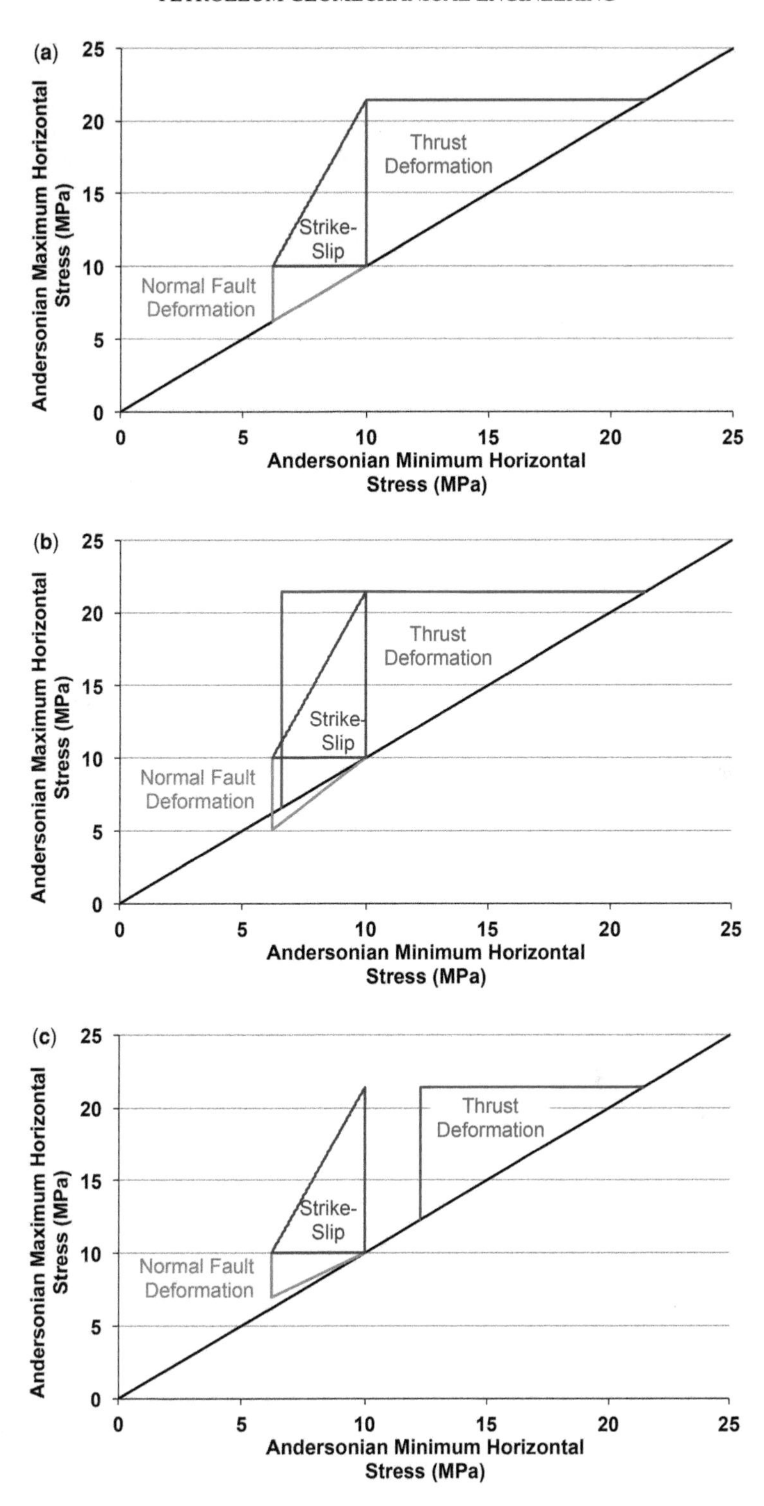

Fig. 8. Comparisons of stress polygons with (**a**) standard Andersonian faulting for stress conditions leading up to fault development; (**b**) possible stress conditions post-fault development in formations with low Poisson's ratios (=0.1); and (**c**) possible post-fault conditions in formations with high Poisson's ratios (=0.35).

Cusiana field, where the bounding Cusiana fault was originally a margin extensional fault, which was later reactivated during the Andes compression. Consequently, the inclination of the Cusiana fault may not be consistent with the typical angular relationship of failure planes relative to the maximum horizontal stress ($\beta = 45^\circ - \phi/2$) for thrust fault development, but may be more representative of normal faulting. To account for the differences in the dip between the actual fault plane after fault formation and the expected fault orientation resulting from rock failure, Wu *et al.* (1998) refined the earlier calculations for stresses resulting from normal and thrust faults for plane strain conditions. These equations represent a simplified analytical approach to estimate the impact of stress inversion on the stresses acting in fault blocks away from the immediate vicinity of the fault planes.

As a further consideration, the normal assumption of vertical and horizontal stresses, while a common assumption for flat-lying formations, does not apply in these compressional environments, where faulted and folded formations predominate. The subvertical principal stress orientation in these folded environments is, as a first approximation, taken to be normal to the bedding planes in moderately inclined and folded beds. The presence of faulting provides additional complexity and local rotations occur based on the orientation of the fault, the relative movements on the fault blocks and the fault friction angles which are best evaluated using numerical models (Thornton & Crook 2014). These folded formations and rotated stress systems have a significant impact on wellbore stability for inclined wells in these compressional environments (Last & McLean 1996).

Other geological processes have a significant impact altering the stress systems acting in relatively passive environments, most notably associated with salt intrusions, where stress rotations in the horizontal plane are observed from breakout analysis (Yassir & Zerwer 1997) and those in the vertical plane are calculated using numerical methods (Peric & Crook 2004).

To conclude, the stress regimes described by Anderson for the onset of different faulting styles, specifically normal and thrust faulting, are calculated to persist in relatively undeformed fault blocks if the formations have a high Poisson's ratio and/or large fault plane friction. The stress states acting in fault blocks deviate from the pre-faulting Andersonian stress states for formations with low Poisson's ratios and for faults with relatively low friction, either initially or as a result of strain weakening and low residual friction angles. This suggests that at depths within the zone of interest for hydrocarbon exploitation, fault blocks containing younger, less compact and less cemented formations, which are more likely to behave in an elastoplastic manner, and bound by faults with high fault friction could be expected to have stress systems compatible with the Andersonian system of stresses required for the formation of the faulting styles. In contrast, older or more competent formations, bound by faults with low fault friction (drained or undrained), may have post-faulting stress systems that are not consistent with the original pre-faulting stress states.

This has significant implications for stress analysis in petroleum field development and for the selection and interpretation of stress estimation methods. These estimation methods are often lithologically constrained, e.g. breakout analysis predominates in more shale-rich formations, whereas minifrac and differential strain curve analysis are more common in sandstones and limestones. There are also implications for strain-based indicators of fault movement, such as seismic moment analysis 'beach balls', which may not always represent the existing stress field throughout these active fault systems.

Neotectonic natural fracture development accompanying compression and extension

Evidence of strain accompanying faulting on a geological timeframe includes the occurrence of extensional vertical fractures or steep hybrid fractures in compressional environments. These neotectonic fractures are oriented perpendicular to the deformation front, with the minimum horizontal stress aligned parallel to the strike of the thrust fault plane ($\sigma_h = \sigma_{//}$), as described by Hancock & Bevan (1987) for a number of foreland basins. The formation of these near-vertical natural fractures oriented parallel to the maximum horizontal stress direction is again inconsistent with a common interpretation of compressional thrust stress regimes, where the minimum stress is expected to be vertical.

The rotation of the minimum stress from vertical at the onset of thrust faulting to horizontal during the development and displacement of the thrust fault, discussed in the preceding sections, may provide a prerequisite condition for the development of the subvertical neotectonic natural fractures. These are observed over hundreds of kilometres in the forelands and hinterlands of orogenic belts and at distances greater >500 km from the compressional deformation front (Hancock & Bevan 1987). Numerous examples are presented by Hancock & Bevan (1987), from southern England, northern France and the Arabian platform, along with the vertical J_2 joint set of the Marcellus Shale in the Appalachian basin reported by Engelder *et al.* (2009). In the absence of this post-faulting stress rotation, these neotectonic fractures have been considered to be restricted to shallow depths and attributed to significant uplift and elevated fluid pressures, or lateral elongation in the forelands parallel or sub-parallel to the fault strike or orogenic margin, in order to produce the necessary tensile effective horizontal stresses to form the fractures.

For normally faulted environments, similar observations have been discussed by Kattenhorn *et al.* (2000), where vertical natural extensional fractures occur oriented perpendicular to the strike of the normal fault plane, again contrary to the common interpretation of stress states associated with normal faults. Kattenhorn *et al.* (2000) discuss the field occurrence of these fractures and assess the conditions required for their formation, primarily considering the elastic stress perturbation around faults. However, one prerequisite for their formation is considered to be $\sigma_h = \sigma_{//}$, which Kattenhorn *et al.* (2000) refer to as a condition where $\delta > 1$; δ is redefined here (where compressional stresses are considered positive) as the ratio of the fault-perpendicular to fault-parallel horizontal stresses ($\sigma_{\perp}/\sigma_{//}$) acting remote from the fault plane (Fig. 7). The stress estimations presented here and discussed in Addis *et al.* (1994) enable us to predict which formations are likely contain such fractures and why these may not extend into adjacent formations of different mechanical properties.

The lithological control on the minimum horizontal stress becoming parallel to the fault strikes is consistent with observations of neotectonic fractures being more prolific in low Poisson's ratio formations, e.g. sandstones and limestones (Hancock & Engelder 1989), and less clay-rich formations (Engelder *et al.* 2009).

In the formulations presented, the minimum horizontal stress does not become tensile, a requirement for the development of new tectonic joints as described by Hancock & Bevan (1987), Hancock & Engelder (1989) and Engelder *et al.* (2009) for compressional regimes and by Kattenhorn *et al.* (2000) for normal faulting environments. Mechanical conditions such as elevated fluid pressures and/or high deviatoric stresses leading to 'extension fractures' (Engelder *et al.* 2009) or extension parallel to the fault plane may be required to make this minimum effective horizontal stress tensile.

Stress evolution during field production and the impact on infill drilling

The compressional stress regime discussed in the previous sections for the Cusiana field considers the relative total stress magnitudes present at the start of production, which have developed over a geological timeframe and are considered to be relatively static, although fault movements are known to perturb this static condition. Production and injection in a field cause the total stresses to vary with the changes in the reservoir pore pressure on an almost daily basis. This dynamic stress environment not only exists in the reservoir, but also in the surrounding formations. It has a significant technical and economic impact on field planning, drilling, stimulation and production of these fields, especially for depleting high pressure–high temperature (HPHT) fields.

The effect of production and reservoir pressure depletion on the minimum total stress magnitude was first documented by Salz (1977), who showed a linear relationship between the change in the total minimum horizontal stress magnitude and the change in the average reservoir pressure ($d\sigma_h/dP_p$) in the Vicksburg Formation in south Texas, where the stress depletion response was shown to be:

$$\frac{d\sigma_h}{dP_p} = \gamma_h = 0.53$$

A large number of subsequent studies on different fields have described similar decreases in the magnitude of the minimum horizontal stress accompanying depletion (Teufel *et al.* 1991; Engelder & Fischer 1994; Addis 1997*a*, *b*; Hillis 2003). Little attention has been given to the changes in the magnitude of the maximum horizontal total stress with pore pressure, which is normally assumed to change with the same depletion ratio as the minimum horizontal total stress ($\gamma_H = \gamma_h$). The change in the vertical total stress with pore pressure (γ_v) has received more attention as a result of the increasing use of field-wide geomechanical numerical modelling and the cross-correlation with four-dimensional seismic velocity changes, which are used to monitor the vertical stress and strain changes resulting from depletion-driven reservoir compaction and subsidence (Kenter *et al.* 2004; Molenaar *et al.* 2004).

Geological factors which influence the dynamic response of the minimum horizontal total stresses to reservoir pressure changes include:

(1) the reservoir dimensions;
(2) the reservoir structure (anticlinal, inclined, flat-lying);
(3) the mechanical property (elastic) contrasts between the reservoir and overburden formations;
(4) pore pressure depletion – the radius of influence, drainage radius or reservoir compartmentalization;
(5) pressure cycles, depletion followed by injection (Santarelli *et al.* 1998, 2008);
(6) faulting style and stress regime;
(7) location on the structure.

These factors contribute to the range of the stress depletion ratios observed globally, which typically vary between $\gamma_h = 0.4$ and 1.0, with the most common being in the range $\gamma_h = 0.6$–0.8. Teufel *et al.* (1991) showed for the Ekofisk field that the horizontal stresses on this domal field decrease at the same rate for wells located at the crest and the flanks of the field, at a rate of *c.* $\gamma_h = 0.8$. The nearby chalk fields show similar stress depletion ratios for the minimum horizontal stress. Geological factors also control the vertical stress changes accompanying depletion, ranging from $d\sigma_v/dP_p = \gamma_v = 0$ for flat-lying, laterally extensive reservoirs to $\gamma_v = 0.2$ for anticlinal or highly inclined reservoirs (Molenaar *et al.* 2004). Molenaar *et al.* (2004) predict that the vertical total stress changes accompanying the depletion of the Shearwater field, in the high pressure–high temperature region of the central North Sea, are dependent on the location of the well on the structure of the inclined reservoir blocks.

The variation of the maximum horizontal stress with depletion is not known and is only estimated through analytical (Addis 1997*a*) or numerical modelling, given the challenges of determining the magnitude of the maximum horizontal stress from field measurements. The assumption that γ_H varies in a similar manner to the minimum horizontal stress might be a reasonable assumption for flat-lying reservoirs exhibiting a passive basin type depletion response. This is unlikely to be the case for anticlinal or faulted reservoirs.

Case study 2. Brent field, North Sea: the impact of depletion on infill drilling

The difficulty in estimating the magnitude of the stress depletion coefficient for the minimum horizontal stress (γ_h) manifested itself during the analysis of drilling challenges which were encountered during the infill drilling of the Brent field during the early 2000s. Brent had been on production since 1976, with the later stages of production benefiting from pressure support through water injection. In January 1998, the water injection was halted over the majority of the field and the reservoir pressure allowed to decrease through reservoir blow-down aimed at recovering remaining bypassed oil and gas. The average pressure depletion rate of the field was 3.4 MPa a^{-1} (500 psi/year).

Infill drilling of high-angle sidetracks from existing wellbores during 1999 began to encounter severe mud losses, which had not been observed with the earlier drilling. A post-well review identified the cause of the mud losses as a reduction in the fracture gradient of the wells, resulting from the decrease in the minimum horizontal stress accompanying the depletion (Addis *et al.* 2001). The losses were stress-related; the most severe losses corresponded to the lowest reservoir pressures and with wellbores drilled in the direction of the maximum horizontal stress direction i.e. predominantly NW–SE (Figure 9).

The infill drilling sidetracks targeted small pockets of bypassed oil on the eastern flanks of Brent, which were isolated from the main reservoir by

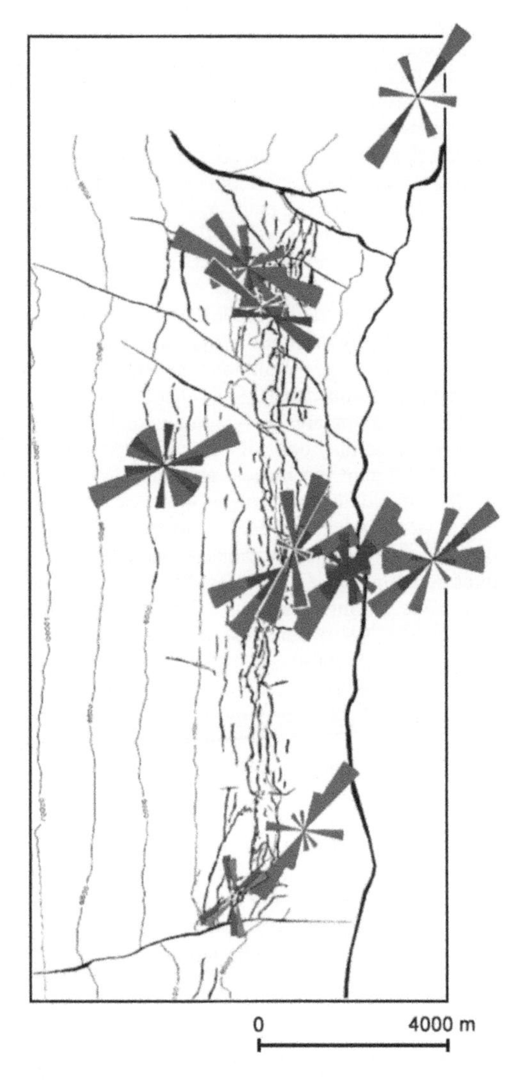

Fig. 9. Breakout (minimum horizontal stress) directions for the Brent field. Orange and pink denote borehole elongation in vertical wells and deviated wells, respectively; blue and green denote possible borehole breakouts in vertical wells and deviated wells, respectively. Rose diagrams are not to scale. After Boylan & Williams (1998). © 1998 Society of Petroleum Engineers. Reproduced with permission of SPE. Further reproduction prohibited without permission.

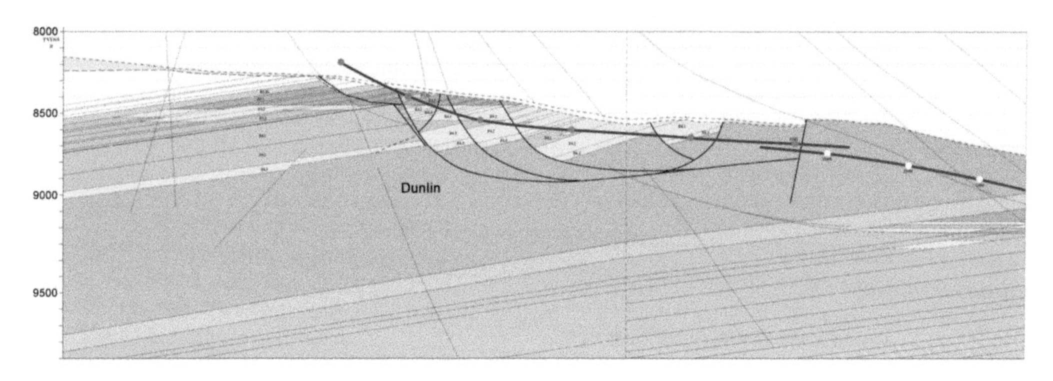

Fig. 10. Cross-section through the Brent reservoir along the BA05 S1 well path (in blue). After Addis *et al.* (2001). © 2001, Society of Petroleum Engineers. Reproduced with permission of SPE. Further reproduction prohibited without permission.

faulted compartments – the crestal slump faults (Struijk & Green 1991) (Fig. 10). These wells were low-cost sidetracks, *c.* £1 million per sidetrack, designed to access 1–2 million bbls of oil per well, but the mud losses and other non-productive time were making the sidetracks uneconomic (Davison *et al.* 2004).

As a result of the mud loss review, a number of operational changes were adopted to continue drilling the sidetracks, as outlined in Addis *et al.* (2001) and Davison *et al.* (2004). An early priority was to establish the stress depletion constant for Brent.

Stress depletion response

During the planning stages, the depletion coefficient for the minimum horizontal total stress was assumed to be $\gamma_h = 0.7–0.75$, based on data obtained from laboratory tests of the Brent sandstone and the simple passive basin assumption for horizontal stress decrease. Such estimates are commonly used as a first approximation, but they ignore the geology and the geological controls of stress and stress evolution accompanying depletion and injection.

The relationships used for estimating the horizontal stress magnitudes for faulted environments, post-faulting, illustrated in Figures 4 and 7 for thrust and normal faulting regimes, allow the horizontal stress depletion responses (γ_h) to be estimated for these different faulted environments (Addis 1997*a*). Given that the eastern flank of Brent was normally faulted, with slump faults, the stress depletion response might be controlled by the stresses acting in the presence of the normal faults. However, the estimates for these normally faulted conditions did not significantly differ from the passive basin values. This simple approach also does not account for the present day stress regime. The maximum horizontal stress direction, oriented perpendicular to the breakouts (Fig. 9), indicates that the horizontal stresses have rotated since the initiation of the faults because they are currently aligned oblique to the strike of the faults.

The dynamic response of the reservoir to pressure changes does not respond in isolation. The reservoir has a predominantly shale overburden, providing a contrast in the average elastic material properties between the reservoirs and the shale overburden. The impact of this modulus contrast on the reservoir stress depletion response was estimated using a semi-analytical model based on the Eshelby ellipsoidal inclusion approach (V. Dunayevsky, pers. comm. 2001), which resulted in an estimate of the stress depletion coefficient of $\gamma_h = 0.61–0.62$ for the horizontal total stresses and $\gamma_v = 0.01–0.02$ for the vertical total stress changes. The initial estimate of the stress depletion response and this update used core-based mechanical property measurements and considered uniform reservoir pressure changes across the field. Subsequently, an elastic finite element model was built and populated with the calculated reservoir pressures from a dynamic reservoir engineering model of the Brent field, leading again to stress depletion estimates for the minimum horizontal total stress in the region of $\gamma_h = 0.60–0.61$ over the majority of the field, but with lower values towards the flanks of the field (J. Emmen pers. comm. 2001).

These stress depletion estimates showed a reduction in the minimum horizontal total stress and in the fracture gradient of between 60 and 70% of the pore pressure decline for these high-angle sidetracks. However, predicting the observed stress depletion responses of reservoirs globally for different geological conditions had proved difficult at the time due to the range of factors affecting the depletion response (Addis 1997*b*).

Consequently, the approach taken to estimate the depletion constant, and the reduction in the fracture gradient with depletion, was based on an analysis of the field data. The only available data reflecting the change in the fracture gradient and the minimum horizontal total stress during the depletion stage of the Brent field development since 1998 were the mud loss occurrences. These are imprecise measurements, but in the absence of actual stress measurements, such as mini-fracture tests, they can be used as an operational estimate.

Business impact

The reservoir pressure estimates at the time of the mud loss, the differences in well trajectory and operational effects such as temperature, swabbing and surging were analysed, which led to an operational bound of $\gamma_h = 0.4$ for the decline in the minimum horizontal total stress (fracture gradient) with depletion. This much slower rate of minimum horizontal stress decline formed the basis for new well planning and, along with operational improvements, resulted in continued infill drilling in the Brent reservoir and significant improved recovery of reserves (Davison *et al.* 2004).

This Brent case study demonstrates the shortcomings of using simple model assumptions about stress and stress variations and the need to include geologically realistic models calibrated to field data measurements. If the early, higher, stress depletion estimates had been used for planning purposes, the infill well drilling could have been prematurely curtailed, with a loss of tens of millions of barrels of oil production and reserves recovery.

Strength differences resulting from cementation and facies variations

The previous sections have focused on the relative magnitudes of the *in situ* stresses for different faulted geological environments and the variation in the stress magnitudes in the reservoir accompanying depletion. Field development must also address the second element of geomechanical analysis: formation mechanical properties and, foremost among these, the strength of the reservoir rock.

Numerous relationships are used by the industry to estimate the strengths and mechanical properties of different lithologies from log-based geophysical measurements, as collated and summarized by Khaksar *et al.* (2009). These generic trends are invaluable, because measurements on reservoir formations are often limited by the available core. For a typical exploration well, core taken across the hydrocarbon-bearing interval may range from 20 m to over 100 m for larger reservoirs. Mechanical testing of 15–20 core sample intervals from the reservoir formation would be a typical sampling density per well. Given that the determination of mechanical rock properties on cores obtained from more than two to three exploration and appraisal wells is uncommon in a field, gives a total number of sample points of 40–60. More testing may occur in larger fields, but this is rarely done.

Table 2 shows calculations of sampling densities for typical geomechanical reservoir characterization for different well lengths or for a reservoir sector. A sampling rate for one well in a reservoir sector of 3 km radius corresponds to a sampling density $<1 \times 10^{-12}$ (<1 ppt). This very low sampling rate illustrates why the use of petrophysical well logs and three-dimensional seismic data, in combination with calibrated rock strength trends, are crucial in interpolating the measured rock properties across the entire reservoir and overburden, and underpin any three-dimensional geomechanical analysis of the field.

The use of these generic strength trends, however, introduces considerable uncertainty when estimating reservoir rock strength profiles for detailed geomechanical analysis because the correlations are generally 'one-parameter' correlations (e.g. unconfined compressive strength v. porosity). By contrast, Coates & Denoo (1980) and Bruce (1990) presented an equation to describe the dependence of sandstone strength on two primary factors (stiffness and clay content) based on a collation of earlier test data:

$$C_0 = 0.026 \times 10^{-6} \frac{2\cos\phi}{1-\sin\phi} EK(0.008V_{\text{clay}} + 0.0045[1 - V_{\text{clay}}])$$

where C_0 = unconfined compressive strength (psi), ϕ = angle of internal friction, E = Young's

Table 2. *Illustration of the sampling densities for geomechanical analysis in wells and reservoir sectors*

	No. of sampling points	Well length (m)	Reservoir radius (m)	Sampling density
Reservoir interval sampling	15–20	100		0.011–0.015
Total well depth sampling	15–20	3000		3×10^{-4}–5×10^{-4}
Reservoir (sector) sampling	15–20	100	3000	0.4×10^{-12}–0.6×10^{-12}

Based on standard mechanical samples 7.5 cm long.

modulus (psi), K = bulk modulus (psi) and V_{clay} = fractional volume of clay minerals

Plumb (1994) followed this approach, attributing the primary petrophysical controls of sandstone strength to porosity and clay content. Plumb (1994) described a number of characteristics of empirical strength curves for sandstones, including a transition from a grain load-bearing skeleton throughout the sandstones at low porosities and clay contents, to a predominantly matrix load-bearing structure at porosities >30–35%. This forms an upper bound for the unconfined compressive strengths of relatively clean load-bearing sandstones. For the lower porosity grain load-bearing sandstones, Plumb (1994) showed that the strength reduces with increasing clay content. Additional developments to define multi-parameter strength correlations include the work of Tokle *et al.* (1986) and Raaen *et al.* (1996).

Improved accuracy of both the strength, and any subsequent analysis, requires strength and mechanical property measurements on the reservoir core materials to calibrate these generic correlations. Ideally, mechanical property characterization tests are performed on each reservoir facies or reservoir layer to identify potential strength variations. In practice, the number of tests possible is limited by the available length of cores cut through the reservoir.

Case study 3: Lunskoye Sakhalin Island, Eastern Russia: the impact of strength variations on completion design

An example of how detailed lithological and strength variations impact geomechanical design is shown in the completion design for the reservoir sandstones in the Lunskoye field, offshore Sakhalin Island, eastern Russia. The Lunskoye gas field consists of 10 layers of the Daghinsky sandstone (layers I–X) with porosities of 15–30% and permeabilities ranging between 2 and 3000 mD. The Daghinsky Formation, of Miocene age, was deposited under cyclic repetitions of transgressive and regressive episodes along a fluvial-dominated delta system. The Upper and Middle Daghinsky sandstones form the Lunskoye reservoir, with the Upper Daghinsky (I–IV) described as a non-coal-bearing shallow marine, inter- to mid-shelf formation with minor deltaic influence and shoreline-parallel sand bodies. The Middle Daghinsky (V–XII) has been described as being deposited in a coal-bearing delta plain environment. The sandstone reservoir has a thickness of *c.* 400 m true vertical depth, each Daghinsky layer being separated by thin, fine-grained siltstone layers. Between Daghinsky layers IV and V is a field-wide siltstone layer *c.* 20 m thick (Ross *et al.* 2006).

The wells in this reservoir were planned as open-hole completions using either slotted or pre-drilled liners. However, the likelihood of sand production was high with these completions and would have led to significant production delays. The wells were re-designed and completed using a selective perforation technique, which involves shooting perforations across the higher strength sandstones and avoiding the high porosity, weaker sand-prone intervals. However, sufficient interval has to be perforated to deliver the 8.5 million m^3 per day per well (300 mmscf per day per well) of gas to meet the liquefied natural gas contract (Addis *et al.* 2008; Gunningham *et al.* 2008). This approach to the lower sandface completion design requires a rock testing programme to define a rock strength profile through the reservoir as the basis of the selective perforation design.

Strength v. porosity correlations

For the Lunskoye development and completion design, rock mechanics tests consisting of unconfined compressive strength tests, thick walled cylinder tests and petrophysical characterization tests were performed on the available cores from three reservoir layers. Even though the data were sparse, two strength trends were identified, with samples from the deeper sandstone layers exhibiting higher strengths than those from the shallower layers.

The strength v. porosity trends are shown in Figure 11, indicating that the deeper sandstones are *c.* 10 MPa (100 bar) stronger than the shallower sandstones for porosities between 15 and 25%. The differences in the strength were not readily explained by differences in texture or clay content. However, following the drilling of a new appraisal well, a spectral gamma ray log indicated that the cements in the different layers differed in geochemical composition.

The different geochemical compositions of the cements identified in the Lunskoye reservoir were used to explain the strength variations observed from the laboratory mechanical testing programme and the two rock strength (thick walled cylinder v. porosity) trends. The final selective perforated completion design was based on the two strength correlations applied to the different reservoir layers contained in the subsurface model. The model was discretized down to intervals 2 m true vertical depth thick in the three-dimensional static reservoir model, resulting in significantly more of the lower Daghinsky layers being selectively perforated in the completion design and contributing to greater production than would have been possible by using the lower strength or an average strength trend.

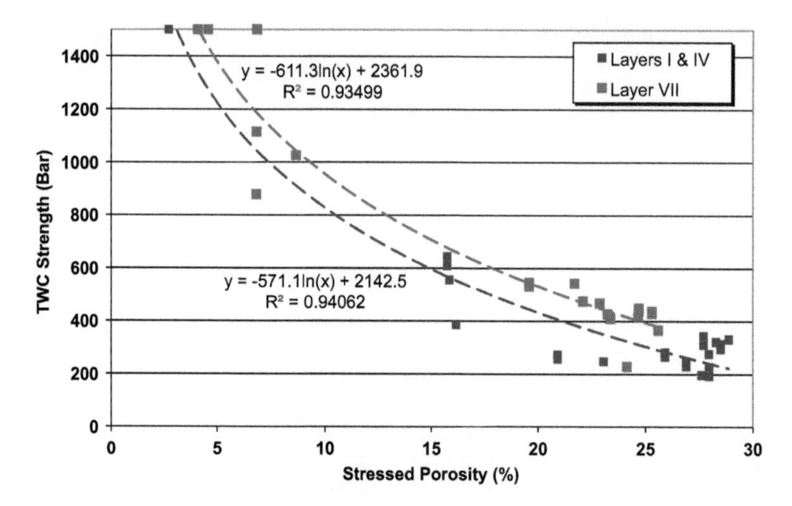

Fig. 11. Two strength trends identified for the Daghinsky sandstones of the Lunskoye field: stronger red trend line for the deeper layers; weaker blue trend line for shallower layers.

Business impact

The first phase of well drilling on Lunskoye used this selective perforating scheme for completing the wells, based on real-time reservoir data obtained from logging while drilling (LWD) to update the reservoir model, which enabled a customized, well-specific, selective perforating design for each of the wells drilled. As a result, the completions successfully delivered the required gas production sand-free for the start-up of the liquefied natural gas development (Zerbst & Webers 2011).

From this example we demonstrate that identifying strength differences in reservoir sandstones can lead to significantly less conservative completion designs and larger operational limits for production, leading to increased production and recovery. The higher strength sandstones are able to withstand larger drawdown pressures and depletions, with improved production and recovery.

Scratch testing of core with subtle lithological variations

Similar variations in rock strength have been observed in other reservoirs around the world, resulting from subtle geological influences. For example, reservoir facies differences in a North Sea field, observable in core, but not discernible from well logs, required two strength correlations to mechanically characterize the reservoir, with the higher strength facies on average 45 MPa (450 bar) stronger than the weaker strength facies for 15–20% porosity sandstones.

The sensitivity of strength to both depositional and diagenetic factors has been based on coarse sampling using plugs from cores, which introduces sampling bias into any analysis. Recent developments with the use of scratch tests, which provide a continuous measure of strength along the entire core, together with ultrasonic measurements, provide a means to use strength to identify different facies in far more detail than with sporadic core plugs (Germay & Lhomme 2016).

Figure 12 shows an example of scratch test based strength estimates over a 400 m long core section, showing a large scatter of data when plotted against the corresponding log porosity. When re-analysed using a clustering scheme, four different facies are identified, enabling unique strength correlations to be established for the different facies. Both standard strength measurements on core plugs and the more detailed measurements of strength obtained from the scratch tests provide independent evidence for the use of facies-based strength correlations for detailed completion design.

Concluding remarks

The contribution of geomechanical engineering to a range of field planning issues, from exploration through to the development and abandonment of oil and gas fields, has seen a continued increase since the 1980s. The uptake of geomechanics has relied on analytical models and, more recently, sophisticated numerical models to address the design optimization of field developments. These range from pore pressure and stress estimation, wellbore stability, sand production and completion design, reservoir management issues of compaction and subsidence and four-dimensional field monitoring to hydraulic fracturing and natural fracture

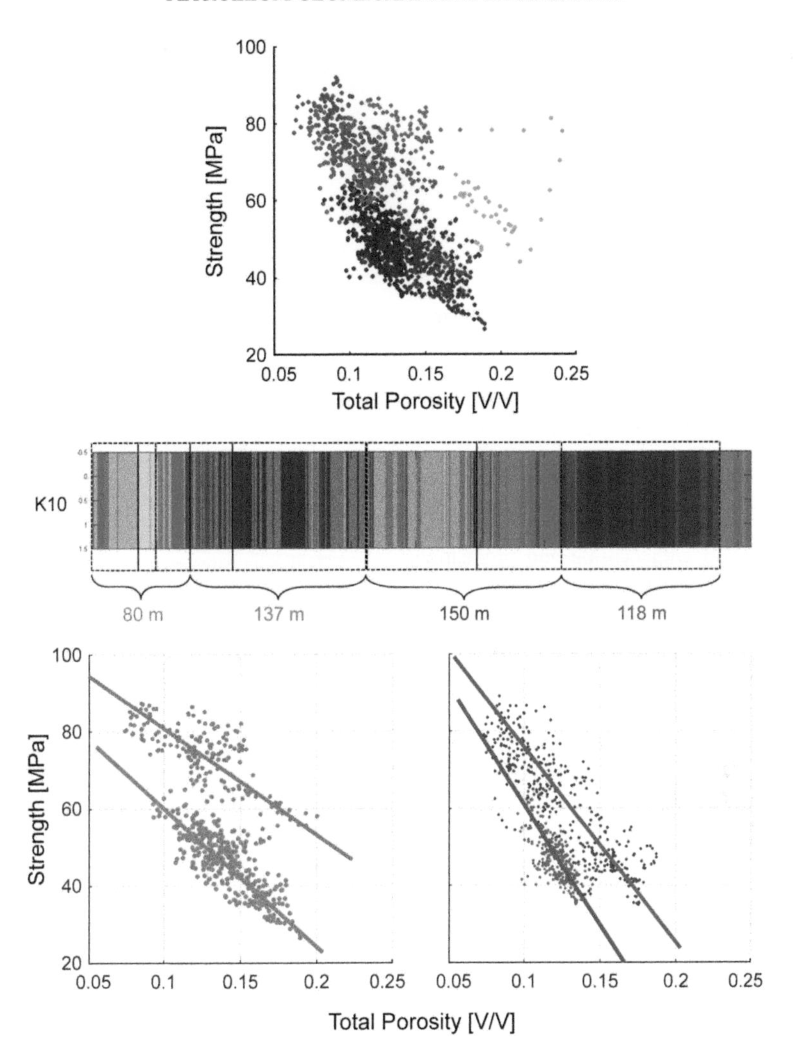

Fig. 12. Identification of different facies using strength v. porosity cross-plots based on continuous strength and p-wave velocity measurements on cores. After Noufal *et al.* (2015). © 2015, Society of Petroleum Engineers. Reproduced with permission of SPE. Further reproduction prohibited without permission.

stimulation. The use of analytical models has been effective, but the availability of field data in the case studies discussed in this paper highlights the limitations of the models, which can oversimplify the industry's view of the subsurface. What is required is both improved modelling and increased geological input into the models.

The use of field data to help characterize the stress state in both the Cusiana and Brent fields, for both the initial field conditions and during production and depletion, has led to novel explanations of the initial stress state with respect to the faulting style and the stress depletion response of the reservoirs. These data, along with improved models and geomechanical subsurface characterization, have led to significant improvements in the field developments.

Core sampling and mechanical testing strategies commonly result in very low sampling rates, down to <1 ppt of the reservoir volume. This highlights the reliance on establishing strength- and property-based correlations to enable petrophysical and geophysical extrapolation of the core measurements away from the well and across the field for subsurface geomechanical characterization. This low sampling density increases the uncertainty of any analysis, which can be managed, but not eliminated, with intelligent calibration of the data based on sound geological models. Nevertheless, analysis still benefits from adequate sampling and laboratory measurements, as shown in the Lunksoye completion design, and from the more detailed strength measurements obtained from scratch tests.

Field-wide porosity distributions derived from the three-dimensional seismic attribute analysis used to generate three-dimensional reservoir strength and mechanical property distributions typically rely on generic correlations for the translation from porosity to strength. The Lunskoye case study shows that significant strength differences measured on similar sandstones within the same reservoir arise from subtle geological changes. As such, future three-dimensional reservoir characterization would benefit from a facies level or unit description for any detailed geomechanical analysis and completion design.

Advanced numerical modelling is a practical approach to improving our visualization and understanding of subsurface geomechanical conditions and their evolution with drilling and production. This move away from the more simplistic models used in the industry to date is facilitated by the availability of more complex material models and increased computational power. However, models need data. This paper has shown how additional measurements throughout the lifetime of a field, from early characterization to the monitoring and surveillance of developments, allow asset teams to assess the validity of the subsurface models and react in a timely manner to optimize field development plans.

Understanding the geomechanical issues addressed here has led to operational changes with considerable financial impact on field developments. It has meant the difference between stable wells and unplanned drilling costs, continued drilling v. field shutdown, and between sub-optimum and improved reserves recovery. As we move into increasingly challenging environments, geomechanics is proving to be an essential key to economically unlocking additional reserves.

The field studies presented here, and the field improvements implemented as a result of these analyses and recommendations, rely on entire asset teams. The ideas and observations presented here are also the result of numerous discussions with supportive colleagues and I acknowledge the contributions of Mike McLean, Nigel Last, Dick Plumb, Philippe Charlez, Mike Cauley, Chris Kuyken, Victor Dunayevsky, Mark Davison, Mike Gunningham, Philippe Brassart, Jeroen Webers, Cor Kenter, Nick Barton, Axel Makurat and Najwa Yassir and the numerous researchers, operational and asset engineers with whom I have had the pleasure of working. The GSL paper reviewers, Paul Gillespie and Miltiadis Parotidis, also made excellent recommendations, helping to improve the paper, and in pointing out the possible impact of stress rotations during faulting on the formation of neotectonic fractures. I acknowledge the support of Rockfield Software Limited in preparing this paper and the support and invaluable suggestions of Najwa Yassir while reviewing and editing this paper.

References

AADNOY, B.S. & LOOYEH, R. 2011. *Petroleum Rock Mechanics: Drilling Operations and Well Design*. Gulf Professional Publishing, Houston, TX.

ADDIS, M.A. 1997*a*. Reservoir depletion and its effect on wellbore stability evaluation. *International Journal of Rock Mechanics and Mining Sciences*, **34**, 423.

ADDIS, M.A. 1997*b*. The stress depletion response of reservoirs. Paper SPE 38720-MS, paper presented at the 72nd SPE Annual Technical Conference & Exhibition, 5–8 October 1997, San Antonio, TX, USA.

ADDIS, M.A., LAST, N.R., BOULTER, D., RAMISA-ROCA, L. & PLUMB, R.A. 1993. The quest for borehole stability in the Cusiana field, Colombia. *Oilfield Review*, **5**, 33–43.

ADDIS, M.A., LAST, N.C. & YASSIR, N.A. 1994. Estimation of horizontal stresses at depth in faulted regions, and their relationship to pore pressure variations. Paper SPE 28140, presented at the 1994 SPE/ISRM Rock Mechanics in Petroleum Engineering Conference, 29–31 August 1994, Delft, the Netherlands. Reproduced in *SPE Formation Evaluation*, **11**, 11–18.

ADDIS, M.A., CAULEY, M.B. & KUYKEN, C. 2001. Brent in-fill drilling programme: lost circulation associated with drilling depleted reservoirs. Paper SPE/IADC 67741, presented at the SPE/IADC Drilling Conference, 27 February–1 March 2000, Amsterdam, the Netherlands.

ADDIS, M.A., GUNNINGHAM, M.C., BRASSART, Ph., WEBERS, J., SUBHI, H. & HOTHER, J.A. 2008. Sand quantification: the impact on sandface completion selection and design, facilities design and risk evaluation. Paper SPE 116713, presented at the 2008 SPE Annual Technical Conference and Exhibition, 21–24 September 2008, Denver, CO, USA.

ALBERTY, M.W. & MCLEAN, M.R. 2001. Fracture gradients in depleted reservoirs – Drilling wells in late reservoir life. Paper SPE 67740, presented at the SPE/IADC Drilling Conference, 27 February–1 March, Amsterdam, Netherlands.

ANDERSON, E.M. 1905. The dynamics of faulting. *Transactions of the Edinburgh Geological Society*, **8**, 387–402, https://doi.org/10.1144/transed.8.3.387

ANDERSON, E.M. 1951. *The Dynamics of Faulting and Dyke Formation with Applications to Britain*. Oliver and Boyd, Edinburgh.

ANTHEUNIS, D., VRIEZEN, P.B., SCHIPPER, B.A. & VAN DER VLIS, A.C. 1976. Perforation collapse: failure of perforated friable sandstones. Paper SPE 5750, presented at the SPE European Spring Meeting, 8–9 April, Amsterdam, Netherlands.

BAILEY, A.H.E., KING, R.C., HOLFORD, S.P. & HAND, M. 2016. Incompatible stress regimes from geological and geomechanical datasets: can they be reconciled? An example from the Carnarvon Basin, Western Australia. *Tectonophysics*, **683**, 405–416.

BELL, J.S. & GOUGH, D.I. 1979. Northeast–southwest compressive stress in Alberta – evidence from oil wells. *Earth and Planetary Science Letters*, **45**, 475–482.

Biddle, K.T. & Rudolph, K.W. 1988. Early Tertiary structural inversion in the Stord Basin, Norwegian North Sea. *Journal of the Geological Society, London*, **145**, 603–611, https://doi.org/10.1144/gsjgs.145.4.0603

Biot, M.A. 1941. General theory of three dimensional consolidation. *Journal of Applied Physics*, **12**, 155–164.

Bishop, A.W. 1974. The strength of crustal materials. *Engineering Geology*, **8**, 139–153.

Bol, G.M., Wong, S.-W., Davidson, C.J. & Woodland, D.C. 1994. Borehole stability in shales. *SPE Drilling & Completion*, **9**, 87–94.

Boylan, A. & Williams, C. 1998. *Brent Breakout Study*. Z&S Geology Report **ZSL-97-462**.

Bradley, W.B. 1979*a*. Failure of inclined boreholes. *Journal of Energy Resource Technology*, **101**, 232–239.

Bradley, W.B. 1979*b*. Mathematical stress cloud – stress cloud can predict borehole failure. *Oil & Gas Journal*, **77**, 92–102.

Bruce, S. 1990. A mechanical stability log. Paper SPE19942, presented at the SPE/IADC Drilling Conference, 27 February–2 March 1990, Houston, TX, USA.

Buchanan, J.G. & Buchanan, P.G. 1995. *Basin Inversion*. Geological Society, London, Special Publications, **88**, http://sp.lyellcollection.org/content/88/1

Charlez, P.A. 1991. *Rock Mechanics: Theoretical Fundamentals*. 1st edn. Editions Technip, Paris.

Charlez, P.A. 1997. *Rock Mechanics: Petroleum applications*. 2nd edn. Editions Technip, Paris.

Coates, G.R. & Denoo, S.A. 1980. Log derived mechanical properties and rock stress. Paper presented at the SPWLA-1980-U. SPWLA 21st Annual Logging Symposium, 8–11 July 1980, Lafayette, LA, USA.

Cooper, M.A. & Williams, G.D. 1989. *Inversion Tectonics*. Geological Society, London, Special Publications, **44**, http://sp.lyellcollection.org/content/44/1

Crook, A.J.L., Willson, S.M., Yu, J.G. & Owen, D.R.J. 2006. Predictive modelling of structure evolution in sandbox experiments. *Journal of Structural Geology*, **28**, 729–744.

Davison, J.M., Leaper, R. *et al.* 2004. Extending the drilling operating window in Brent: solutions for infill drilling in depleting reservoirs. Paper SPE 87174, presented at the IADC/SPE Drilling Conference, 2–4 March 2004, Dallas, TX, USA.

Eaton, B.A. 1969. Fracture gradient prediction and its application in oilfield operations. Paper SPE 2163. *Journal of Petroleum Technology*, October, 1353–1360.

Engelder, T. & Fischer, M.P. 1994. Influence of poroelastic behavior on the magnitude of minimum horizontal stress, S_h, in overpressured parts of sedimentary basins. *Geology*, **22**, 949–952.

Engelder, T., Lash, G.G. & Uzcategui, R.S. 2009. Joint sets that enhance production from Middle and Upper Devonian gas shales of the Appalachian Basin. *American Association of Petroleum Geologists Bulletin*, **93**, 857–889.

Fjaer, E., Holt, R.M., Horsrud, P., Raaen, A.M. & Risnes, R. 2008. *Petroleum Related Rock Mechanics*. 2nd edn. Developments in Petroleum Science, **53**. Elsevier, Amsterdam.

Geertsma, J. 1957. The effect of fluid pressure decline on volumetric changes of porous rocks. *Transactions AIME*, **210**, 331.

Germay, C. & Lhomme, T. 2016. Upscaling of Rock (Mechanical) Properties Measured on Core Plugs. Epslog Technical Note, internal report.

Gunningham, M.C., Addis, M.A. & Hother, J.A. 2008. Applying sand management process on the Lunskoye high gas-rate platform using quantitative risk assessment. Paper SPE 112099, presented at the 2008 SPE Intelligent Energy Conference and Exhibition, 25–27 February 2008, Amsterdam, the Netherlands.

Hancock, P.L. & Bevan, T.G. 1987. Brittle modes of foreland extension. *In*: Coward, M.P., Dewey, J.F. & Hancock, P.L. (eds) *Continental Extensional Tectonics*. Geological Society, London, Special Publications, **28**, 127–137, https://doi.org/10.1144/GSL.SP.1987.028.01.10

Hancock, P.L. & Engelder, T. 1989. Neotectonic joints. *Geological Society of America Bulletin*, **101**, 1197–1208.

Hatchell, P.J., van den Beukel, A. *et al.* 2003. Whole Earth 4D: reservoir monitoring geomechanics. Paper SEG-2003-1330, presented at the SEG Annual Meeting, 26–31 October, Dallas, Texas.

Hennig, A., Yassir, N., Addis, M.A. & Warrington, A. 2002. Pore pressure estimation in an active thrust region and its impact on exploration and drilling. *In*: Huffman, A.R. & Bowers, G.L. (eds) *Pressure Regimes in Sedimentary Basins and their Prediction*. AAPG Memoirs, **76**, 89–105.

Hillis, R.R. 2003. Pore pressure/stress coupling and its implications for rock failure. *In*: Van Rensbergen, P., Hillis, R.R., Maltman, A.J. & Morley, C.K. (eds) Geological Society, London, Special Publications, **216**, 359–368, https://doi.org/10.1144/GSL.SP.2003.216.01.23

Hottman, C.E., Smith, J.H. & Purcell, W.R. 1979. Relationship among Earth stresses, pore pressure, and drilling problems offshore Gulf of Alaska. *Journal of Petroleum Technology*, **31**, 1477–1484.

Hubbert, M.K. & Willis, D.G. 1957. Mechanics of hydraulic fracturing. *Petroleum Transactions, AIME*, **210**, 153–168.

Hubbert, M.K. & Rubey, W.W. 1959. Role of fluid pressure in mechanics of overthrust faulting: I. Mechanics of fluid-filled porous solids and its application to overthrust faulting. *Geological Society of America Bulletin*, **70**, 115–166.

Kattenhorn, S.A., Aydin, A. & Pollard, D.D. 2000. Joints at high angles to normal fault strike: an explanation using 3-D numerical models of fault-perturbed stress fields. *Journal of Structural Geology*, **22**, 1–23.

Kenter, C.J., van den Beukel, A.C. *et al.* 2004. Geomechanics and 4d: evaluation of reservoir characteristics from timeshifts in the overburden. Paper ARMA-04-627, presented at the Gulf Rocks 2004, the 6th North America Rock Mechanics Symposium (NARMS), 5–9 June, Houston, Texas.

Khaksar, A., Taylor, P.G., Fang, Z., Kayes, T.J., Salazar, A. & Rahman, K. 2009. Rock strength from core and logs, where we stand and ways to go. Paper SPE 121972, presented at the EUROPEC/EAGE

Conference and Exhibition, 8–11 June 2009, Amsterdam, the Netherlands.

Kirsch, E.G. 1898. Die Theorie der Elastizität und die Bedürfnisse der Festigkeitslehre. *Zeitschrift des Vereines deutscher Ingenieure*, **42**, 797–807.

Last, N.C. & McLean, M.R. 1996. Assessing the impact of trajectory on wells drilled in an overthrust region. *Journal of Petroleum Technology*, SPE 30465, 620–626.

Last, N.C., Plumb, R.A., Harkness, R.M., Charlez, P., Alsen, J. & McLean, M.R. 1995. An integrated approach to evaluating and managing wellbore instability in the Cusiana Field, Colombia, South America. Paper SPE 30464, presented at the SPE Annual Technical Conference and Exhibition, 22–25 October 1995, Dallas, TX, USA.

Last, N.C., Harkness, R.M. & Plumb, R.A. 1998. From theory to practice: evaluation of the stress distribution for wellbore stability analysis in an overthrust regime by computational modelling and field calibration. Paper SPE/ISRM 47209, presented at the SPE/ISRM Eurock '98 Conference, 8–10 July 1998, Trondheim, Norway.

Mandl, G. 2000. *Faulting in Brittle Rocks: An Introduction to the Mechanics of Tectonic Faults*. Springer, Berlin.

Molenaar, M.M., Hatchell, P.J., van den Beukel, A.C., Jenvey, N.J., Stammeijer, J.G.F., van der Velde, J.J. & de Haas, W.O. 2004. Applying geomechanics and 4D: '4D In-situ Stress' as a complementary tool for optimizing field management. Paper ARMA/NARMS 04-639, presented at Gulf Rocks 2004, the 6th North America Rock Mechanics Symposium (NARMS): Rock Mechanics Across Borders and Disciplines, 5–9 June 2004, Houston, TX, USA.

Moos, D. 2014. The future of geomechanics – where we are, how we got here, and where we're going. Paper presented at the SPE Workshop, Applying Geomechanics in the E&P Industry: Best Practices and Recent Technological Developments. 29–30 April 2014, Guadalajara, Mexico.

Morita, N., Black, A.D. & Guh, G.-F. 1990. Theory of lost circulation pressure. Paper SPE 20409, presented at the SPE Annual Technical Conference and Exhibition, New Orleans, LA, 43–58.

Noufal, A., Germay, C., Lhomme, T., Hegazy, G. & Richard, T. 2015. Enhanced core analysis workflow for the geomechanical characterisation of reservoirs in a giant offshore oilfield, Abu Dhabi. Paper SPE 175412, presented at ADIPEC, Abu Dhabi International Petroleum Exhibition and Conference, 5–9 November 2015, Abu Dhabi, UAE.

Osorio, J.G. & Lopez, C.F. 2009. Geomechanical factors affecting the hydraulic fracturing performance in a geomechanically complex, tectonically active area in Colombia. Paper SPE 122315, presented at the SPE Latin American and Caribbean Petroleum Engineering Conference, 31 May–3rd June 2009, Cartagena, Colombia.

Osorio, J.G., Penuela, G. & Otalora, O. 2008. Correlation between microseismicity and reservoir dynamics in a tectonically active area of Colombia. Paper SPE 115715, presented at the SPE Annual Technical Conference and Exhibition, 21–24 September 2008, Denver, CO, USA.

Peric, D. & Crook, A.J.L. 2004. Computational strategies for predictive geology with reference to salt tectonics. *Computer Methods in Applied Mechanics and Engineering*, **193**, 5195–5222.

Plumb, R.A. 1994. Influence of composition and texture on the failure properties of clastic rocks. Paper SPE 28022, presented at Rock Mechanics in Petroleum Engineering, 29–31 August 1994, Delft, the Netherlands.

Raaen, A.M., Hovem, K.A., Joranson, H. & Fjaer, E. 1996. FORMEL: a step forward in strength logging. Paper SPE 36533, presented at the SPE Annual Technical Conference and Exhibition, 6–9 October 1996, Denver, CO, USA.

Ross, L., King, K. *et al.* 2006. Seismically based integrated reservoir modelling, Lunskoye Field, offshore Sakhalin, Russian Federation. Paper SPE 102650, presented at the Russian Oil and Gas Technical Conference and Exhibition, 3–6 October 2006, Moscow, Russia.

Salz, L.B. 1977. Relationship between fracture propagation pressure and pore pressure. Paper SPE 6870, presented at the 52nd SPE Annual Technical Conference, 9–12 October 1977, Denver, CO, USA.

Santarelli, F.J., Tronvoll, J.T., Svennekjaier, M., Skeie, H., Henriksen, R. & Bratli, R.K. 1998. Reservoir stress path: the depletion and the rebound. Paper SPE 47350, presented at the SPE/ISRM Rock Mechanics in Petroleum Engineering, 8–10 July 1998, Trondheim, Norway.

Santarelli, F.J., Havmoller, O. & Naumann, M. 2008. Geomechanical aspects of 15 years water injection on a field complex: an analysis of the past to plan the future. Paper SPE 112944, presented at the SPE North Africa Technical Conference & Exhibition, 12–14 March 2008, Marrakech, Morocco.

Sibson, R.H. 1977. Fault rocks and fault mechanisms. *Journal of the Geological Society, London*, **133**, 191–213, https://doi.org/10.1144/gsjgs.133.3.0191

Sibson, R.H. 2003. Brittle-failure controls on maximum sustainable overpressure in different tectonic regimes. *American Association of Petroleum Geologists Bulletin*, **87**, 901–908.

Skelton, J., Hogg, T.W., Cross, R. & Verheggen, L. 1995. Case history of directional drilling in the Cusiana Field in Colombia. Paper IADC/SPE 29380, presented at the IADC/SPE Drilling Conference, 28th February–2nd March, 1995, Amsterdam, the Netherlands.

Skempton, A.W. 1954. The pore pressure coefficients A and B. *Geotechnique*, **4**, 143–147.

Struijk, A.P. & Green, R.T. 1991. The Brent Field, Block 211/29, UK North Sea. *In*: Abbotts, I.L. (ed.) *United Kingdom Oil and Gas Fields – 25 Years Commemorative Volume*. Geological Society, London, Memoirs, **14**, 63–72, https://doi.org/10.1144/GSL.MEM.1991.014.01.08

Terzaghi, K. 1925. *Erdbaumechanik auf Bodenphysikalischer Grundlage*. F. Deuticke.

Terzaghi, K. 1943. *Theoretical Soil Mechanics*. Wiley, Chichester.

Teufel, L.W., Rhett, D.W. & Farrell, H.E. 1991. Effect of reservoir depletion and pore pressure drawdown on in situ stress and deformation in the Ekofisk Field, North Sea. Paper ARMA-91-063, presented at the 32nd US Symposium on Rock Mechanics (USRMS), 10–12 July 1991, Norman, OK, USA.

Thornton, D.A. & Crook, A.J.L. 2014. Predictive modeling of the evolution of fault structure: 3-D modeling and coupled geomechanical/flow simulation. *Rock Mechanics and Rock Engineering*, **47**, 1533–1549.

Tokle, K., Horsrud, P. & Bratli, R.K. 1986. Predicting uniaxial compressive strength from log parameters. Paper SPE 15645, presented at the 61st SPE Annual Technical Conference and Exhibition, 5–8 October 1986, New Orleans, LA, USA.

Willson, S.M., Last, N.C., Zoback, M.D. & Moos, D. 1999. Drilling in South America: a wellbore stability approach for complex geologic conditions. Paper SPE 53940, presented at the Latin American and Caribbean Petroleum Engineering Conference, 21–23 April 1999, Caracas, Venezuela.

Woodland, D.C. & Bell, J.S. 1989. In situ stress magnitudes from mini-frac records in Western Canada. PETSOC-89-05-01. *Journal of Canadian Petroleum Technology*, **28**, 22–31.

Wu, B., Addis, M.A. & Last, N.C. 1998. Stress estimation in faulted regions: the effect of residual friction. Paper SPE 47210, presented at Eurock '98 Conference, 8–10 July 1998, Trondheim, Norway.

Yassir, N.A. 1989. Undrained shear characteristics of clay at high total stresses. Paper ISRM-IS-1989-114, presented at Rocks at Great Depth, ISRM International Symposium, 30 August–2 September, Pau, France.

Yassir, N.A. & Zerwer, A. 1997. Stress regimes in the Gulf coast, offshore Louisiana: data from wellbore breakout analysis. *American Association of Petroleum Geologists Bulletin*, **81**, 293–307.

Zerbst, C. & Webers, J. 2011. Completing the first bigbore gas wells in Lunskoye – a case history. *SPE Drilling & Completion*, December, 462–471.

Zoback, M.D. 2008. *Reservoir Geomechanics*. Cambridge University Press, Cambridge.

In situ stress distribution and mechanical stratigraphy in the Bowen and Surat basins, Queensland, Australia

EMMA TAVENER[1], THOMAS FLOTTMANN[2]* & SAM BROOKE-BARNETT[2]

[1]*Santos Ltd, Santos Place, 32 Turbot Street, Qld 4000, Australia*

[2]*Origin Energy, 339 Coronation Drive, Qld 4064, Australia*

**Correspondence: thomas.flottman@originenergy.com.au*

Abstract: We present regional *in situ* stress analyses based on publicly available log and pressure data from coal seam gas developments in the Permian Bowen basin, Australia. Together with earlier data from the eastern part of the Jurassic Surat basin, our results show a broad, but systematic, rotation of S_{Hmax} azimuths in this part of eastern Australia as well as systematic changes in stress state with depth. Overall, the geomechanical state of the region appears to reflect the interplay between basin-controlling structures and a complex far-field stress regime. At the reservoir level, within and between Permian coal seams, this stress complexity is reflected in highly variable stress states both vertically and laterally. Stress data, including direct pressure measurements and observations of borehole failure in image logs, have been used to calibrate sonic-derived one-dimensional wellbore stress models that consistently exhibit a change in tectonic stress regime with depth. Shallow depths (<600 m) are characterized by a reverse-thrust stress regime and deeper levels are characterized by a strike-slip regime. Changes in the stress state with depth influence the mechanical stratigraphy of rocks with widely contrasting mechanical attributes (coals and clastic sediments). Our results highlight the interdependency between regional tectonic, local structural and detailed rheological influences on the well scale geomechanical conditions that have to be taken into consideration in drilling and completion designs.

Supplementary material: Database of additional wells with image log data are available at https://doi.org/10.6084/m9.figshare.c.3785849

The Australian continent is characterized by significant variability in the magnitude and orientation of *in situ* stresses (Coblentz *et al.* 1995, 1998; Hillis *et al.* 1999; Hillis & Reynolds 2000, 2003; Reynolds *et al.* 2002, 2003) at both the continental and regional scale. This is well documented from oil-field data, particularly in the eastern–central interior basins (Reynolds *et al.* 2005; Nelson *et al.* 2007). The *in situ* stress distribution in the Australian plate is controlled by plate boundary forces acting on the Australian plate (Fig. 1). The key plate tectonic elements bracketing the Australian plate include the divergent southern margin between Australia and Antarctica, transpressional convergence at the southeastern plate margin, compression along the northern and northwestern plate margin (particularly the Papua New Guinea fold–thrust belt and the Himalayan collision zone) and subduction at the northeastern margin (Indonesian Arc; Reynolds *et al.* 2002, 2003; Sandiford *et al.* 2004).

The Bowen basin (Figs 1 & 2) is interpreted as a Permian to Triassic back-arc basin (Holcombe *et al.* 1997*a*, *b*; Korsch & Totterdell 2009) and is one of a series of rift basins that developed across eastern Australia from the Early Permian (Korsch *et al.* 2009*a*). Between *c.* 265 and 230 Ma, the Bowen basin and the New England Orogen were subjected to contractional and strike-slip deformation known as the Hunter Bowen Orogeny (Holcombe *et al.* 1997*b*; Korsch *et al.* 2009*b*). This deformation initiated basin-bounding fault systems (Fig. 2). The Bowen basin is broadly characterized by two north–south-trending depocentres, the Denison and Taroom troughs (Fig. 2), with internal half-graben structures that initiated during the Early Permian. The early Permian basin-fill is dominated by fluvio-lacustrine clastic successions.

The deposition of thick, mid–late Permian coal measures (2–15 m), particularly in the eastern part of the basin, occurred during thermal sag, which was followed by significant late Permian inversion of local half-graben. In the Bowen basin, one to three coal seams of economic interest can be developed at the well scale. Late Permian and Triassic contraction led to complex reactivation structures and the deposition of late Permian–Early Triassic fluvio-marine clastics; deposition ceased during a mid–late Triassic contractional event. The development of permeability in coal seam gas fields is loosely associated with structural highs.

From: Turner, J. P., Healy, D., Hillis, R. R. & Welch, M. J. (eds) 2017. *Geomechanics and Geology*. Geological Society, London, Special Publications, **458**, 31–47.
First published online May 24, 2017, https://doi.org/10.1144/SP458.4

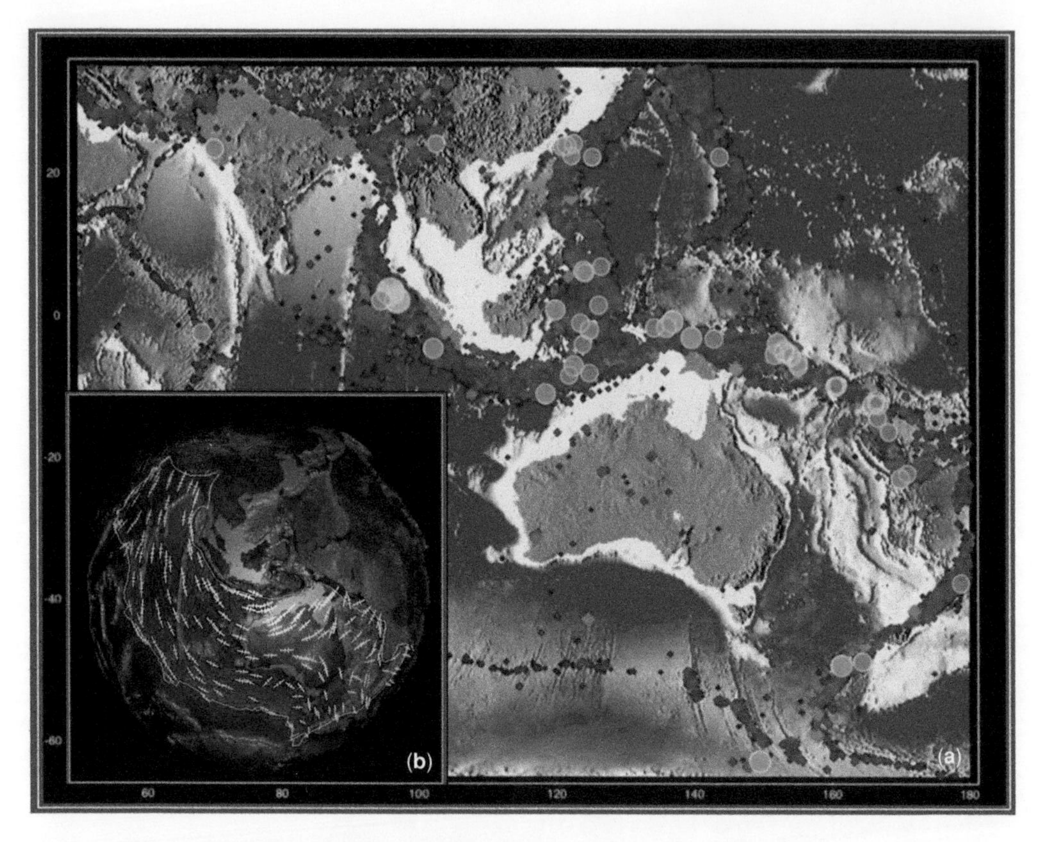

Fig. 1. (**a**) Indo-Australian plate and surrounding plate margins. Orange circles indicate earthquakes at plate margins; the diameter of the circles corresponds to the relative earthquake magnitude. (**b**) Plate motion vectors, Indo-Australian plate. Adapted from Sandiford (2016).

The southeastern part of the Bowen basin is overlain by the Surat basin (grey coloured area in Figs 2 & 3), which formed a broad intracontinental depression during the Jurassic. Surat basin sediments are dominated by siltstones and mudstones with minor sandstones, all of which were deposited in a fluvio-lacustrine depositional environment. All the clastic lithologies contain significant volcanic components, resulting in low porosity and permeability. The Surat basin contains multiple coal seams (on average 60) with an average thickness of 0.4 m. These coal seams form the Walloons fairway (Fig. 2).

Structuring in the Surat basin is subtle and reflects the more intense deformation of the underlying reactivated inversion structures in the Permian Bowen basin. Triassic structural highs set up gentle, low-amplitude highs in the Surat basin, which are characterized by exceptional permeabilities ranging from hundreds of millidarcies (mD) to multidarcies (e.g. Undulla Nose, Fig. 2). High permeability regions are characterized by coals with multiple fracture orientations readily identifiable on image logs (i.e. no preferred fracture orientation).

Coal seam methane has been produced in SE Queensland for over 20 years. The industry saw a marked acceleration of drilling and broad data acquisition in the early 2000s, when several world-scale coal seam gas to liquefied natural gas projects commenced construction and production. Early production was mainly from vertical wells with or without hydraulic fracture completions. Ongoing optimization of both drilling design and hydraulic fracture completions has fostered a wealth of diagnostic data acquisition and some design changes (Flottmann *et al.* 2013; Kirk-Burnnand *et al.* 2015). Initial work on the geomechnical framework (Brooke-Barnett *et al.* 2015) has shown the variability of stress magnitudes and the potential influence of fundamental basement and basin structures.

The focus of this study is the *in situ* stress state in the context of the geology and geomechanics of the Permian Bowen basin and the Roma Shelf region of the Surat basin (Fig. 2). The wireline data utilized in this study are publicly available at QDEX (2016). The results presented here complement earlier studies conducted in the eastern part of the Jurassic Surat

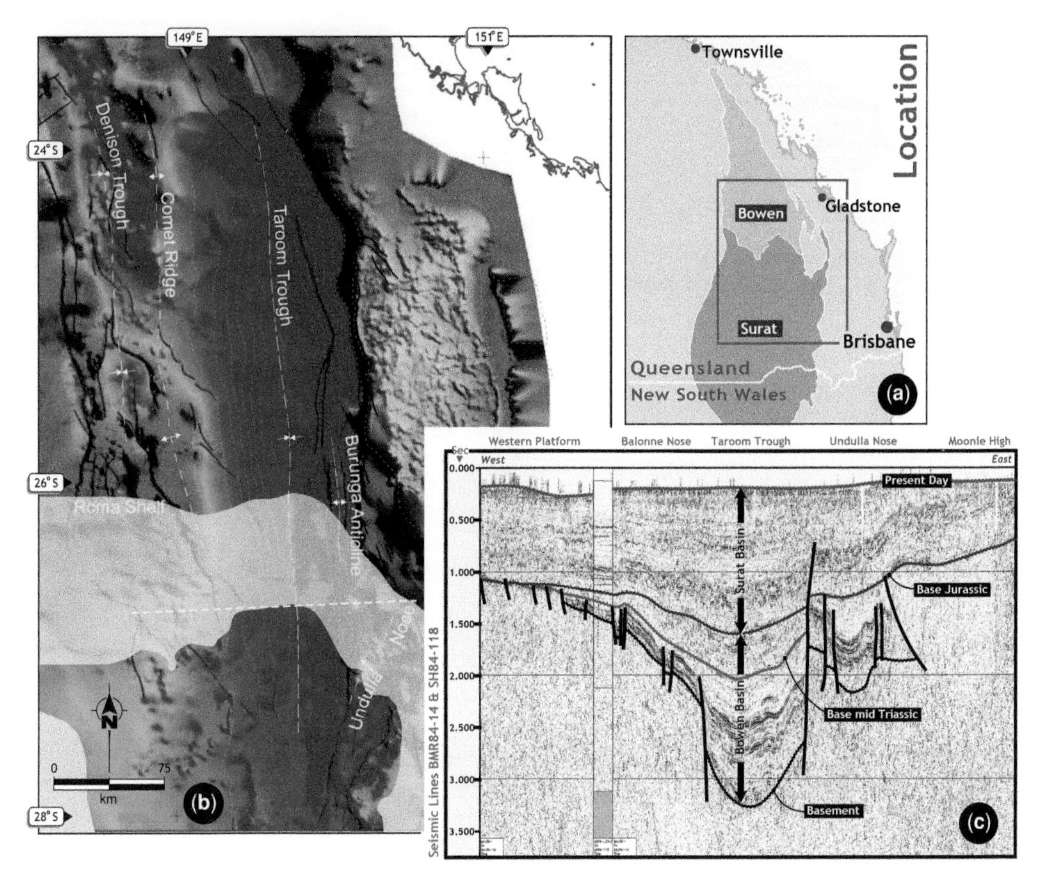

Fig. 2. (**a**) Location of Bowen and Surat basins in Australia. (**b**) SEEBASE map showing the structure of the top basement underlying the Bowen basin (see text for further details); grey area shows extent of the Walloons (production) fairway of the Surat Basin; dotted line is the location of the regional seismic line in part (c). (**c**) Seismic line showing basic structural elements of Bowen and Surat basins

basin (excluding the Roma Shelf region) to the south of the Bowen basin (Figs 2 & 3; Flottmann *et al.* 2013; Brooke-Barnett *et al.* 2015). The data presented here show significant three-dimensional complexity and granularity in the stress tensor (the relative magnitude of the principal stresses) and their plan view orientation in an area with hitherto sparse datasets; the World Stress Map (Heidbach *et al.* 2008, 2010) shows a comparatively uniform distribution of the maximum horizontal stress (S_{Hmax}) in eastern Queensland. Similarly, both the magnitude of differential stresses and the Andersonian stress state (reverse/strike-slip/normal; Anderson 1951) varies significantly with depth. The geomechanics of the Bowen and Surat basins are uniquely influenced by the geological setting because the stratigraphic column vertically juxtaposes lithologies with starkly contrasting rheological properties (e.g. clastic sediments v. coals).

This paper has four objectives:

(1) Documenting the significant plan view variability of the S_{Hmax} orientation in an intracontinental basin setting utilizing a comprehensive and regionally extensive *in situ* stress dataset based on >180 wells, of which 145 wells present new data.
(2) Establishing systematic variations of stress geometry with depth from representative examples of one-dimensional wellbore stress models based on log-derived strain-based stress calculations.
(3) Discussing the implications of both the lateral and vertical stress variability on the bulk mechanical stratigraphy which, in turn, influence completion strategies, such as hydraulic fracture stimulation and inclined and horizontal drilling.

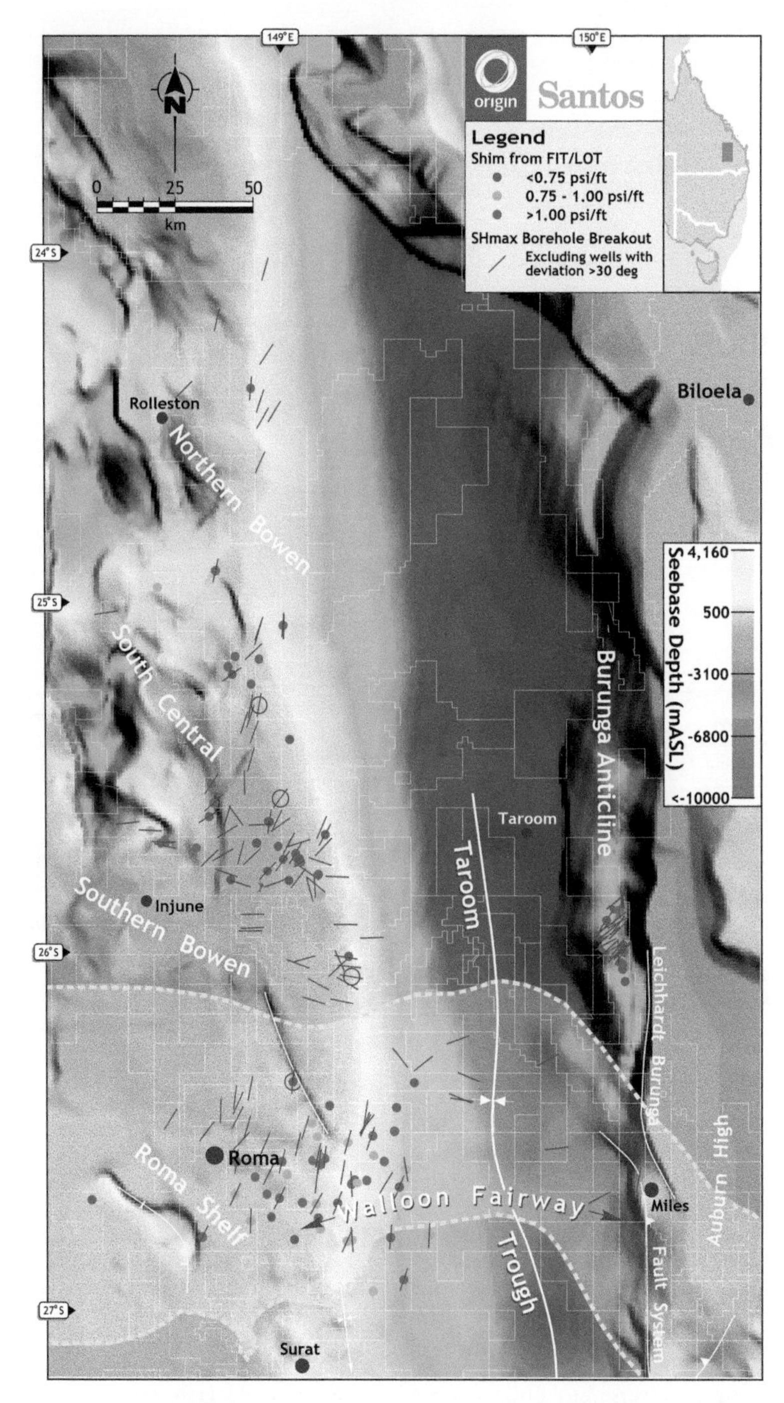

Fig. 3. Azimuths of S_{Hmax} from breakout interpretation on image logs on the SEEBASE map (see Fig. 2b). All orientations are from breakouts deeper than 450 m and wells with a wellbore inclination <30°. Note the variability of S_{Hmax} azimuths in the different domains (see text for discussion). Coloured circles represent the S_{hmin} magnitude as derived from LOTs (below 450 m TVD RT). The map illustrates that the north and central Bowen regions appear to be highly stressed, the Roma Shelf and Burunga Anticline are at intermediate stress and the south Bowen is in the lowest present day stress. Open circles refer to wells presented in Figures 5–8 (from north to south).

(4) Assessing the *in situ* stress and geomechanical implications based on data from the Permian Bowen basin (presented here) with existing data and interpretations from the Jurassic Surat basin.

Data, conditioning and calibration

The basic stratigraphic correlations are from standard well logs (gamma ray, density, resistivity, two-arm caliper; for stratigraphic overview, see Cook & Jell 2013). Integrating the density log and extrapolating the trend to the surface allows calculation of the vertical stress magnitude. The integral of the density logs with reference to depth gives the gradient of the vertical stress (S_V), usually around 1 psi/ft (*c.* 19.2 ppg, *c.* 22.6 kPa m^{-1}) (equation 1).

Vertical (overburden) stress at depth *z* (Pa):

$$S_V = \int_0^z \rho(z) g \, dz \tag{1}$$

where z is the depth below ground level, ρ is the density in kg m^{-3} and g is the acceleration due to gravity (assumed to be 9.81 m s^{-2}).

The orientation of S_{Hmax} (the maximum horizontal stress) is derived from borehole breakouts and the drilling-induced tensile fractures (DITFs) observed in image logs. Breakouts form due to conjugate microfracturing and shear failure at the wellbore wall in response to hoop stresses (Kirsch 1898). Breakout is a product of far-field stresses interacting with the wellbore wall and elongating the wellbore in the direction of S_{hmin} (the minimum horizontal stress; see Bell 1990, 1996*a*, *b*), leading to an overall oval shape of the wellbore. DITFs form in the azimuth of S_{Hmax}; they form sharp, usually linear, fractures where the tensile hoop stresses are greater than the tensile rock strength at the wellbore wall. Both breakouts and DITFs are sensitive to mud-weight changes and the magnitude of the differential stresses (in vertical wellbores, $S_{Hmax} - S_{hmin}$) applied to the wellbore wall. For a given far-field stress state, elevated mud-weights can suppress the initiation of borehole breakout, whereas high mud-weights can, in turn, initiate DITFs. The wells used here are usually drilled either under balance or slightly over balance with respect to the hydrostatic gradient of.433 psi/ft (*c.* 8.3 ppg; 9.8 kPa m^{-1}). Wireline logging is undertaken after the wellbore is filled with a 3% KCl brine under slightly overbalanced conditions. All logs are referenced and corrected to true north. Stress measurements were ranked according the classification scheme of the World Stress Map (Tingay *et al.* 2008). The stress data are displayed on a basin structure map generated using the SEEBASE method of integrating various datasets to generate a best approximate of a 'depth-to-basement' structure image (SEEBASE 2005).

One-dimensional wellbore stress models are used to constrain the relative magnitude of S_{hmin} and S_{Hmax} (in relation to S_V) with depth. One-dimensional wellbore stress models are based on Poisson's ratio (equation 2) and Young's modulus (equation 3), which are derived from dipole sonic and density wireline data (dynamic data). A dynamic to static conversion of Poisson's ratio and Young's modulus was derived regionally from rock mechanics laboratory measurements (equations 4–7). Minimum and maximum horizontal stresses were then calculated using poroelastic stress equations (Eaton 1968, 1972, 1975; Thiercelin & Plumb 1994; equations 8 & 9), which incorporate the static Poisson's ratio, vertical stress, pore pressure, the static Young's modulus and Biot's coefficient, as well as tectonic strain in the minimum (ε_{min}) and maximum (ε_{max}) horizontal stress directions. Nominal tectonic strain values of $\varepsilon_{max} = 0.0009$ and $\varepsilon_{min} = 0.0003$ were used for the initial calculations before calibration (Brooke-Barnett *et al.* 2015). Pore pressure was calculated based on a freshwater hydrostatic gradient, which is commonly observed in undepleted coal reservoirs. Biot's coefficient was not independently constrained and was set as 1 to ensure consistency across the basin, thus true variation in poroelastic strain constants can be assessed.

Dynamic Poisson's ratio (ν_{dyn}) (no units):

$$\nu_{dyn} = \frac{(V_p/V_s)^2 - 2}{2[(V_p/V_s)^2 - 1]} \tag{2}$$

where V_p is the compressional sonic velocity in m s^{-1} and V_s is shear sonic velocity in m s^{-1}.

Dynamic Young's modulus (E_{dyn}) (Pa):

$$E_{dyn} = \frac{\rho V_s^2 (3V_p^2 - 4V_s^2)}{V_p^2 - V_s^2} \tag{3}$$

where ρ is the density in kg m^{-3}, V_p is the compressional sonic velocity in m $s^{-1}$and V_s is the shear sonic velocity in m s^{-1}.

Sandstone static Poisson's ratio (ν_{stat}) (no units):

$$\nu_{stat} = 0.7\nu_{dyn} + 0.06 \tag{4}$$

where ν_{dyn} is the dynamic Poisson's ratio.

Siltstone static Poisson's ratio (v_{stat}) (no units):

$$v_{stat} = 0.7v_{dyn} + 0.08 \quad (5)$$

where v_{dyn} is the dynamic Poisson's ratio.

Sandstone static Young's modulus (E_{stat}) (Pa):

$$E_{stat} = 0.32E_{dyn} \quad (6)$$

where E_{dyn} is the dynamic Young's modulus.

Siltstone static Young's modulus (E_{stat}) (Pa):

$$E_{stat} = 0.3E_{dyn} \quad (7)$$

where E_{dyn} is the dynamic Young's modulus.

Strain-derived S_{Hmax} (Pa):

$$\underbrace{S_{Hmax} - \alpha P_p}_{\sigma_{Hmax}} = \underbrace{\frac{v_{stat}}{1 - v_{stat}}(S_V - \alpha P_p)}_{\text{Vertical component}} + \underbrace{\frac{E_{stat}}{(1 - v_{stat}^2)}(\varepsilon_{max} + v_{stat}\varepsilon_{min})}_{\text{Tectonic component}}$$

$$\Rightarrow S_{Hmax} = \frac{v_{stat}}{1 - v_{stat}}(S_V - \alpha P_p) + \frac{E_{stat}}{(1 - v_{stat}^2)}(\varepsilon_{max} + v_{stat}\varepsilon_{min}) + \alpha P_p \quad (8)$$

Strain-derived S_{hmin} (Pa):

$$\underbrace{S_{hmin} - \alpha P_p}_{\sigma_{hmin}} = \underbrace{\frac{v_{stat}}{1 - v_{stat}}(S_V - \alpha P_p)}_{\text{Vertical component}} + \underbrace{\frac{E_{stat}}{(1 - v_{stat}^2)}(\varepsilon_{min} + v_{stat}\varepsilon_{max})}_{\text{Tectonic component}}$$

$$\Rightarrow S_{hmin} = \frac{v_{stat}}{1 - v_{stat}}(S_V - \alpha P_p) + \frac{E_{stat}}{(1 - v_{stat}^2)}(\varepsilon_{min} + v_{stat}\varepsilon_{max}) + \alpha P_p \quad (9)$$

where σ_{Hmax} is the maximum effective horizontal stress, σ_{hmin} is the minimum effective horizontal stress, v_{stat} is the static Poisson's ratio, S_v is the vertical stress, α is Biot's coefficient, P_p is the formation pressure, E_{stat} is the static Young's modulus, ε_{max} is the strain in the maximum horizontal stress direction and ε_{min} is the strain in the minimum horizontal stress direction.

The initial calibration of the stress profiles was undertaken using the stress polygon method (Moos & Zoback 1990; Zoback 2007). This method uses the incidence of borehole failure (breakout and drilling-induced tensile failure) to estimate the stress conditions required to induce failure within the rock, using the frictional limit (defined by the friction angle) and compressional and tensional strength of the rock at the point of failure occurrence. The friction angle was calculated using the method defined by Lal (1999; equation 10). The unconfined compressive rock strength was defined based on empirical relationships listed in Chang *et al.* (2006). The equation defined by McNally (1987), based on data from the Bowen basin, was used for sandstones (equation 8). Where available, fracture closure pressures derived from pressure data such as leak-off tests (LOTs), diagnostic fracture injection tests and modular formation dynamic tester minifracs and pre-injection minifrac tests were used to constrain the minimum principal stress (Barree *et al.* 2007, 2009). Table 1 gives the wells and the specific calibration method(s) used.

Friction angle (°) (Lal 1999):

$$\varphi = \sin^{-1}\left(\frac{V_p - 1000}{V_p + 1000}\right) \quad (10)$$

where V_p is compressional sonic velocity in m s^{-1}.

Sandstone compressive rock strength (MPa) (McNally 1987):

$$UCS = 1200e^{(-0.036\Delta t)} \quad (11)$$

where Δt is the compressional sonic slowness in μs/ft and *e* is the base of natural logarithm.

Stress orientation

In situ stress data covering an area of 350 × 210 km (>70 000 km^2; for comparison, an area more than half the size of England) show significant variation in the orientation of S_{Hmax}. Based on the results, six distinct domains can be identified: the north Bowen, south Bowen and Burunga Anticline regions in the Bowen basin and the Roma Shelf and Taroom Trough regions in the Surat basin (Fig. 3).

Regional maps of the S_{Hmax} orientation and S_{hmin} magnitude have been constructed using all the available data from open file wells to December 2015. The S_{Hmax} orientation has been determined from observations of breakout or drilling-induced fractures on image logs or from breakout observed on four- or six-arm caliper logs. The S_{Hmax} orientation represented on the map in Figure 3 is derived from breakout data. The DITF data give the same

Table 1. *Wells (including offset wells) used in this study and calibration method for one-dimensional wellbore stress models*

Well name	Latitude (S)	Longitude (E)	LOT	FIT	ISIP	Pc	IL	RST
Durham Ranch 164	26° 03′ 38.10″	149° 13′ 32.72″		•			•	
FV03-15-1	25° 33′ 25.31″	148° 59′ 51.29″		•	•		•	•
Arcadia Branch 5	25° 17′ 29.35″	148° 55′ 41.88″		•	•	•	•	
Hermitage 14	26° 20′ 44.63″	149° 0′ 59.09″		•	•	•	•	•
Ironbark Gully 4	25° 31′ 45.57″	148° 55′ 58.90″	•					
Hermitage 11	26° 20′ 42.63″	149° 2′ 55.79″		•				
Sunnyholt 2	25° 17′ 37.66″	148° 50′ 44.84″	•		•		•	
Sunnyholt 3	25° 17′ 17.98″	148° 51′ 06.91″		•	•	•		
Sunnyholt 4	25° 17′ 47.61″	148° 51′ 12.36″	•		•	•		
Sunnyholt 11	25° 17′ 50.82″	148° 51′ 36.55″	•					
Arcadia Valley 2	25° 24′ 15.00″	148° 50′ 57.82″	•				•	•
Mount Kingsley 1	25° 13′ 55.22″	148° 54′ 6.84″		•			•	•
FV17-35-1DW1	25° 47′ 19.97″	149° 01′ 29.01″	•		•	•	•	•

FIT, formation integrity test; IL, image log; ISIP, instantaneous shut-in pressure; LOT, leak-off test; Pc, closure pressure; RST, rock strength testing.

result, but there are fewer data points and they are not discussed further herein. The mean S_{Hmax} azimuth is represented by the straight lines given in Figure 3. To avoid ambiguity, all data presented are from depths >450 m and from wells with <30° wellbore inclination.

This study builds on the findings of Brooke-Barnett *et al.* (2015) by applying the statistical methodology outlined by Hillis & Reynolds (2000, 2003) over both the Surat and Bowen basins. Consequently, the Rayleigh test was applied to the stress orientation data to determine the confidence of stress orientations over the study area (Mardia 1972; Table 2). Wells were also grouped into regions based on the underlying SEEBASE topography and S_{Hmax} orientation (Fig. 3) and the Rayleigh test was applied separately to each of these regions. The regions were then classified into six types using the following criteria: a type 1 region can reject the null hypothesis that stress orientations are random at the 99.9% confidence interval; a type 2 region can reject the null hypothesis at the 99% confidence interval; a type 3 region can reject the null hypothesis at the 97.5% interval; a type 4 region can reject the null hypothesis at the 95% interval; a type 5 can reject the null hypothesis at the 90% interval; and a type 6 region suggests that the null hypothesis cannot be rejected at the 90% interval (Hillis & Reynolds 2000, 2003). As per the methodology of Hillis & Reynolds (2000, 2003), Table 2 shows the results of the Rayleigh test as applied to the mean S_{Hmax} orientations from A to C quality borehole breakouts and DITF measurements. However, the mean statistics were also calculated using all borehole breakouts and DITF measurements (A to E quality) as well as the average S_{Hmax} for each well as per the methodology of Brooke-Barnett *et al.* (2015). Over the entire study area, the mean S_{Hmax} orientation is *c.* 42° N with a standard deviation of *c.* 36° (Table 2), giving the area a type 1 stress ranking. However, the orientation and quality of the *in situ* stresses varies significantly between the six domains that make up the whole area.

(1) The northern Bowen domain is dominated by a NNE-trending S_{Hmax} orientation; the dominant S_{Hmax} orientation here is *c.* 22° N with a standard deviation of *c.* 19° (Table 2). There are individual diversions from the dominant orientation in the very west of the study area; in the far north individual easterly trending as well as one southeasterly trending outlier are recorded. This region is designated as a type 2 stress region.
(2) The south-central Bowen domain of the Bowen basin forms a transitional corridor of variable S_{Hmax} orientations; both NE and NW trends as well as easterly trends occur, which gradually changes to an east–west to WNW–ESE trend of S_{Hmax} further south. Overall, this region exhibits a mean S_{Hmax} orientation of *c.* 70° N with a standard deviation of *c.* 34° (Table 2). Despite the higher spread in orientation in this region, the sheer amount of reliable stress indicators (21 A–C type measurements) enable this region to have a type 1 stress ranking.
(3) The southern Bowen domain (immediately north of the Roma Shelf on Fig. 3) exhibits a very consistent east–west S_{Hmax} orientation with a standard deviation of *c.* 4° (Table 2). Similar to the north Bowen domain, the consistency of orientation between wells, despite

Table 2. *Results of Raleigh analysis*

Region	Count						A–C type		Statistics A–C						Statistics A–E				Statistics well average S_{hmax}			
	A	B	C	D	E	Total	Borehole breakouts	DITFs	Count	Mean (°N)	*Rn*	SD (°)	Conf.	Type	Count	Mean (°N)	*Rn*	SD (°)	Count	Mean (°N)	*Rn*	SD (°)
North Bowen	1	3	1	32	21	58	5	0	5	12.1	0.9	10.7	>99	2.0	58.0	25.2	0.7	22.1	29.0	22.3	0.8	18.9
South-central Bowen	2	8	11	55	18	94	5	16	21	88.2	0.6	26.7	>99.9	1.0	94.0	67.8	0.4	37.5	47.0	69.5	0.5	34.2
Southern Bowen	0	1	4	3	2	10	4	1	5	93.3	1.0	3.8	>99	2.0	10.0	87.4	1.0	9.1	5.0	90.7	1.0	4.4
Roma Shelf	0	0	1	55	22	78	0	1	1	36.0	1.0	0.0	<90	–	78.0	17.8	0.9	13.0	39.0	18.0	0.9	13.8
Burunga Anticline	2	8	7	9	12	38	15	2	17	53.0	0.9	14.3	>99.9	1.0	38.0	50.9	0.8	17.2	19.0	48.7	0.9	13.8
Taroom Trough	1	1	2	41	21	66	4	0	4	60.2	0.6	27.8	<90	6.0	66.0	82.4	0.3	41.6	33.0	77.9	0.3	42.9
New England Orogen	0	0	1	10	7	18	1	0	1	23.9	1.0	0.0	<90	–	18.0	56.4	0.7	22.8	9.0	52.1	0.8	18.9
Total	6	21	27	205	103	362	34	20	54	66.2	0.5	33.1	>99.9	1.0	362.0	43.8	0.4	38.1	181.0	41.5	0.4	36.3

The Rayleigh test was applied to the entire area and the sub-regions defined in this paper based on A to C quality stress indicators (Statistics A–C). Mean statistics have also been calculated on A–E quality stress indicators (Statistics A–E) and the mean stress orientation for each well (Statistics well total S_{Hmax}) based on input well data. Mean is the mean S_{Hmax} calculated for a region, *Rn* is the length of the vector resulting from the sum of all S_{Hmax} orientations within a region (Mardia 1972), SD is the standard deviation of the calculated mean S_{Hmax}, Conf. is the confidence level at which the null hypothesis that the stress orientation is random can be rejected and the type is the stress province type as per the methodology of Hillis & Reynolds (2000, 2003). DITFs, drilling-induced tensile fractures.

the small sample size, means this domain also has a type 2 ranking.

(4) The Roma Shelf domain, in which most stress measurements are from wells in the Jurassic Walloons sequence, is again dominated by NNE- (*c.* 18°) trending S_{Hmax} orientations. Note that although the orientations between wells are relatively consistent (standard deviation of *c.* 14°) the quality of stress indicators in this region is low, meaning there are insufficient data to assign a ranking to this area (Table 2).

(5) The Burunga Anticline on the eastern margin of the Taroom Trough displays a consistent NE trend of S_{Hmax} (*c.* 51°) with a standard deviation of *c.* 14° (Table 2). The low standard deviation and high quality of stress indicators give this region a type 1 stress ranking.

(6) The Taroom Trough domain of the Surat basin (which overlies the depocentre of the Permian Bowen basin; Figs 2 & 3) to the east of the Roma Shelf shows a complex S_{Hmax} orientation. In the basin centre the stress azimuths are dominated by broadly easterly trends (both ENE and ESE), but show a number of significant deviations from a clear overall trend. At the flanks of the depocentre the S_{Hmax} orientation swings into parallelism with the depocentre boundaries (a fault system in the east and a ramp in west, Brooke-Barnett *et al.* 2015). Other datasets (Brooke-Barnett *et al.* 2015) show a NE trend to the east, where the Walloons depocentre is underlain by a basement high (New England Orogen Region; Table 2). Despite the inclusion of additional data, this region retains the type 6 designation from Brooke-Barnett *et al.* (2015).

One-dimensional wellbore stress models (mechanical Earth models)

A one-dimensional wellbore stress model (also called a mechanical Earth model) is a numerical representation of the geomechanical state of the subsurface over a given interval and combines known pore pressures, the stress state (vertical, minimum and maximum horizontal) and rock mechanical properties (uniaxial compressive strength, Young's modulus and Poisson's ratio). The one-dimensional wellbore stress models herein have been generated using RokDoc623 software.

The one-dimensional wellbore stress models have been compiled for four example wells within the Bowen basin (circled well locations, Fig. 3). The one-dimensional wellbore stress models are created using shear and p-wave sonic velocities and elastic models to produce estimates of stress and rock properties. The strainless one-dimensional wellbore stress models are validated and refined with known data, including leak-off data, DFIT or mini-frac closure pressure data, which gives an estimate of the minimum horizontal stress at a particular depth. Alternatively, laboratory-based rock strength testing or frictional limits based on stress indicators from image log analyses were used to calibrate the horizontal stress magnitudes (for method applied, see figure captions). One-dimensional wellbore stress models show systematic variations of Andersonian stress geometry with depth in the Bowen and Surat basins (see Flottmann *et al.* 2013; Brooke-Barnett *et al.* 2015).

At depths shallower than around 500–600 m TVD (total vertical depth, i.e. the depth below the surface measured from the drill rig floor), the stress state is characterized by a reverse stress regime ($S_V < S_{hmin} < S_{Hmax}$, where S_V = vertical stress, S_{hmin} = minimum horizontal stress and S_{Hmax} = maximum horizontal stress). Image log data in reverse stress regimes are dominated by borehole breakouts (DITFs are largely absent) and breakout orientations indicate significant variability in the azimuth of S_{Hmax} (Fig. 4). The one-dimensional wellbore stress models presented in Figures 5–8 indicate low horizontal differential stresses (i.e. the difference between S_{Hmax} and S_{hmin}), in particular at depths where reverse stress regimes are dominant. The low differential stress is a likely cause of the scatter in the breakout orientations at shallow depths. Below 500–600 m TVD the stress geometry is typically of a strike-slip stress regime ($S_{hmin} < S_V < S_{Hmax}$). Image log data in strike-slip stress regimes show both borehole breakouts and DITFs (Fig. 4a), both of which occur dominantly in shale/siltstone units.

Figure 4b and c show the wellbore stress conditions at 400 and 800 m, respectively, using the stress and rock strength data given in Figure 6 for those depths. The stress/rock strength conditions at 400 m allow for broad compressive failure where the maximum horizontal stress exceeds the compressive failure. This condition is represented by the occurrence of scattered borehole breakouts. The stress conditions do not reach tensile failure, resulting in the sparse development of DITFs at this depth (Fig. 4a). At 800 m depth both the differential stresses and the uniaxial compressive strength are higher than at 400 m (Fig. 6). This results in the preferential development of DITFs (as the stress conditions exceed the tensile rock strength). Both DITFs and borehole breakouts occur in a well-defined narrow band at this depth. The data compilation in Figure 4a shows the dominance of DITFs at depths >1000 m; the dominance of DITFs at greater depths appears to be related to increasing differential stress with depth in this part of the Bowen basin.

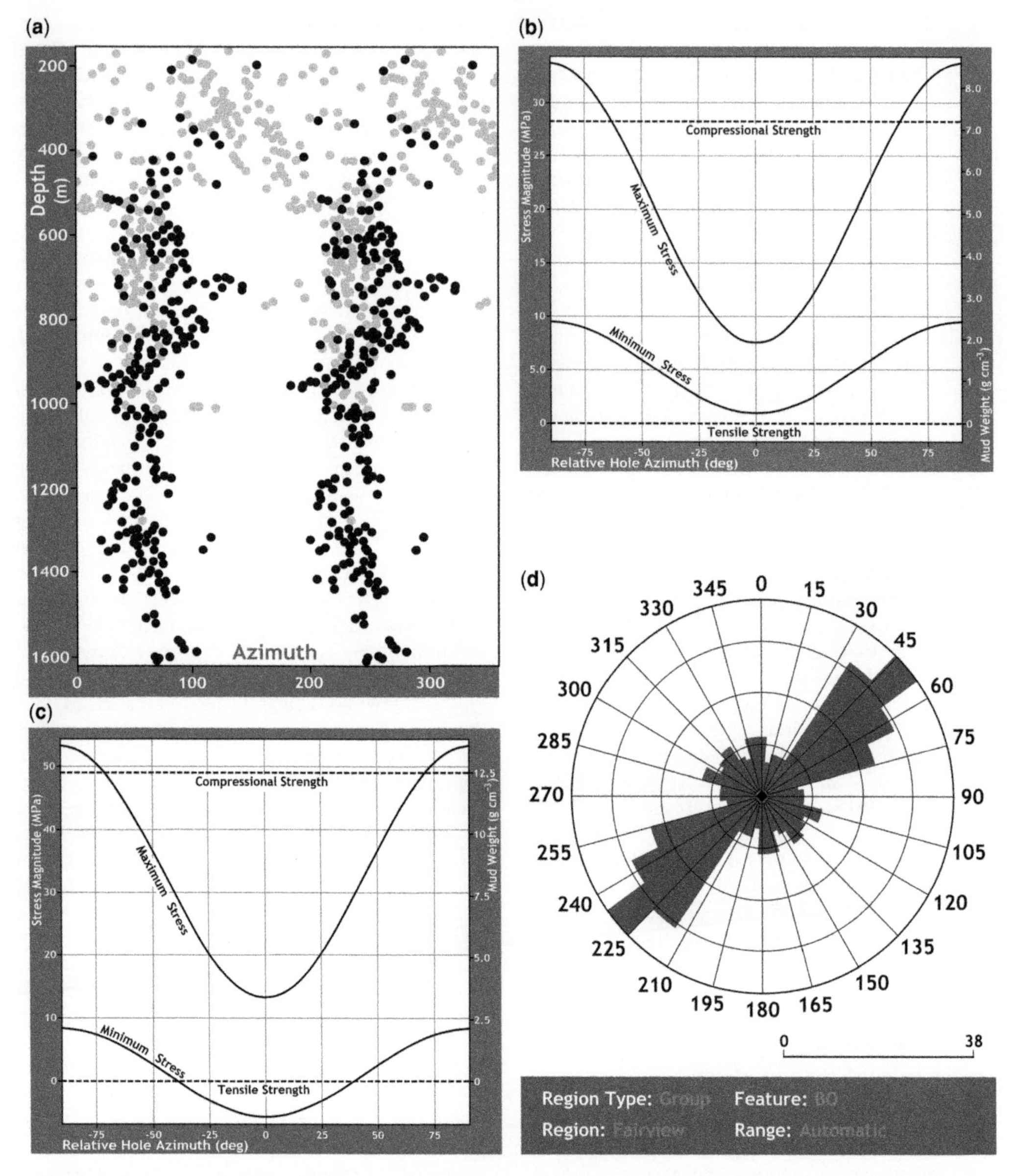

Fig. 4. (**a**) Compilation of wellbore breakout data (grey circles) and DITFs (black circles) from the central Bowen basin region. Note the variability of breakout azimuth at shallow depth and the progressive domination of DITFs at greater depth. (**b**) Circumferential stress distribution at 400 m depth using rock strength and stress data given in Figure 6; note that the minimum stress does not allow the generation of tensile fractures (i.e. no DITFs), but the maximum stress exceeds the compressive strength, resulting in the wide-ranging development of borehole breakouts (BO). (**c**) Same as part (b), but at 800 m; note that the minimum stress exceeds the tensile strength, resulting the preferential development of DITFs at greater depths. (**d**) Rose plot of combined azimuth of S_{Hmax} data from wellbore analysis presented in part (a).

The transition between reverse stress regimes and strike-slip stress regimes is also well documented in tiltmeter data acquired during hydraulic stimulations in the Jurassic Surat basin (Flottmann *et al.* 2013). The dataset presented here shows a similar stress regime transition at the same depth range, but in the Permian Bowen basin. The co-occurrence of the transition from a reverse to a strike-slip stress

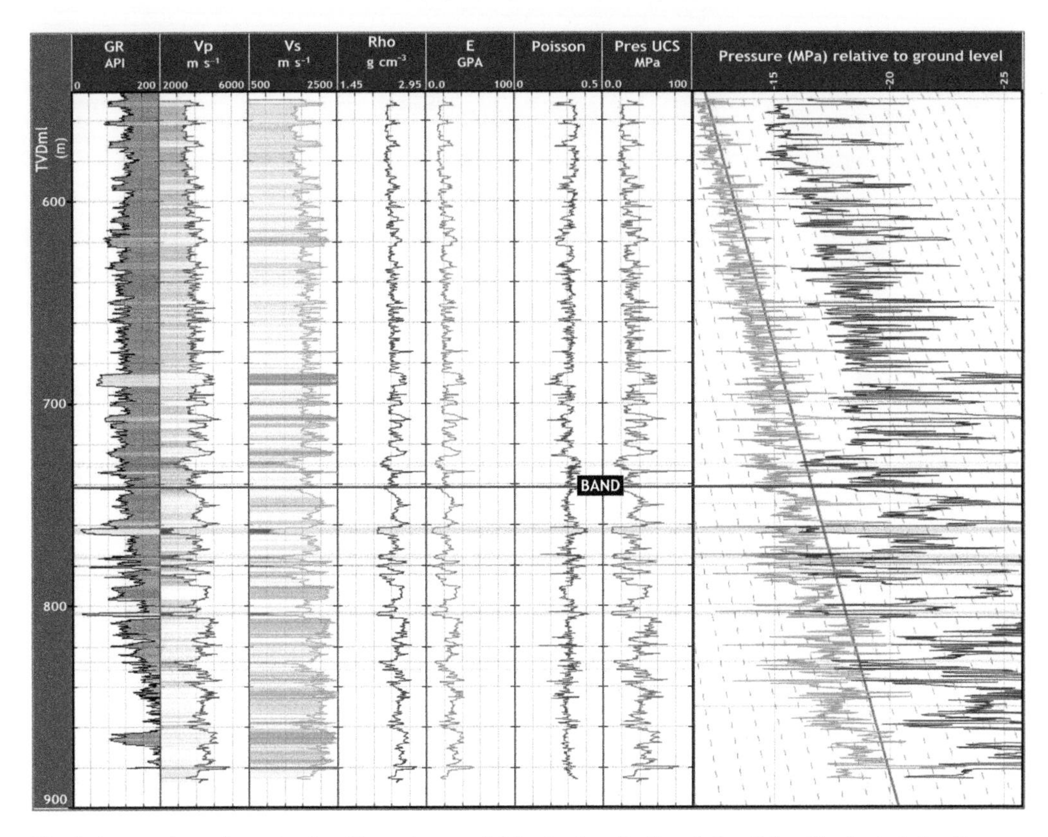

Fig. 5. Logs and one-dimensional wellbore stress model for the Arcadia Branch 5 well (see Fig. 3, circled wells show well locations of Figs 5–8 from north to south). Left-hand side: input log data used to calculate S_{hmin} and S_{Hmax} on right-hand side. S_{hmin} generally smaller than S_{V}, but S_{Hmax} greater than S_{V}, indicating a strike-slip stress regime. Colour scheme: *Vs* and *Vp* logs, blue colours slow (e.g. coals, carbonaceous shales), red colours fast (hard sand and shales). Formation tops: BAND denotes Bandanna Formation, above Rewan Formation (for stratigraphic detail, see Cook & Jell 2013).

state at a similar depth range in two different basins suggests that the transition in stress state is controlled by the present day depth rather being controlled by geological or stratigraphic parameters.

A second transition from a strike-slip to a normal stress regime ($S_{\mathrm{hmin}} < S_{\mathrm{Hmax}} < S_{\mathrm{V}}$) has been documented at *c.* 650–800 m in some areas in the Surat basin. This transition does not occur in the Bowen basin. The magnitude of differential stresses in the Surat basin also show significant variability. This appears to be attributed to both the nature of the basement and/or the thickness of the sedimentary section underlying the Surat basin (Brooke-Barnett *et al.* 2015).

Mechanical stratigraphy

Rock properties (Young's modulus, the uniaxial compressive strength and Poisson's ratio) are constrained by rock strength testing from offset wells in both the interburden and the coals. S_{hmin} has been constrained using DFIT, minifrac and leak-off data. Results from one-dimensional wellbore stress models display some key contrasts in mechanical stratigraphy. In principle, (non-coal) interburden rocks have a higher rock strength than coals, which display a consistently low rock strength based on a high Poisson's ratio and low Young's modulus. Coals are dominated by a normal stress regime, regardless of whether the surrounding rocks are in a reverse, strike-slip or normal overall stress regime. Importantly, in reverse and strike-slip stress regimes coals exhibit generally lower overall stresses than the surrounding country rock (Fig. 9a, b). This has been established by numerous systematic DFIT tests in numerous wells in both the Bowen and Surat basins (Fig. 9a, b; Flottmann *et al.* 2013). The same result is achieved by establishing frictional limits theory based on image log analyses.

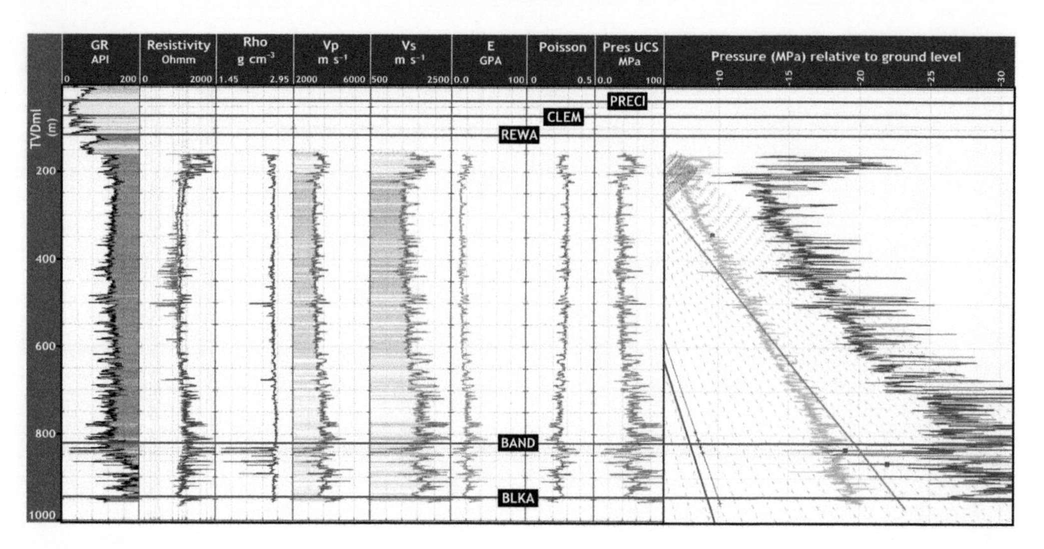

Fig. 6. Logs and one-dimensional wellbore stress model for the Fairview F-V 03-15-1 well. Colour scheme as in Figure 5. Note distinct change from a reverse to a strike-slip stress regime around 500 m. Black squares show DFIT calibration points for stress model. Formation tops: PRECI, Precipice Sandstone; CLEM, Clematis Formation; REW, Rewan Formation; BAND, Bandanna Formation; BLKA, Black Alley Shale.

The relationships given in equations (8) and (9) suggest that rocks with a high Poisson's ratio (e.g. coals) are more susceptible to accommodating high stress in tectonic scenarios dominated by vertical 'loading' (overburden); conversely, horizontal 'loading' (i.e. the tectonic component) is dominantly accommodated in rocks with a high Young's modulus. Based on the one-dimensional wellbore stress models, the essential elements impacting the mechanical stratigraphy in the Bowen and Surat basins can be reduced to three key components: (1) coals with a comparatively high Poisson's ratio and low Young's modulus; (2) interburden rocks with a comparatively low Poisson's ratio and a high Young's modulus; and (3) a stress regime dominated by horizontal (tectonic) components.

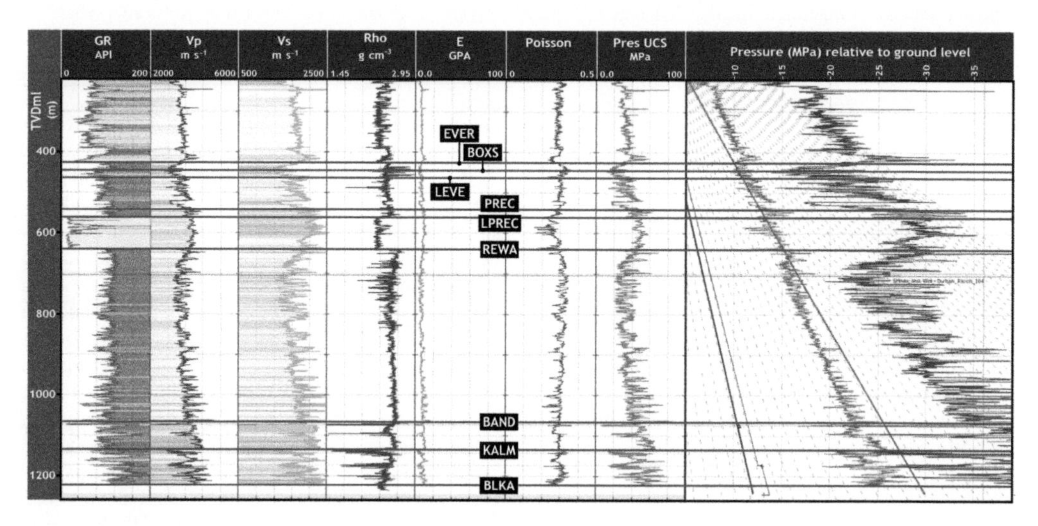

Fig. 7. Logs and one-dimensional wellbore stress model for the Durham Ranch DM 164 well. Colour scheme same as Figure 5. Note distinct change from reverse stress regime above 400 m to strike-slip stress regime below 650 m. Formation tops: EVER, Evergreen Formation; BOXS, Boxvale Sandstone; LEVE, Lower Evergreen; PRECI, Precipice Sandstone; LPREC, Lower Precipice Sandstone; REW, Rewan Formation; BAND, Bandanna Formation; KALM, Kaloola Member; BLKA, Black Alley Shale.

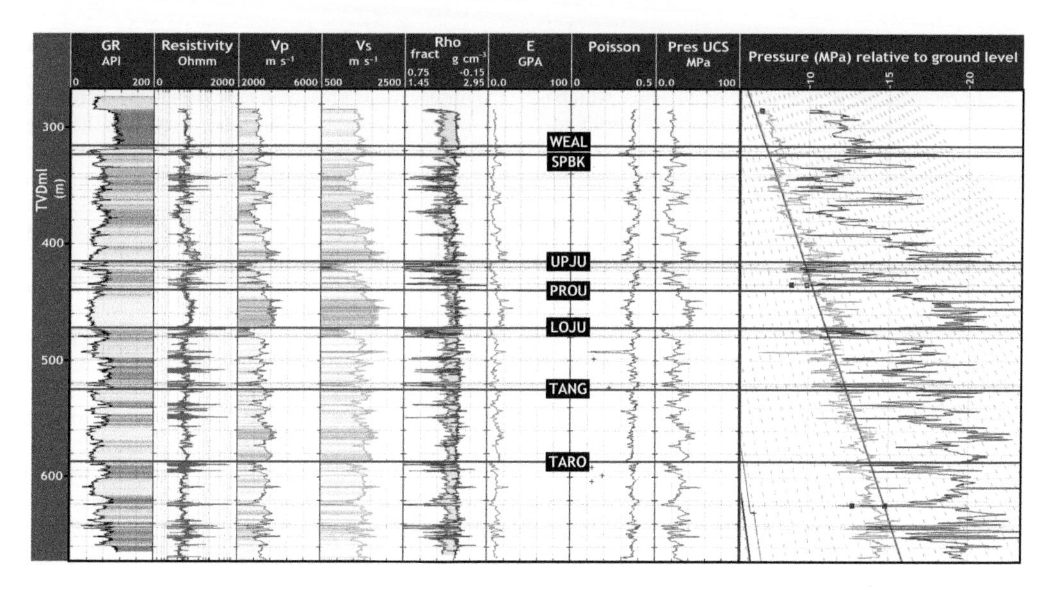

Fig. 8. Logs and one-dimensional wellbore stress model for the Hermitage 14 well (location Fig. 3). Colour scheme same as Figure 5. Note numerous transitions from a reverse to a strike-slip regime. Black squares show DFIT calibration points for stress model. Formation tops: WEAL, Weald Sandstone; SPBK, Springbok Formation; UPJU, Upper Juandah coal measures; PROU, Proud Sandstone; LOJU, Lower Juandah coal measures; TANG, Tangalooma Sandstone; TARO, Taroom coal measures.

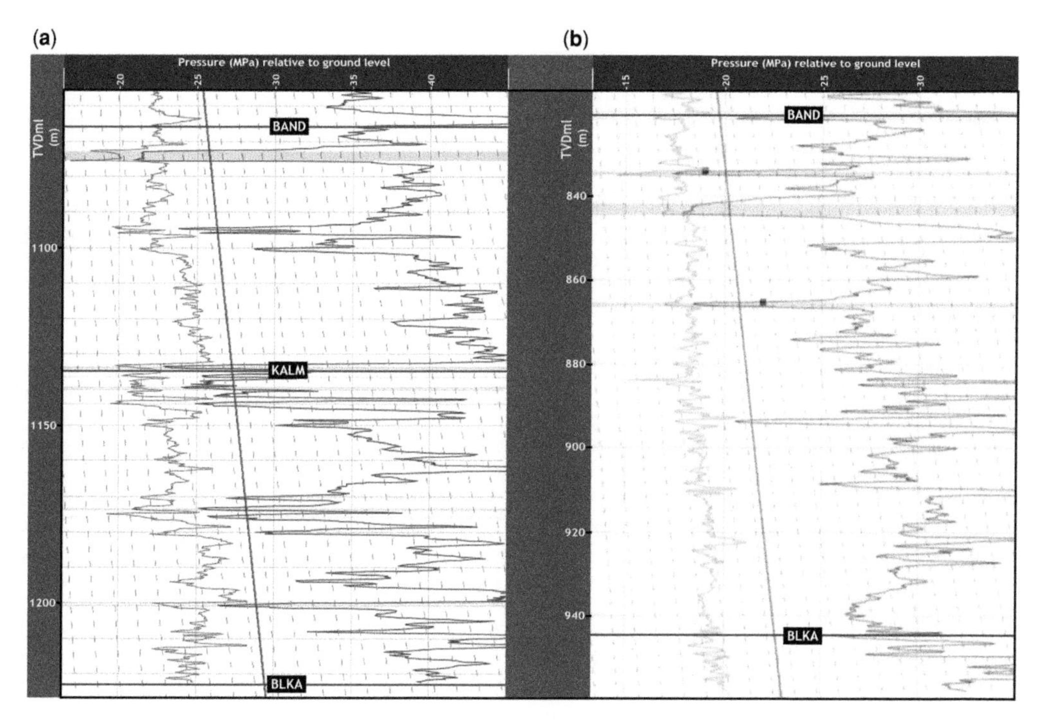

Fig. 9. (**a**) Close up of DM 164 log over coal interval (grey shading); note that both S_{Hmax} and S_{hmin} are smaller than S_V (straight brown line); coals are in normal stress regime. (**b**) Same observations as for well FV-03-15-1.

In the following discussion we explore the oilfield impacts of the key elements of the mechanical stratigraphy in the Bowen and Surat basins, which is relevant to basins with similar conditions worldwide.

Discussion

The data presented in this paper show variability in the plan view stress azimuths and Andersonian stress geometry with depth. The data and interpretations from the Permian Bowen basin presented here, in combination with similar data from the Jurassic Surat basin (Brooke-Barnett *et al.* 2015), show similar depths for the transition from a reverse to a strike-slip regime (400–600 m). This suggests that the first-order influence for the occurrence of this transition in stress regime is depth (TVD) rather than local conditions, such as the basin-specific stratigraphy. The transition from a reverse to a strike-slip stress state at *c.* 400–600 m depth thus appears to be typical for eastern Queensland. This is in contrast with interior parts of Australia (e.g. the Cooper basin), where a transition from a strike-slip stress regime at shallow depths to a reverse stress regime is reported to occur at depths in excess of *c.* 2.5 km (e.g. Reynolds *et al.* 2006).

Previous datasets presented in the World Stress Map (Heidbach *et al.* 2008, 2010) suggest an overall NNE–NE orientation of S_{Hmax} in the Bowen basin. Our data support this observation over the entire region, including both the Bowen and Surat basins (Table 2). However, our data show significant local deviations from the inferred regional trend. In particular, the south Bowen domain of the Bowen basin and the Taroom Trough domain of the Surat basin both display variable S_{Hmax} orientations and the Denison Trough domain displays an east–west S_{Hmax} which, although consistent, deviates significantly from the regional S_{Hmax} orientation.

The reasons why the orientation of S_{Hmax} azimuths is subject to significant local variations remains, at this stage, a matter of speculation. Brooke-Barnett *et al.* (2015) showed that S_{Hmax} azimuths in the Surat basin are significantly influenced by the underlying basement structures. However, available SEEBASE data show no obvious major structural trends that could guide stress rotations of the severity seen in the southern Bowen basin. The easterly stress orientations in the south-central and southern domains may be guided by deep-seated easterly trending fault systems and volcanism related to the opening of the Coral Sea during the Tertiary (Cook & Jell 2013). The changing composition of the deeper basement could result in changes in the bulk rock strength, which, in turn, could contribute to the stress rotations observed – in fact, the stress rotations may provide a guide for future remote deep data acquisition. Regardless, the observations presented here suggest that stresses in intracontinental basins can vary significantly and without any visible first-order manifestation in complementary datasets, such as faults or basin structure.

However, from an oilfield perspective, it is important to take local variations of S_{Hmax} into consideration. For horizontal and/or deviated wells, for example, the S_{Hmax} azimuths can have significant implications with regard to optimizing wellbore stability in deviated/horizontal wells in the context of rock strength and mud-weight optimization to prevent wellbore collapse. Wellbore stability is highest where the differential stresses around wellbores are at a minimum. This is particularly true in deviated/horizontal wells (regardless of the overall stress magnitudes). In many strike-slip regimes, which are common around the world, this often entails deviating the wellbore into S_{Hmax} as the differential stresses between S_{hmin} and S_V are less than, for example, the differential stress between S_V and S_{Hmax}.

In the context of fracture stimulation completions, a well deviated into S_{Hmax} may not be optimum because hydraulic fractures propagate preferentially into the azimuth of S_{Hmax}, resulting in a longitudinal fracture stimulation. Generally, transverse stimulations (i.e. the fracture propagates perpendicular to the wellbore) are more advantageous, particularly in low-permeability formations, because they access greater reservoir rock volumes. Consequently, the wellbore has to be deviated into S_{hmin} and, during drilling, the mud-weight window has to be adjusted to deliver a stable wellbore under high differential stresses acting perpendicular to the wellbore axis. Regionally varying stress azimuths as documented here highlight the need to acquire a complete stress dataset, allowing full stress tensor and stress azimuth descriptions to optimize both drilling and completion options. The local variability of the S_{Hmax} azimuths in our dataset demonstrates that it may be insufficient to rely on regional datasets alone; local conditions have to be established to allow for optimum drilling and completion designs.

Stress state transitions with depth, and between lithologies, are of primary practical importance. Away from the wellbore-influenced hoop stresses, hydraulic fractures open against the regional minimum principal stress. Consequently, hydraulic fractures are expected to be vertical in both strike-slip and normal stress regimes; however, in a reverse stress regime fracture opening is expected to be horizontal and leads to a horizontal hydraulic fracture. The occurrence of horizontal fracturing at shallow depth is corroborated by tiltmeter data in the Surat basin, which show a predominance of horizontal components (up to 100%) at shallow depths. At depths >400–500 m (Flottmann *et al.* 2013), tiltmeter data show a predominance of vertical fracture components.

Three-dimensional stress characterization is a key requirement for the planning and implementation of drilling and completions such as hydraulic fracture stimulations. Hydraulic stimulations propagate perpendicular to the lowest principal stress in the plane defined by the intermediate and maximum principal stresses; hydraulic fractures tend to grow in the azimuth of S_{Hmax}. The mechanical stratigraphy and stress magnitudes are of particular importance for the vertical containment of fracture growth. Fractures initiate and grow preferentially in formations with low S_{hmin}. Formations with low S_{hmin} have a low mean stress $[(S_V + S_{Hmax} + S_{hmin})/3]$; conversely, formations with a high S_{hmin} and a high mean stress tend to act as barriers to the vertical growth of hydraulic stimulations, thereby containing the stimulation to the target zone.

However, different Andersonian stress regimes result in contrasting stress accommodation and stress intensity in rocks with different rheological properties. Lithologies with a high Poisson's ratio and a low Young's modulus (such as coals) tend to accommodate high stresses in a normal stress regime ($S_V > S_{Hmax} > S_{hmin}$), where the maximum principal stress is vertical and the vertical component of stress is governed by Poisson's ratio (equations 8 & 9; see also Herwanger *et al.* 2015). Conversely, in tectonic regimes where the maximum principal stress is horizontal (reverse and strike-slip tectonic regimes), the stress distribution is governed by rocks with a high Young's modulus and a low Poisson's ratio (i.e. the tectonic component in equations 8 & 9). Consequently, in reverse and strike-slip regimes the interburden rocks (high Young's modulus and low Poisson's ratio) will accommodate high stresses and the coals will be comparatively less stressed.

The interaction of the Andersonian stress state and rocks of contrasting rheological properties thus has significant implications for the propagation of hydraulic fractures. In a normal stress regime, for example, where coals are comparatively highly stressed, hydraulic fractures will initiate in coals, but will tend to grow into (interburden) formations dominated by a lower mean stress (Fig. 10a). Our interpretation of one-dimensional wellbore stress models shows that, in the Bowen basin, the actual Andersonian stress states are reverse and strike slip and the coals tend to be the least stressed members of the mechanical stratigraphy. Interburden rocks with a high Young's modulus accommodate the horizontal tectonic stress component (equations 8 & 9) and are the most highly stressed components of the mechanical stratigraphy in the Bowen basin (Fig. 10b). Consequently, hydraulic fracture completions targeting coals are well contained in the low stress coals (Fig. 10b). Similar observations are documented in the Surat basin, where tracer logs show containment in coal during hydraulic treatments (Kirk-Burnnand *et al.* 2015).

Different Andersonian stress regimes can result in potentially stark contrasts in rock-specific stress states (Fig. 10a, b). This highlights the need for the full three-dimensional characterization of stress parameters to achieve the appropriate conditioning of one-dimensional geomechanical models, which are key (software) inputs for planning fracture stimulations or well planning. The three-dimensional variation of stress states in the Bowen basin

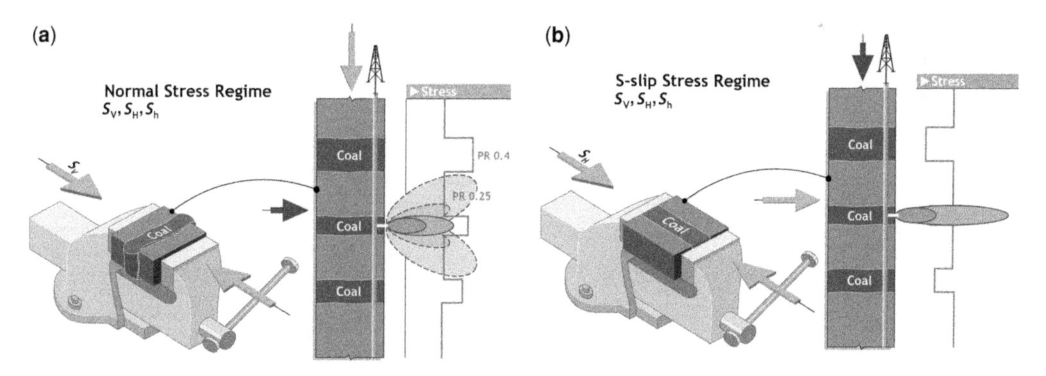

Fig. 10. (**a**) Compression in vice represents normal stress regime (maximum stress vertical, i.e. perpendicular to layering), resulting in high stress in coal (Poisson's ratio (PR) *c.* 0.4) and low stress in clastic interburden rocks (Poisson's ratio *c.* 0.25); hypothetical fracture stimulation indicated by oval shapes; initiation in coal (solid outline), but progressive growth into low stress interburden. (**b**) Compression in vice represents strike-slip/reverse stress regime (maximum stress horizontal, i.e. parallel to layering), resulting in high stress in interburden and low stress in coal. Hypothetical fracture stimulation (ovals) stays contained in coal.

highlight the interdependency between the plate tectonics boundary conditions, the local structural geological setting and rheological parameters, all of which ultimately contribute to the geomechanical conditions at individual wellbores.

Conclusions

This paper documents broad, but systematic, changes in the *in situ* stress orientations and *in situ* stress state with depth in the intracontinental Bowen and Surat basins of eastern Queensland, Australia. Both basins show a transition from a reverse stress regime at depths shallower than 400–600 m; at greater depths strike-slip stress geometries are dominant. The (Andersonian) stress regime transition is primarily depth-controlled, independent of the basin and/or lithology. The spatial variability of the *in situ* stress distribution results from the interaction of regional intra-plate stresses with basin-scale structures and basement rheology. Stress geometries and rock properties materially influence drilling and completion considerations. In particular, hydraulic fracture completions have to be designed to take into account the rock-specific geomechanical conditions established from log-derived one-dimensional wellbore stress models. The data and analyses presented suggest that first-order regional stresses increase in complexity at a local scale; ultimately, the resolved geomechanical state at the wellbore level reflects a scale-dependent interplay of plate tectonic forces that are geologically modulated by the local structure and the mechanical properties of individual rock packages.

We thank Santos Ltd and Origin Energy for permission to publish this paper. We acknowledge the efforts of countless colleagues and field personnel who contributed to data acquisition and discussions leading to the results presented here. Particular thanks to the convenors of the GSL conference Geology of Geomechanics for the opportunity to present and encouragement to publish. Tony Addis and Joe English prepared thorough and insightful reviews, which greatly improved the manuscript.

References

ANDERSON, M.E. 1951. *The Dynamics of Faulting*. Oliver and Boyd, Edinburgh.

BARREE, R.D., BARREE, V.L. & CRAIG, D.P. 2007. Holistic fracture diagnostics. Paper SPE 107877, presented at the SPE Rocky Mountain Oil & Gas Technology Symposium, 16–18 April, Denver, CO, USA.

BARREE, R.D., GILBERT, J.V. & CONWAY, M.W. 2009. Stress and rock property profiling for unconventional reservoir stimulation. Paper SPE 118703, presented at the SPE Hydraulic Fracturing Convention, 19–21 January, The Woodlands, TX, USA.

BELL, J.S. 1990. The stress regime of the Scotian Shelf offshore eastern Canada to 6 kilometres depth and implications for rock mechanics and hydrocarbon migration. *In*: MAURY, V. & FOURMAINTRAUX, D. (eds) *Rock at Great Depth*. Balkema, Rotterdam, 1243–1265.

BELL, J.S. 1996*a*. Petro Geoscience 1. *In situ* stresses in sedimentary rocks (part 1): measurement techniques. *Geoscience Canada*, **23**, 85–100.

BELL, J.S. 1996*b*. Petro Geoscience 2. *In situ* stresses in sedimentary rocks (part 2): applications of stress measurements. *Geoscience Canada*, **23**, 135–153.

BROOKE-BARNETT, S., FLOTTMANN, T. *ET AL.* 2015. Influence of basement structures on in-situ stresses over the Surat Basin, southeast Queensland. *Journal of Geophysical Research: Solid Earth*, **120**, 4946–4965.

CHANG, C., ZOBACK, M.D. & KHAKSAR, A. 2006. Empirical relations between rock strength and physical properties in sedimentary rocks. *Journal of Petroleum Science and Engineering*, **51**, 223–237.

COBLENTZ, D.D., SANDIFORD, M., RICHARDSON, R.M., ZHOU, S. & HILLIS, R. 1995. The origins of the intraplate stress field in continental Australia. *Earth and Planetary Science Letters*, **133**, 299–309.

COBLENTZ, D.D., ZHOU, S., HILLIS, R.R., RICHARDSON, R.M. & SANDIFORD, M. 1998. Topography, boundary forces, and the Indo-Australian intraplate stress field. *Journal of Geophysical Research*, **103**, 919–931.

COOK, A.G. & JELL, J.S. 2013. Paleogene and Neogene. *In*: JELL, P.A. (ed.) *Geology of Queensland*. Geological Survey of Queensland, Brisbane, 577–685.

EATON, B.A. 1968. Fracture gradient prediction and its application in oilfield operations. Paper SPE-2163-PA, https://doi.org/10.2118/2163-PA

EATON, B.A. 1972. The effect of overburden stress on geopressures prediction from well logs. Paper SPE-3719-PA, https://doi.org/10.2118/3719-PA

EATON, B.A. 1975. The equation for geopressure prediction from well logs. Paper SPE-5544-MS, https://doi.org/10.2118/5544-MS

FLOTTMANN, T., BROOKE-BARNETT, S. *ET AL.* 2013. Influence of in-situ stresses on fracture stimulations in the Surat Basin, southeast Queensland. Paper SPE 167064-MS, presented at the SPE Unconventional Resources Conference and Exhibition-Asia Pacific, 11–13 November 2013, Brisbane, Australia, https://doi.org/10.2118/167064-MS

HEIDBACH, O., TINGAY, M., BARTH, A., REINECKER, J., KURFESS, D. & MÜLLER, B. 2008. The World Stress Map Database Release 2008, https://doi.org/10.1594/GFZ.WSM.Rel2008

HEIDBACH, O., TINGAY, M., BARTH, A., REINECKER, J., KURFESS, D. & MÜLLER, B. 2010. Global crustal stress pattern based on the World Stress Map database release 2008. *Tectonophysics*, **482**, 3–15.

HERWANGER, J.V., BOTTRILL, A.D. & MILDREN, S.D. 2015. Uses and abuses of the brittleness index with application to hydraulic stimulations. Paper URTEC-2172545-MS, presented at the URTeC Conference, 20–22 July 2015, San Antonio, TX, USA, https://doi.org/10.15530/URTEC-2015-2172545

HILLIS, R.R. & REYNOLDS, S.D. 2000. The Australian stress map. *Journal of the Geological Society, London*, **157**, 915–921, https://doi.org/10.1144/jgs.157.5.915

HILLIS, R.R. & REYNOLDS, S.D. 2003. *In situ* stress field of Australia. *In*: HILLIS, R.R. & MÜLLER, R.D. (eds) *Evolution and Dynamics of the Australian Plate*. Geological Society of America, Boulder, 49–58.

HILLIS, R.R., ENEVER, J.R. & REYNOLDS, S.D. 1999. *In situ* stress field of eastern Australia. *Australian Journal of Earth Sciences*, **46**, 813–825.

HOLCOMBE, R.J., STEPHENS, C.J. *ET AL.* 1997*a*. Tectonic evolution of the northern New England fold belt: Carboniferous to Early Permian transition from active accretion to extension. *In*: ASHLEY, P.M. & FLOOD, P.G. (eds) *Tectonics and Metallogenesis of the New England Orogen: Alan H. Voisey Memorial Volume*. Geological Society of Australia, Special Publications, **19**, 66–79.

HOLCOMBE, R.J., STEPHENS, C.J. *ET AL.* 1997*b*. Tectonic evolution of the northern New England fold belt: the Permian–Triassic Hunter-Bowen event. *In*: ASHLEY, P.M. & FLOOD, P.G. (eds) *Tectonics and Metallogenesis of the New England Orogen: Alan H. Voisey Memorial Volume*. Geological Society of Australia, Special Publications, **19**, 52–65.

KIRK-BURNNAND, E., PANDEY, V.J., FLOTTMAN, T. & TRUBSHAW, R.L. 2015. Hydraulic fracture design optimization in low permeability coals. Paper SPE 176895-MS, presented at the SPE Unconventional Resources Conference and Exhibition-Asia Pacific, 9–11 November 2015, Brisbane, Australia, https://doi.org/10.2118/176895-MS

KIRSCH, E.G. 1898. Die Theorie der Elastizität und die Bedürfnisse der Festigkeitslehre. *Zeitschrift des Vereines deutscher Ingenieure*, **42**, 797–807.

KORSCH, R.J. & TOTTERDELL, J.M. 2009. Evolution of the Bowen, Gunnedah and Surat Basins, eastern Australia. *Australian Journal of Earth Sciences*, **56**, 271–272.

KORSCH, R.J., TOTTERDELL, J.M., CATHRO, D.L. & NICOLL, M.G. 2009*a*. Early Permian East Australian Rift System. *Australian Journal of Earth Sciences*, **56**, 381–400.

KORSCH, R.J., TOTTERDELL, J.M., CATHRO, D.L. & NICOLL, M.G. 2009*b*. Contractional structures and deformational events in the Bowen, Gunnedah and Surat Basins, eastern Australia. *Australian Journal of Earth Sciences*, **56**, 477–499.

LAL, M. 1999. Shale stability: drilling fluid interaction and shale strength. Paper presented at the SPE Latin American and Caribbean Petroleum Engineering Conference, 21–23 April 1999, Caracas, Venezuela.

MARDIA, K.V. 1972. *Statistics of Directional Data*. Academic Press, New York.

MCNALLY, G.H.N. 1987. Estimation of coal measures rock strength using sonic and neutron logs. *Geoexploration*, **24**, 381–395.

MOOS, D. & ZOBACK, M.D. 1990. Utilization of observations of well bore failure to constrain the orientation and magnitude of crustal stresses: application to continental, Deep Sea Drilling Project and Ocean Drilling Program boreholes. *Journal of Geophysical Research*, **95**, 9305–9325, https://doi.org/10.1029/JB095iB06p09305

NELSON, E.J., CHIPPERFIELD, S.T., HILLIS, R.R., GILBERT, J., MCGOWEN, J. & MILDREN, S.D. 2007. The relationship between closure pressures from fluid injection tests and the minimum principal stress in strong rocks. *International Journal of Rock Mechanics & Mining Sciences*, **44**, 787–801.

QDEX 2016. www.business.qld.gov.au/industry/mining/mining-online-services/qdex-data

REYNOLDS, S.D., COBLENTZ, D.D. & HILLIS, R.R. 2002. Tectonic forces controlling the regional intraplate stress field in continental Australia: results from new finite element modeling. *Journal of Geophysical Research*, **107**(B7), https://doi.org/10.1029/2001JB000408

REYNOLDS, S.D., COBLENTZ, D.D. & HILLIS, R.R. 2003. Influences of plate-boundary forces on the regional intraplate stress field of continental Australia. *In*: HILLIS, R.R. & MÜLLER, R.D. (eds) *Evolution and Dynamics of the Australian Plate*. Geological Society of America, Boulder, 59–70.

REYNOLDS, S.D., MILDREN, S.D., HILLIS, R.R., MEYER, J.J. & FLOTTMANN, T. 2005. Maximum horizontal stress orientations in the Cooper Basin, Australia: implications for plate-scale tectonics and local stress sources. *Geophysical Journal International*, **160**, 331–343.

REYNOLDS, S.D., MILDREN, S.D., HILLIS, R.R. & MEYER, J.J. 2006. Constraining stress magnitudes using petroleum exploration data in the Cooper–Eromanga Basins, Australia. *Tectonophysics*, **415**, 123–140.

SANDIFORD, M. 2016. http://jaeger.earthsci.unimelb.edu.au/Images/images.html

SANDIFORD, M., WALLACE, M. & COBLENTZ, D.C. 2004. Origin of the *in situ* stress field in south-eastern Australia. *Basin Research*, **16**, 325–338.

SEEBASE 2005. Public domain report to Shell Development Australia by FrOG Tech Pty Ltd, edited.

THIERCELIN, M.J. & PLUMB, R.A. 1994. A core-based prediction of lithologic stress contrast in east texas formations. *SPE Formation Evaluation*, **9**, 251–258.

TINGAY, M., REINECKER, J. & MÜLLER, B. 2008. *Borehole Breakout and Drilling-induced Fracture Analysis from Image Logs*. World Stress Map Project Guidelines: Image Logs. World Stress Map Project.

ZOBACK, M.D. 2007. *Reservoir Geomechanics*. Cambridge University Press, Cambridge.

Contemporary stress and neotectonics in the Otway Basin, southeastern Australia

DAVID R. TASSONE[1], SIMON P. HOLFORD[1]*, ROSALIND KING[2], MARK R. P. TINGAY[1] & RICHARD R. HILLIS[1,3]

[1]*Australian School of Petroleum, Centre for Tectonics, Resources and Exploration, The University of Adelaide, North Terrace, Adelaide, SA 5005, Australia*

[2]*Department of Earth Sciences, Centre for Tectonics, Resources and Exploration, The University of Adelaide, North Terrace, Adelaide, SA 5005, Australia*

[3]*Deep Exploration Technologies CRC, 26 Butler Boulevard, Burbridge Business Park, Adelaide Airport, SA 5950, Australia*

**Correspondence: simon.holford@adelaide.edu.au*

Abstract: Geomechanical and geological datasets from fold–thrust belts and passive margins that have been subject to neotectonic activity often provide contradictory evidence for the state of contemporary stress. Southeastern Australia has relatively high levels of neotectonic activity for a so-called stable continental region. In the eastern Otway Basin, this neotectonic activity consists of compressional deformation and uplift, indicating a reverse fault stress regime. However, this is inconsistent with the stress magnitudes estimated from petroleum exploration data, which indicate normal or strike-slip fault stress regimes. A new wellbore failure analysis of 12 wells indicates that the maximum horizontal stress azimuth in this basin is *c.* 135° N, consistent with neotectonic structural trends. Our results indicate that the lithology and variations in structural style with depth exert important controls on horizontal stress magnitudes. The observed partitioning of stress regimes and deformation styles with depth within the basin may reflect the contrasting mechanical properties of the basin-fill. There is an overall increase in the minimum horizontal stress gradient of *c.* 1–2 MPa km^{-1} from west to east, corresponding to a change in structural style across the basin. In the central Otway Basin, rift-related faults strike near-parallel to the maximum horizontal stress azimuth and there are comparatively low levels of neotectonic activity, whereas in the eastern Otway Basin, where rift-related faults strike near-orthogonal to the maximum horizontal stress azimuth, the level of neotectonic faulting and uplift is much higher. Our results show that the integration of structural geology with geomechanical datasets can lead to improved interpretations of contemporary stresses, consistent with neotectonic observations.

The Otway Basin in southeastern Australia formed as a result of the Late Jurassic to Early Palaeogene separation of Australia from Antarctica (Norvick & Smith 2001; Krassay *et al.* 2004; MacDonald *et al.* 2013). A significant NW–SE compressional deformation event during the mid-Cretaceous interrupted this rifting (Fig. 1), causing regional exhumation and subsequent rifting to be focused along the Tasman Fault Zone, effectively creating a failed rift system through the Bass Strait (Fig. 2; Hill *et al.* 1995; Cayley 2011). Following final continental separation at *c.* 43 Ma, post-rift subsidence in the Otway Basin was interrupted by several periods of uplift, exhumation and compressional deformation, with the most recent phase beginning during the late Miocene to early Pliocene and continuing to the present day (Fig. 1; Cooper & Hill 1997; Krassay *et al.* 2004; Sandiford & Quigley 2009; Holford *et al.* 2014). This complex tectonic history makes the Otway Basin a suitable natural laboratory to help understand the causes and effects of compressional deformation and neotectonics in a post-rift, passive margin setting.

Neogene neotectonic compressional deformation is widespread over much of southeastern Australia (Dickinson *et al.* 2002), apparently consisting of two separate phases of deformation: an initial late Miocene–early Pliocene phase and a subsequent late Pliocene–Holocene phase (Fig. 1; Wallace *et al.* 2005). Within the on- and offshore Gippsland Basin (e.g. the Strezleki Ranges) and in the vicinity of the Otway Ranges and adjacent offshore areas across the eastern Otway Basin, there is strong evidence for local Neogene basin inversion (Cooper

From: Turner, J. P., Healy, D., Hillis, R. R. & Welch, M. J. (eds) 2017. *Geomechanics and Geology*. Geological Society, London, Special Publications, **458**, 49–88.
First published online May 25, 2017, https://doi.org/10.1144/SP458.10

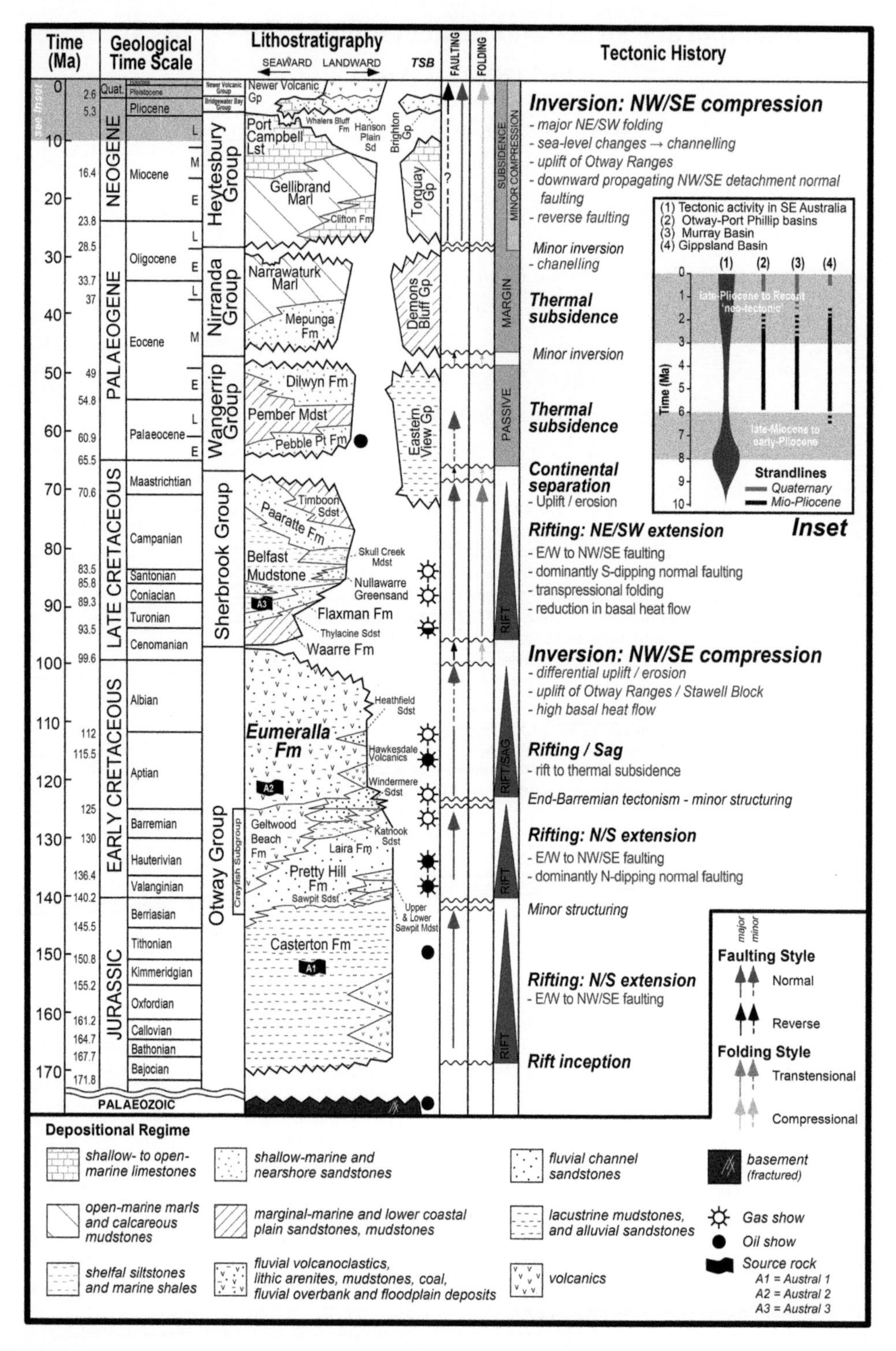

Fig. 1. Stratigraphy of the Otway Basin showing the depositional regimes and major tectonic stages that caused folding and faulting. Modified after Duddy (2003) and Geary & Reid (1998). Inset shows the suggested tectonic activity across southeastern Australia in the last 10 myr interpreted from strandlines. Modified after Wallace *et al.* (2005).

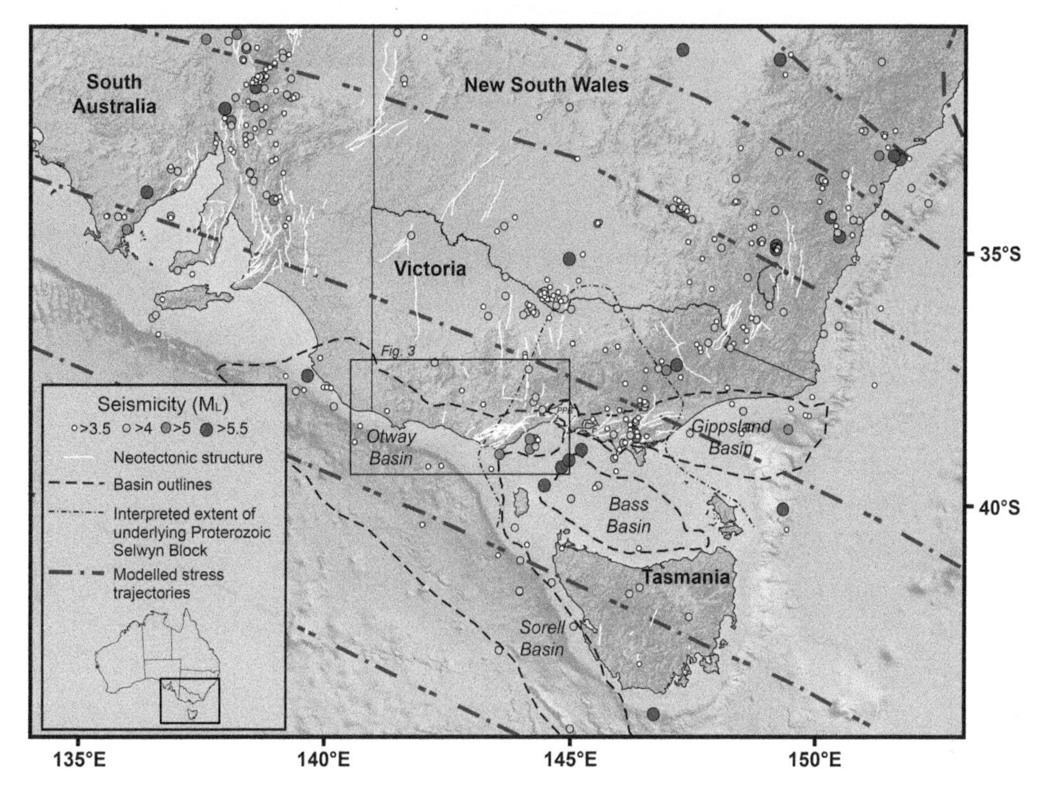

Fig. 2. Topographic map of southeastern Australia modified after Holford *et al.* (2011*a*) showing Meso-Cenozoic basin outlines, neotectonic faults and structural features (after Clark *et al.* 2011), the distribution of earthquakes with $M_L > 3.5$ (after Sandiford *et al.* 2004; Holford *et al.* 2011*b*) and modelled stress trajectories (after Reynolds *et al.* 2003*a*; Nelson *et al.* 2006*a*). Also shown is the interpreted extent of the underlying Proterozoic Selwyn Block in central Victoria (after Cayley 2011).

& Hill 1997; Holford *et al.* 2011*a*) and an abundance of neotectonic features, such as fault scarps, monoclines and anticlines. These neotectonic features collectively point towards a reverse fault stress regime (Fig. 2; Sandiford 2003*a*, *b*; Hillis *et al.* 2008; Clark *et al.* 2011, 2012). High levels of seismicity in the Otway Basin and over much of southeastern Australia (relative to other stable intraplate settings; e.g. Johnston 1994) provide further evidence that neotectonic deformation continues to the present day (Fig. 2). Although still ongoing, geological observations from within the Otway Basin indicate that the intensity of this activity has declined since a peak during the late Miocene to early Pliocene (Sandiford 2003*b*; Tassone *et al.* 2012).

A number of contemporary stress studies conducted in Australian basins (e.g. Hillis & Williams 1992, 1993; Hillis *et al.* 1995; Hillis & Reynolds 2000; Mildren & Hillis 2000; Reynolds & Hillis 2000; Reynolds *et al.* 2003*a*, *b*; Nelson *et al.* 2006*a*; Bailey *et al.* 2014, 2016) have shown that observed maximum horizontal stress orientations are consistent with the predictions of plate-scale finite-element stress modelling, which suggest that plate boundary forces exert a first-order control on the state of stress in the Australian continent (Fig. 2; Reynolds *et al.* 2003*a*; Dyksterhuis & Müller 2008). Across both the Otway Basin and broader regions of southeastern Australia, stress modelling studies suggest that the observed *c.* NW–SE horizontal stress orientations primarily reflect the increased coupling of the Australian and Pacific plate boundaries associated with the formation of the southern Alps in New Zealand since the late Miocene to early Pliocene (Sandiford *et al.* 2004). However, although there is consensus regarding the controls on contemporary stress orientations in most Australian margin basins, contemporary stress magnitudes constrained using petroleum exploration data appear to be inconsistent with those inferred from geological observations (Hillis *et al.* 2008; King *et al.* 2012). For example, in the central and eastern Otway Basin, some studies (e.g. Bérard *et al.* 2008) have constrained a normal fault *in situ* stress regime (i.e. $S_V > S_{Hmax} > S_{hmin}$) (e.g. Nelson *et al.* 2006*a*; Bailey *et al.* 2014) indicate

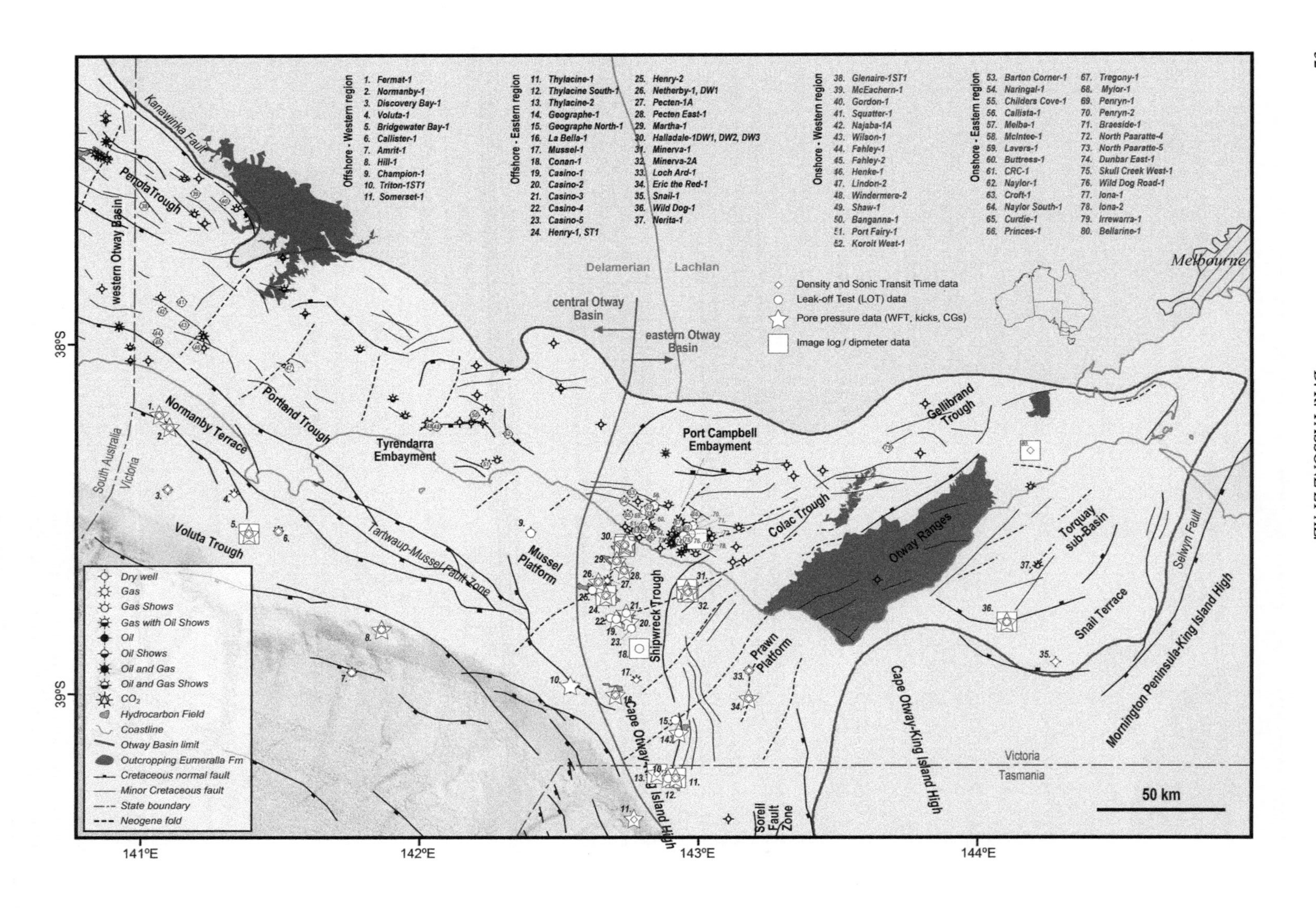
Offshore - Western region
1. Fermat-1
2. Normanby-1
3. Discovery Bay-1
4. Voluta-1
5. Bridgewater Bay-1
6. Callister-1
7. Amrit-1
8. Hill-1
9. Champion-1
10. Triton-1ST1
11. Somerset-1
Offshore - Eastern region
11. Thylacine-1
12. Thylacine South-1
13. Thylacine-2
14. Geographe-1
15. Geographe North-1
16. La Bella-1
17. Mussel-1
18. Conan-1
19. Casino-1
20. Casino-2
21. Casino-3
22. Casino-4
23. Casino-5
24. Henry-1, ST1
25. Henry-2
26. Netherby-1, DW1
27. Pecten-1A
28. Pecten East-1
29. Martha-1
30. Halladale-1DW1, DW2, DW3
31. Minerva-1
32. Minerva-2A
33. Loch Ard-1
34. Eric the Red-1
35. Snail-1
36. Wild Dog-1
37. Nerita-1
Onshore - Western region
38. Glenaire-1ST1
39. McEachern-1
40. Gordon-1
41. Squatter-1
42. Najaba-1A
43. Wilson-1
44. Fahley-1
45. Fahley-2
46. Henke-1
47. Lindon-2
48. Windermere-2
49. Shaw-1
50. Banganna-1
51. Port Fairy-1
52. Koroit West-1
Onshore - Eastern region
53. Barton Corner-1
54. Naringal-1
55. Childers Cove-1
56. Callista-1
57. Melba-1
58. McIntee-1
59. Lavers-1
60. Buttress-1
61. CRC-1
62. Naylor-1
63. Croft-1
64. Naylor South-1
65. Curdie-1
66. Princes-1
67. Tregony-1
68. Mylor-1
69. Penryn-1
70. Penryn-2
71. Braeside-1
72. North Paaratte-4
73. North Paaratte-5
74. Dunbar East-1
75. Skull Creek West-1
76. Wild Dog Road-1
77. Iona-1
78. Iona-2
79. Irrewarra-1
80. Bellarine-1
Density and Sonic Transit Time data
Leak-off Test (LOT) data
Pore pressure data (WFT, kicks, CGs)
Image log / dipmeter data
Melbourne
Delamerian
Lachlan
central Otway Basin
eastern Otway Basin
Kanawinka Fault
Penola Trough
western Otway Basin
Portland Trough
Normanby Terrace
Tyrendarra Embayment
Voluta Trough
Tartwaup-Mussel Fault Zone
Mussel Platform
Port Campbell Embayment
Shipwreck Trough
Colac Trough
Gellibrand Trough
Otway Ranges
Torquay sub-Basin
Snail Terrace
Selwyn Fault
Mornington Peninsula-King Island High
Prawn Platform
Cape Otway-King Island High
Sorell Fault Zone
South Australia
Victoria
Tasmania
50 km
38°S
39°S
141°E
142°E
143°E
144°E
Dry well
Gas
Gas Shows
Gas with Oil Shows
Oil
Oil Shows
Oil and Gas
Oil and Gas Shows
CO_2
Hydrocarbon Field
Coastline
Otway Basin limit
Outcropping Eumeralla Fm
Cretaceous normal fault
Minor Cretaceous fault
State boundary
Neogene fold

a strike-slip fault *in situ* stress regime (i.e. $S_{Hmax} > S_V > S_{hmin}$). These conflicting results may be due in part to the application of different techniques used to constrain horizontal stress magnitudes (Vidal-Gilbert *et al.* 2010). Establishing the causes of the incompatibility between petroleum data and neotectonic evidence, which clearly indicates a reverse fault stress regime (e.g. Cooper & Hill 1997; Hillis *et al.* 2008; Holford *et al.* 2010, 2011*a*; Tassone *et al.* 2012, 2014), remains a challenging and outstanding issue (Couzens-Schultz & Chan 2010; King *et al.* 2012; Bailey *et al.* 2016).

In this study, we use recently available petroleum exploration datasets, including petrophysical wireline and drilling data (Fig. 3), to constrain the contemporary stress states within the central and eastern Otway Basin. Recent offshore hydrocarbon discoveries and renewed exploration interest in the unconventional hydrocarbon potential of the onshore basin has provided a wealth of new data within the Victorian sector of the Otway Basin, which can be used to tackle the problem of the incompatible stress regimes indicated by petroleum datasets and geological evidence (cf., Couzens-Schultz & Chan 2010). In contrast with previous contemporary stress studies in the Otway Basin, which have mainly been concerned predominantly with constraining the state of stress (e.g. Hillis *et al.* 1995; Nelson *et al.* 2006*a*), we integrate a variety of complementary structural geology observations and geophysical datasets. These observations and datasets help to define the roles of mechanical rock properties, pre-existing basement structures and intra-basin structural heterogeneities in influencing the nature of vertical and horizontal stress magnitudes throughout the basin. We show that petroleum exploration datasets, when integrated with other structural geology observation and geophysical datasets, can help us to understand and reconcile the state of stress in sedimentary basins.

Geological background

The Otway Basin is a large, broadly NW–SE-trending extensional basin encompassing onshore and offshore parts of South Australia and Victoria, and Tasmanian waters (Fig. 2). Rifting in the Otway Basin commenced in the late Jurassic–early Cretaceous (Fig. 3), first in the western region (i.e. the South Australian sector), then progressing to the central and eastern regions (i.e. the Victorian sector), resulting in several distinct graben and half-graben with varying geometries and orientations (Fig. 3: Perincek & Cockshell 1995; Krassay *et al.* 2004). The early Cretaceous rift axis is located onshore and consists of *c.* W–NW-trending depocentres in the western and central regions (e.g. the Penola Trough) and *c.* NE-trending depocentres in the eastern region (e.g. the Colac Trough; Krassay *et al.* 2004). The near-orthogonal change in rift axis and fault strike occurs at *c.* 143° E longitude and is primarily the result of episodic rifting at different times and in different directions (Hill *et al.* 1994, 1995), probably related to substantial rheological differences in the basement across different Proterozoic–Palaeozoic boundaries (i.e. Delamerian and Lachlan; Miller *et al.* 2002; Cayley 2011). This boundary broadly continues southwards through the Shipwreck Trough towards the Tasman Fracture Zone (i.e. the Sorell Fault Zone), effectively separating the passive margin in the central parts and a failed rift zone in the eastern parts (Fig. 3; Hill *et al.* 1995; Miller *et al.* 2002; Cayley 2011). Initial rift-fills were dominated by carbonaceous lacustrine shales with minor interbedded sandstones and volcanics (Casterton Formation); as the rate of extension increased into the Berriasian and Barremian, the synrift accommodation space was filled by amalgamated fluvial and lacustrine facies (Crayfish Subgroup; Krassay *et al.* 2004). A thick mudstone-rich volcaniclastic succession (Eumeralla Formation) was deposited during a decrease in tectonic activity during the Aptian and Albian (Fig. 1; Krassay *et al.* 2004).

The basin experienced a pulse of *c.* NW–SE-directed shortening, uplift and exhumation during the late Albian–Cenomanian (Hill *et al.* 1995; Duddy 2003), resulting in a regional mid-Cretaceous unconformity (Fig. 1). Major uplift in the eastern Otway Basin effectively isolated the Torquay Sub-Basin (Krassay *et al.* 2004), the late

Fig. 3. Structural elements map of the Victorian Otway Basin (modified after Totterdell 2012) showing: the distribution of wells with available wireline formation test (WFT) data to determine formation pore pressure (P_P) values; wells with bulk density, sonic velocity and/or caliper data to estimate vertical stress (S_V) magnitudes; wells with reported leak-off tests (LOTs) to estimate minimum horizontal stress (S_{hmin}) magnitudes; and wells with available image log and caliper/dipmeter data to determine the orientation of maximum horizontal stress (S_{Hmax}). The Victorian Otway Basin has been divided into central and eastern regions at *c.* 142.6° E, where pronounced Neogene compressional deformation occurs and the structural trend, due predominantly to episodic Cretaceous rifting and crustal rheological differences, changes from *c.* NW–SE (western region) to *c.* east–west and NE–SW (eastern region). Neogene folds after Geary & Reid (1998), Messent *et al.* (1999) and Boult *et al.* (2008).

Cretaceous–Cenozoic stratigraphy of which shares a closer affinity with the Bass Basin to the east (Messent *et al.* 1999).

The western and central parts of the Otway Basin experienced renewed, *c.* north–south-directed to *c.* NE–SW-directed rifting in the late Cretaceous (Perincek *et al.* 1994), with some workers arguing for a significant oblique (sinistral) component (Schneider *et al.* 2004). The locus of extension shifted to the south during the late Cretaceous, resulting in a series of *c.* NW–SE-trending depocentres (e.g. the Voluta Trough) located to the south of the Tartwaup–Mussel fault zone (Fig. 3; Bernecker *et al.* 2003). The Upper Cretaceous Sherbrook Group generally consists of fluvial–deltaic and nearshore to shallow marine siliciclastic deposits (Fig. 1).

The end of rifting in the Otway Basin is marked by a regional intra-Maastrichtian unconformity (Fig. 1; Krassay *et al.* 2004), but the evidence for seafloor spreading off the Otway Basin (i.e. the first appearance of oceanic crust) and thus continental separation between Australia and Antarctica is dated as middle Eocene (Norvick & Smith 2001). Sedimentary successions in the Otway Basin become progressively more marine-influenced and calcareous throughout the Cenozoic, reflecting the progressive establishment of open marine circulation (McGowran *et al.* 2004; Blevin & Cathro 2008). Subsidence following intra-Maastrichtian break-up initiated a major transgression across the Otway Basin, resulting in the deposition of the siliciclastic Wangerrip Group (Fig. 1; Bernecker *et al.* 2003). Small growth wedges bounded by basinwards-dipping reactivated late Cretaceous normal faults in the Portland Trough (Fig. 3; Krassay *et al.* 2004), where the Wangerrip Group reaches a maximum thickness >1200 m (Holdgate & Gallagher 2003), imply that an extensional stress regime continued into the Palaeogene (Holford *et al.* 2014). It is separated by a major intra-Lutetian unconformity associated with significant localized erosion from the overlying prograding nearshore to offshore marine clastics and carbonates of the late Eocene–early Oligocene Nirranda Group (Holdgate & Gallagher 2003; Krassay *et al.* 2004). This unconformity also correlates with the onset of fast seafloor spreading in the Southern Ocean at *c.* 43 Ma (Veevers 2000; McGowran *et al.* 2004). Since the mid-Eocene onset of fast spreading, the Australian continent has been subjected to a largely compressional stress field resulting from the configuration of the Indo-Australian plate boundaries (Holford *et al.* 2014).

The Nirranda Group reaches its maximum thickness of *c.* 200 m in two major depocentres, in the Port Campbell Embayment and Portland Trough (Holdgate & Gallagher 2003), and is separated from the overlying Heytesbury Group by a regional intra-Oligocene unconformity (Bernecker *et al.* 2003). The late Oligocene–late Miocene Heytesbury Group (maximum thickness >1600 m) consists of marls and limestones deposited under fully marine conditions (Krassay *et al.* 2004). A regional late Miocene–Pliocene unconformity (Dickinson *et al.* 2002; Holford *et al.* 2011*b*) separates the Heytesbury Group from the late Neogene succession of the Otway Basin, which is characterized by relatively thin and localized mixed siliciclastic–carbonate sediments and basaltic volcanic rocks that usually unconformably or disconformably overlie Heytesbury Group strata (Dickinson *et al.* 2002; Tassone *et al.* 2011).

The main source rocks in the Otway Basin are of early Cretaceous age: the Casterton Formation and Crayfish Subgroup in the onshore western and central parts and the Eumeralla Formation in the onshore and offshore central and eastern parts of the basin (Fig. 1; Bernecker *et al.* 2003). The Pretty Hill Formation is the major reservoir unit in the onshore western and central parts and the basal Waarre Formation acts as the major regional reservoir interval in the onshore and offshore eastern parts of the Otway Basin (Fig. 1; Bernecker *et al.* 2003). The Eumeralla Formation seals the Pretty Hill Formation reservoirs, whereas the Flaxman Formation and Belfast Mudstone (Fig. 1) seal the Waarre Formation reservoirs. Traps are structurally controlled and generally fault-bound; they have been affected by periods of deformation during the Eocene, Oligocene–early Miocene and late Miocene–Pliocene (Dickinson *et al.* 2002). The most significant compressional and neotectonic structural features occur onshore in the eastern Otway Basin, in and around the Otway Ranges, and in the adjacent Torquay Sub-Basin (Fig. 3; Hill *et al.* 1994, 1995; Cooper & Hill 1997; Geary & Reid 1998; Messent *et al.* 1999; Dickinson *et al.* 2002; Miller *et al.* 2002; Sandiford 2003*b*, Sandiford *et al.* 2004; Holford *et al.* 2010, 2011*a*, 2014; Tassone *et al.* 2012, 2014). Similar to the Gippsland Basin, these structures consist of a combination of *c.* NE-trending anticlines resulting from the reactivation of synrift normal faults (Fig. 2). Boult *et al.* (2008) has nevertheless shown that Neogene neotectonic structures do extend into the central Otway Basin (Fig. 3).

Determination of contemporary stresses within the central and eastern Otway Basin

Previous determinations of contemporary stress in the Otway Basin have utilized petroleum exploration data dominantly from the western Otway Basin in South Australia (Hillis *et al.* 1995; Nelson *et al.* 2006*a*; Bailey *et al.* 2014), although a few have focused within the Port Campbell Embayment (Bérard *et al.* 2008; Rogers *et al.* 2008) and offshore regions of Victoria (Nelson *et al.* 2006*a*). We note that there is often significant contemporary stress variation (i.e. orientation and magnitude) within basins (Bell 1996; Tingay *et al.* 2006; Heidbach *et al.* 2007, 2010). As a result of this, and the observation that many of the neotectonic features lie in the eastern Otway Basin, we have divided the Victorian sector of the Otway Basin, where less contemporary stress information is present, into two regions: the central and eastern regions either side of *c.* 142.6° E longitude. In addition to the onshore parts, the central region encompasses the Voluta Trough, the Normanby Terrace, the Mussel Platform and deeper water wells near the shelf break, whereas the eastern region encompasses Torquay Sub-Basin, the Prawn Platform and the Shipwreck Trough wells NE of the Tartwaup–Mussel hingeline (Fig. 3). This subdivision enables us to better assess the influence of underlying structural trends on the distribution of contemporary stresses within the basin.

Assuming that one of the principal stress directions is vertical, and utilizing a large dataset from across the central and eastern regions, the five components that make up the simplified contemporary stress tensor in sedimentary rocks were determined. This includes the formation pore pressure (P_P), the vertical stress (S_V) magnitude, the minimum horizontal stress (S_{hmin}) magnitude and the maximum horizontal stress (S_{Hmax}) magnitude and orientation. For each component of the stress tensor described in the following sections, the way in which the tensor component is constrained is first discussed, followed by a distribution of the applicable available data and then the outcomes of the analyses.

Maximum horizontal stress orientations

Constraining S_{Hmax} *orientations.* The maximum horizontal stress (S_{Hmax}) orientation is constrained in this study by identifying wellbore failure features such borehole breakouts and drilling-induced tensile fractures (DITFs) (Zoback *et al.* 2003). Breakouts occur when the wellbore circumferential stress ($\sigma_{\theta\theta}$) exceeds the compressive rock strength of the wellbore wall, causing compressional shear failure surfaces to intersect and break off into the open wellbore (Fig. 4a; Gough & Bell 1982). DITFs occur when the pressure of the drilling mud exceeds the wellbore circumferential stress (Fig. 4a), typically resulting in vertical tensile fractures in the wellbore wall (Bell 1990; Zoback 2010). Breakouts form broad, flat enlargements of the borehole within the spalled wellbore aligned parallel with the minimal horizontal stress (S_{hmin}) orientation and orthogonal to the S_{Hmax} orientation (Fig. 4b), whereas DITFs form thin tensile cracks parallel to the S_{Hmax} orientation and are usually insensitive to caliper data (Fig. 4c; Hillis & Williams 1992; Zoback 2010).

The interpretation of resistivity or acoustic image log data is considered to be a more reliable approach to constrain S_{Hmax} orientations rather than using four- or six-arm caliper data. This is because image logs not only clearly identify borehole breakouts, but also reveal DITFs, natural fractures, bedding planes and other sedimentary features (Heidbach *et al.* 2010). Analysis of wellbore failure features using either image log or caliper data requires additional data in the form of dipmeter data. The dipmeter tool records the azimuth of wellbore drift relative to horizontal and magnetic north, the inclination of the wellbore from vertical, the azimuth of the reference pad in the wellbore plane relative to magnetic north and the bearing of the first pad relative to the high side of the wellbore (Reinecker *et al.* 2003). Whereas image log tools measure and image resistivity and acoustic variations, caliper data require the width of the wellbore to be measured across two orthogonal pairs of caliper arms (four-arm caliper; Reinecker *et al.* 2003) or six independent arms positioned evenly around the wellbore circumference (six-arm; Wagner *et al.* 2004). As the four- or six-arm caliper tool is pulled up the wellbore, the tool rotates about a semi-vertical axis due to cable torque (Hillis & Williams 1992). In zones of wellbore enlargement due to breakout, one caliper pair gets stuck in the enlargement direction and the tool stops rotating until another interval of round hole is encountered (Plumb & Hickman 1985).

Successful discrimination of stress-induced wellbore borehole breakouts from other wellbore enlargement and elongation features that result from drilling, such as washout, mudcake and key seats (asymmetrical abrasions due to wear by the drill string on the side of the wellbore that deviates from the vertical), is ensured by the use of other wireline logging parameters (e.g. the azimuth of the reference pad in the wellbore plane relative to magnetic north and the bearing of the first pad relative to the high side of the wellbore; Reinecker *et al.* 2003). Six-arm caliper data provide detailed information regarding the shape of the elongation wellbore because the six arms move independently of one another, allowing the tool to decentralize and record radius lengths. For this reason, wellbore borehole breakout analysis of six-arm caliper tools is more complicated (cf. Wagner *et al.* 2004).

DITFs are generally not associated with wellbore elongation. However, if orthogonal trends are observed using only caliper/dipmeter data, then it is impossible to differentiate the horizontal stress orientations (Hillis & Williams 1992). Borehole

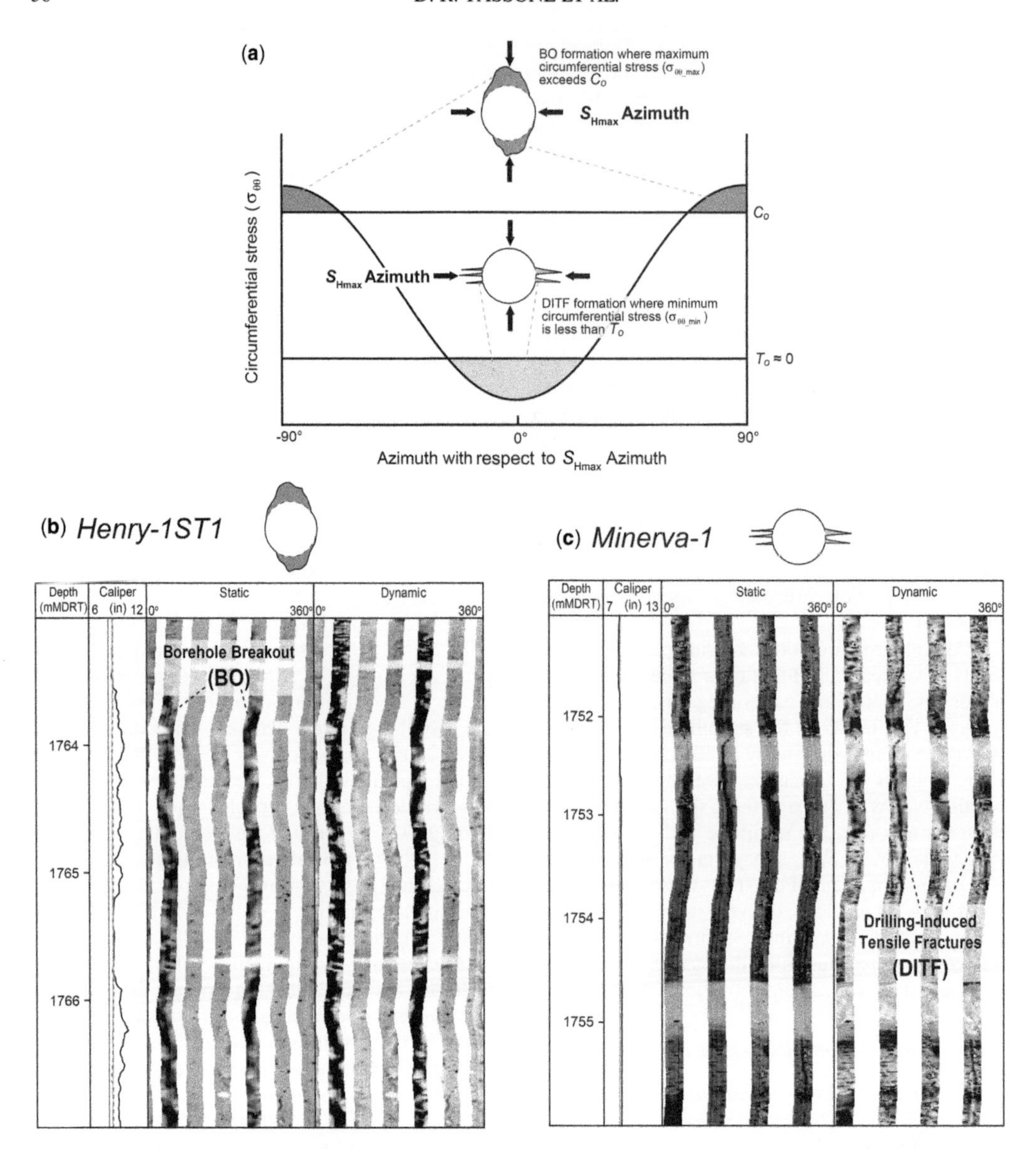

Fig. 4. (**a**) Schematic diagram of development of stress-induced wellbore failure features when the circumferential stress around a vertical wellbore exceeds the compressive rock strength or is less than the tensile rock strength, in which breakouts and DITFs form, respectively (modified after Reynolds *et al.* 2005). DITFs form parallel to the orientation of S_{Hmax}, whereas breakouts form orthogonal to the orientation of S_{Hmax} (i.e. parallel to the orientation of S_{hmin}). (**b**) Example of breakouts observed in Henry-1ST1. (**c**) Example of DITF observed in Minerva-1. Note the caliper response in (b) due to breakouts compared with the lack of caliper response in (c) where DITFs were observed.

breakouts and DITFs from deviated wells may not be reliable indicators of S_{Hmax} orientation because their orientations are controlled by the entire stress tensor (Mastin 1988; Hillis & Reynolds 2000). However, within a strike-slip fault stress regime, Mastin (1988) concludes that a wellbore deviated at least 35° from vertical is required before the horizontal projection of a borehole breakout differs by more than 10° from the S_{hmin} direction.

Data availability and distribution. Image log data and caliper data from 12 petroleum wells in the central and eastern Otway Basin were available for this study to enable S_{Hmax} orientations to be determined

(Fig. 3). Eight wells were from the Shipwreck Trough in the eastern Otway Basin, of which three wells used Schlumberger's Formation MicroImager (FMI) tool (Conan-1, Thylacine-1and 2), three wells used Baker Atlas' SimulTaneous Acoustic and Resistivity Imager (STAR) tool (Henry-1ST1, Halladale-1DW1 and DW2) and two wells used Schlumberger's Formation MicroScanner (FMS) tool (Minerva-1 and 2A; Table 1). Wild Dog-1 in the Torquay Sub-Basin, also in the eastern Otway Basin, ran a six-arm caliper log (the tool was assumed to be centralized given that all three caliper arms were measured as diameters), while the only well located in the central Otway Basin, Bridgewater Bay-1 in the Voluta Trough, ran four-arm caliper logs (Table 1 and Fig. 3). In addition to the ten offshore wells, two onshore wells from the eastern Otway Basin (Bellarine-1 and Wild Dog Road-1) ran FMI image tool data that were also available to constrain S_{Hmax} orientations. The maximum horizontal stress orientation at Bellarine-1 has been reported by Tassone *et al.* (2014) and is included in this study of wellbore failure statistics.

The depth intervals in which these tools ran are listed in Table 1, along with other relevant parameters such as the maximum wellbore deviation (with respect to vertical), the bit size used and imaged and logged intervals for specific formations. In addition to Dunbar-1 in the Port Campbell Embayment and Eric the Red-1 in the Prawn Platform, Minerva-1 and Minerva-2A have been previously interpreted by Nelson *et al.* (2006*a*), but are reinterpreted in this study to provide length-weighted statistics (not provided by Nelson *et al.* 2006*a*). Bérard *et al.* (2008) published the S_{Hmax} orientation in CRC-1 within the Port Campbell Embayment.

Table 1. *Victorian Otway Basin wells with image log and dipmeter data available to constrain the maximum horizontal stress (S_{Hmax}) orientations in this study, including maximum well deviations encountered over corresponding depth ranges and stratigraphy in which these were logged*

Well	Tool type	Bit size (inches)	Maximum deviation (°)	Data depth range			Images
				Top (m bKB)	Bottom (m bKB)	Interval (m)	
Bellarine-1	FMI	8.5	1.00	278	978	700	Tertiary/Eumeralla/Pretty Hill?
Bridgewater Bay-1	HDT	17.5	1.23	487	1607	1120	Heytesbury/Nirranda/Wangerrip/Timboon/Paaratte
		12.25	1.25	1586	3536	1950	Timboon/Paaratte/Belfast
		8.5	1.25	3850	4183	333	Belfast/Flaxman/Waarre
Conan-1	FMI	12.25	0.40	1675	1949	274	Belfast/Flaxman/Waarre/Eumeralla
Halladale-1DW1	STAR	8.5	20.54	775	1916	1141	Pember/Pebble Point/Massacre/Paaratte/Skull Creek/Nullawarre/Belfast/Flaxman/Waarre/Eumeralla
Halladale-1DW2	STAR	8.5	21.72	1668	1927	259	Belfast/Flaxman/Waarre/Eumeralla
Henry-1ST1	STAR	8.5	1.00	1725	2015	290	Skull Creek/Waarre/Eumeralla
Minerva-1	FMS	8.5	8.80	1189	2024	835	Belfast/Flaxman/Waarre
		6.0	9.00	2110	2424	314	Waarre/Eumeralla
Minerva-2A	FMS	12.25	3.90	1526	2170	644	Belfast/Flaxman/Waarre
Thylacine-1	FMI	8.5	2.97	2022	2503	481	Belfast/Flaxman/Waarre
Thylacine-2	FMI	8.5	3.51	2104	2530	426	?/Flaxman/Waarre
			1.25	2245	2350	105	?/Flaxman/Waarre
Wild Dog-1	SED	8.5	1.07	724	1244	500	Torquay/Demons Bluff/Eastern View/Eumeralla
Wild Dog Road-1	FMI	8.5	34.5	1200	1676	476	Skull Creek/Nullawarre/Belfast/Waarre

FMI, Schlumberger Formation MicroImager; FMS, Formation MicroScanner; HDT, Schlumberger high-resolution dipmeter tool (four-arm); SED, Halliburton six-arm dipmeter tool; STAR, Baker Atlas SimulTaneous Acoustic and Resistivity Imager. m bKB, metres below Kelly bushing.

Table 2. *Wellbore failure analysis of image log data showing the mean orientations of borehole breakout (BO) and DITF failure features and their corresponding standard deviations (i.e. circular statistics of Mardia 1972) when weighted by number (N) and length (in m)*

Well	Indicator	Number-weighted			Length-weighted			S_{Hmax} orientation (°)	World Stress Map quality rating
		N	Mean borehole breakout/DITF orientation (°)	SD (°)	Length (m)	Mean borehole breakout/DITF orientation (°)	SD (°)		
Bellarine-1	BO	4	47.2	3.81	3.69	46.9	2.77	137.1	D
Bellarine-1	DITF	55	143.6	11.68	67.94	144.6	11.18	144.1	B
Bellarine-1	DITF_INC	40	237.5	17.55				147.5	
Conan-1	BO	32	40.8	6.88	58.06	44.6	5.95	132.7	B
Halladale-1DW1	BO	20	34.3	3.91	194.41	35.3	2.98	124.5	A
Halladale-1DW1	DITF	2	123.3	2.06	2.43	122.6	2.03	123.0	D
Halladale-1DW2	BO	18	43.0	7.74	78.57	41.5	6.73	132.3	B
Henry-1ST1	BO	23	38.0	5.36	107.24	40.5	2.63	129.3	A
Minerva-1	BO	14	51.9	5.01	554.27	53.2	1.74	142.6	A
Minerva-1	DITF	7	131.3	7.45	36.52	131.5	7.80	131.4	C
Minerva-2A	BO	4	51.2	2.09	280.32	52.8	0.46	141.0	C
Minerva-2A	DITF	5	138.1	2.43	6.44	138.0	2.56	138.1	D
Thylacine-1	BO	3	42.2	4.49	3.67	40.8	4.64	131.5	D
Thylacine-2	BO	15	42.3	9.45	7.89	46.2	9.62	134.3	D
Wild Dog Road-1	BO	22	51.5	3.26	187.90	48.9	5.30	140.2	A
Wild Dog Road-1	DITF_INC	18	132.4	18.14				132.4	

Also listed is the corresponding orientation of S_{Hmax} (average of number- and length-weighted estimates) and World Stress Map quality rating after Heidbach *et al.* (2010).

S_{Hmax} *orientations within the central and eastern Otway Basin.* Our analysis has positively identified a number of drilling-induced wellbore failure features. In total, 155 borehole breakouts and 69 DITFs were interpreted from image logs (i.e. FMI/FMS/STAR) and 18 borehole breakouts were interpreted from caliper logs. Examples of borehole breakouts and DITFs from Henry-1ST1 and Minerva-1can be seen in Figure 4b and c, respectively. Cumulatively, the borehole breakouts identified a combined length of 1981.5 m and the DITFs a total of 138.0 m. Tables 2 and 3 show the number of borehole breakouts and DITFs in each well, the total length, the mean orientation and standard deviation of the wellbore failure features, and the mean S_{Hmax} orientation for image logs and dipmeter data, respectively. In a number of wells, borehole breakouts had developed over 30% of the interpretable image log and *c.* 70% of borehole breakouts interpreted by length, or *c.* 1244 m in total, occurred within the Upper Cretaceous Belfast Mudstone (Fig. 5). The distribution of interpreted

Table 3. *Wellbore failure analysis of caliper/dipmeter data showing the mean orientations of borehole breakout (BO) failure features and their corresponding standard deviations (i.e. circular statistics of Mardia 1972) when weighted by number (N) and length in (m)*

Well	Tool	Indicator	Number-weighted			Length-weighted			S_{Hmax} orientation (°)	World Stress Map quality rating
			N	Mean BO orientation (°)	SD (°)	Length	Mean BO orientation (°)	SD (°)		
Bridgewater Bay-1	HDT	BO	15	49.0	10.51	486.00	52.2	8.04	140.6	A
Wild Dog-1	FMS	BO	3	57.7	16.60	19.46	60.6	15.30	149.1	D

HDT, Schlumberger high-resolution dipmeter tool (four-arm); FMS, Formation MicroScanner.
Also listed is the corresponding orientation of S_{Hmax} (average of number- and length-weighted estimates) and World Stress Map quality rating after Heidbach *et al.* (2010).

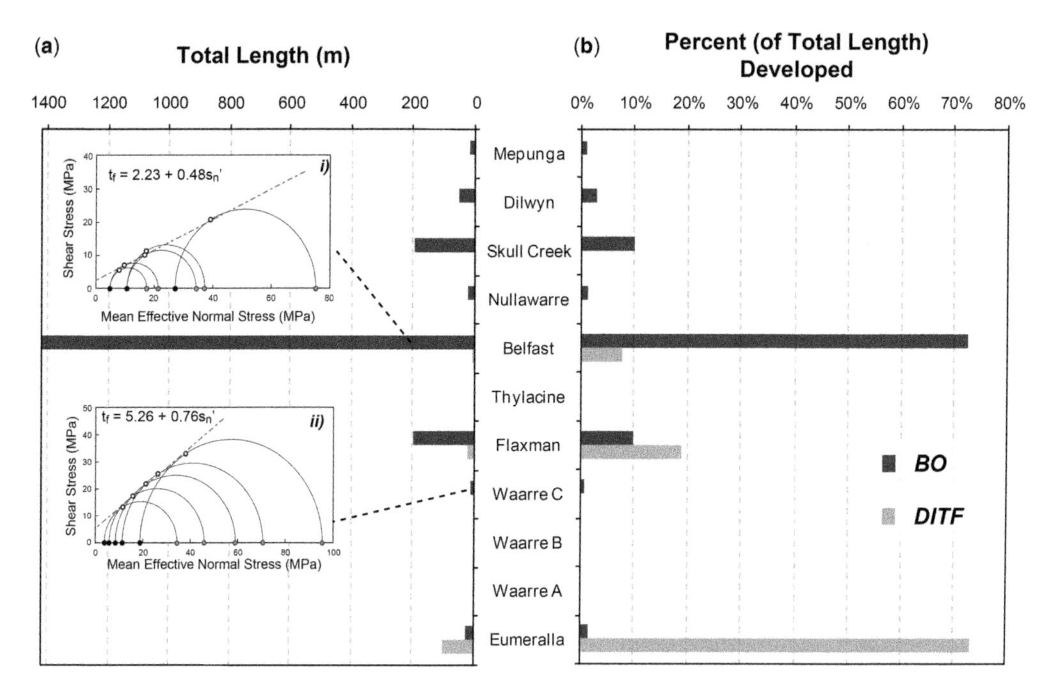

Fig. 5. Wellbore failure analysis statistics calculated for the 12 wells listed in Table 1. (**a**) The cumulative total length of interpreted breakouts (BOs) and DITFs for each stratigraphic formation logged. It can be seen that significantly more breakouts were confidently interpreted than DITFs and that these predominantly occurred within finer grained rocks. In particularly, the Upper Cretaceous Belfast Mudstone (seal), which triaxial testing shows (inset (i), after Dewhurst *et al.* in press) to be much weaker than coarser grained rocks such as the Waarre C reservoir unit of the Upper Cretaceous Waarre Formation (inset (ii)). (**b**) The percentage of breakouts and DITFs, with respect to the total lengths, developed within each stratigraphic formation logged showing that the breakouts formed in the Belfast Mudstone account for >70% of all breakouts confidently interpreted.

borehole breakouts and DITFs, represented as both number-weighted and length-weighted statistics, are shown in Figure 6a and b. Figure 6c shows the resulting S_{Hmax} orientations compared with those determined from previously published and unpublished studies. This indicates consistent orientations in the central and eastern Otway Basin of *c.* 135 ± 15° N. The determined S_{Hmax} orientations are predominantly of A–C quality according to the World Stress Map criteria (Heidbach *et al.* 2010), although lower D quality measurements also yield a consistent NW–SE trend. There is, however, a regional clockwise rotation of *c.* 10° in S_{Hmax} orientation from the western to central Otway Basin (Nelson *et al.* 2006*a*). Although the regional S_{Hmax} orientations are remarkably consistent across the entire Otway Basin, it is worth mentioning the distribution of DITFs with depth interpreted within Bellarine-1 (see Tassone *et al.* 2014, fig. 23). At *c.* 450 m there is a local stress rotation from the regional orientation (i.e. *c.* 140° N) to *c.* 10° N and then abruptly back to the regional orientation, possibly indicating a perturbation due to a proximal structural feature, such as a fault. We also note that Bellarine-1, which Tassone *et al.* (2014) showed to have been exhumed by *c.* 1.8 km, contains a large number of DITFs and relatively few breakouts. This is consistent with the notion of exhumed rocks being strong and unlikely to develop breakouts even with low mud weights.

In the West Tuna Field in the neighbouring Gippsland Basin, Nelson *et al.* (2006*b*) observed from image log data and triaxial testing results that strong, cemented sandstones units act as a stress-bearing framework in this highly horizontally stressed region. Consequently, stresses were partitioned between ‘strong’ interbedded sandstones and ‘weaker’ shales, such that borehole breakouts only occurred in the stronger sandstones. A similar observation was interpreted within siliciclastic rocks in the Perth Basin (King *et al.* 2008). These results stand in contrast to wellbore failure analyses in the Otway Basin, which show that borehole breakouts preferentially develop in weaker finer grained rocks, not sandstones. For example, triaxial testing indicates that the Belfast Mudstone is relatively weak (Fig. 5a; Dewhurst *et al.* in press). This observation is more consistent with

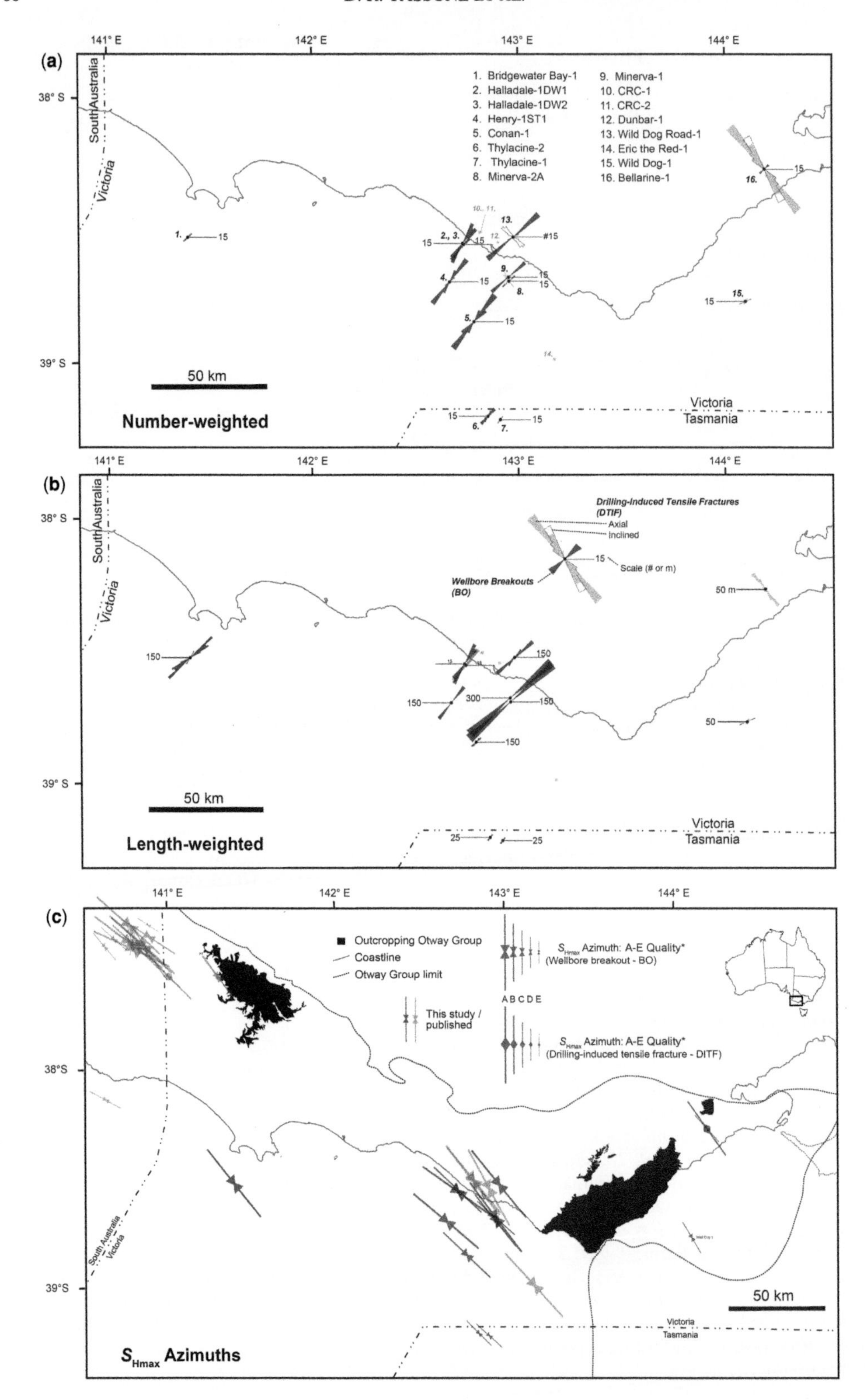
(a)
141° E
142° E
143° E
144° E
38° S
39° S
South Australia
Victoria
1. Bridgewater Bay-1
2. Halladale-1DW1
3. Halladale-1DW2
4. Henry-1ST1
5. Conan-1
6. Thylacine-2
7. Thylacine-1
8. Minerva-2A
9. Minerva-1
10. CRC-1
11. CRC-2
12. Dunbar-1
13. Wild Dog Road-1
14. Eric the Red-1
15. Wild Dog-1
16. Bellarine-1
50 km
Number-weighted
Victoria
Tasmania
(b)
Drilling-Induced Tensile Fractures (DTIF)
Axial
Inclined
Scale (# or m)
Wellbore Breakouts (BO)
50 km
Length-weighted
Victoria
Tasmania
(c)
Outcropping Otway Group
Coastline
Otway Group limit
S_{Hmax} Azimuth: A-E Quality*
(Wellbore breakout - BO)
A B C D E
This study / published
S_{Hmax} Azimuth: A-E Quality*
(Drilling-induced tensile fracture - DITF)
38°S
39°S
South Australia
Victoria
50 km
Victoria
Tasmania
S_{Hmax} Azimuths

conventional wellbore failure analysis (Nelson *et al.* 2006*b*), where the difference may largely be due to the strength of the sandstones in question. For example, the quartz-rich Waarre C unit has a relatively low C_0 (uniaxial compressive strength) of *c.* 21.21 MPa and a μ_i (coefficient of friction) value of *c.* 0.76 (Fig. 5a) compared with the strong, cemented sandstone in the West Tuna Field C_0 of *c.* 60 MPa and μ_i of *c.* 0.9; (Nelson *et al.* 2006*b*). Thus the Waarre C unit in the Otway Basin is unlikely to provide a similar stress-bearing framework.

Pore pressure magnitudes

Constraining magnitudes of P_P. The magnitudes of formation P_P at depth have a significant impact on the effective stress magnitudes and hence the complete stress tensor. Several wells located in the offshore eastern Otway Basin have P_P measurements from potential reservoir intervals determined using wireline formation tools (WFTs), such as repeat formation tests, modular dynamic tests and reservoir characterization instruments. WFTs estimate P_P by isolating a section of wellbore with packers and measuring the rate at which the pressure stabilizes in the isolated section as the wellbore pressure is reduced, drawing fluids from the formation (Nelson & Hillis 2005). The success of these measurements depends on the permeability of the formation and, as a result of time and cost constraints, such tests are mainly conducted in reservoirs and rarely performed in low-permeability rocks such as shales. If such data are unavailable, drilling mud weight or equivalent circulating density (ECD) data can also be used as proxies for P_P because the pressure of the drilling mud is often kept just in excess of P_P to avoid drilling problems such as fluid influxes (i.e. 'kicks') and to maximize drilling efficiency (Van Ruth & Hillis 2000). This approach can only be used as a proxy for P_P when the increases in mud density that occur prior to any fracture-induced mud losses are due to changes in P_P and not for other reasons, such as to improve wellbore stability or overcome tight (i.e. undergauged) hole sections (Van Ruth & Hillis 2000; Reynolds *et al.* 2006). Increases in background gas during drill pipe connections – that is, connection gas – can also be an indicator of increasing P_P if the drilling mud weight remains unchanged (Mouchet & Mitchell 1989; Sagala & Tingay 2012).

Data availability and distribution. Wells that contain direct and indirect measurements of P_P are shown in (Fig. 3). In the central Otway Basin, two wells contained WFTs, four wells reported kicks and two wells reported connection gas. Mud weight and ECD data were also available for six wells. In the eastern Otway Basin, 19 wells contained WFT data and many of these wells also contained mud weight data, but only three wells were used to highlight their gradients as they were all similar.

P_P *gradients.* Formation pore pressure data (i.e. WFT, kicks, connection gas, mud weight) are plotted against true vertical depth below seabed or mudline (m TVDML) in Figure 7. Previous contemporary stress studies have typically assumed that P_P gradients are hydrostatic (i.e. 10.0 MPa km^{-1}) across the Otway Basin (e.g. Nelson *et al.* 2006*a*).

Permeable units from wells in both the Voluta and Shipwreck troughs provide evidence for both hydrostatic P_P and abnormal P_P indicative of overpressure. WFT data in the Voluta Trough from Fermat-1 indicate near hydrostatic pressures at depths shallower than *c.* 3000 m TVDML, although water kicks with associated mud pit gains, together with WFT data at Normanby-1, located only *c.* 5 km to the SE, indicate a P_P gradient of >14 MPa km^{-1} at depths >3000 m TVDML (Fig. 7a). Drilling and mud log data record several water and gas kicks, together with connection gas at depths below *c.* 3300 m TVDML at Callister-1, indicating high P_P gradients of *c.* 14–17 MPa km^{-1} (Fig. 7a). Petrophysical log data in Bridgewater Bay-1 suggest P_P gradients >16 MPa km^{-1}, consistent with estimates from drilling mud weight data, but also indicate that the top of overpressure within Upper Cretaceous fine-grained Belfast Mudstone strata begins at *c.* 2900 m TVDML, indicating that drilling was actually underbalanced for *c.* 500 m (Fig. 7a; Tassone *et al.* 2011). As a result of the lack of hydrocarbon shows in Belfast Mudstone units and the rapid deposition of thick Upper Pliocene sediments within Miocene submarine canyons at Bridgewater Bay-1 well, Tassone *et al.* (2011) suggested a disequilibrium undercompaction mechanism for the generation of overpressures (Fig. 7a) in this area.

The discovery of hydrocarbons within Upper Cretaceous strata (i.e. the Waarre Formation and Thylacine Member; Fig. 1), and the subsequent development of numerous gas fields in the Shipwreck Trough, have led to the acquisition of many

Fig. 6. Stress-induced wellbore failure features plotted as rose diagrams weighted by (**a**) number and (**b**) length in metres. (**c**) The corresponding orientations of S_{Hmax} calculated as the average between the number- and length-weighted orientations. The size of the symbols relates to the quality of the measurement as defined by the World Stress Map (Heidbach *et al.* 2010). It can be seen that the S_{Hmax} orientations determined in this study correlate well with previously published and unpublished estimates (e.g. Hillis *et al.* 1995; Nelson *et al.* 2006*a*; Bérard *et al.* 2008).

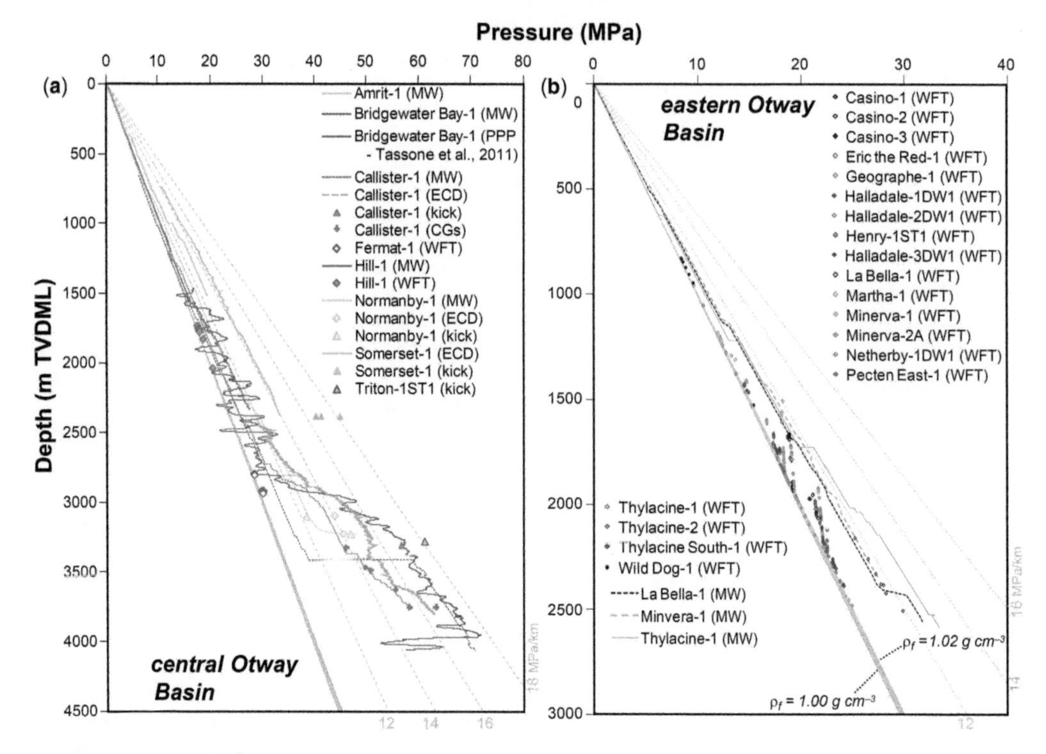

Fig. 7. Available wireline formation test (WFT), kick, connection gas (CG) and mud weight data (mud weight, or equivalent circulation density, ECD), which measures and estimates P_P magnitudes, respectively, plotted against true vertical depth (below mudline) for (**a**) the central Otway Basin and (**b**) the eastern Otway Basin. The distribution of P_P data can be seen in Fig. 3. Included in (a) is the predicted pore pressure (PPP) for Bridgewater Bay-1 based on petrophysical wireline log data indicating high overpressures matching increases in mud weight possibly in excess of c. 16 MPa km^{-1} (Tassone *et al.* 2011). ρ_f, pore fluid density.

WFT data measurements in this part of the eastern Otway Basin. Changes in the P_P gradient occur in a number of wells, indicating variations in the formation fluid density related to the occurrence of gas. Gas accumulations at Minerva-1 and Thylacine-1 occur at depths of *c.* 1565 and 1920 m TVDML, respectively, and hydrocarbon buoyancy forces at these locations cause P_P gradients equivalent of up to *c.* 12 MPa km^{-1} (Figs 7b & 8a). Moderate overpressures observed in La Bella-1 do not appear to be associated with hydrocarbon buoyancy forces, however, because the WFT data indicate two isolated overpressured intervals that are not in pressure communication with P_P gradients of *c.* 10.5 and 11.7 MPa km^{-1} in the water-saturated zones of the Waarre C and Waarre A units, respectively (Fig. 8a). The cause of these moderate overpressures is unclear, although they may be associated with enhanced lateral transfer pressures at the crest of the tilted structure (Yardley & Swarbrick 2000; Tassone *et al.* 2011). A number of other gas fields show P_P discrepancies within vertical stratigraphic sequences or across faults. The Waarre B unit in the Casino Field separates two gas reservoirs (i.e. Waarre C and A units) with a pressure differential of *c.* 200 psi (*c.* 1.379 MPa; Sharp & Wood 2004). Although WFT data from Minerva-1 and Minerva-2A show pressure communication across a SW-dipping fault (Fig. 8b, d), other gas fields such as the Halladale/Blackwatch and Thylacine do not show pressure communication across faults at Waarre Formation levels (Fig. 8b, c and e, f). A fault seal analysis is required to determine whether the differing fault geometries between the Minerva and Halladale/Blackwatch Fields is the cause of the contrasting pressure compartmentalization (i.e. structurally permeable and critically stressed) or if there is sand-on-sand juxtaposition across the fault (e.g. Lyon *et al.* 2005). It can be seen in Figure 8c that the Waarre C and Waarre A unit reservoirs display varying water P_P gradients – for example *c.* 10.02 MPa km^{-1} at Thylacine-1 compared with *c.* 9.91 MPa km^{-1} at Geographe-1, indicating that these reservoirs are not in pressure communication. Hence P_P magnitudes may be compartmentalized by faults in which reservoirs are dissipating P_P

gradients differentially or, alternatively, they may have differing water densities.

Also noteworthy in the westernmost Shipwreck Trough is a saltwater kick encountered at Somerset-1 at *c.* 2380 m TVDML and recorded connection gas as well as gas kick at Triton-1ST1 in the Belfast Mudstone at *c.* 3283 m TVDML, which indicate significantly high overpressures, both with P_P gradients of *c.* 18.5 MPa km^{-1} (Fig. 7a). More work is required to understand the mechanism(s) of the significant overpressure experienced at these locations.

Vertical stress magnitudes

Constraining S_V *magnitudes.* For offshore wells, the total vertical stress (S_V) magnitude is defined as the pressure exerted by the weight of the water column from the surface to the seabed, plus the pressure exerted by the weight of overlying rocks at a specified true vertical depth below mean sea-level (*Z*, m TVDSS) (Tingay *et al.* 2003*a*). It is expressed as:

$$Sv = (\rho_w \times g \times Z_w) + \int_{Z_w}^{Z} [\rho_b(Z) \times g] \cdot dZ \quad (1)$$

where $\rho_b(Z)$ is the bulk density of the overlying rock column at depth *Z* (m TVDSS), *g* is the acceleration due to gravity, ρ_w is the density of seawater and Z_w is the water depth (Engelder 1993). Petrophysical wireline well log data are typically acquired over large intervals of the well at high resolution (typically *c.* 15 cm) and thus magnitude of S_V at any depth can be constrained with a high degree of confidence because the vertical lithological variation of the sampled formations is taken into account in an unbiased way.

To calculate accurate S_V magnitudes, the bulk density log data were carefully filtered to remove spurious data that resulted from poor contact between the tool and the wellbore wall, caused by rugose wellbore conditions (Tingay *et al.* 2003*a*). The filtering process followed the procedure outlined by Tingay *et al.* (2003*a*), whereby bulk density data were assumed to be affected by rugosity if the density error log was greater than ± 0.1 g cm^{-3} and the caliper log was $\geq 5\%$ of the bit or hole size. Any wells that did not have all or some of these additional data were still analysed, but were considered to be less reliable. Filtered (and non-filtered) logs were then also manually edited and de-spiked to remove anomalous measurements prior to calculating the vertical stress (Tingay *et al.* 2003*a*). As the depth of water has a significant influence on the magnitude of S_V with depth (Reynolds *et al.* 2003*b*), we have corrected for water pressure at the seabed and referenced the depth to TVDML. Furthermore, because the principal stresses acting in the vicinity of a deviated wellbore wall are not aligned with the wellbore axis, a horizontal stress component may also be assumed and therefore only density data from near-vertical (<10°) wellbore intervals were considered (Zoback *et al.* 2003).

Vertical stress magnitude calculations require that the bulk rock densities are integrated from the seabed. As it is uncommon for bulk density wireline logs to be logged from the seabed, well check-shot velocity data (or from an offset well) were used to determine the average rock density from the seabed to the top of the density log using the Nafe-Drake (Ludwig *et al.* 1970) velocity–density relationships.

Data availability and distribution. Twenty-four offshore wells that ran near-vertical density, sonic and caliper wireline well log data were available for this study to calculate S_V magnitudes. Of these 24 wells, 20 wells had check-shot velocity data acquired that were used to estimate average density values from the seabed to the top of the bulk density log. For wells that did not have check-shot velocity data, nearby wells with requisite data were used as an approximation. These wells are shown in Figure 3.

S_V *profiles.* The S_V magnitude–depth profiles for all the wells in this study are presented in Figure 9. Overall, there is not a significant variation in S_V gradients. The lower and upper bounds of S_V magnitudes with depth are well described by the power law relationships:

$$\text{Lower bound: } 19.82 \times Z^{1.0747} \quad (2)$$

$$\text{Upper bound: } 21.70 \times Z^{1.0680} \quad (3)$$

where S_V is calculated in MPa and *Z* is the TVDML. At a depth of *c.* 2500 m TVDML, the difference in S_V magnitude within these bounds is *c.* 4.66 MPa (Fig. 9). This variation is *c.* 2.25 and 4.34 MPa less than that reported for the same depth in the Cooper Basin (cf. Reynolds *et al.* 2006) and Carnarvon Basin (cf. King *et al.* 2010), respectively. The larger variation in S_V magnitudes in the Cooper and Carnarvon basins are probably due to localized uplift and erosion (*c.* 900–1000 m; Densley *et al.* 2000; Mavromatidis & Hillis 2005) that has brought denser sedimentary rocks to shallower depths (King *et al.* 2010). The Lower Cretaceous Eumeralla Formation is predominantly near the maximum post-depositional burial depths in the offshore central and eastern Otway Basin (Tassone *et al.* 2014), which may explain the small amount of variation observed across this area. Where significant uplift and erosion has occurred in the Otway Basin – for example, at Bellarine-1 in the eastern part of the study area (*c.* 1800 m; Tassone *et al.* 2014) – S_V gradients are as high as *c.* 26 MPa km^{-1} at 2000 m TVDML (Fig. 9). This is *c.* 3 MPa km^{-1}

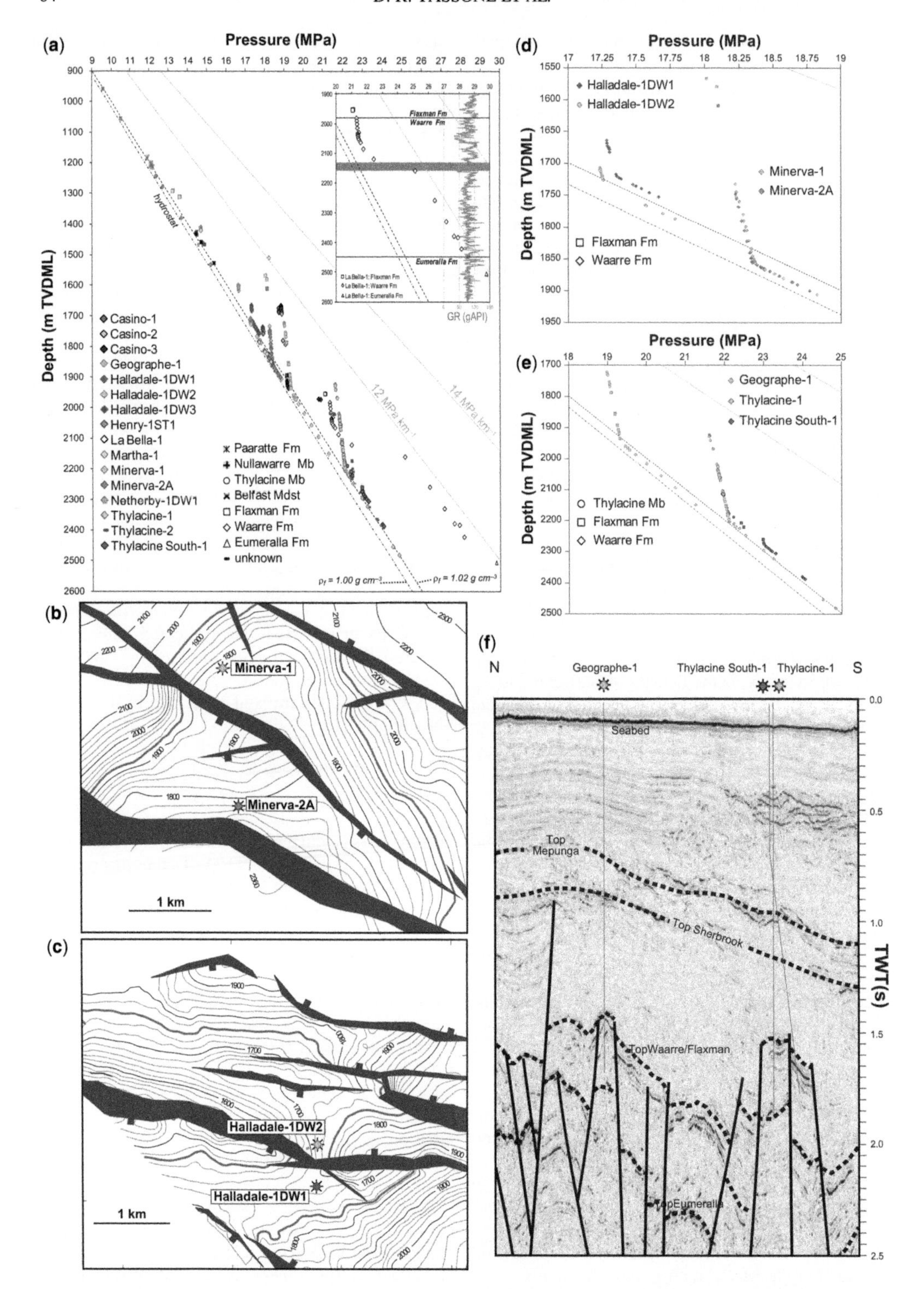

Fig. 8. (**a**) WFT data showing P_P magnitudes plotted against TVDBM for wells in the Shipwreck Trough differentiated by stratigraphic formations and units. The higher (steeper) gradient with respect to the hydrostat indicates that the pores are gas-saturated, creating hydrocarbon (i.e. gas) column buoyancy force overpressures.

greater than the upper limit S_V magnitude–depth relationships defined in equation (3) (Fig. 9).

Minimum horizontal stress magnitudes

Constraining S_{hmin} *magnitudes.* The minimum horizontal stress (S_{hmin}) magnitudes were determined from leak-off tests (LOTs), which are commonly acquired during drilling operations after the casing is cemented in place (Zoback 2010). More reliable estimates of the S_{hmin} magnitudes are often attained from the instantaneous shut-in pressure (ISIP), or preferably the fracture closure pressure during mini-frac or extended LOTs (XLOTs; Fig. 10; Addis *et al.* 1998; Zoback 2010). Unlike the Cooper Basin in central Australia, where the exploration and development of low-permeability hydrocarbon plays has resulted in a plethora of mini-frac data (cf. Reynolds *et al.* 2006; Nelson *et al.* 2007), these tests have never been routinely conducted in wells drilled in the Otway Basin. Leak-off pressures (LOPs) during LOTs represent the creation of a supposedly new hydraulic fracture in the surrounding host formation when increasing wellbore pressures suddenly drop. This causes a deviation from linearity in a pressure v. time plot (Fig. 10). A fundamental assumption of LOTs is that hydraulic fractures will develop axial to vertical wellbores and are caused by tensile failure of the host formation (Brudy & Zoback 1999). Unfortunately, image logs are generally not run after LOTs (e.g. Evans *et al.* 1989) and therefore the orientation, nature and vertical extent of hydraulic fractures in vertical wells is often not verified.

Although XLOTs, which are repeated cycles of LOTs, provide superior estimates of S_{hmin} magnitudes, lower bound LOPs from LOTs are generally considered to provide good approximations of S_{hmin} magnitude (Breckels & Van Eekelen 1982; Addis *et al.* 1998). This is especially true in near-vertical wellbore sections because they are often similar to fracture propagation pressures when flow rates and fluid viscosities are sufficiently low enough (Haimson & Fairhurst 1967; Zoback 2010). Furthermore, they do not impose an S_V component (and thus shear component), as would be the case in highly deviated sections (Brudy & Zoback 1993). Observations of low LOPs (*c.* 30–60% less than S_V) in active thrust belts, where a reverse fault stress regime dominates, has led Couzens-Schultz & Chan (2010) to propose an alternate interpretation of LOT results. In their alternative interpretation, the magnitude of S_{hmin} is constrained by assuming that the LOP causes shear failure along (cohesionless) pre-existing planes of weaknesses rather than the traditional assumption of failure via new tensile fractures. This approach leads to slightly higher estimates of S_{hmin} magnitude (King *et al.* 2012). Given that there is no way to determine whether or not leak-offs were caused by shear or tensile failure as image log data were not acquired before and after LOTs were conducted, we did not apply this alternative technique and acknowledge that our S_{hmin} gradients maybe underestimated.

Minimum horizontal stress magnitudes are arguably the least well-constrained component of the *in situ* stress tensor that can be directly measured in a single wellbore. This is because they are often reliant on one or perhaps two LOTs at shallow depths below the surface and intermediate casing shoes. These are then often linearly extrapolated to depths of interest or used as calibration data points to extrapolate the developed fracture gradient (i.e. S_{hmin}) prediction algorithms to reservoirs. A common factor in all these fracture gradient prediction algorithms (e.g. Matthews & Kelly 1967; Eaton 1969; Breckels & Van Eekelen 1982; Daines 1982) is P_P. An increase in P_P (i.e. overpressure) causes an increase in S_{hmin} and a decrease in P_P causes a decrease in S_{hmin}. This coupling relationship

Fig. 8. (*Continued*) ρ_f, pore fluid density. It can be seen that small columns of gas are reservoired in the Nullawarre Greensand Member and larger gas columns are typically reservoired in the Thylacine Sandstone Member, Flaxman and Waarre formations. The top right-hand corner shows WFT data from La Bella-1, indicating two zones of moderate overpressures separated by an intra-formational sealing unit characterized by a high gamma ray response. (**b**, **c**) Top-Waarre Formation isochron maps (in milliseconds) along with interpreted faults for the Minerva (after O'Brien *et al.* 2006) and Halladale/Blackwatch (after Constantine *et al.* 2009) gas fields, respectively. (**d**) WFT data from within the Flaxman and Waarre formations across faults in the Minerva and Halladale/Blackwatch gas fields, showing that the Minera-1 and Minerva-2A wells are in pressure communication (indicative of across-fault hydrocarbon migration), whereas the Halladale-1DW1 and Halladale-1DW2 wells are not in pressure communication, with overpressures in Halladale-1DW1 suggesting a sealing fault. (**e**) WFT data from within the Thylacine Sandstone Member, Flaxman and Waarre formations from Geographe-1, Thylacine-1 and Thylacine South-1. Similar to the Halladale/Blackwatch gas field, the Thylacine South-1 well has slight overpressures compared with the Thylacine-1 well, which are separated by a south-dipping fault. Compared with the Geographe-1 well, water-saturated pore fluid pressures are slightly higher, which, along with (a), suggests the P_P is compartmentalized across the entire Shipwreck Trough or the pore fluid has varying fluid densities as a result of salinity changes. (**f**) Interpreted regional composite seismic profile modified after O'Brien *et al.* (2006) showing rift-related faulting between the Geographe-1 and Thylacine-1 wells in which there is likely pressure compartmentalization.

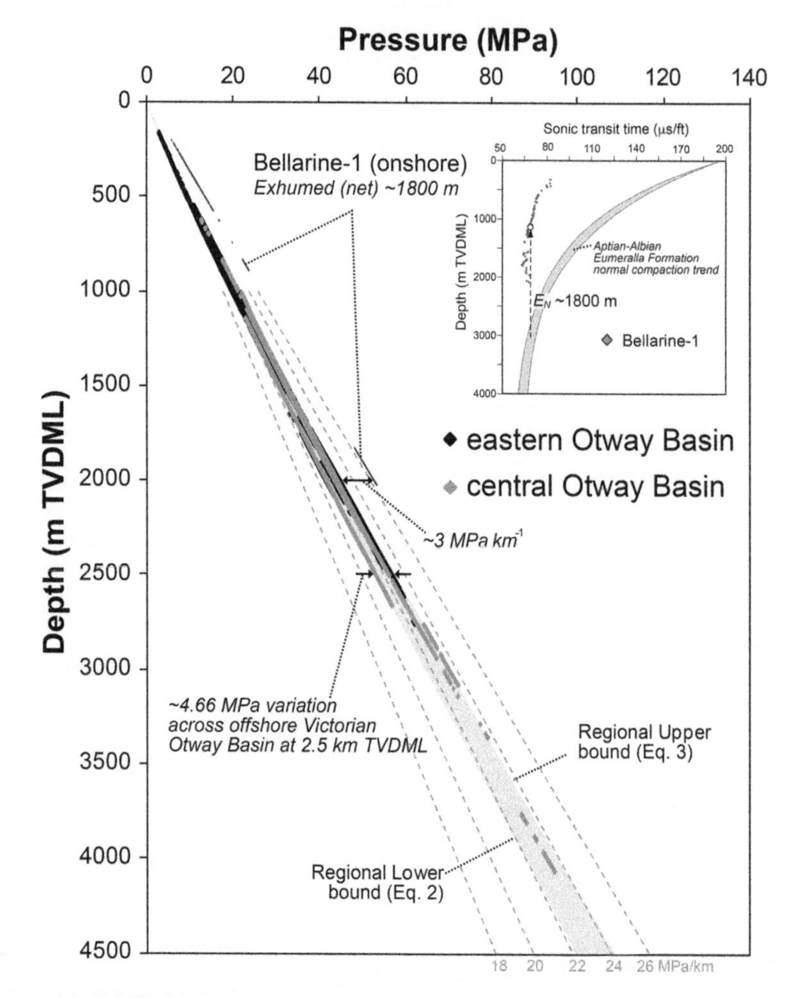

Fig. 9. S_V magnitude data plotted against true vertical depth (below mudline) for the central (nine wells) and eastern (15 wells) Otway Basin showing similar amounts of variation across the entire offshore Victorian Otway Basin, in which the lower (equation 2) and upper (equation 3) bounds were estimated with a power law magnitude–depth relationship. At 2500 m below mudline, there is *c.* 4.66 MPa difference between the upper and lower bounds. Also shown is the S_V magnitude–depth profile of the Bellarine-1 well (after Tassone *et al.* 2014) located in the eastern onshore Otway Basin (see Fig. 3). Here the Lower Cretaceous Eumeralla Formation rocks have been exhumed (net) by >1800 m (Tassone *et al.* 2014), which has resulted in denser rocks closer to the surface and an S_V magnitude–depth profile at 2000 m below mudline *c.* 3 MPa km^{-1} greater than the upper bound to the offshore S_V magnitudes.

between P_P/S_{hmin} has been characterized over both hydrocarbon field and basin-wide scales (e.g. Addis 1997; Hillis 2001; Goulty 2003; Tingay *et al.* 2003*b*) and it is worth noting that such a phenomenon cannot be discounted in overpressured rocks within the Otway Basin.

Data availability and distribution. The majority of LOT data used in this study were reported from basic data or completion reports and did not contain the actual pressure v. time data (i.e. they are ranked as D quality, cf. King *et al.* 2008). Therefore we assume that the LOPs were correctly interpreted and that the reported LOPs are a good estimation of the S_{hmin} magnitude (Haimson & Fairhurst 1967). Comparing reported LOPs with actual pressure v. time data was beyond the scope of this study and we acknowledge that this may affect the veracity of our estimates of S_{hmin} magnitudes. We note that in a previous study of contemporary stress in the Otway Basin by Nelson *et al.* (2006*a*), S_{hmin} was determined using 13 reported LOPs and two XLOTs, which are considered to be reliable estimates of S_{hmin}. Nelson *et al.* (2006*a*) found that

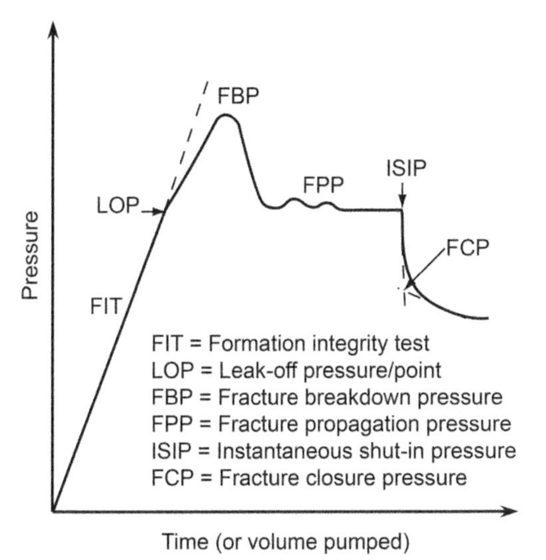

Fig. 10. Schematic plot of an extended leak-off test (modified after Zoback 2010).

closure pressures determined from the XLOTs were consistent with the reported LOPs, thereby lending confidence to the reliability of reported LOPs in constraining S_{hmin} in the Otway Basin.

It has long been known that lithology exerts a strong influence on both ISIPs during mini-frac tests and LOPs during XLOTS, which accordingly influences the magnitude of S_{hmin} (Bush & Meyer 1988; Evans *et al.* 1989; Warpinski & Teufel 1989, 1991; Addis *et al.* 1998; Reynolds *et al.* 2006). Consequently, we have made use of the substantial amount of LOT data available from the Otway Basin. Eighty-nine LOTs from 73 near-vertical (<10° inclination) wellbores within the offshore and onshore central and eastern Otway Basin (Fig. 3) have been used to estimate S_{hmin} magnitudes. These tests have been acquired since early exploration and encompass differing formations (and lithologies) at various depths within sections believed to be close to the maximum post-depositional burial depths (cf. Corcoran & Doré 2007; Tassone *et al.* 2014). In contrast with previous contemporary stress studies in the Otway Basin, we have grouped the LOT data by the dominant lithology of the formation in which the test was run and by geographical location (i.e. central v. eastern Otway Basin). Table 4 shows the formation bulk lithology groupings used to differentiate the LOT data, as well as information regarding the depositional environments of the corresponding formations. The Heytesbury and Nirranda groups are mainly marl and carbonate-dominated sections that are the most shallow stratigraphic sequences in the Otway Basin (Fig. 1; *c.* <1000–1500 m TVDML). These stratigraphic units are often favoured for cementing-in-place surface casing, thereby accounting for a large proportion (*c.* 47%) of the available reported LOT data in this study.

S_{hmin} gradients. Unlike petroleum exploration, which requires conservative fracture gradient estimates for well design and safe drilling practice, we are interested in best-fit estimates of S_{hmin} to understand the regional contemporary *in situ* stresses. Figure 11a, b shows the relationships between LOPs with depth, differentiated by lithology (i.e. marl/carbonate, sand, shale, sand/shale) and location within the Otway Basin (i.e. central and eastern). Although there is still considerable variation within each lithology, a comparison of different lithologies shows that the highest best-fit S_{hmin} gradients occur within marl and carbonate-dominated formations, with best-fit S_{hmin} gradients of *c.* 20.3 and 21.2 MPa km^{-1} for the central and eastern Otway Basin, respectively (Fig. 11a). The shale-dominated formations have best-fit S_{hmin} gradients of *c.* 18.8 and 20.9 MPa km^{-1} for the central and eastern Otway Basin, respectively, while the sand-dominated formations have the lowest best-fit S_{hmin} gradients of *c.* 15.9 and 17.0 MPa km^{-1} for the central and eastern Otway Basin, respectively (Fig. 11a). Undifferentiated sand/shale-dominated formations have best-fit S_{hmin} gradients generally constrained between the shale (upper) and sand (lower) dominated formations (Table 4), although the central Otway Basin best-fit S_{hmin} gradient (*c.* 15.4 MPa km^{-1}) is slightly lower than the sand-dominated best-fit S_{hmin} gradients in the same region (Fig. 11a). This may reflect the fact that LOT data for undifferentiated sand/shale-dominated formations in the eastern part of the study area were obtained in more shale-rich units, whereas LOT data in the central Otway Basin were obtained in more sand-rich units.

Considering the difference between sandstone- and shale-dominated best-fit S_{hmin} gradients at typical reservoir/seal depths (e.g. *c.* 2000 m TVDML in the Shipwreck Trough), the difference in S_{hmin} magnitudes between these lithologies for the central and eastern regions is *c.* 5.72 and 7.74 MPa, respectively. These differences in LOPs, and thus estimates of S_{hmin} magnitudes, in siliciclastic rocks across both regions are consistent with the findings of Warpinski (1989), who showed, based on 60 mini-frac tests within a single well, that reservoir (sand) units typically have S_{hmin} magnitudes 5–12 MPa lower than non-reservoir (shale) units. A similar conclusion was reached by Addis *et al.* (1998), who showed that XLOTs and LOTs in shale- and mudstone-dominated formations generally indicate higher stresses and fracture gradients than sand-dominated units. The observation of

Table 4. *Best-fit S_{hmin} gradients differentiated by lithology and region derived from LOT data assuming LOPs were caused by tensile failure.*

Lithology	Formation/Group	Region	Best-fit S_{hmin} magnitude gradient (MPa km^{-1})	N	Negative (MPa km^{-1})	Positive (MPa km^{-1})	R^2	Equation No.
Marl/carbonate	Whales Bluff Formation/Port Campbell Limestone Gellibrand Marl/Jan Juc Formation/Puebla Formation/Narrawaturk Marl/undiff. Heytesbury Group/undiff. Nirranda Group	Eastern	$S_{hmin\_Marl/Carbonate\_East} \approx 21.19 \times Z$	36	5.40	7.09	0.7947	4
		Central	$S_{hmin\_Marl/Carbonate\_West} \approx 20.32 \times Z$	11	6.52	3.48	0.8059	5
Sand	Mepunga Formation/Dilwyn Formation/Pebble Point Formation/Timboon Sandstone/Paaratte Formation	Eastern	$S_{hmin\_Sand\_East} \approx 17.01 \times Z$	6	4.04	2.73	0.7441	6
		Central	$S_{hmin\_Sand\_West} \approx 15.93 \times Z$	13	3.09	3.83	0.9094	7
Shale	Clifton Formation/Pember Mudstone/Belfast Mudstone/Laira Formation	Eastern	$S_{hmin\_Shale\_East} \approx 20.89 \times Z$	9	7.88	0.48	0.9275	8
		Central	$S_{hmin\_Shale\_West} \approx 18.79 \times Z$	6	2.54	1.64	0.9599	9
Sand/shale	Eumeralla Formation/Wangerrip Group/Sherbrook Group	Eastern	$S_{hmin\_Sand/Shale\_East} \approx 19.30 \times Z$	5	1.45	1.15	0.9829	10
		Central	$S_{hmin\_Sand/Shale\_West} \approx 15.37 \times Z$	5	2.25	0.81	0.9938	11

Also listed is the number of data points (N), range of values from the best-fit S_{hmin} gradient derived from the lower and upper limits, respectively, and coefficient of determination (R^2). S_{hmin} gradients in MPa km^{-1} with depth referenced to the mudline. Lithology classifications are based on general characteristics of stratigraphic formations and groups.

wide natural fractures in the sandstones abruptly terminating at shale-rich interfaces also led Warpinski & Teufel (1989) to emphasize that stress contrasts exist between shale- and sand-dominated units, with higher stresses in the shales. This reflects a contrast in the mechanical properties between the different lithologies, especially Poisson's ratio, which is usually higher in shales. Thus fracture gradients determined in shale and mudstone lithology rocks should not be directly extrapolated to reservoir sandstones, particularly if production has depleted P_P magnitudes (Addis *et al.* 1998; Hillis 2001).

For all the lithology end-members (i.e. marl/carbonate, shale and sand-dominated formations), best-fit S_{hmin} gradients in the eastern Otway Basin are *c.* 1–2 MPa km^{-1} higher than in the central Otway Basin (Fig. 11a, b). Although there is a lot of overlap at shallower depths, especially within marl and carbonate-dominated formations at depths deeper than *c.* 800 m TVDML, there appears to be a more noticeable differentiation of LOPs within siliciclastic formations from across the basin (Fig. 11b). This clearly demonstrates that lithology has a major control on LOPs and thus the magnitude of S_{hmin} with depth. It also indicates a relative increase in S_{hmin} gradients from west to east across the central to eastern parts of the basin.

Constraining maximum horizontal stress magnitudes

The magnitude of S_{Hmax} is the most difficult component of the stress tensor to constrain using petroleum data because there is no way to measure it directly (White & Hillis 2004). It is well accepted that the Earth's crust is in a state of incipient, albeit slow, frictional failure equilibrium; seismicity in the brittle upper crust within stable continental regions provides strong evidence for this notion (e.g. Stein *et al.* 1992; Townend & Zoback 2000). Assuming that one principal stress is vertical, the differences in magnitude between the maximum and minimum principal stresses at depth will be limited by the frictional strength of planar discontinuities such as faults (Jaeger & Cook 1979). Frictional limit theory states that stresses in the Earth's upper crust cannot be such that they exceed the frictional strength of pre-existing, optimally oriented faults (Sibson 1974 1995; Jaeger & Cook 1979).

The presence of wellbore failure features, such as DITFs and/or borehole breakouts, can provide additional constraints on S_{Hmax} magnitudes if the mechanical rock properties, such as compressive and tensile rock strengths, are known (Bell & Gough 1979; Brudy & Zoback 1999). However, because this study focuses on linking contemporary stress data with neotectonic observations rather than modelling the state of stress (i.e. S_{Hmax} magnitude) for geomechanical applications (e.g. wellbore stability, hydraulic fracturing, sand production), S_{Hmax} magnitudes were not constrained in this study. We did not attempt to constrain S_{Hmax} magnitudes using wellbore observations because we did not have access to rock strength data and often the image logs we used did not preserve breakouts with enough accuracy to provide an accurate estimate of their width.

Possible states of stress constrained from petroleum exploration data

During normal burial of sedimentary rocks, mechanical and thermochemical compaction (i.e. the reduction of porosity with depth) processes are generally non-linear (Athy 1930). It is for this reason that Warpinski (1989) emphasizes the perils of fitting linear stress–depth regressions because this maintains the misconception that magnitudes of S_{hmin} always increase with depth regardless of parameters in addition to lithology, such as changes in stress states, pore pressures, temperatures and diagenesis throughout time (i.e. elastic/viscoelastic behaviour; Warpinski 1989). Unfortunately, however, there is insufficient data from within individual Otway Basin wells to constrain horizontal pressure–depth trends other than linear gradients at this stage.

Assuming that LOPs constrain S_{hmin} magnitudes, marl- and carbonate-dominated formations exhibit higher S_{hmin} gradients than $S_{V,Lower}$ over all depths for where these rocks occur (i.e. <1600 m TVDM; Fig. 11). Furthermore, these S_{hmin} gradients are about equal to or higher than $S_{V,Upper}$ at depths less than *c.* 380 and 710 m TVDML for the central and eastern parts of the basin, respectively (Fig. 11c, d). This implies that the shallow, post-rift Heytesbury and Nirranda groups are within a strike-slip to reverse faulting stress regime or, potentially, a pure reverse faulting stress regime if the LOPs actually represent S_V (i.e. S_3) rather than S_{hmin} (i.e. S_2) (Evans & Engelder 1989). In the latter instance, the amount by which S_{hmin} magnitudes exceed S_V magnitudes is therefore indeterminable using LOT data (Evans *et al.* 1989).

Throughout the central and eastern Otway Basin, the LOPs obtained in sand-dominated formations are generally less than the S_V magnitudes for all depths tested (Fig. 11c, d), implying either a strike-slip or normal fault stress regime. In the central Otway Basin, the LOPs from sandstones define a best-fit S_{hmin} gradient of *c.* 15.9 MPa km^{-1}. Likewise, LOPs from shale-dominated formations in the central Otway Basin are also less than $S_{V,Lower}$ at all depths, providing further support for either a strike-slip or normal fault stress regime. However, the LOPs from shale-dominated formations in the

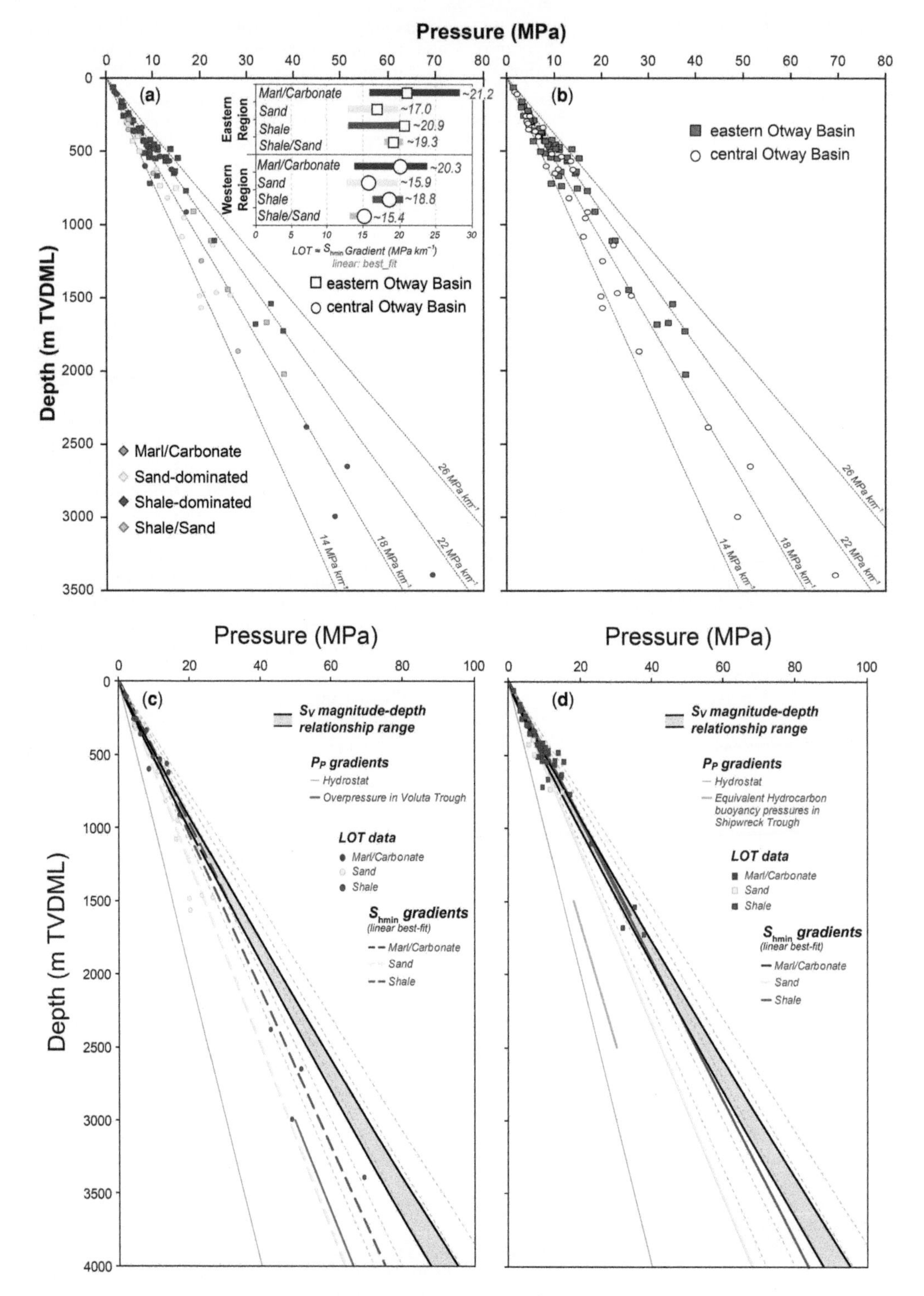

Fig. 11. (**a**) Available LOT data ($N = 89$) plotted against TVDBM differentiated by region (central or eastern; see Fig. 3) and lithology, whereby LOPs are used as estimates of S_{hmin} magnitudes. Lithology has been generalized based on the stratigraphy (Fig. 1) as listed in Table 4 into either marl/carbonate, sand, shale or sand/

eastern Otway Basin indicate variably greater and lower S_{hmin} magnitudes compared with S_V magnitudes over various depths, implying that all faulting regimes may be possible (i.e. normal, strike-slip and reverse).

In addition to S_V and S_{hmin} stresses, Figure 11 also shows the maximum P_P gradients observed across the basin. In particular, it highlights the shortcomings of linearly extrapolating S_{hmin} gradients calibrated at shallow depths to deeper depths if pore pressure–stress coupling is not taken into account. Alternatively, it may highlight potential drilling challenges if sand-dominated S_{hmin} gradients are indeed lower than overpressured mudstone P_P gradients if the reservoir unit is laterally drained.

A best-fit S_{hmin} gradient determined for eastern Otway Basin shales is *c.* 20.9 MPa km^{-1}, which is less than $S_{V,Lower}$ at depths >2010 m TVDML. This indicates a potential partitioning of stress regimes with depth, with strike-slip faulting stress regimes at depths >2 km TVDML and borderline strike-slip to reverse faulting stress conditions at shallower depths (e.g. <2 km TVDML; Fig. 11d). In a previous study, Nelson *et al.* (2006*a*) constrained a best-fit S_{hmin} gradient of *c.* 18.5 MPa km^{-1} in the eastern Otway Basin, which is generally lower than our estimates of the S_{hmin} gradient for marl, carbonate- and shale-dominated formations throughout the basin (Table 4). Nelson *et al.* (2006*a*) suggested that the simplest explanation for the increase in S_{hmin} gradients from *c.* 15.5 MPa km^{-1} in the western Otway Basin (the South Australian sector) to *c.* 20 MPa km^{-1} in the Gippsland Basin is largely due to the closer proximity to the New Zealand compressional boundary. However, the observation of S_{hmin} gradients greater than *c.* 20 MPa km^{-1} in marl- and carbonate-dominated stratigraphic units throughout the basin and in shale-dominated stratigraphic units in the eastern Otway Basin implies that horizontal stress magnitudes are not purely related to the proximity to a plate boundary.

State of stress in underlying basement rocks

Wellbore failure analysis of 12 petroleum wells has constrained the S_{Hmax} azimuth in the eastern Otway Basin to be broadly *c.* 135 ± 15° N, which is consistent with previous studies in this region (Fig. 6c; Nelson *et al.* 2006*a*; Bérard *et al.* 2008). The remarkable consistency of S_{Hmax} orientations stands in contrast with many basins, which typically exhibit local stress variations near geological structures such as faults, fractures or igneous intrusions (Bell 1996; Yale 2003; Tingay *et al.* 2010; Holford *et al.* 2016). Several wells for which we have determined S_{Hmax} orientations are situated close to large faults (Fig. 8), suggesting that these faults do not act as weak or strong mechanical discontinuities (Bell 1996).

However, such data only provide constraints on the orientation of stresses within the shallow upper crust (depths <4 km). In an attempt to obtain a more comprehensive understanding of the state of stress in the Otway Basin, here we combine our results with independent datasets that constrain the state of stress in the underlying basement. This includes earthquake focal mechanism solutions (Denham *et al.* 1981, 1985; Bock & Denham 1983; McCue *et al.* 1990; McCue & Paull 1991; Leonard *et al.* 2002; Allen *et al.* 2005; Clark 2009). We also consider the implications of Pliocene–Holocene volcanic eruption point alignments (Lesti *et al.* 2008).

Earthquake focal mechanisms represent the pattern of seismic radiation resulting from slip on a fault. Table 5 and Figure 12a show 19 published focal mechanisms for 17 earthquakes in the vicinity of the Otway Basin, with magnitudes ranging from M_L 3.2 to 5.5, which occurred at depths between 5 and 21 km. These are superimposed over basement discontinuities identified from both field mapping and geophysical datasets (e.g. magnetic, radiometric, gravity and seismic; Vandenberg *et al.* 2000). All the focal mechanism solutions were determined

Fig. 11. (*Continued*) shale-dominated formations. The distribution of LOT data can be seen in Fig. 3. The best-fit linear magnitude–depth gradients for each lithology with respect to region, as well as their corresponding range, are shown in the top right-hand corner of (a). It can be seen that marl/carbonate-dominated formations have the highest gradients followed shale-dominated formations and then sand-dominated formations. (**b**) The same LOT data differentiated only by region, showing higher gradients in the eastern Otway Basin than in the central Otway Basin. Although there is smaller difference in magnitude within the shallower marl and carbonate rocks across both regions, there is more appreciable difference in siliciclastic rocks (sand and shale-dominated formations) at depths deeper than *c.* 800 m TVDML from the central (lower gradients) to the eastern Otway Basin (higher gradients). (**c**) Stress gradients estimated using petroleum exploration data in the central Otway Basin. (**d**) Stress gradients in the eastern Otway Basin. It can be seen that LOT data in shallow marl/carbonate-dominated formations in both (c) and (d) have LOPs generally $\geq S_V$ magnitudes at the equivalent depth, suggesting a strike-slip to reverse *in situ* faulting stress regime. The LOT data in the central Otway Basin (c) define lower S_{hmin} gradients within siliciclastic rocks (i.e. sand and shale-dominated formations) compared with the eastern Otway Basin (d) by *c.* 1–2 MPa km^{-1}. This may suggest a normal to strike-slip contemporary fault stress regime in the central Otway Basin and possibly either a normal, strike-slip or strike-slip to reverse contemporary fault stress regime in the eastern Otway Basin within these deeper rocks.

Table 5. *Published focal mechanism solutions for earthquakes in southeastern Australia, with corresponding date, location, magnitude, depth, axis of compression (P-axis ≡ S_{Hmax} orientation), interpreted stress regime, World Stress Map quality rating after Zoback (1992) and publication reference*

Location	Date	Latitude (°S)	Longitude (°E)	M_L	Depth (km)	P-axis (°)	Stress regime	*N* (Figure 12a)	World Stress Map quality	Reference
Berridale	18 May 1959	−36.22	148.64	5.2	15	327	TF	15	D	Denham *et al.* (1981), Leonard *et al.* (2002)
Mt Hotham	3 May 1966	−37.042	147.158	5.5	15	219	NF	11	B	Denham *et al.* (1985), Leonard *et al.* (2002)
Middlingbank	21 June 1971	−36.22	148.78	4	5	136	SS	16	??	Bock & Denham (1983), Leonard *et al.* (2002)
Murrumbateman	6 May 1974	−35.06	149.02	3.8	5	271	SS/TF	17	D	Denham *et al.* (1981), Leonard *et al.* (2002), Nelson *et al.* (2006*a*)
The Pilot	8 September 1976	−36.72	148.24	3.8	6	326	TF	13	D	Bock & Denham (1983), Leonard *et al.* (2002)
Balliang	2 December 1977	−37.88	144.27	4.2	21	292	TF	2a	??	Denham *et al.* (1981), Leonard *et al.* (2002)
						286	SS	2b	??	Denham *et al.* (1981), Leonard *et al.* (2002)
Suggan Buggan	30 November 1981	−36.09	148.33	3.7	7	95	SS	14	D	Bock & Denham (1983), Leonard *et al.* (2002)
Woonangatta	21 November 1982	−37.205	146.956	5.4	17	113	TF	10	B	Denham *et al.* (1985), Leonard *et al.* (2002), Nelson *et al.* (2006*a*)
Nhill	22 December 1987	−36.107	141.539	4.9	6	264	SS/NF	1a	B	McCue *et al.* (1990), Leonard *et al.* (2002), Nelson *et al.* (2006*a*)
						136	TF	1b	B	Denham *et al.* (1981), Leonard *et al.* (2002)
Bunnaloo	3 July 1988	−35.73	144.49	4	10	118	SS	3	C	McCue & Paull (1991), Leonard *et al.* (2002)
Thomson Reservoir	25 September 1996	−37.863	146.422	5	11	141	TF	9	C	Allen *et al.* (2005)
Tatong	27 June 1997	−36.781	146.094	4.2	6	354.2	TF	6	C	Allen *et al.* (2005), Nelson *et al.* (2006*a*)
Corryong	17 July 1998	−36.441	148.005	4.7	20	332.2	SS	12	C	Allen *et al.* (2005), Nelson *et al.* (2006*a*)
Boolarra South	29 August 2000	−38.402	146.245	4.7	15	322	TF	7	D	Allen *et al.* (2005)
Dumbalk	30 October 2000	−38.56	146.05	3.2	16	161	SS	4	D	Leonard *et al.* (2002)
Boolarra South	4 July 2001	−38.74	146.34	3.4	7	286	SS	8	D	Leonard *et al.* (2002)
Korumburra	24 April 2009	??	??	4.6	8	??	SS	5	??	Clark (2009)

TF, thrust fault; SS, strike slip; NF, normal fault.

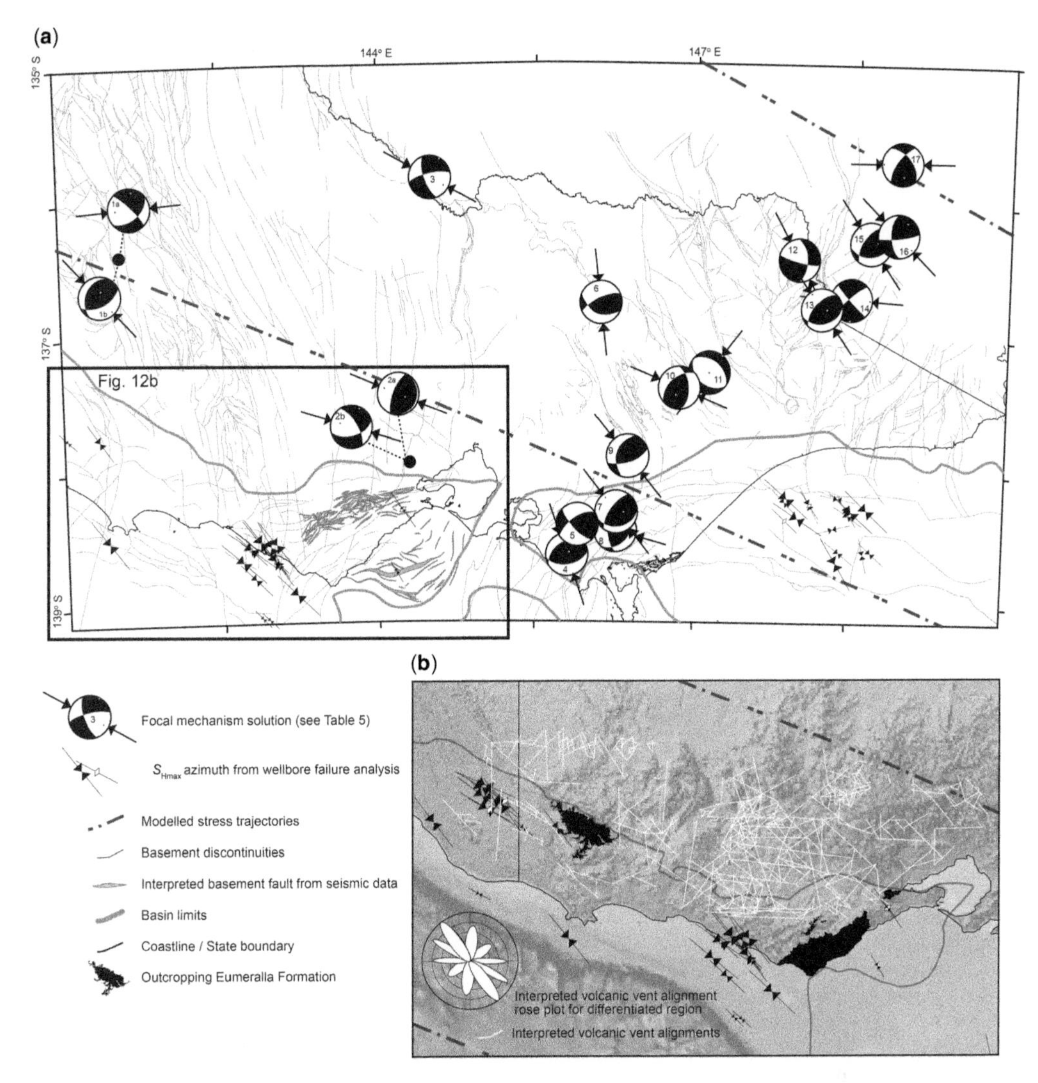

Fig. 12. (**a**) Published focal mechanism solutions – –for earthquakes occurring in southeastern Australia with magnitudes ranging between M_L 3.2 and 5.5 at depths between 5 and 21 km (Denham *et al.* 1981, 1985; Bock & Denham 1983; McCue *et al.* 1990; McCue & Paull 1991; Leonard *et al.* 2002; Allen *et al.* 2005; Nelson *et al.* 2006*a*; Clark 2009) superimposed over Palaeozoic basement discontinuities identified from field mapping and geophysical datasets, including magnetic, radiometric, gravity and seismic data (Messent *et al.* 1999; Vandenberg *et al.* 2000). Also shown are estimates of S_{Hmax} orientations derived from wellbore failure analyses (this study; Nelson *et al.* 2006*a*; Bérard *et al.* 2008) and modelled stress trajectories (after Reynolds *et al.* 2003*a*). The focal mechanism solution numbers correspond to those listed Table 5. (**b**) Mapping of Pliocene–Holocene volcano alignments (i.e. Newer Volcanics) in western Victoria (after Lesti *et al.* 2008) showing that the dominant NW–SE trend probably reflects the opening (or reopening) of deep fractures favourably oriented parallel to the S_{Hmax} orientation (as determined from wellbore failure features and focal mechanisms), while the north–south and east–west trends probably reflect reactivated Palaeozoic discontinuities (Lesti *et al.* 2008).

from onshore earthquakes that occurred within basement rocks.

No focal mechanisms solution exists underlying the Otway Basin to directly compare with our estimates of S_{Hmax} orientation from wellbore failure analysis (Fig. 6). The two most proximal earthquakes to the central and eastern Otway Basin for which focal mechanism solutions are available are both poorly constrained, allowing for alternative interpretations. McCue *et al.* (1990) suggested that the 1987 Nhill earthquake in western Victoria was consistent with a strike-slip focal mechanism solution (1a in Fig. 12a). However, Leonard *et al.* (2002) showed that an alternative focal mechanism solution is also possible with similar amounts of uncertainty: a thrust regime (1b in Fig. 12a) in which the P-axis (assumed to be close to the S_{Hmax} orientation) is roughly aligned with nearby wellbore failure azimuths defined by borehole breakouts and DITFs (Hillis *et al.* 1995; Nelson *et al.* 2006*a*). The only other focal mechanism solution in close proximity to the Victorian Otway Basin is the 1977 Balliang earthquake near Geelong, just north of the northern Otway Basin boundary. This focal mechanism solution is also poorly constrained (Denham *et al.* 1981) and two alternative solutions have been presented: the original published solution by Denham *et al.* (1981), indicating a thrust fault regime, and strike-slip faulting on a pre-existing normal fault (Leonard *et al.* 2002). Both solutions yield a similar P-axis azimuth rotated by *c.* 30° towards the north with respect to the S_{Hmax} orientation determined for the Lower Cretaceous rocks in Bellarine-1.

Four of the 17 focal mechanism solutions are for intra-basement earthquakes underlying the Mesozoic–Cenozoic sedimentary rocks of the Gippsland Basin, where wellbore failure analysis defined by borehole breakouts and DITFs also indicate consistent S_{Hmax} orientations (Fig. 12a). Apart from the Mt Hotham earthquake in 1966, which indicates a normal faulting focal mechanism solution (P-axis *c.* 219°; Denham *et al.* 1985) and a possible focal mechanism solution for the Nhill earthquake in 1987 (P-axis *c.* 264°; McCue *et al.* 1990; Leonard *et al.* 2002), all other focal mechanism solutions listed in Table 5 have a P-axis azimuth that trends broadly NW–SE. It is thus clear from the focal mechanism solutions that earthquakes within the basement rocks in southeastern Australia are dominantly the result of strike-slip and reverse faulting. The broad consistency between stress orientations determined from both petroleum data and earthquake focal mechanisms suggests some degree of coupling between the basement and the overlying sedimentary cover, consistent with observations from a number of seismogenic zones across Australia (Clark & Leonard 2003). It is difficult, however, to correlate specific small-magnitude earthquakes with known basement-involved faults (Clark 2009).

The alignment of Pliocene–Holocene volcanoes in western Victoria, which overlaps the northern margin of the Otway Basin, provides additional evidence of the state of stress within the underlying basement (Fig. 12b; Lesti *et al.* 2008). Mapping the spatial density and alignment of deeply sourced eruption points (Fig. 12b) led Lesti *et al.* (2008) to suggested that the dominant NW–SE trend reflects the opening (or reopening) of deep fractures favourably oriented parallel to the S_{Hmax} azimuth, while subsidiary north–south and east–west trends reflect reactivated basement discontinuities (Fig. 12a). Hence the alignment of eruption points highlights the influence of synvolcanic tectonics and the *in situ* stress field during Pliocene–Holocene times.

Regional S_{Hmax} orientations estimated from Pliocene–Holocene volcanic eruption point alignments, together with petroleum well data estimates and focal mechanism solutions across southeastern Australia, agree well with plate-scale finite-element stress modelling (Walcott 1998; Hillis & Reynolds 2000; Reynolds *et al.* 2003*a*). This modelling indicates that the regional S_{Hmax} orientation in the Otway Basin, as well as southeastern Australia, primarily reflects the increased coupling of the Australian and Pacific plate boundary along New Zealand's Alpine Fault (Sandiford 2003*b*; Sandiford *et al.* 2004).

Evidence for neotectonic compressional deformation

The onset of Neogene neotectonic compressional deformation across southeastern Australia is marked by a regional late Miocene–early Pliocene angular unconformity. This unconformity is well developed in the Otway Basin, particularly over localized neotectonic structures (Dickinson *et al.* 2002), and corresponds to a transition in the nature of basin-fill from carbonates to siliciclastics around 6–8 myr ago (Sandiford 2003*b*). Although this deformation occurs throughout the entire Otway Basin (Geary & Reid 1998; Messent *et al.* 1999; Boult *et al.* 2008), neotectonic compressional deformation is far more pronounced in the eastern Otway Basin (Fig. 3). The main structural style for neotectonic deformation takes the form of folding and reverse faulting within upper Miocene and older strata in the Torquay Sub-Basin (e.g. Nerita Anticline; Fig. 13) and around the Otway Ranges (Fig. 14b; Cooper & Hill 1997; Dickinson *et al.* 2002; Holford *et al.* 2011*a*). The *c.* 40 m high cliff near Port Campbell in the eastern Otway Basin reveals a low-angle fault within the middle to upper Miocene Port Campbell Limestone section (Fig. 14b). This fault exhibits reverse motion and slightly offsets an unconformable surface at the top of the cliff.

In addition to the Torquay Sub-Basin and Otway Ranges, there is also clear evidence from seismic data for earlier compressional deformation in parts of the Shipwreck Trough and Mussel and Prawn platforms within the offshore eastern Otway Basin (Fig. 15). Within these regions, seismic mapping of the post-rift mid-Oligocene units

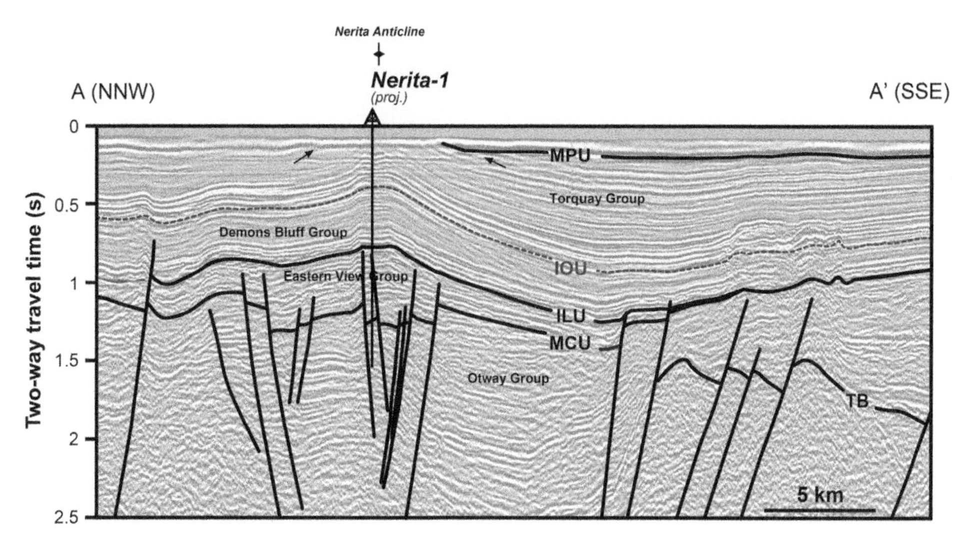

Fig. 13. Interpreted seismic profile O40-21 perpendicular to the fold axes of the Nerita Anticline in the Torquay Sub-Basin. Modified after Messent *et al.* (1999) and Holford *et al.* (2014). The Nerita Anticline is interpreted as an inversion structure with erosional truncation (black arrows) at the late Miocene–early Pliocene unconformity, which dips away from the anticlinal crest, perhaps suggesting neotectonic deformation. TB, Top Basement; MCU, Mid-Cretaceous unconformity; ILU, Intra-Lutetian unconformity; IOU, Intra-Oligocene unconformity; MPU, Miocene-Pliocene unconformity.

(i.e. the base-Heytesbury Group isochron; Fig. 1) shows that neotectonic compressional deformation is dominantly accommodated by low-amplitude (*c.* 100–500 m; Fig. 15b) folds with wavelengths of *c.* 10–25 km that plunge towards the SW (Geary & Reid 1998; Tuitt *et al.* 2011) with little evidence of faulting (Fig. 15a). A number of these folds, such as the Miverva, Pecten and Crowes anticlines, strike parallel and are contiguous to onshore neotectonic structures (Fig. 15b). The onset of neotectonic deformation in the eastern Otway Basin also correlates well with apatite fission track data that show cooling of Cenozoic–Cretaceous rocks beginning between 10 and 5 Ma as result of up to 400–1500 m of uplift and erosion (Cooper & Hill 1997; Dickinson *et al.* 2002; Green *et al.* 2004; Holford *et al.* 2011*a*; Tassone *et al.* 2012).

Pliocene and Quaternary strata overlying the regional late Miocene–early Pliocene angular unconformity in the onshore eastern Otway Basin also reveal evidence for neotectonic deformation (Sandiford, 2003*a*, *b*; Clark *et al.* 2011). For this reason, we categorize the onset of regional Neogene compressional deformation that resulted in the late Miocene–early Pliocene angular unconformity as early neotectonic deformation, and subsequent deformation as late neotectonic deformation.

Based on the displacement of Miocene strata, Pliocene strandlines and Pliocene–Quaternary volcanic flows, Clark *et al.* (2011) have estimated late neotectonic deformation slip rates of between *c.* 4 and 58 m Ma^{-1} for *c.* 2–1 Ma faults, monoclinal and anticlinal structural features in and adjacent to the Otway Ranges (Figs 14a & 15a; Sandiford 2003*a*, *b*). Similar analyses of neotectonic structures in the Gippsland Basin reveal slightly higher late neotectonic deformation slip rates (>63 m Ma^{-1}), which complements the observation of higher levels of present day seismicity in this part of southeastern Australia (Figs 2 & 12a). Excluding the Ferguson Hill Anticline (which is probably linked to an underlying basement fault), the basement-linked Selwyn and Rowsley faults that bound the northeastern and southeastern limits of the eastern Otway Basin (Fig. 16a), respectively, have been shown to have the highest late neotectonic deformation slip rates in the basin (Clark *et al.* 2011). Determining longer term (i.e. >10 Ma) neotectonic deformation slip rates in the Otway Basin is more challenging, although the conformable nature of most Pliocene strandlines across Victoria has been taken as evidence for very little deformation having occurred in the interval between 6 and 3 Ma. This demonstrates that most late neotectonic deformation occurred post-3 Ma (Fig. 1: Wallace *et al.* 2005; Clark *et al.* 2011), probably linked to an abrupt change in the Australian–Pacific plate interactions from a steady strike-slip motion across the Alpine Fault exposed on New Zealand's South Island to a steady transpressional regime (Walcott 1998; Sandiford 2003*b*). Through detailed analysis of Pliocene–Quaternary drainage patterns on the

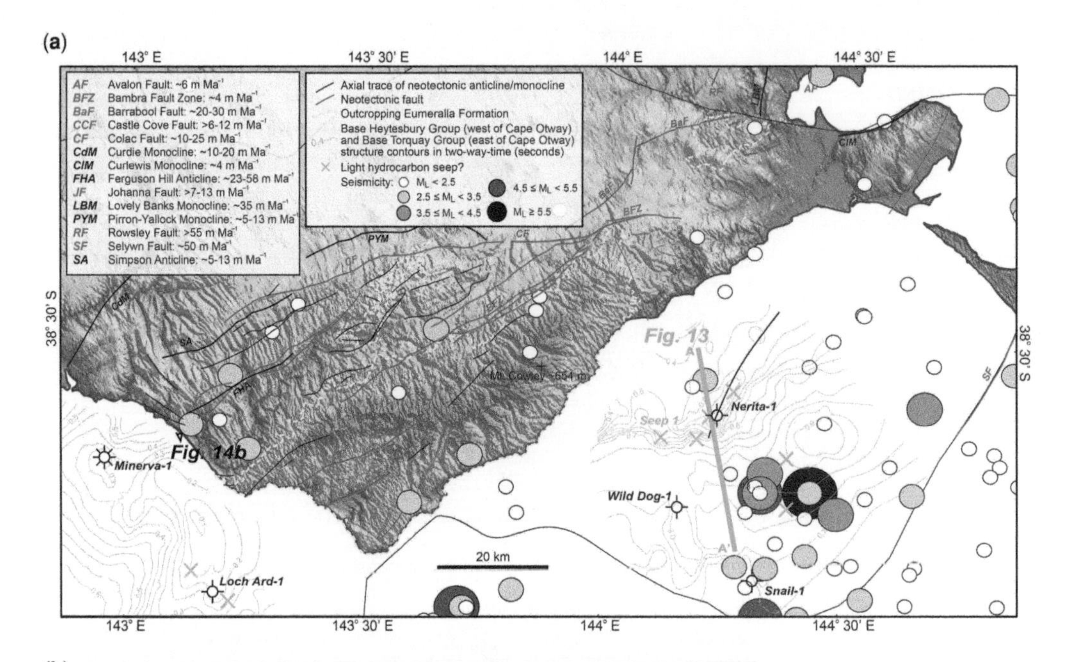

Fig. 14. (**a**) Topographic map of the Colac Trough and Otway Ranges (after Constantine & Liberman 2001) showing neotectonic features (e.g. faults, anticlines and monoclines), some with estimated slip/deformation rates (Clark *et al.* 2011). Also illustrated are isochrons (two-way time in seconds) of the intra-Oligocene unconformable surface mapped in the Shipwreck Trough and Prawn Platform (i.e. base-Heytesbury Group; Geary & Reid 1998) as well as in the Torquay Sub-Basin (i.e. base-Torquay Group; after Holford *et al.* 2011*a*, *b*) showing the similar trend in late Miocene–early Pliocene structural features. Green line represents location of seismic profile O40-21 in Fig. 13. Also shown are suspected light hydrocarbon seeps interpreted to be derived from a gas/condensate or dry thermogenic gas 'source' (after Bishop *et al.* 1992; Messent *et al.* 1999) and earthquakes (after Sandiford *et al.* 2004; Holford *et al.* 2011*a*, *b*), with Seep 1 apparently associated with a seabed pockmark (O'Brien *et al.* 1992). (**b**) Evidence of post-rift reverse faulting near the Otway Ranges. Small apparent low-angle fault with reverse sense is observed in the mid- to upper Miocene Port Campbell Limestone cliffs at the Twelve Apostles site in the Port Campbell National Park. The fault slightly displaces the top of the cliff (i.e. late Miocene–early Pliocene unconformity), indicating that the fault must have formed in the past *c.* 5–10 myr. Photographer is facing *c.* north–NNW, suggesting that the fault dips apparently broadly towards the east. Cliff face is *c.* 40 m high and the people at the top of the cliff can be used for scale.

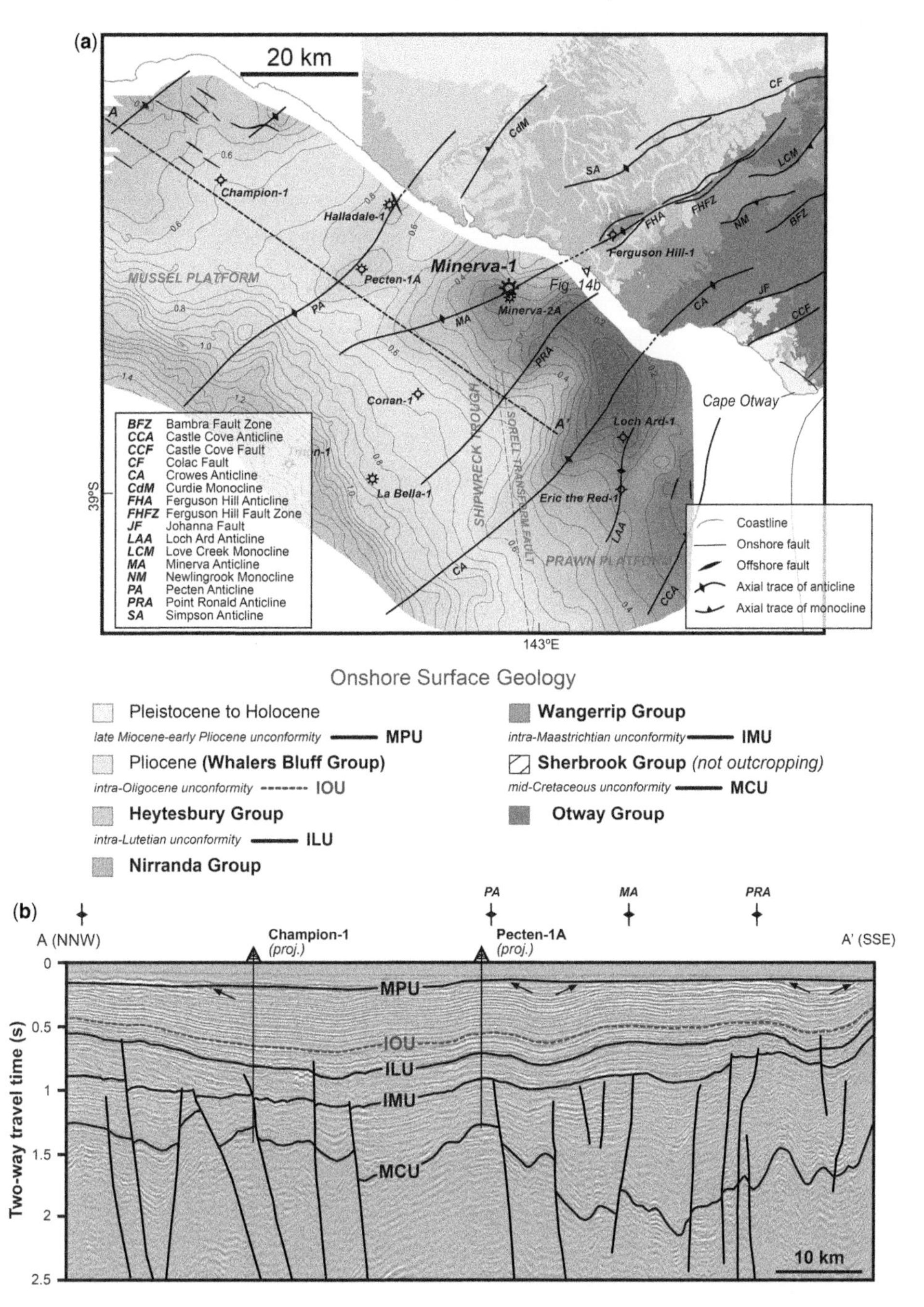

Fig. 15. (**a**) Surface geology map of the Port Campbell Embayment and northwestern margin of the Otway Ranges (after Edwards *et al.* 1996) and isochrons (two-way time in seconds) of the intra-Oligocene unconformable surface (i.e. base-Heytesbury Group) mapped in the Shipwreck Trough, Mussel and Prawn platforms (Geary & Reid 1998). Superimposed are the axes of offshore Neogene folds (after Geary & Reid 1998) that can be correlated to onshore neotectonic features (Clark *et al.* 2011). Note that the Ferguson Hill Anticline has Paleocene rocks outcropping surrounded by upper Oligocene to upper Miocene rocks, potentially suggesting appreciable amounts of Neogene erosion. (**b**) Interpreted seismic profile OH91-113, perpendicular to the fold axis of low-amplitude Neogene anticlines that plunge towards the SW within wavelengths of *c.*15–25 km (after Geary & Reid 1998; Holford *et al.* 2014). Black arrows indicate erosional truncation of strata beneath the late Miocene–early Pliocene unconformity.

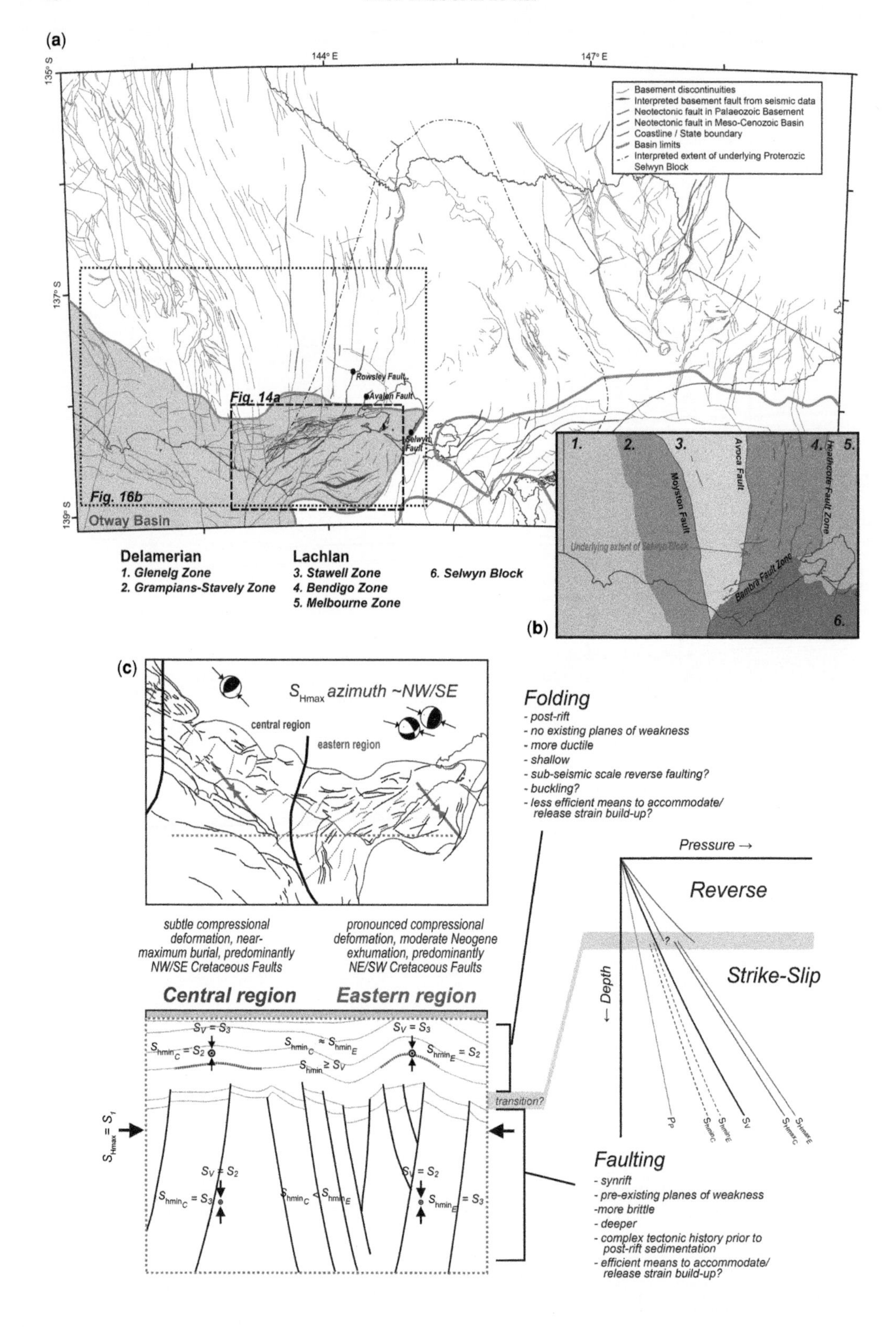
(a)
144° E
147° E
135° S
137° S
139° S
Basement discontinuities
Interpreted basement fault from seismic data
Neotectonic fault in Palaeozoic Basement
Neotectonic fault in Meso-Cenozoic Basin
Coastline / State boundary
Basin limits
Interpreted extent of underlying Proterozic Selwyn Block
Rowsley Fault
Avalon Fault
Selwyn Fault
Fig. 14a
Fig. 16b
Otway Basin
Delamerian
1. Glenelg Zone
2. Grampians-Stavely Zone
Lachlan
3. Stawell Zone
4. Bendigo Zone
5. Melbourne Zone
6. Selwyn Block
(b)
Moyston Fault
Avoca Fault
Heathcote Fault Zone
Bambra Fault Zone
Underlying extent of Selwyn Block
(c)
S_{Hmax} azimuth ~NW/SE
central region
eastern region
subtle compressional deformation, near-maximum burial, predominantly NW/SE Cretaceous Faults
pronounced compressional deformation, moderate Neogene exhumation, predominantly NE/SW Cretaceous Faults
Central region
Eastern region
Folding
- post-rift
- no existing planes of weakness
- more ductile
- shallow
- sub-seismic scale reverse faulting?
- buckling?
- less efficient means to accommodate/ release strain build-up?
Pressure →
Reverse
Strike-Slip
← Depth
transition?
Faulting
- synrift
- pre-existing planes of weakness
-more brittle
- deeper
- complex tectonic history prior to post-rift sedimentation
- efficient means to accommodate/ release strain build-up?

northern flanks of the Otway Ranges, Sandiford (2003*a*) constrained the most recent episode of reverse fault related uplift and incision to occurring between *c.* 2 and 1 Ma.

Although no focal mechanism solution has been determined in the eastern Otway Basin, recorded seismicity supports active brittle deformation at mid-crustal depths (Fig. 12). Potential evidence of thermogenic hydrocarbon seeps above seabed pockmarks in the vicinity of neotectonic compressional deformation structures (e.g. the Nerita Anticline; Fig. 13: Bishop *et al.* 1992; O'Brien *et al.* 1992) may also provide additional evidence of brittle deformation at shallow depths (<1 km) if indeed valid (cf. Logan *et al.* 2010). It is important to highlight that although seismic profile data reveals less evidence of major brittle deformation, faulting comparable with the fault displacement observed in the cliffs near Port Campbell would be beyond that interpretable at typical scales of seismic resolution (the minimum vertical resolution for the data used in this study is generally *c.* 30 m).

Figure 16a, b shows the distribution of late neotectonic features across the broader southeastern Australian region (Clark *et al.* 2011). This encompasses the Mesozoic–Cenozoic age Otway and Gippsland basins, the Cenozoic Murray–Darling Basin and outcropping basement, which mostly comprise the remnants of the early Palaeozoic Delamerian and Lachlan orogenies (Glen 2005). This map also highlights the distribution of basement discontinuities identified from field mapping and geophysical datasets (e.g. magnetic, radiometric, gravity and seismic; Vandenberg *et al.* 2000), including the interpreted underlying extent of the Proterozoic Selwyn Block microcontinent (Fig. 16a, b; Cayley 2011). There appears to be some spatial and geometric association between late neotectonic deformational structures and the underlying basement discontinuities. This suggests that basement structures may exert first-order control on the locus of late neotectonic deformation in the Otway Basin. Most late neotectonic structures in the eastern Otway Basin also strike parallel to early neotectonic inversion structures (e.g. the Nerita Anticline; Fig. 13, Fig. 14b & 16a) and are near-orthogonal to the contemporary S_{Hmax} orientation identified from the wellbore failure analysis of sedimentary rocks (Fig. 6c), as well as estimates of basement stress orientation (Fig. 12). This implies that both late and early neotectonic compressional deformation structures formed under comparable stress fields (Sandiford 2003*b*).

Linking contemporary stresses determined from petroleum data with geological and geophysical observations

In the previous sections we presented contemporary stress magnitudes using petroleum exploration data and also evidence of neotectonic compressional deformation from geological observations and focal mechanism solution data, emphasizing the key uncertainties associated with constraining the contemporary state of stress. Although S_V magnitude–depth profiles exhibit relatively little variation across the central and eastern offshore Otway Basin, an important finding is that LOT data, which was used to estimate the magnitude of S_{hmin}, appears to be sensitive, with depth, to both the lithology and structural province (Fig. 11). The aim of this section is to reconcile our new results with existing geological and geophysical observations to present a consistent model for the contemporary state of stress in the Otway Basin.

Fig. 16. (**a**) Neotectonic faults (after Clark *et al.* 2011) in Palaeozoic basement rocks and Meso-Cenozoic basins superimposed over Palaeozoic basement discontinuities identified from field mapping and geophysical datasets (Messent *et al.* 1999; Vandenberg *et al.* 2000) showing a clear relationship with the trends of neotectonic faults and underlying basement discontinuities. This suggests that the basement faults have a strong control, with the locus of neotectonic deformation in the Meso-Cenozoic passive margin sedimentary basins. Also shown is the interpreted extent of the underlying Proterozoic Selwyn Block in central Victoria (after Cayley 2011). (**b**) Interpreted pre-Permian basement terranes (after Vandenberg *et al.* 2000) showing that neotectonic deformation in the eastern onshore region occurs in the vicinity of the northwestern boundary of the Selwyn Block. (**c**) A summary of regional contemporary stresses constrained from petroleum exploration data integrated with geological observations that incorporate underlying structural trends across the basin, the mode of deformation with depth and lithology at a larger scale. Post-rift shallow Oligocene to Miocene marl and carbonate-dominated formations are folded (more ductile) with little evidence of pre-existing planes of weakness and have high horizontal stresses indicative of a reverse *in situ* fault stress regime. By contrast, deeper synrift siliciclastic rocks are heavily faulted (many pre-existing planes of weakness) and have lower horizontal stresses indicative of a strike-slip contemporary fault stress regime. There is also an increase in horizontal stresses from the central Otway Basin, where faults typically strike *c.* NW–SE, to the eastern Otway Basin, where faults typically strike *c.* NE–SW. Horizontal stress orientations are similar across both regions, are uniform over the entire basin sedimentary infill (excluding local stresses near faults) and agree well under focal mechanism solutions that sample basement stresses, which also indicate reverse and strike-slip faulting movements.

Influence of neotectonic compressional deformation on contemporary stresses in shallow post-rift sequences

Petroleum exploration data from the shallow, post-rift Heytesbury and Nirranda groups indicate a strike-slip to reverse faulting stress regime (i.e. $S_{hmin} < S_V$ to $S_{hmin} \approx S_V$) or, potentially, a pure reverse faulting stress regime (i.e. $S_{hmin} \geq S_V$), with less variation in the LOP across the central and eastern Otway Basin compared with the deeper synrift siliciclastic sequences (Fig. 11). The observation of a low-angle reverse fault in the cliff near Port Campbell (Fig. 14b), as well as neotectonic compressional structures around the topographically prominent Otway Ranges (Fig. 14a) – which contain over-steepened creek profiles due to recent uplift (Hill *et al.* 1995) – all point towards a reverse faulting stress regime. Consequently, we infer that a pure reverse fault stress regime predominately exists within the shallow, post-rift Heytesbury and Nirranda groups preserved at depths within 1–2 km below mudline.

Seismic data generally reveal little evidence for faulting within the post-rift marls and carbonates of the Heytesbury Group, despite plentiful evidence for low-amplitude folding associated with late Miocene to Holocene inversion structures (Fig. 15). Therefore, as there is no significant pre-existing weakness in these rocks (i.e. faults generated from the initial rifting), S_{Hmax} magnitudes should be considered to be minimum estimates if frictional limit theory is used. This implies that these shallow, post-rift marls and carbonate-dominated formations have predominantly accommodated compression through ductile deformation. Any fault displacement associated with these inversion structures is probably below the resolution of seismic data (i.e. *c.* 30 m), similar to the small, low-angle reverse faults seen in outcrop (Fig. 14b).

We suggest that compressional deformation accommodated by folding in these post-rift Miocene marls and carbonate-dominated formations has been a less efficient means of releasing strain compared with fault slip, potentially explaining the higher LOPs. It should be noted that there are other potential explanations for LOPs higher than S_V at these shallow depths, such as the wellbore walls being lined with mud during the test (Evans *et al.* 1989) or gravimetric reduction due to regional erosion (Bush & Meyer 1988; Warpinski & Teufel 1989). Gravimetric reduction due to regional uplift and erosion can result in high horizontal-to-vertical stress ratios at shallow depths (Bush & Meyer 1988). This is particularly possible if mudstones (or potentially ductile marls as is the case here) behave viscoelastically and effectively preserve a small portion of the higher stress state from the maximum palaeoburial depth (Warpinski & Teufel 1989).

Reconciling horizontal stress magnitudes constrained from petroleum exploration data in synrift rock sequences with geological observations

In contrast with the highly stressed, shallow post-rift marl and carbonate-dominated formations, petroleum exploration data indicate that deeper synrift, siliciclastic-dominated formations have lower S_{hmin} stresses. The LOT data show an overall increase in S_{hmin} gradients of *c.* 1–2 MPa km^{-1} at depths below *c.* 800 m TVDML within the deeper synrift units from west to east (Fig. 11). A major difference between the post-rift and synrift sequences in the Otway Basin is that the latter contain significant faulting. The siliciclastic units within the Paleocene Wangerrip Group, which witness the transition to post-rift conditions, generally record only minor faulting, but the underlying Cretaceous sequence contains a thick sequence of fault-controlled, synrift sedimentary rocks (Fig. 1). Minimum horizontal stress magnitudes are lower in the deeper synrift siliciclastic formations in the central Otway Basin, where faults generally strike near-parallel to the S_{Hmax} orientation, and are higher in the eastern Otway Basin where faults generally strike near-orthogonal to the S_{Hmax} orientation (i.e. broadly NE–SW). This reflects a similar change in structural trends within the underlying basement (Fig. 16a). A normal to strike-slip fault stress regime is evident in the central Otway Basin, where there is generally minor neotectonic (i.e. folding) compressional deformation (Fig. 3; Boult *et al.* 2008), while a strike-slip to reverse fault stress regime is observed in the eastern Otway Basin, where there is pronounced neotectonic compressional deformation. Given that there is less variation in LOPs within the shallower and folded post-rift marl and carbonate-dominated formations across both regions compared with the deeper and faulted synrift sequences, we suggest that the subsurface structural style and mode of compressional deformation exert an important control on horizontal stress gradients in the central and eastern Otway Basin (Fig. 16c).

Eastern Otway Basin. Changes in the horizontal stress gradient (at a larger scale than that caused by slight lithological variations) with depth resulting from varying structural styles with depth has been documented previously. For example, both Evans *et al.* (1989) and Couzens-Schultz & Chan (2010) have documented LOPs that change from near-lithostatic pressure (i.e. S_V) at shallow depth intervals (<1 km), abruptly becoming much less than lithostatic pressure (*c.* 60–90% S_V) in rocks at depths >1 km, in regions where independent geological evidence indicates reverse fault stress conditions. A possible explanation for the occurrence of lower LOPs in structural settings, where geological evidence indicates reverse fault stress conditions, but petroleum exploration data indicate strike-slip fault stress conditions, is an instantaneous stress drop following brittle deformation (Couzens-Schultz & Chan 2010). In the central and eastern Otway Basin, compressional strain in the deep, faulted synrift sections (particularly sandstones) may have been efficiently released by intermittent slip along faults since the late Miocene to present day, thus lowering horizontal stresses. Thus LOTs tend to yield lower LOPs and indicate a strike-slip fault stress regime as opposed to a reverse slip fault stress regime, despite the clear geological evidence favouring the latter (Fig. 14b; e.g. Mildren & Hillis 2000; Reynolds *et al.* 2003*b*; Mildren *et al.* 2004; White & Hillis 2004; Nelson *et al.* 2006*a*; King *et al.* 2008). If an LOP database is predominantly composed of LOTs performed within deeper synrift sequences with lower horizontal stress, then S_{hmin} gradients within shallower, post-rift sequences may be underestimated in a compressional deformation regime, or vice versa. We acknowledge that if the LOPs reflect shear failure along pre-existing fractures rather than the creation of new tensile failures (which may be possible in older, more deformed synrift packages), this could result in the underestimation of synrift S_{hmin} gradients.

Central Otway Basin. A normal to strike-slip faulting stress regime has been deduced for the central Otway Basin based on petroleum exploration data. Although there is evidence of small-amplitude folding within the Miocene marl and carbonate sequences that extend laterally westwards towards the South Australian Otway Basin with consistent NE–SW trends (Fig. 3; cf., Boult *et al.* 2008), there is less evidence for substantial inversion structures westwards away from the Otway Ranges. Hence a strike-slip faulting stress state in deeper synrift sequences, combined with a lack of suitably oriented structures on which inversion could occur, can adequately describe geological observations in the central Otway Basin. This may also indicate a buckling style of compressional deformation within the shallower, highly stressed ductile marl and carbonate-dominated formations in the central Otway Basin, where the inverted synrift faults are non-parallel to the shallower anticlinal trends. A *c.* 20 m surface height 'pop-up' structure that formed within the last 0.8 myr in an inter-dune coastal flat near the South Australia–Victoria border (Boult *et al.* 2008) may also be indicative of a strike-slip faulting stress regime. There is evidence for recent hydrocarbon leakage along reactivated basement faults that witness neotectonic compressional deformation (Boult *et al.* 2008), resulting in palaeohydrocarbon columns in the South Australian Otway Basin (Lyon *et al.* 2007).

Alternative explanations for conflicting geomechanical and geological datasets

We note that if LOPs reflect shear failure along (cohesionless) pre-existing planes of weakness, rather than the traditional assumption of failure via new tensile fractures, then this may lead to an underestimation of the magnitude of S_{hmin} (Couzens-Schultz & Chan 2010; King *et al.* 2012). Given that there is no way to determine whether leak-offs were caused by shear or tensile failure because the image log data were not acquired before and after LOTs were conducted, we acknowledge that our S_{hmin} gradients may well be underestimated within the deeper synrift sections. Under this scenario, particularly in the eastern Otway Basin, a reverse fault stress regime might be predominant throughout both the post- and synrift successions.

If the state of stress in the basin has changed since the youngest neotectonic features were formed, this could also account for conflicting assessments from geomechanical and geological datasets. The youngest, most reliable evidence for a neotectonic reverse fault stress regime appears to be provided by the analysis of drainage patterns on the northern flank of the Otway Ranges by Sandiford (2003*a*). This analysis suggests that the present relief in the Curdies and Gellibrand drainage basins occurred as a consequence of the tilting of fault blocks in the interval 2–1 Ma (Sandiford 2003*a*). If horizontal stresses have relaxed since *c.* 1 Ma, the stress state may have switched from a reverse fault to strike-slip fault regime, consistent with the geomechanical data. Further investigation of this possibility may yield useful insights into the magnitude and tempo of fluctuating horizontal stress magnitudes in intra-plate regions.

Conclusions

In fold–thrust belts and passive margins that have been subject to neotectonic activity, geomechanical and geological datasets often provide contradictory evidence for the state of contemporary stress. We have investigated the state of contemporary stress in the Otway Basin using newly available petroleum exploration data, with a focus on reconciling observations from geomechanical datasets with independent geological observations and geophysical datasets. In contrast with many previous contemporary stress studies in basins around Australia, we have highlighted the importance of factors such as the underlying structural fabrics and variations in the mechanical properties of syn- and post-rift basin-fill on stress magnitudes.

Our main conclusions are as follows:

(1) Wellbore failure analysis indicates that S_{Hmax} orientations in the central and eastern region of the Otway Basin are *c.* 135 ± 15° N, consistent with previous investigations using petroleum exploration data. Independent S_{Hmax} orientation constraints from focal mechanism solutions of intra-basement earthquakes and Pliocene–Holocene volcanic vent alignments are also broadly consistent with the *c.* NW–SE orientations determined from petroleum data. This implies a coupling of stress orientations between the Mesozoic–Cenozoic sedimentary basin-fill and the underlying basement.

(2) Shallow, post-rift, upper Oligocene to upper Miocene age marl and carbonate-dominated formations generally report the highest LOP gradients in the basin, which are typically in excess of the S_V gradients (>20 MPa km^{-1}) above *c.* 800 m TVDML. Although other reasons may explain the high LOPs across the entire basin (e.g. viscoelasticity due to recent uplift and erosion), the outcropping rocks show evidence of reverse neotectonic faulting and folding, in particularly in the eastern Otway Basin, indicating a reverse faulting stress regime.

(3) Deeper siliciclastic synrift sequences generally exhibit a strike-slip fault stress regime based on petroleum exploration data. The observation that S_{hmin} gradients in shales are typically *c.* 3–4 MPa km^{-1} greater than those in sandstones confirms that the lithology exerts a strong control on horizontal stress magnitudes in this basin.

(4) There is a *c.* 1–2 MPa km^{-1} increase in S_{hmin} gradients within synrift rocks from west to east in the basin. This coincides with both a change in structural trend and an increase in the intensity of neotectonic compressional deformation, implying that horizontal stresses are probably influenced by the orientation of the underlying structural fabrics with respect to the contemporary horizontal stress field.

(5) Neotectonic deformation styles vary with depth. Neotectonic compressional deformation in the post-rift, marl and carbonate-dominated formations is primarily accommodated by folding. There is little evidence for faulting within post-rift sequences, in contrast with deeper synrift sequences (which mostly consist of siliciclastic lithologies) that are highly faulted and are associated with lower horizontal stresses than the folded post-rift sequences. We suggest that compressional deformation accommodated through folding represents a less efficient means of releasing regional strain than faulting (of optimally oriented, reactivated planes of weakness). This could potentially explain the observed stress partitioning with depth, whereby petroleum data from post-rift sequences indicate higher horizontal stresses and a reverse stress state, but data from synrift sequences indicate lower horizontal stresses and a strike-slip stress state, in apparent conflict with geological evidence.

(6) Alternatively, the lower S_{hmin} gradients observed in deeper synrift sequences may reflect the underestimation of S_{hmin}. This might occur if LOPs reflect shear failure along (cohesionless) pre-existing planes of weakness, rather than the traditional assumption of failure via new tensile fractures. In basins where neotectonic activity is observed, we recommend that more reliable XLOT data are regularly acquired to better constrain S_{hmin} magnitudes. A further scenario that might explain aspects of our results is a change in stress regime from dominantly reverse faulting to strike-slip faulting over the past *c.* 1 myr.

This work forms part of ARC Discovery Projects DP0879612 and DP160101158 and ASEG Research Foundation Project RF09P04. We thank Will Jones, Adam Smith, Ian Brown and Huw Edwards of PGS for the provision of SAMDA, and gratefully acknowledge Geoscience Australia, DPI Victoria, Occam Technologies and Lakes Oil NL for access to seismic and well data. We also thank Jerry Meyer and Scott Mildren of Ikon Geomechanics for technical assistance and the provision of their propriety software JRS Suite for wellbore analyses, and IHS for provision of their Kingdom Suite seismic interpretation software. We thank Ian Duddy and Paul Green of Geotrack International for valuable discussions regarding the geology of the Otway Basin. We also thank Dave Dewhurst and an anonymous referee for their helpful reviews, and Dan Clark for his comments on an earlier version of this paper.

References

Addis, M.A. 1997. The stress-depletion response of reservoirs. Paper SPE-38720-MS, presented at the SPE Annual Technical Conference and Exhibition, 5–8 October 1997, San Antonio, TX, USA, https://doi.org/10.2118/38720-MS

Addis, M.A., Hanssen, TH., Yassir, N., Willoughby, D.R. & Enever, J. 1998. A comparison of leak-off test and extended leak-off test data for stress estimation. Paper SPE-47235-MS, presented at the SPE/ISRM Rock Mechanics in Petroleum Engineering Conference, 8–10 July 1998, Trondheim, Norway, https://doi.org/10.2118/47235-MS

ALLEN, I., GIBSON, G. & CULL, J.P. 2005. Stress-field constraints from recent intraplate seismicity in southeastern Australia. *Australian Journal of Earth Sciences*, **52**, 217–229.

ATHY, L.F. 1930. Density, porosity and compaction of sedimentary rocks. *American Association of Petroleum Geologists Bulletin*, **14**, 7675–7708.

BAILEY, A., KING, R., HOLFORD, S., SAGE, J., BACKÉ, G. & HAND, M. 2014. Remote sensing of subsurface fractures in the Otway Basin, South Australia. *Journal of Geophysical Research: Solid Earth*, **119**, 6591–6612.

BAILEY, A.H.E., KING, R.C., HOLFORD, S.P. & HAND, M. 2016. Incompatible stress regimes from geological and geomechanical datasets: can they be reconciled? An example from the Carnarvon Basin, Western Australia. *Tectonophysics*, **683**, 405–416.

BELL, J.S. 1990. Investigating stress regimes in sedimentary basins using information from oil industry wireline logs and drilling records. *In*: HURST, A., LOVELL, M. & MORTON, A. (eds) *Geological Applications of Wireline Logs*. Geological Society, London, Special Publications, **48**, 305–325, https://doi.org/10.1144/GSL.SP.1990.048.01.26

BELL, J.S. 1996. Petro Geoscience 1. In situ stresses in sedimentary rocks (part 2): applications of stress measurements. *Geoscience Canada*, **23**, 135–153.

BELL, J.S. & GOUGH, D.I. 1979. Northeast-southwest compressive stress in Alberta: evidence from oil wells. *Earth and Planetary Science Letters*, **45**, 475–482.

BÉRARD, T., SINHA, B.K., VAN RUTH, P., DANCE, T., JOHN, Z. & TAN, C.P. 2008. Stress estimation at the Otway CO_2 storage site, Australia. Paper SPE 116422, presented at the 2008 SPE Asia Pacific Oil & Gas Conference & Exhibition, 20–22 October, Perth, Australia, https://doi.org/10.2118/116422-MS

BERNECKER, T., SMITH, M.A., HILL, K.A. & CONSTANTINE, A.E. 2003. Oil and gas, fuelling Victoria's economy. *In*: BIRCH, W.D. (ed.) *Geology of Victoria*. Geological Society of Australia, Special Publications, **23**, 469–487.

BISHOP, J.H., BICKFORD, G.P. & HEGGIE, D.T. 1992. *South-eastern Australia Surface Geochemistry II: Light Hydrocarbon Geochemistry in Bottom Waters of the Gippsland Basin, Eastern Otway Basin, Torquay Sub-Basin and the Durroon Sub-Basin*. Vols 1 and 2. Bureau of Mineral Resources Record **1992/54**.

BLEVIN, J. & CATHRO, D. 2008. *Australian Southern Margin Synthesis*. Client report to Geoscience Australia by FrOG Tech Pty Ltd, Project GA707.

BOCK, G. & DENHAM, D. 1983. Recent earthquake activity in the Snowy Mountain region and its relationship to major faults. *Journal of the Geological Society of Australia*, **30**, 423–429.

BOULT, P.J., LYON, P., CAMAC, B., HUNT, S. & ZWINGMANN, H. 2008. Unravelling the complex structural history of the Penola Trough – revealing the St George Fault. *In*: BLEVIN, J.E., BRADSHAW, B.E. & URUSKI, C. (eds) *Eastern Australasian Basins Symposium III*. Petroleum Exploration Society of Australia, Special Publications, 81–93.

BRECKELS, I.M. & VAN EEKELEN, H.A.M. 1982. Relationships between horizontal stress and depth in sedimentary basins. *Journal of Petroleum Technology*, **34**, 2191–2198.

BRUDY, M. & ZOBACK, M.D. 1993. Compressive and tensile failure of boreholes arbitrarily-inclined to principal stress axes – application to the KTB boreholes, Germany. *International Journal of Rock Mechanics and Mining Sciences & Geomechanics Abstracts*, **30**, 1035–1038.

BRUDY, M. & ZOBACK, M.D. 1999. Drilling-induced tensile wall-fractures: implications for determination of in situ stress orientations and magnitudes. *International Journal of Rock Mechanics and Mining Sciences & Geomechanics Abstracts*, **36**, 191–215.

BUSH, D.D. & MEYER, B.S. 1988. In situ stress magnitude dependency on lithology. Paper SPE 88-0729, presented at the 29th US Symposium of Rock Mechanics (USRMS), 13–15 June 1988, Minneapolis, MN, USA.

CAYLEY, R.A. 2011. Exotic crustal block accretion to the eastern Gondwanaland margin in the Late Cambrian – Tasmania, the Selwyn Block, and implications for the Cambrian–Silurian evolution of the Ross, Delamerian, and Lachlan orogens. *Gondwana Research*, **19**, 628–649.

CLARK, D.J. 2009. What is an 'active' fault in the Australian intraplate context? A discussion with examples from eastern Australia. *Australian Earthquake Engineering Society Newsletter,* June, 3–6.

CLARK, D.J. & LEONARD, M. 2003. Principal stress orientations from multiple focal plane solutions: new insight in to the Australian intraplate stress field. *In*: HILLIS, R.R. & MULLER, D. (eds) *Evolution and Dynamics of the Australian Plate*. Geological Society of Australia, Special Publications, **22**, 91–105 and Geological Society of America, Special Papers, **372**.

CLARK, D.J., MCPHERSON, A. & COLLINS, C.D.N. 2011. *Australia's Seismogenic NeotectonicRrecord: a Case for Heterogeneous Intraplate Deformation*. Geoscience Australia, Records, 2011/11.

CLARK, D., MCPHERSON, A. & VAN DISSEN, R. 2012. Long-term behaviour of Australian stable continental region (SCR) faults. *Tectonophysics*, **266**, 1–30.

COOPER, G.T. & HILL, K.C. 1997. Cross-section balancing and thermochronological analysis of the Mesozoic development of the eastern Otway Basin. *Journal of the Australian Petroleum Production and Exploration Association*, **37**, 390–414.

CONSTANTINE, A.E. & LIBERMAN, N. 2001. *Hydrocarbon Prospectivity for VIC/O-01(1), VIC/O-01(2) and VIC/O-01(3), Eastern Onshore Otway Basin, Victoria, Australia: 2001 Acreage Release*. Victorian Initiative for Minerals and Petroleum Report **70**. Department of Natural Resources and Environment, Victoria.

CONSTANTINE, A., MORGAN, G. & TAYLOR, R. 2009. The Halladale and Black Watch gas fields – drilling AVO anomalies along Victoria's Shipwreck Coast. *The APPEA Journal*, **49**, 101–128.

CORCORAN, D.V. & DORÉ, A.G. 2007. Top seal assessment in exhumed basin settings – some insights from Atlantic Margin and borderland basins. *In*: KOESTLER, A.G. & HUNSDALE, R. (eds) *Hydrocarbon Seal Quantification*. Norwegian Petroleum Society, Special Publications, **11**, 89–107.

COUZENS-SCHULTZ, B.A. & CHAN, A.W. 2010. Stress determination in active thrust belts: an alternative leak-off pressure interpretation. *Journal of Structural Geology*, **22**, 1061–1069.

DAINES, S.R. 1982. Aquathermal pressuring and geopressure evaluation. *American Association of Petroleum Geologists Bulletin*, **66**, 931–939.

DENHAM, D., WEEKES, J. & KRAYSHEK, C. 1981. Earthquake evidence for compressive stress in the southeast Australian crust. *Journal of the Geological Society of Australia*, **28**, 323–332.

DENHAM, D., GIBSON, G., SMITH, R.S. & UNDERWOOD, R. 1985. Source mechanisms and strong ground motion from the 1982 Wonnangatta and the 1966 Mount Hotham earthquakes. *Australian Journal of Earth Sciences*, **32**, 37–46.

DENSLEY, M.R., HILLIS, R.R. & REDFEARN, J.E.P. 2000. Quantification of Tertiary uplift in the Carnarvon Basin based on interval velocities. *Australian Journal of Earth Sciences*, **47**, 111–122.

DEWHURST, D.N., DELLE PIANE, C., ESTEBAN, L., SAROUT, J., JOSH, M., PERVUKHINA, M. & CLENNELL, M.B. In press. Microstructural, geomechanical and petrophysical characterisation of shale caprocks. *In*: VIALLE, S., AJO-FRANKLIN, J. & CAREY, J.W. (eds) *Caprock Integrity in the Context of Geological Carbon Storage*. Wiley.

DICKINSON, J.A., WALLACE, M.W., HOLDGATE, G.R., GALLAGHER, S.J. & THOMAS, L. 2002. Origin and timing of the Miocene–Pliocene unconformity in southeast Australia. *Journal of Sedimentary Research*, **72** , 288–303.

DUDDY, I.R. 2003. Mesozoic, a time of change in tectonic regime. *In*: BIRCH, W.D. (ed.) *Geology of Victoria*. Geological Society of Australia, Special Publications, **23**, 239–286.

DYKSTERHUIS, S. & MÜLLER, R.D. 2008. Cause and evolution of intraplate orogeny in Australia. *Geology*, **36**, 495–498.

EATON, B.A. 1969. Fracture gradient prediction and its application in oil field operations. *Journal of Petroleum Technology*, **21**, 1353–1360.

EDWARDS, J., TICKELL, S.J., WILCOCKS, A.J., EATON, A.R., CRAMER, M.L., KING, R.L., BOURTON, S.M. 1996. *Colac 1:250000 Geological Map*. Geological Survey of Victoria, Melbourne.

ENGELDER, T. 1993. *Stress Regimes in the Lithosphere*. Princeton University Press, Princeton, NJ.

EVANS, K. & ENGELDER, T. 1989. Some problems in estimating horizontal stress magnitudes in 'thrust' tregimes. *International Journal of Rock Mechanics and Mining Sciences & Geomechanics Abstracts*, **26**, 647–660.

EVANS, K., ENGELDER, T. & PLUMB, R.A. 1989. Appalachian stress study 1: a detailed description of in situ stress variation in Devonian shales of the Appalachian Plateau. *Journal of Geophysical Research*, **94**, 1729–1754.

GEARY, G.C. & REID, I.S.A. 1998. *Hydrocarbon Prospectivity of the Offshore Eastern Otway Basin, Victoria, for the 1998 Acreage Release*. Victorian Initiative for Minerals and Petroleum Report **55**, Department of Natural Resources and Environment, Melbourne.

GLEN, R.A. 2005. The Tasmanides of eastern Australia. *In*: VAUGHAN, A.P.M., LEAT, P.Y. & PANKHURST, R.J. (eds) *Terrane Processes at the Margins of Gondwana*. Geological Society, London, Special Publications, **246**, 23–96, https://doi.org/10.1144/GSL.SP.2005.246.01.02

GOUGH, D.I. & BELL, J.S. 1982. Stress orientations from borehole wall fractures with examples from Colorado, east Texas, and northern Canada. *Canadian Journal of Earth Sciences*, **19**, 1358–1370.

GOULTY, N.R. 2003. Reservoir stress path during depletion of Norwegian chalk oilfields. *Petroleum Geoscience*, **9**, 233–241, https://doi.org/10.1144/1354-079302-545

GREEN, P.F., CROWHURST, P.V. & DUDDY, I.R. 2004. Integration of AFTA and (U-Th)/He thermochronology to enhance the resolution and precision of thermal history reconstruction in the Anglesea-1 well, Otway Basin, SE Australia. *In*: BOULT, P.J., JOHNS, D.R. & LANG, S.C. (eds) *Eastern Australasian Basins Symposium II*. Petroleum Exploration Society of Australia, Special Publications, 117–131.

HAIMSON, B.C. & FAIRHURST, C. 1967. Initiation and extension of hydraulic fractures in rocks. *Society of Petroleum Engineers Journal*, **7**, 310–318.

HEIDBACH, O., REINECKER, J., TINGAY, M., MÜLLER, B., SPERNER, B., FUCHS, K. & WENZEL, F. 2007. Plate boundary forces are not enough: second- and third-order stress patterns highlighted in the World Stress Map database. *Tectonics*, **26**, TC6014, https://doi.org/10.1029/2007TC002133

HEIDBACH, O., TINGAY, M.R.P., BARTH, A., REINECKER, J., KURFESS, D. & MÜLLER, B. 2010. Global crustal stress patterns based on the 2008 World Stress Map database release. *Tectonophysics*, **482**, 3–15.

HILL, K.C., HILL, K.A., COOPER, G.T., RICHARDSON, M.J. & LAVIN, C.J. 1994. Structural framework of the eastern Otway Basin: inversion and interaction between two major structural provinces. *Exploration Geophysics*, **25**, 79–87.

HILL, K.C., HILL, K.A., COOPER, G.T., O'SULLIVAN, A.J., O'SULLIVAN, P.B. & RICHARDSON, M.J. 1995. Inversion around the Bass Basin, SE Australia. *In*: BUCHANAN, J.G. & BUCHANAN, P.G. (eds) *Basin Inversion*. Geological Society, London, Special Publications, **88**, 525–547, https://doi.org/10.1144/GSL.SP.1995.088.01.27

HILLIS, R.R. 2001. Coupled changes in pore pressure and stress in oil fields and sedimentary basins. *Petroleum Geoscience*, **7**, 419–425, https://doi.org/10.1144/petgeo.7.4.419

HILLIS, R.R. & REYNOLDS, S.D. 2000. The Australian stress map. *Journal of the Geological Society, London*, **157**, 915–921, https://doi.org/10.1144/jgs.157.5.915

HILLIS, R.R. & WILLIAMS, A.F. 1992. Borehole breakouts and stress analysis in the Timor Sea. *In*: HURST, A., GRIFFITHS, C.M. & WORTHINGTON, P.F. (eds) *Geological Applications of Wireline Logs II*. Geological Society, London, Special Publications, **65**, 157–168, https://doi.org/10.1144/GSL.SP.1992.065.01.11

HILLIS, R.R. & WILLIAMS, A.F. 1993. The contemporary stress field of the Barrow-Dampier Sub-basin and its implications for horizontal drilling. *Exploration Geophysics*, **24**, 567–576.

HILLIS, R.R., MONTE, S.A., TAN, C.P. & WILLOUGHBY, D.R. 1995. The contemporary stress field of the Otway Basin, South Australia: implications for hydrocarbon exploration and production. *Journal of the Australian Petroleum Production and Exploration Association*, **31**, 494–506.

HILLIS, R.R., SANDIFORD, M., REYNOLDS, S.D. & QUIGLEY, M.C. 2008. Present-day stresses, seismicity and Neogene-o-Recent tectonics of Australia's 'passive' margin: intraplate deformation controlled by plate boundary forces. *In*: JOHNSON, H., DORÉ, A.G., GATLIFF, R.W., HOLDSWORTH, R., LUNDIN, E.R. & RITCHIE, J.D. (eds) *The Nature and Origin of Compression in Passive Margins*. Geological Society, London, Special Publications, **306**, 71–90, https://doi.org/10.1144/SP306.3

HOLDGATE, G.R. & GALLAGHER, S.J. 2003. Tertiary, a period of transition to marine basin environments. *In*: BIRCH, W.D. (ed.) *Geology of Victoria*. Geological Society of Australia, Special Publications, **23**, 289–335.

HOLFORD, S.P., HILLIS, R.R., DUDDY, I.R., GREEN, P.F., TUITT, A.K. & STOKER, M.S. 2010. Impacts of Neogene–Recent compressional deformation and uplift on hydrocarbon prospectivity of the passive southern Australian margin. *Journal of the Australian Petroleum Production and Exploration Association*, **50**, 267–286.

HOLFORD, S.P., HILLIS, R.R., DUDDY, I.R., GREEN, P.F., TASSONE, D.R. & STOKER, M.S. 2011*a*. Palaeothermal and seismic constraints on late Miocene–Pliocene uplift and deformation in the Torquay sub-basin, southern Australian margin. *Australian Journal of Earth Sciences*, **58**, 543–562.

HOLFORD, S.P., HILLIS, R.R., HAND, M. & SANDIFORD, M. 2011*b*. Thermal weakening localizes intraplate deformation along the southern Australian continental margin. *Earth and Planetary Science Letters*, **305**, 207–214.

HOLFORD, S.P., TUITT, A.K., HILLIS, R.R., GREEN, P.F., STOKER, M.S., DUDDY, I.R. & SANDIFORD, M. 2014. Cenozoic deformation in the Otway Basin, southern Australian margin: implications for the origin and nature of post-breakup compression at rifted margins. *Basin Research*, **26**, 10–37, https://doi.org/10.1111/bre.12035

HOLFORD, S.P., TASSONE, D.R., STOKER, M.S. & HILLIS, R.R. 2016. Contemporary stress orientations in the Faroe–Shetland region. *Journal of the Geological Society, London*, **173**, 142–152, https://doi.org/10.1144/jgs2015-048

JAEGER, J.C. & COOK, N.G.W. 1979. *Fundamentals of Rock Mechanics*. Chapman and Hall, London.

JOHNSTON, A.C. 1994. Seismotectonic interpretations and conclusions from the stable continental region seismicity database. *In*: JOHNSTON, A.C., COPPERSMITH, K.J., KANTER, L.R. & CORNELL, C.A. (eds) *The Earthquakes of Stable Continental Region – Vol. 1: Assessment of Large Earthquake Potential*. Report **TR-102261-V1**. Electric Power Research Institute, Palo Alto, CA.

KING, R.C., HILLIS, R.R. & REYNOLDS, S.D. 2008. In situ stresses and natural fractures in the Northern Perth Basin, Australia. *Australian Journal of Earth Sciences*, **55**, 685–701.

KING, R.C., NEUBAUER, M., HILLIS, R.R. & REYNOLDS, S.D. 2010. Variation in vertical stress in the Carnarvon Basin, NW Shelf, Australia. *Tectonophysics*, **482**, 73–81.

KING, R.C., HOLFORD, S.P. *ET AL*. 2012. Reassessing the in situ stress regimes of Australia's petroleum basins. *Journal of the Australian Petroleum Production and Exploration Association*, **52**, 415–426.

KRASSAY, A.A., CATHRO, D.L. & RYAN, D.J. 2004. A regional tectonostratigraphic framework for the Otway Basin. *In*: BOULT, P.J., JOHNS, D.R. & LANG, S.C. (eds) *Eastern Australasian Basins Symposium II*. Petroleum Exploration Society of Australia, Special Publications, 97–116.

LEONARD, M., RIPPER, I.D. & YUE, L. 2002. *Australian Fault Plane Solutions*. Geoscience Australia Records, **2002/1**.

LESTI, C., GIORDANO, G., SALVINI, F. & CAS, R. 2008. Volcano tectonic setting of the intraplate, Pliocene–Holocene, Newer Volcanic Province (southeast Australia): role of crustal fracture zones. *Journal of Geophysical Research*, **113**, B07407, https://doi.org/10.1029/2007JB005110

LOGAN, G.A., JONES, A.T., KENNAED, J.M., RYAN, G.J. & ROLLET, N. 2010. Australian offshore natural hydrocarbon seepage studies, a review and re-evaluation. *Marine and Petroleum Geology*, **27**, 26–45.

LUDWIG, W.J., NAFE, J.E. & DRAKE, C.L. 1970. Seismic refraction. *In*: MAXWELL, A.E. (ed.) *The Sea*, Vol. **4**. Wiley-Interscience, New York, 53–84.

LYON, P.J., BOULT, P.J., HILLIS, R.R., MILDREN, S.D. 2005. Sealing by shale gouge and subsequent seal breach by reactivation: a case study of the Zema Prospect, Otway Basin. *In*: BOULT, P.J. & KALDI, J. (eds) *Evaluating Fault and Cap Rock Seals*. American Association of Petroleum Geologists, Hedberg Series, **2**, 179–197.

LYON, P.J., BOULT, P.J., HILLIS, R.R. & BIERBRAUER, K. 2007. Basement controls on fault development in the Penola Trough, Otway Basin, and implications for fault-bounded hydrocarbon traps. *Australian Journal of Earth Sciences*, **54**, 675–689.

MACDONALD, J.D., HOLFORD, S.P., GREEN, P.F., DUDDY, I.R., KING, R.C. & BACKÉ, G. 2013. Detrital zircon data reveal the origin of Australia's largest delta system. *Journal of the Geological Society, London*, **170**, 3–6, https://doi.org/10.1144/jgs2012-093

MARDIA, K., K.V. 1972. *Statistics of Directional Data*. Academic Press, London.

MASTIN, L. 1988. Effect of borehole deviation on breakout orientations. *Journal of Geophysical Research*, **93**, 9187–9195.

MATTHEWS, W.R. & KELLY, J. 1967. How to predict formation pressure and fracture gradient from electric and sonic logs. *Oil and Gas Journal*, 92–106.

MAVROMATIDIS, A. & HILLIS, R.R. 2005. Quantification of exhumation in the Eromanga Basin and its implications for hydrocarbon exploration. *Petroleum Geoscience*, **11**, 79–92, https://doi.org/10.1144/1354-079304-621

MCCUE, K. & PAULL, E. 1991. *Australian Seismological Report 1988*. Australian Bureau of Mineral Resources Report **304**.

McCue, K., Gibson, G. & Wesson, V. 1990. The earthquake near Nhill, western Victoria, on 22 December 1987 and the seismicity of eastern Australia. *Journal of Australian Geology and Geophysics*, **11**, 415–420.

McGowran, B., Holdgate, G.R., Li, Q. & Gallagher, S.J. 2004. Cenozoic stratigraphic succession in south-eastern Australia. *Australian Journal of Earth Sciences*, **51**, 459–496.

Messent, B.E., Collins, G.I. & West, B.G. 1999. *Hydrocarbon Prospectivity of the Offshore Torquay Sub-Basin, Victoria*. Gazettal Area V99-1. Victorian Initiative for Minerals and Petroleum Report **60**.

Mildren, S.D. & Hillis, R.R. 2000. In situ stresses in the Bonaparte Basin, Australia: implications for first- and second-order controls on stress orientations. *Geophysical Research Letters*, **27**, 3413–3416.

Mildren, S.D., Hillis, R.R., Kivior, T. & Kaldi, J.G. 2004. Integrated seal assessment and geologic risk with application to the Skua Field, Timor Sea, Australia. *In*: Ellis, G.K., Baillie, P.W. & Munson, T.J. (eds) *Timor Sea Petroleum Geoscience*. Proceedings of the Timor Sea Symposium, Darwin, Australia. Northern Territory Geological Survey, Special Publications, **1**, 275–294.

Miller, J.M., Norvick, M.S. & Wilson, C.J.L. 2002. Basement controls on rifting and the associated formation of ocean transform faults – Cretaceous continental extension of the southern margin of Australia. *Tectonophysics*, **359**, 131–155.

Mouchet, J.P. & Mitchell, A. 1989. *Abnormal Pressures While Drilling*. Elf Aquitaine, Boussens.

Nelson, E.J. & Hillis, R.R. 2005. In situ stresses of the West Tuna area, Gippsland Basin. *Australian Journal of Earth Sciences*, **52**, 299–313.

Nelson, E.J., Hillis, R.R., Sandiford, M., Reynolds, S.D. & Mildren, S.D. 2006*a*. Present-day state-of-stress of southeast Australia. *Journal of the Australian Petroleum Production and Exploration Association*, **46**, 283–305.

Nelson, E.J., Hillis, R.R. & Mildren, S.D. 2006*b*. Stress partitioning and wellbore failure in the West Tuna Area, Gippsland Basin. *Exploration Geophysics*, **37**, 215–221.

Nelson, E.J., Chipperfield, S.T., Hillis, R.R., Gilbert, J. & McGowen, J. 2007. Using geological information to optimize fracture stimulation practices in the Cooper Basin, Australia. *Petroleum Geoscience*, **13**, 3–16, https://doi.org/10.1144/1354-079306-700

Norvick, M.S. & Smith, M.A. 2001. Mapping the plate tectonic reconstruction of southern and southeastern Australia and implications for petroleum systems. *Journal of the Australian Petroleum Production and Exploration Association*, **41**, 15–35.

O'Brien, G.W., Heggie, D.T., Hartman, B., Bickford, G. & Bishop, J.H. 1992. *Light Hydrocarbon Geochemistry of the Gippsland, North Bass, Bass, Otway and Stansbury Basins and the Torquay Sub-basin, South-Eastern Australia*. Bureau of Mineral Resources Record **1992/52**.

O'Brien, G.W., Bernecker, T., Thomas, J.H., Driscoll, J.P. & Rikus, L. 2006. *An Assessment of the Hydrocarbon Prospectivity of Areas VIC/O-06(1), VIC/O-06(2), VIC/O-06(3) and V06-1, Eastern Onshore and Offshore Otway Basin, Victoria, Australia*. Victorian Initiative for Minerals and Petroleum Report **87**

Perincek, D. & Cockshell, C.D. 1995. The Otway Basin: early Cretaceous rifting to Neogene inversion. *The APPEA Journal*, **35**, 451–466.

Perincek, D., Simons, B. & Pettifer, G.R. 1994. The tectonic framework and associated play types of the Western Otway Basin, Victoria, Australia. *The APPEA Journal*, **34**, 460–478.

Plumb, R.A. & Hickman, S.H. 1985. Stress-induced borehole enlargement: a comparison between the four-arm dipmeter and the borehole televiewer in the Auburn geothermal well. *Journal of Geophysical Research*, **90**, 5513–5521.

Reinecker, J., Tingay, M. & Müller, B. 2003. *Borehole Breakout Analysis from Four-arm Caliper Logs. Guidelines for Interpreting Stress Indicators: World Stress Map Online Publication*, http://www.world-stress-map.org

Reynolds, S. & Hillis, R.R. 2000. The in situ stess field of the Perth Basin, Australia. *Geophysical Research Letters*, **27**, 3421–3424.

Reynolds, S., Coblentz, D.D., Hillis, R.R. 2003*a*. Influences of plate-boundary forces on the regional intraplate stress field of continental Australia. *In*: Hillis, R.R. & Müller, R.D. (eds) *Evolution and Dynamics of the Australian Plate*. Geological Society of Australia, Special Publications, **22** and Geological Society of America, Special Papers, **372**, 53–64.

Reynolds, S., Hillis, R.R. & Paraschivoiu, E. 2003*b*. In situ stress field, fault reactivation and seal integrity in the Bight Basin, South Australia. *Exploration Geophysics*, **34**, 174–181.

Reynolds, S., Mildren, S.D., Hillis, R.R., Meyer, J.J. & Flottmann, T. 2005. Maximum horizontal stress orientations in the Cooper Basin, Australia: implications for plate-scale tectonics and local stress sources. *Geophysical Journal International*, **160**, 331–343.

Reynolds, S., Mildren, S.D., Hillis, R.R. & Meyer, J.J. 2006. Constraining stress magnitudes using petroleum exploration data in the Cooper-Eromanga Basins, Australia. *Tectonophysics*, **415**, 123–140.

Rogers, C., van Ruth, P., Hillis, R.R. 2008. Fault reactivation in the Port Campbell Embayment with respect to carbon dioxide sequestration, Otway Basin, Australia. *In*: Johnson, H. & Doré, A.G., Gatliff, R.W., Holdsworth, R., Lundin, E.R. & Ritchie, J.D. (eds) *The Nature and Origin of Compression in Passive Margins*. Geological Society, London, Special Publications, **306**, 201–214, https://doi.org/10.1144/SP306.10

Sagala, A.J.I. & Tingay, M.R.P. 2012. Analysis of overpressure and its generating mechanisms in the Northern Carnarvon Basin from drilling data. *Journal of the Australian Petroleum Production and Exploration Association*, **52**, 375–390.

Sandiford, M. 2003*a*. Geomorphic constraints on the late Neogene tectonics of the Otway Ranges. *Australian Journal of Earth Sciences*, **50**, 69–80.

Sandiford, M. 2003*b*. Neotectonics of south-eastern Australia: linking the Quaternary faulting record with seismicity and in situ stress. *In*: Hillis, R.R. & Muller, R.D. (eds) *Evolution and Dynamics of the Australian*

Plate. Geological Society of Australia, Special Publications, **22**, 107–119.

SANDIFORD, M. & QUIGLEY, M. 2009. TOPO-OZ: insights into the various modes of intraplate deformation in the Australian continent. *Tectonophysics*, **474**, 405–416.

SANDIFORD, M., WALLACE, M. & COBLENTZ, D.D. 2004. Origin of the in situ stress field in south-eastern Australia. *Basin Research*, **16**, 325–338.

SCHNEIDER, C.L., HILL, K.C. & HOFFMAN, N. 2004. Compressional growth of the Minerva Anticline, Otway Basin, southeast Australia – evidence of oblique rifting. *Journal of the Australian Petroleum Production and Exploration Association*, **44**, 463–480.

SHARP, N.C. & WOOD, G.R. 2004. Casino Gas Field, offshore Otway Basin, Victoria – the appraisal story and some stratigraphic enlightenment. *In*: BOULT, P.J., JOHNS, D.R. & LANG, S.C. (eds) *Eastern Australasian Basins Symposium II*. Petroleum Exploration Society of Australia, Special Publications, 97–116.

SIBSON, R.H. 1974. Frictional constraints on thrust, wrench and normal faults. *Nature*, **249**, 542–544.

SIBSON, R.H. 1995. Selective fault reactivation during basin inversion: potential for fluid redistribution through fault-valve action. *In*: BUCHANAN, J.G. & BUCHANAN, P.G. (eds) *Basin Inversion*. Geological Society, London, Special Publications, **88**, 3–19, https://doi.org/10.1144/GSL.SP.1995.088.01.02

STEIN, R.S., KING, G.C. & LIN, J. 1992. Change in failure stress on the southern San Andreas fault system caused by the 1992 magnitude = 7.4 Landers earthquake. *Science*, **258**, 1328–1332.

TASSONE, D.R., HOLFORD, S.P., TINGAY, M.R.P., TUITT, A.K., STOKER, M.S. & HILLIS, R.R. 2011. Overpressures in the central Otway Basin: the result of rapid Pliocene–Recent sedimentation? *Journal of the Australian Petroleum Production and Exploration Association*, **51**, 439–458.

TASSONE, D.R., HOLFORD, S.P., HILLIS, R.R. & TUITT, A.K. 2012. Quantifying Neogene plate-boundary controlled uplift and deformation of the southern Australian margin. *In*: HEALY, D., BUTLER, R.W.H., SHIPTON, Z.K. & SIBSON, R.H. (eds) *Faulting, Fracturing and Igneous Intrusion in the Earth's Crust*. Geological Society, London, Special Publications, **367**, 91–110, https://doi.org/10.1144/SP367.7

TASSONE, D.R., HOLFORD, S.P., DUDDY, I.R., GREEN, P.F. & HILLIS, R.R. 2014. Quantifying Cretaceous–Cenozoic exhumation in the Otway Basin using sonic transit time data: implications for conventional and unconventional hydrocarbon prospectivity. *American Association of Petroleum Geologists Bulletin*, **98**, 67–117.

TINGAY, M., MÜLLER, B., REINECKER, J. & HEIDBACH, O. 2006. State and origin of the present-day stress field in sedimentary basins: new results from the World Stress Map Project. Paper ARMA-06-1049, presented at Golden Rocks 2006, The 41st U.S. Symposium on Rock Mechanics (USRMS), 17–21 June 2006, Golden, CO, USA.

TINGAY, M.R.P., HILLIS, R.R., SWARBRICK, R.E. & OKPERE, E.C. 2003*a*. Variation in vertical stress in the Baram Basin, Brunei: tectonic and geomechanical implications. *Marine and Petroleum Geology*, **20**, 1201–1212.

TINGAY, M.R.P., HILLIS, R.R., MORLEY, C.K., SWARBRICK, R.E. & OKPERE, E.C. 2003*b*. Pore pressure/stress coupling in Brunei Darussalam – implications for shale injection. *In*: VAN RENSBERGEN, P., HILLIS, R.R., MALTMAN, A.J. & MORLEY, C.K. (eds) *Subsurface Sediment Mobilization*. Geological Society, London, Special Publications, **216**, 369–379, https://doi.org/10.1144/GSL.SP.2003.216.01.24

TINGAY, M.R.P., MORLEY, C., HILLIS, R.R. & MEYER, J.J. 2010. Present-day stress orientation in Thailand's basins. *Journal of Structural Geology*, **32**, 235–248.

TOTTERDELL, J. 2012. New exploration opportunities along Australia's southern margin. *Journal of the Australian Petroleum Production and Exploration Association*, **52**, 29–44.

TOWNEND, J. & ZOBACK, M.D. 2000. How faulting keeps the crust strong. *Geology*, **28**, 399–402.

TUITT, A.K., HOLFORD, S.P. *ET AL*. 2011. Continental margin compression: A comparison between compression in the Otway Basin of the southern Australian margin and the Rockall–Faroe area in the north east Atlantic margin. *Journal of the Australian Petroleum Production and Exploration Association*, **51**, 241–258.

VAN RUTH, P. & HILLIS, R.R. 2000. Estimating pore pressure in the Cooper Basin, South Australia: sonic log method in an uplifted basin. *Exploration Geophysics*, **31**, 441–447.

VANDENBERG, A.H.M., WILLMAN, C.E. *ET AL*. 2000. *The Tasman Fold Belt System in Victoria*, Geological Survey of Victoria, Special Publications.

VEEVERS, J.J. 2000. Change of tectono-stratigraphic regime in the Australian plate during the 99 Ma (mid-Cretaceous) and 43 Ma (mid-Eocene) swerves of the Pacific. *Geology*, **28**, 47–50.

VIDAL-GILBERT, S., TENTHOREY, E., DEWHURST, D., ENNIS-KING, J., VAN RUTH, P. & HILLIS, R.R. 2010. Geomechanical analysis of the Naylor Field, Otway Basin, Australia: implications for CO_2 injection and storage. *International Journal of Greenhouse Gas Control*, **4**, 827–839.

WAGNER, D., MULLER, B. & TINGAY, M.R.P. 2004. Correcting for tool decentralization of oriented six-arm caliper logs for determination of contemporary tectonic stress orientations. *Petrophysics*, **45, 530–539**.

WALCOTT, R.I. 1998. Modes of oblique compression: late Cainozoic tectonics of the South Island of New Zealand. *Reviews of Geophysics*, **36**, 1–26.

WALLACE, M.W., DICKINSON, J.A., MOORE, D.H. & SANDIFORD, M. 2005. Late Neogene strandlines of southern Victoria: a unique record of eustasy and tectonics in southeast Australia. *Australian Journal of Earth Sciences*, **52**, 277–295.

WARPINSKI, N.R. 1989. Elastic and viscoelastic calculations of stresses in sedimentary basins. SPE Formation Evaluation, **4**, SPE 15243, https://doi.org/10.2118/15243-PA

WARPINSKI, N.R. & TEUFEL, L.W. 1989. In-situ stresses in low-permeability, nonmarine rocks. *Journal of Petroleum Technology*, **41**, 405–414.

WARPINSKI, N.R. & TEUFEL, L.W. 1991. In situ stress measurements at Rainier Mesa, Nevada Test Site – influence of topography and lithology on the stress state in tuff. *International Journal of Rock Mechanics and Mining Sciences & Geomechanics Abstracts*, **28**, 143–161.

White, A. & Hillis, R.R. 2004. In-situ stress field and fault reactivation in the Mutineer and Exeter Fields, Australian North West Shelf. *Exploration Geophysics*, **35**, 217–223.

Yale, D.P. 2003. Fault and stress magnitude controls on variations in the orientation of *in situ* stress. *In*: Ameen, S.A. (ed.) *Fracture and In-Situ Stress Characterization of Hydrocarbon Reservoirs*. Geological Society, London, Special Publications, **209**, 55–64, https://doi.org/10.1144/GSL.SP.2003.209.01.06

Yardley, G.S. & Swarbrick, R.E. 2000. Lateral transfer: a source of additional overpressures? *Marine and Petroleum Geology*, **17**, 523–538.

Zoback, M.L. 1992. First- and second-order patterns of stress in the lithosphere: the world stress map project. *Journal of Geophysical Research*, **97**, 11,703–11,728.

Zoback, M.D. 2010. *Reservoir Geomechanics*. Cambridge University Press, New York.

Zoback, M.D., Barton, C.A. *et al.* 2003. Determination of stress orientation and magnitude in deep wells. *International Journal of Rock Mechanics & Mining Sciences*, **40**, 1049–1076.

State of stress in exhumed basins and implications for fluid flow: insights from the Illizi Basin, Algeria

JOSEPH M. ENGLISH[1,2]*, THOMAS FINKBEINER[3], KARA L. ENGLISH[1] & RACHIDA YAHIA CHERIF[4]

[1]*Petroceltic International, 16 Fitzwilliam Place, Dublin 2, Ireland*

[2]*Present address: Stellar Geoscience Limited, Dublin, Ireland*

[3]*King Abdullah University of Science and Technology, Thuwal 23955-6900, Kingdom of Saudi Arabia*

[4]*Direction Coordination Groupe Associations – Sonatrach, Djenane El-Malik, Hydra, Algiers, Algeria*

**Correspondence: je@stellargeoscience.com*

Abstract: The petroleum prospectivity of an exhumed basin is largely dependent on the ability of pre-existing traps to retain oil and gas volumes during and after the exhumation event. Although faults may act as lateral seals in petroleum traps, they may start to become hydraulically conductive again and enable fluid flow and hydrocarbon leakage during fault reactivation. We constrain the present day *in situ* stresses of the exhumed Illizi Basin in Algeria and demonstrate that the primary north–south and NW–SE (vertical strike-slip) fault systems in the study area are close to critical stress (i.e. an incipient state of shear failure). By contrast, the overpressured and unexhumed Berkine Basin and Hassi Messaoud areas to the north do not appear to be characterized by critical stress conditions. We present conceptual models of stress evolution and demonstrate that a sedimentary basin with benign *in situ* stresses at maximum burial may change to being characterized by critical stress conditions on existing fault systems during exhumation. These models are supportive of the idea that the breaching of a closed, overpressured system during exhumation of the Illizi Basin may have been a driving mechanism for the regional updip flow of high-salinity formation water within the Ordovician reservoirs during Eocene–Miocene time. This work also has implications for petroleum exploration in exhumed basins. Fault-bounded traps with faults oriented at a high angle to the maximum principal horizontal stress direction in strike-slip or normal faulting stress regimes are more likely to have retained hydrocarbons in exhumed basins than fault-bounded traps with faults that are more optimally oriented for shear failure and therefore have a greater propensity to become critically stressed during exhumation.

The relative timing of petroleum charge and trap formation in sedimentary basins is a crucial factor in the formation of petroleum accumulations in the subsurface (e.g. Magoon & Dow 1994). In normally subsiding basins, active petroleum generation in mature organic-rich source rock results in the expulsion of petroleum fluids that migrate under hydrodynamic and buoyancy forces within carrier beds to ultimately accumulate in petroleum traps or escape to the surface (e.g. Schowalter 1979). However, many sedimentary basins around the world are no longer under conditions of maximum burial and have been subjected to one or more exhumation events during their history (Japsen 1998; Issler *et al.* 1999; Doré *et al.* 2002; Corcoran & Mecklenburgh 2005; Underdown & Redfern 2007; Hillis *et al.* 2008; Dixon *et al.* 2010; Green & Duddy 2010; Japsen *et al.* 2010, 2012; K.L. English *et al.* 2016*c*). Although low levels of petroleum generation may continue during the initial cooling phase (K.L. English *et al.* 2016*d*), the exhumation and cooling of source rocks generally leads to the cessation of active petroleum generation within the basin. Unless the exhumed basin is situated along a migration route from an adjacent normally subsiding basin that is actively generating hydrocarbons, post-exhumation charging of petroleum traps can only occur via the redistribution of hydrocarbons already existing within the basin (e.g. Doré *et al.* 2002; K.L. English *et al.* 2016*d*) or via late-stage exhumation charge during final depressurization of the source rock (J.M. English *et al.* 2016). Aside from these potential post-exhumation charge or remigration mechanisms, the petroleum prospectivity of an exhumed basin is largely dependent on the ability of pre-existing traps to retain oil and gas volumes during and after the exhumation event.

From: Turner, J. P., Healy, D., Hillis, R. R. & Welch, M. J. (eds) 2017. *Geomechanics and Geology*. Geological Society, London, Special Publications, **458**, 89–111.
First published online May 30, 2017, https://doi.org/10.1144/SP458.6

Fluid flow in the subsurface is driven by gradients in fluid potential, which can result from variations in excess water pressure, the natural buoyancy of lower density petroleum fluids within water-saturated rock and differences in the capillary pressure required for petroleum fluids to displace water from the rock it is trying to penetrate (England *et al.* 1987). In a petroleum trap, the top-seal represents the lithology that prevents the continued vertical migration of petroleum fluids because the capillary entry pressure of the seal exceeds the upwards buoyancy pressure acting on the petroleum phase underneath. In addition, faults may act as lateral seals in petroleum traps if they juxtapose permeable reservoir rocks against sealing lithologies (Allan 1989; Yielding *et al.* 1997) or if the fault zone itself has become impermeable (e.g. due to clay smearing; Bouvier *et al.* 1989). However, these fault zones may start to become conductive again and accommodate fluid flow during fault reactivation (Jones & Hillis 2003). Active and critically stressed faults and fractures can provide high-permeability conduits for fluid flow (e.g. Sibson 1994; Barton *et al.* 1995; Sibson 2000; Finkbeiner *et al.* 2001; Wiprut & Zoback 2002; Rogers 2003).

A recent subsurface study based on fluid inclusion and present day formation water chemistry data from Ordovician sandstones has indicated that long-distance lateral brine migration and mixing have occurred within the Illizi Basin, Algeria (K.L. English *et al.* 2016*b*). The chemistry of the formation water in the Ordovician sandstone is consistent with derivation from a Triassic–Liassic halite-bearing sequence deposited in the Berkine Basin, which is located >400 km to the north (Fig. 1). The fluid inclusion record confirms that an early stage, low-salinity brine in the Ordovician sandstones in the Illizi Basin was displaced by a higher salinity fluid (K.L. English *et al.* 2016*b*, *d*) during *c.* 1.0–1.4 km of exhumation that occurred during

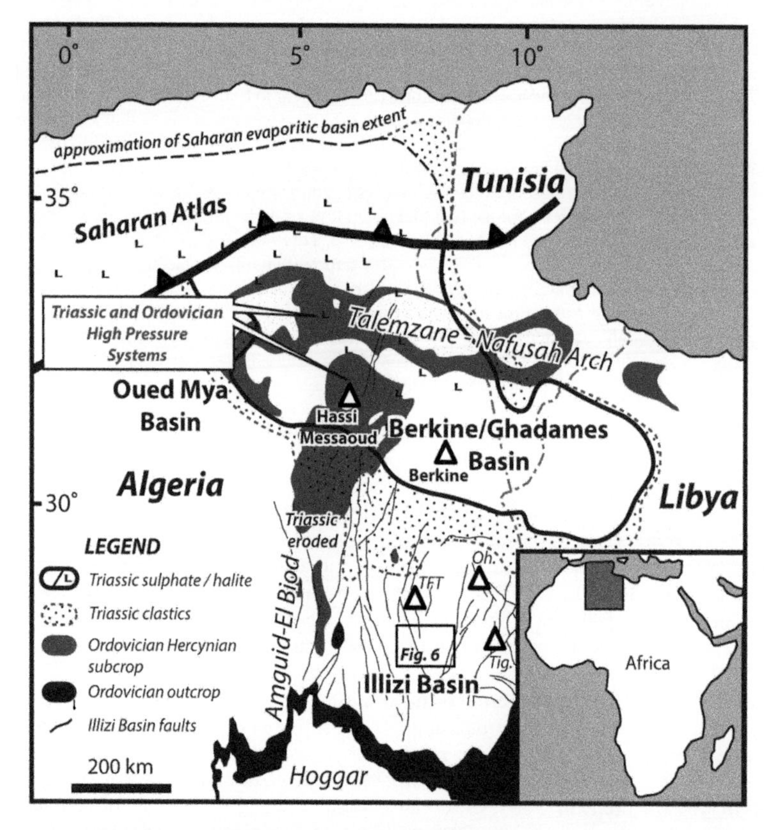

Fig. 1. Location of the study area in the Illizi Basin, Algeria (modified from K.L. English *et al.* 2016*b*). Rectangular black box in the Illizi Basin shows the outline of the study area in Figure 6. The vertical strike-slip faults observed in the Illizi Basin correspond to north–south-, NNW- and NNE-oriented Pan-African basement faults. The white-filled triangles show the locations of the fields discussed in the text. Oh., Ohanet; Tig., Tiguentourine; TFT, Tin Fouyé Tabankort.

the Eocene–Miocene (K.L. English *et al.* 2016*c*) (Fig. 2). Potentiometric maps indicate the presence of an active gravity-flow hydrodynamic regime on the southern flank of the exhumed Illizi Basin and an overpressured system beneath the Triassic salts of the unexhumed Berkine Basin to the north (Chiarelli 1978). It has been proposed that the release of fluid overpressure during exhumation of the Illizi Basin may have been a crucial contributor to updip fluid flux (K.L. English *et al.* 2016*b*).

From a structural perspective, the Illizi and Berkine basins are generally characterized by vertical north–south-, NNW- and NNE-oriented Pan-African basement faults (Fig. 1), whereas NNE- to NE-trending basement-involved high-angle Triassic normal faults are locally important in the Berkine Basin. The long-lived vertical strike-slip faults have been reactivated at various points during the geological evolution of the basin with episodes of transpressional or transtensional deformation depending on the orientation of the faults relative to the far-field tectonic stresses at the time (Galeazzi *et al.* 2010). Herein, we constrain the present day *in situ* stresses of the Illizi Basin and establish whether the primary (vertical strike-slip) fault systems within the basin are likely to be critically stressed. We compare these findings with the *in situ* stresses that have been described from the overpressured Berkine Basin to the north and investigate conceptually whether exhumation of an overpressured sedimentary basin is likely to reactivate pre-existing fault systems and facilitate basin-scale fluid flow during depressurization. The implications for petroleum system analysis will also be discussed.

Approach

Subsurface *in situ* stresses can be described in terms of three orthogonal principal stress components denoted as the maximum principal stress (S_1), the intermediate principal stress (S_2) and the minimum principal stress (S_3), which, by definition, are related as follows: $S_1 \geq S_2 \geq S_3$. When one of the principal stresses is vertical, the three orthogonal principal stress components can also be represented by the vertical stress (S_v), the minimum horizontal stress (S_{hmin}) and the maximum horizontal stress (S_{Hmax}). Anderson (1951) described three stress regimes defined based on relative magnitudes: normal faulting ($S_v \geq S_{Hmax} \geq S_{hmin}$); strike-slip faulting ($S_{Hmax} \geq S_v \geq S_{hmin}$); and reverse faulting ($S_{Hmax} \geq S_{hmin} \geq S_v$). The equal signs correspond to transtensional and transpressive environments. The failure of rock in the subsurface is governed by a relationship between the principal effective stresses and a failure criterion. Terzaghi's effective stress definition is adopted in failure criteria and corresponds to the difference between the externally applied stress (S) and the internal pore pressure (P_p) (e.g. Terzaghi 1943; Jaeger *et al.* 2007; Fjær *et al.* 2008). Herein, we denote the maximum, intermediate and minimum effective stresses as σ_1, σ_2 and σ_3, respectively, and use σ_v, σ_{hmin} and σ_{Hmax} when discussing the vertical, minimum horizontal and maximum horizontal effective stresses. We utilize the Mohr–Coulomb failure criterion, whereby instability or shear failure occurs beyond a limiting ratio of maximum to minimum effective stress. This ratio is governed by the frictional properties of a

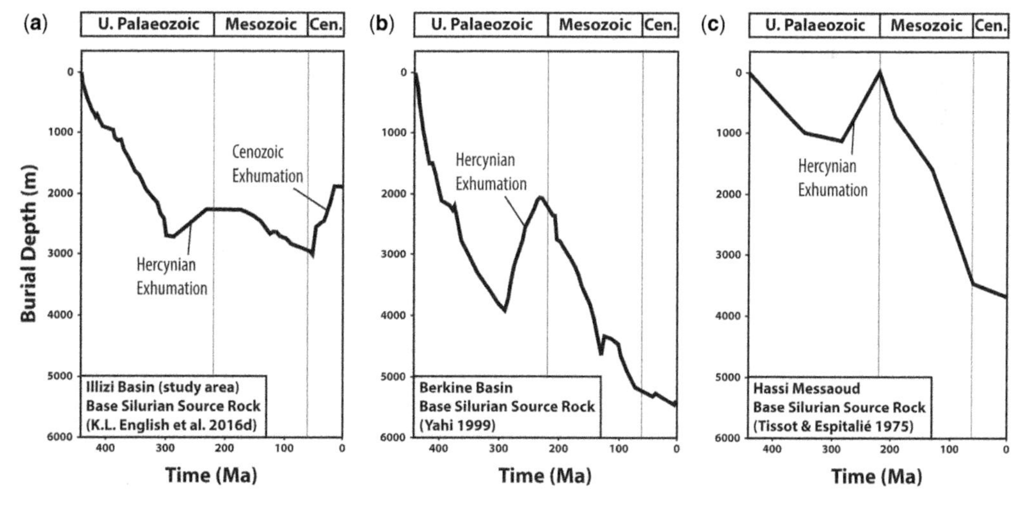

Fig. 2. Burial history models for the Lower Silurian source rock in each of the areas discussed in this study: (**a**) this study, Illizi Basin (K.L. English *et al.* 2016*d*); (**b**) Berkine area (Yahi 1999); and (**c**) Hassi Messaoud field (Tissot & Espitalié 1975). The locations of these areas are shown in Figure 1.

pre-existing, cohesionless and optimally oriented fault zone (referenced and discussed in greater detail in the following sections).

To evaluate the probability of fluid leakage along bounding faults and/or leakage through top-seals, it is important to understand the burial and exhumation history of the basin, in addition to the present day *in situ* pore pressure and stress state. A variety of techniques for estimating the magnitude and timing of exhumation have been studied and are well accepted (e.g. Corcoran & Doré 2002). These techniques provide a reliable understanding of a particular basin's burial history as well as the timing of petroleum generation and charging of petroleum traps (e.g. the Illizi Basin; K.L. English *et al.* 2016*d*). The pore fluid pressure (P_p) and total overburden stress (S_v) are generally well constrained in subsurface wellbores because adequate measurements and logging runs are often available and can be readily interpreted. The orientation of S_{Hmax} is available in many basins through the World Stress Map project (Heidbach *et al.* 2009) and can also be locally constrained through the analysis of well logs that detect characteristic features (e.g. borehole breakouts, drilling-induced tensile wall fractures, acoustic anisotropy) unambiguously induced by the *in situ* far-field stress state.

The most reliable and direct measurements of the *in situ* minimum principal stress (S_3) magnitude are those derived from the analysis of injection tests (e.g. leak-off tests, extended leak-off tests, minifracs, wireline fracs, diagnostic formation injection tests and step-rate tests). When properly conducted, ideally with a downhole pressure gauge, these tests measure the downhole fluid pressures required to create and propagate hydraulically induced fractures and also the pressures under which these newly created fractures close. The instantaneous shut-in pressure (ISIP) is, by definition, the pressure during hydraulic fracturing immediately after shut-in. It is generally lower than the fracture propagation pressure, but greater than the fracture closure pressure (FCP). The ISIP can be considered as a reliable upper bound for the magnitude of S_3. However, this pressure can be significantly in excess of S_3 depending on the treatment and the rock type. In particular, in tight formations such as the Ordovician sandstone in this study area, the ISIPs are generally above the interpreted FCPs. The FCP is the fluid pressure that counteracts the stress in the rock perpendicular to the plane of the induced fracture at the point of closure. Hence the FCP is considered to be equal to, or slightly lower than, the magnitude of S_3; it provides an estimate of S_{hmin} in normal faulting and strike-slip faulting stress regimes and an estimate of S_v in a reverse faulting stress regime. The FCP is interpreted after monitoring the pressure diffusion as a function of time after the well is shut-in (i.e. no flow). A detailed summary of the various regression methods for interpreting the FCP is provided by Barree *et al.* (2007).

In contrast with the minimum principal stress, the magnitude of S_{Hmax} cannot be directly measured. One approach for constraining the magnitude of S_{Hmax} is based on frictional limits theory, which states that the ratio of the maximum effective stress (σ_1) to the minimum effective stress (σ_3) is limited by the magnitude required to cause faulting on pre-existing, optimally oriented fault planes (Zoback & Healy 1984; Townend & Zoback 2000). The Mohr–Coulomb frictional sliding criterion on an optimally oriented fault is given by (e.g. Jaeger *et al.* 2007; Zoback 2007):

$$\frac{\sigma_1}{\sigma_3} = \frac{(S_1 - P_p)}{(S_3 - P_p)} \leq \left(\sqrt{(1 + \mu^2)} + \mu\right)^2 \qquad (1)$$

where μ is the coefficient of friction. Essentially, the Mohr–Coulomb criterion states that for any given magnitude of σ_3, there is a maximum limit to the magnitude of σ_1 established by the frictional strength of the pre-existing faults. Based on a global compilation of *in situ* stress measurements, the ratio of the maximum and minimum effective stresses (σ_1/σ_3) is observed to correspond to a crust in equilibrium with, or at least limited by, frictional failure with μ in the range 0.6–1.0 (Zoback & Healy 1984; Townend & Zoback 2000). This is the same range for the coefficient of friction that has been observed in laboratory experiments (Byerlee 1978). If we assume a typical value of 0.6 for μ, the limiting ratio of the maximum and minimum effective stresses (σ_1/σ_3) is equal to 3.1. Utilizing this hypothesis and Anderson's classification system for faulting, it is possible to construct a stress polygon (Fig. 3) that constrains the range of possible stress states and magnitudes at a particular depth and given pore pressure (Zoback *et al.* 1986; Moos & Zoback 1990). The boundaries of the stress polygon define the maximum allowable differential stress (for a given pore pressure), limited by the strength of pre-existing faults within the crust. The region within the stress polygon captures the range of possible stress states for each particular faulting regime (Fig. 3).

For the purposes of our study, we modify the stress polygon in Figure 3 by plotting the horizontal principal effective stresses normalized by the vertical effective stress and hence construct an equivalent effective stress ratio (ESR) polygon. The ESR polygon enables us to depict stress and pressure data from various depths and locations within the same diagram. As stated earlier, the inherent assumption with the stress polygon approach is that the state of stress is limited by faults that are

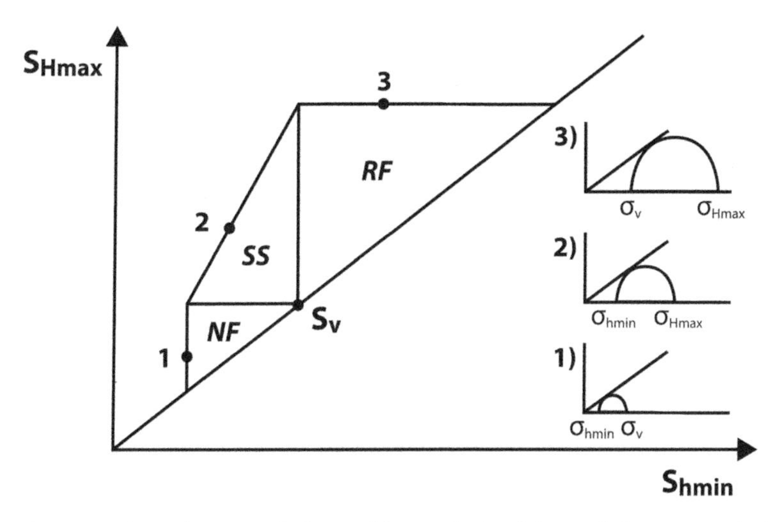

Fig. 3. Stress polygons can be used to constrain the possible magnitudes of S_{hmin} and S_{Hmax} at a particular depth using Anderson's faulting theory and the Coulomb faulting theory for a given coefficient of sliding friction and pore pressure (modified from Zoback *et al.* 2003). The perimeter of the stress polygon delineates the onset of reactivation on optimally oriented faults. NF, normal faulting; RF, reverse faulting; SS, strike-slip faulting.

optimally oriented for frictional shear failure in the present day stress regime. In this study, we also assess how the frictional limit will vary as a function of fault angle. This becomes relevant when the pre-existing faults are sub-optimally oriented for reactivation, but may still act to limit the maximum differential effective stress. Hence, in addition to constructing the perimeter of the ESR polygon as a function of μ, we can also use the angle β (Fig. 4), which is defined as the angle between the normal of the sliding fault plane and the maximum principal stress direction. In fact, the parameters μ and β are closely related via the following equation (e.g. Jaeger *et al.* 2007; Fjær *et al.* 2008):

$$\beta = \frac{\pi}{4} + \frac{\tan^{-1}\mu}{2} \qquad (2)$$

Geomechanical characterization of the Illizi Basin

The Illizi and Berkine basins in southeastern Algeria contain >6000 m of Palaeozoic–Mesozoic strata (Echikh 1998; Dixon *et al.* 2010; Galeazzi *et al.* 2010). This study focuses on geomechanical data collected from a large (80 × 50 km) Ordovician gas condensate field in the southern Illizi Basin (Fig. 1), where intraplate uplift resulted in *c.* 1.0–1.4 km of exhumation during the Eocene–Miocene (K.L. English *et al.* 2016*c*). This field is a large, low-relief, four-way structural closure with a hydrocarbon column in excess of 100 m. The reservoir is Upper Ordovician (Ashgill) glacial to glacio-marine Unit IV sandstone and this sequence is capped by lower Silurian mudrocks, which provide both the source rock and seal (K.L. English *et al.* 2016*d*). Other Ordovician fields in the Illizi Basin include the Tin Fouyé Tabankort, Ohanet and Tiguentourine fields (Fig. 1).

Mechanical properties

Rock strength testing on core is necessary to constrain the *in situ* stress tensor – in particular, the magnitude of S_{Hmax} – when using the stress polygon approach (Moos & Zoback 1990). The results from a series of uniaxial and triaxial mechanical tests conducted on Silurian shale and Ordovician sandstone cores are summarized in Table 1.

The unconfined compressive strength (C_0) of the Ordovician sandstone exhibits values ranging from 90 to 191 MPa with a mean of 144 (±36) MPa. Even though this is a wide range, these sandstones can be described as strong, primarily due to their well-cemented, low-porosity characteristics (see K.L. English *et al.* 2016*a*). It is noteworthy that the weakest sandstone in the dataset (90 MPa) has a relatively high porosity (11%), whereas the relatively stronger samples have lower porosities (<6%). Table 1 also shows a suite of Ordovician IV-3 sandstone samples (Well D, *c.* 1922 m depth) tested at different confining pressures. The resulting peak strength values enable us to derive a value for C_0 and the coefficient of internal friction (μ_i) using regression analysis. As the resulting peak stress at

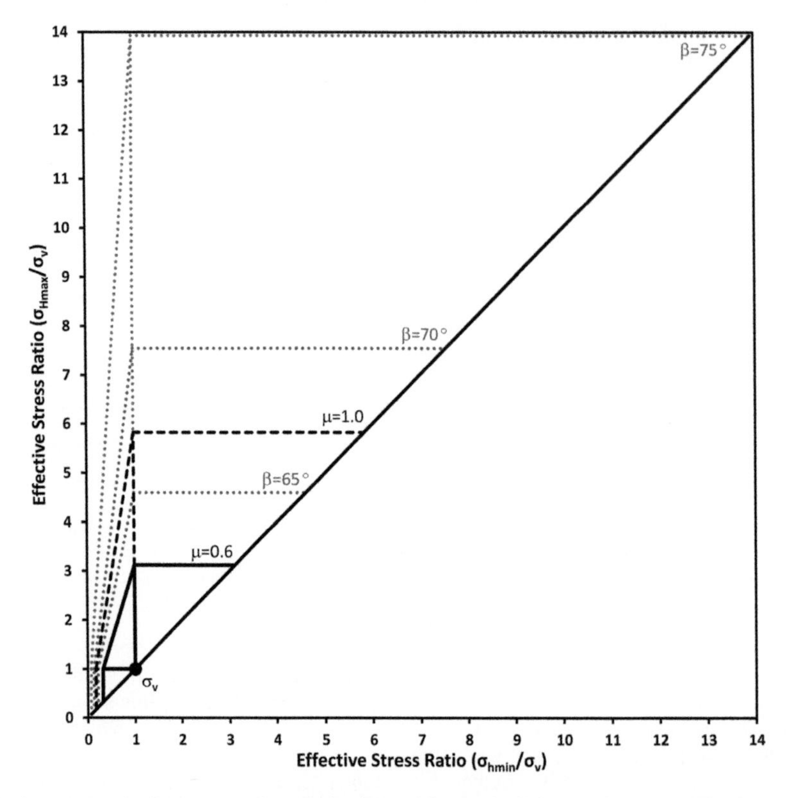

Fig. 4. Effective stress ratio diagrams can be used in place of the standard stress polygon (see Fig. 3) so that data from different depths and with different pore pressures can be plotted together on one diagram. This figure shows several different polygon perimeters: the black solid and dashed lines depict two different coefficients of sliding frictions ($\mu = 0.6$ and $\mu = 1.0$) and the dotted lines show different β angles (measured between the normal of a fault plane and the maximum principal stress direction).

failure for the sample tested at 4.1 MPa confinement appears to be comparatively low, we excluded it from the regression analysis. Using the remaining three samples, we calculated an unconfined compressive strength of $C_0 = 108$ MPa (which is within the range of values obtained from the unconfined tests) and a coefficient of internal friction of $\mu_i = 1.42$ (a typical value for strong sandstones).

By comparison, the confined compressive strength of the Silurian shale is relatively low (29–37 MPa), indicating a much weaker rock. The marked difference in mechanical properties between the Ordovician sandstone and the Silurian shale is also apparent in the measured Young's modulus and Poisson's ratio. Across the uniaxial and triaxial tests (19 measurements), the Young's moduli of the Ordovician sandstone range from 41 to 70 GPa with a mean of 50 (± 7) GPa; Poisson's ratio varies from 0.06 to 0.18 with a mean of 0.12 (± 0.03). By contrast, the Young's moduli of the Silurian shale are much lower, ranging from 6 to 8 GPa, whereas Poisson's ratio is higher, ranging from 0.25 to 0.26 (both based on only two measurements). Hence the Silurian shale is a much softer and more compliant formation than the Ordovician sandstones.

Pore pressure

In general, the Illizi Basin is normally pressured and present day potentiometric maps suggest the northwards hydrodynamic flow of meteoric water from elevated outcrops in the south (Chiarelli 1978). The reservoir pressure within the Ordovician reservoir in the field is 20.2 MPa at a measured depth of 1917 m (−1462 m true vertical depth subsea), which equates to a pressure gradient of 10.6 kPa m^{-1} with respect to the surface. The pore pressure is slightly elevated with respect to a hydrostatic gradient due to the presence of a gas column (Fig. 5) and the presence of high-salinity formation water (K.L. English *et al.* 2016*b*). As the field is not yet in production, virgin pore pressure conditions have been assumed throughout the study.

Table 1. *Rock mechanical data, Illizi Basin, Algeria*

Well	Core depth (m)	Formation	Lithology	Porosity (%)	Type of test	Confining pressure (MPa)	Unconfined compressive strength (MPa)	Confined compressive strength (MPa)	Young's modulus (GPa)	Poisson's ratio
A	1916.32	Ordovician IV-3	Sandstone		Uniaxial	0	130.27		50.06	0.13
A	1916.38	Ordovician IV-3	Sandstone	10.76	Uniaxial	0	90.38		43.10	0.15
A	1916.36	Ordovician IV-3	Sandstone		Triaxial	30		348.53	52.11	0.14
A	1918.40	Ordovician IV-3	Sandstone		Uniaxial	0	191.02		48.71	0.10
A	1918.58	Ordovician IV-3	Sandstone	1.72	Uniaxial	0	155.93		47.82	0.06
A	1918.37	Ordovician IV-3	Sandstone		Triaxial	35		565*	69.50	0.09
A	1927.77	Ordovician IV-2	Sandstone		Uniaxial	0	95.97		41.71	0.13
A	1927.84	Ordovician IV-2	Sandstone	3.68	Uniaxial	0	170.81		50.27	0.08
A	1927.79	Ordovician IV-2	Sandstone		Triaxial	35		370.56	49.00	0.16
A	1947.39	Ordovician IV-2	Sandstone		Uniaxial	0	149.87		44.39	0.13
A	1947.31	Ordovician IV-2	Sandstone	5.72	Uniaxial	0	163.96		49.63	0.18
A	1947.36	Ordovician IV-2	Sandstone		Triaxial	30		452.13	58.42	0.12
D	1920.30	Lower Silurian	Shale		Triaxial	4.1		28.76	7.80	0.26
D	1920.30	Lower Silurian	Shale		Triaxial	10.3		36.73	6.07	0.25
D†	1921.80	Ordovician IV-3	Sandstone		Triaxial	2.1		136.10		0.11
D	1921.83	Ordovician IV-3	Sandstone		Triaxial	4.1		104.73	41.10	0.17
D†	1921.77	Ordovician IV-3	Sandstone		Triaxial	6.9		160.34	43.16	0.11
D†	1921.86	Ordovician IV-3	Sandstone		Triaxial	10.3		220.47	46.47	0.12
D	1927.83	Ordovician IV-2	Sandstone		Triaxial	6.9		94.93	43.16	0.13
D	1935.82	Ordovician IV-1	Sandstone		Triaxial	6.9		230.28	56.12	0.10
D	1942.18	Ordovician IV-1	Sandstone		Triaxial	6.9		290.39	56.81	0.15

*No failure at 35 MPa confining pressure up to a maximum of 576.49 MPa. Failure observed at 565 MPa when confining pressure reduced to 23.57 MPa.
†Using these three samples, we estimate that the unconfined compressive strength (C_0) is 108 MPa and the coefficient of internal friction (μ_i) is 1.42. The sample tested at 4.1 MPa confinement appears comparatively low and was therefore excluded from the regression analysis.

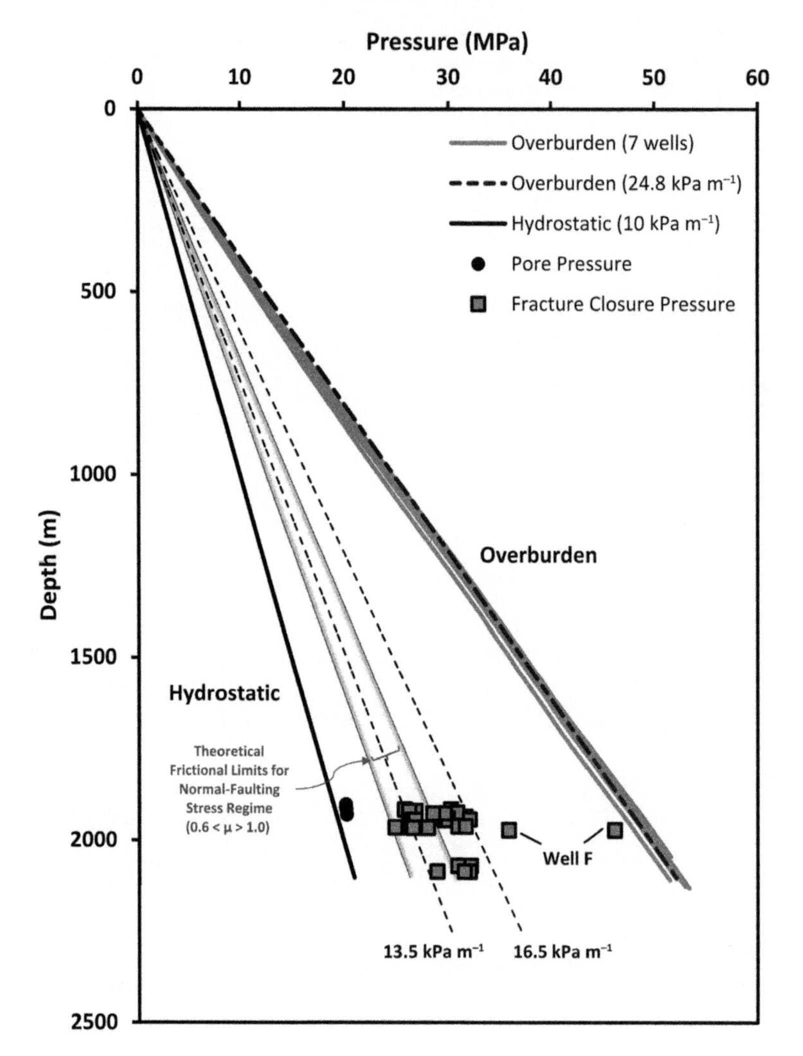

Fig. 5. *In situ* stress and pore pressure data obtained from wells in the study area of the Illizi Basin. The fracture closure pressure and the pore pressure data are all from the Ordovician reservoir section. A detailed breakdown of the fracture closure pressure data is presented in Table 3. Based on the fracture closure pressures, the minimum principal horizontal stress (S_{hmin}) is interpreted to generally be in the 13.5–16.5 kPa m^{-1} range, although some higher magnitudes were observed in Well F. This range for S_{hmin} partially overlaps with the theoretical frictional limits for the normal faulting stress regime assuming that S_v is the maximum principal stress and that μ ranges from 0.6 to 1.0.

Vertical stress magnitude

The total vertical stress or overburden stress (S_v) can be approximated by the weight of the overlying rocks (e.g. Zoback *et al.* 2003; Zoback 2007; Jaeger *et al.* 2007):

$$S_v = \int_0^z \rho g \, dz \tag{3}$$

where ρ is the density of the overburden, z is the depth and g is the gravitational acceleration. Wireline density log measurements from seven vertical wells in the study area were used to estimate the bulk density of the rock column. The bulk density of the shallowest preserved section, where no openhole log measurements were available, was estimated via a linear extrapolation of the underlying section. This was appropriate because this sequence is overcompacted due to deeper burial in the past and no major unconformity exists in the preserved stratigraphic column. The vertical stress profiles obtained for the study area are shown in

Figure 5. The resulting overburden curves are consistent from well to well. At 2 km depth, the average overburden stress gradient is 24.9 kPa m^{-1}, although it predominantly ranges from 24.8 to 25.2 kPa m^{-1} with one outlier at 24.4 kPa m^{-1}. The relatively high magnitudes are a result of the overcompacted nature of the sedimentary rocks due to deeper burial in the past (K.L. English *et al.* 2016*d*). The small variations in vertical stress across the study area may, in part, be a function of the spatial variation in exhumation patterns and in part a function of lateral changes in facies and lithologies through the rock column.

Maximum horizontal stress azimuth

The orientation of the maximum horizontal stress (S_{Hmax}) in the Illizi Basin can be determined using drilling-induced tensile fractures (DITFs) and stress-induced borehole breakouts observed on electrical and/or acoustic image logs. In the study area, image logs are available from 11 wells with coverage exclusively over the basal Silurian shale and Ordovician reservoir sequence. Table 2 presents the statistics (mean and standard error) of the S_{Hmax} azimuth based on both DITFs and breakout orientations from the image logs. The mean azimuths for each well (based solely on the DITFs) are plotted on a Top Ordovician depth structure map in Figure 6. The quality of each data point is assigned based on the criteria of the World Stress Map project (Heidbach *et al.* 2009). The DITFs represent the predominant wellbore failure feature in the Ordovician sandstones. These features can be generated either by a high overbalance in the well during drilling (i.e. mud weights appreciably above the formation pore pressure) or they are indicative of tectonic environments with a large difference between the magnitude of the two horizontal stresses (e.g. a strike-slip faulting stress state; Fig. 3; Zoback 2007). In this study, the wells were only drilled with a slight overbalance (ΔP = mud pressure–pore pressure ≈ 1.3 MPa), and the common presence of DITFs probably reflects a high differential in the horizontal stresses. The orientation of S_{Hmax} is generally NW–SE to NNW–SSE based on the average values for both the observed DITFs (325° N) and borehole breakouts (320° N) (Table 2). Well C is the only anomaly, where the S_{Hmax} orientation is interpreted to be north–south (Fig. 6; Table 2); the explanation for this stress rotation remains poorly understood, but it is possible that it is purely a local feature related to a minor fault. The average 320° N azimuth for S_{Hmax} (based on DITFs) is regionally consistent when integrated with observations from other Ordovician fields in the Illizi Basin (Fig. 1), including Tin Fouyé Tabankort (320° N, Donati *et al.* 2016) and Tiguentourine (336° N,

Table 2. *Maximum horizontal stress orientation in Ordovician Unit IV, Illizi Basin, Algeria*

Well	Based on drilling-induced tensile fractures				Based on borehole breakouts			
	Mean S_{Hmax} orientation (°N)	Standard deviation (°)	Count	Quality	Mean S_{Hmax} orientation (°N)	Standard deviation (°)	Count	Quality
A	323.5	1.4	810	C				
B	312.1	1.3	495	C	308.3	8.7	10	D
C	001.8	1.7	557	C	015.5	5.6	59	B
D	321.0	1.0	1039	A	321.4	6.7	6	D
E	313.5	2.5	143	D				
Ez	326.9	2.0	85	C				
F	324.8	2.3	168	B	327.2	18.2	7	D
G	335.2	1.1	693	C	337.1	5.0	8	D
H	332.6	1.7	434	A	319.9	13.0	8	D
I	321.4	1.3	78	B				
S	339.5	3.3	6	D				
T	321.2	7.1	8	D	306.9	17.3	8	D
	Mean	324.7				320.1		

Quality criteria: A, azimuth believed to be within ±15°, at least 10 distinct events with a combined length ≥100 m and s.d. ≤12°; B, azimuth believed to be within ±15–20°, at least six distinct events with a combined length ≥40 m and s.d. ≤20°; C, azimuth believed to be within ±20–25°; at least four distinct events with a combined length ≥20 m and s.d. ≤25°; D, questionable azimuth (±25–40°); less than four distinct events with a combined length <20 m and s.d. ≤40°; and E, no reliable information (>±40°), no reliable event or s.d. >40°.

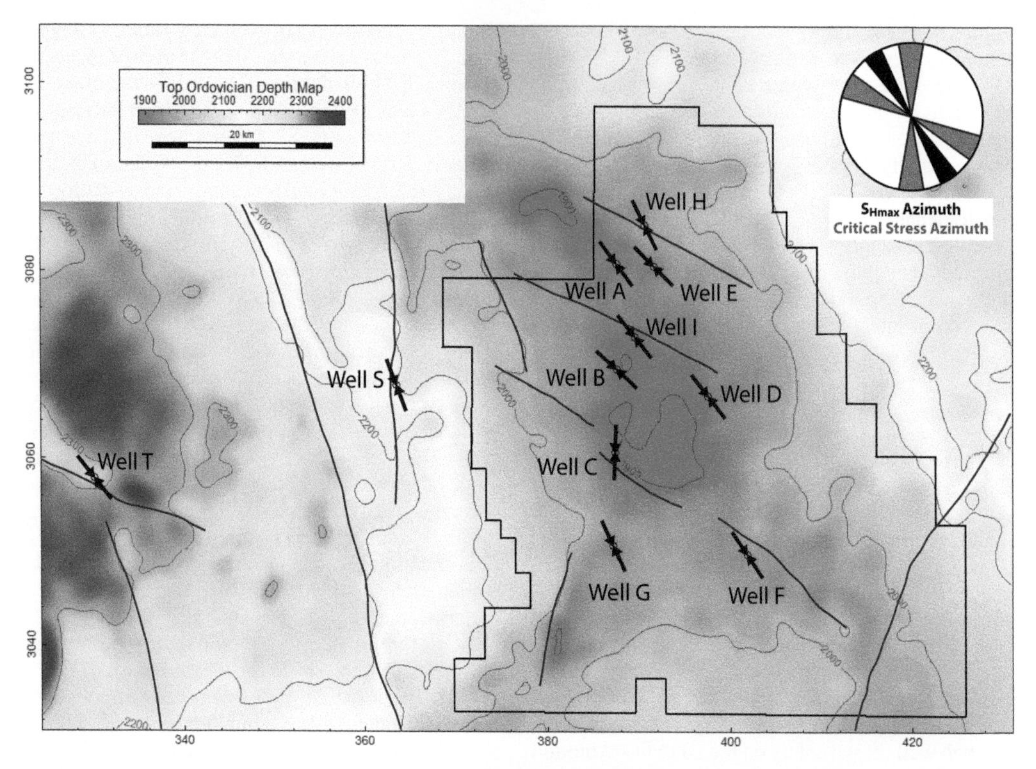

Fig. 6. Depth map of the Top Ordovician within the study area in the Illizi Basin (see map outline in Fig. 1). The field area is situated on a major structural high within the basin and is outlined by the black polygon. The inwardly pointing black arrow pairs mark the mean azimuth of the maximum principal horizontal stress (S_{Hmax}) as determined from drilling-induced tensile fractures in image logs acquired at the well locations labelled A–I and T and S. The map also shows major, interpreted (vertical strike-slip) faults as dark grey lines; these traces indicate north–south and NW–SE trends oriented at *c.* 30° to the average S_{Hmax} azimuth. The rose diagram in the upper right-hand corner illustrates the mean S_{Hmax} azimuth from the image log analysis (Table 2) in black and the corresponding azimuths in grey for fault systems to be critically stressed in the present day stress regime assuming sliding friction values between 0.6 and 1.0, corresponding to β angles between 62 and 67° (see equation 2).

Patton *et al.* 2003), and it is broadly perpendicular to the Atlas Front to the north (Fig. 1). When comparing the observed S_{Hmax} azimuths with the orientation of the major fault systems in the Illizi Basin (Figs 1 & 6), we conclude that these faults often fall within a range of 30–40° from the regional S_{Hmax} azimuth. The significance of this observation will be discussed in the following sections.

Minimum horizontal stress magnitude

A series of 32 injection tests from eight wells were analysed in this study and enable us to constrain S_3 for the Ordovician reservoir in the study area. Note that for normal and strike-slip tectonic environments, the minimum principal stress is the minimum horizontal total stress (i.e. $S_3 = S_{hmin}$). A variety of different types of injection tests were used in the area, including minifracs, step-rate tests, diagnostic injection formation tests (DFITs) and tests run using the Reservoir Characterization Instrument (Table 3). All tests were executed successfully and the acquired data are of satisfactory to very good quality and suitable for detailed analysis. The ISIP, fracture propagation pressure and FCP were interpreted in each of these tests – the latter using both square root (time) and Nolte-G derivative analyses (Fig. 7) (see detailed description of analytical techniques in Barree *et al.* 2007). Maximum and minimum FCP picks were also identified during the G-function and square root (time) derivative analyses of the pressure fall-off data to characterize the range of possible values (Fig. 7) and a quality index was assigned to each data point (Table 3). Based on the FCP points, values for the S_{hmin} gradient exhibit a wide range, from as low as 12.7 kPa m^{-1} to as high as 23.4 kPa m^{-1}

Table 3. *Minimum horizontal stress magnitude in Ordovician Unit IV, Illizi Basin, Algeria*

Well	Depth (m TVD)	Unit	Lithology	Type of test	FPP (MPa)	ISIP (MPa)	FCP (MPa)	S3 (=S_{hmin}) gradient (kPa m^{-1})	FCP range (MPa)	S3 gradient range(kPa m^{-1})	Quality
A	1915–1919	IV-3	Sandstone	SRT	38.8	35.1	30.3	15.8	29.7–31.0	15.5–16.2	A
				Minifrac	39.7	34.1	25.9	13.5	24.3–27.8	12.7–14.5	B
B	1920–1922	IV-3	Sandstone	DFIT	n/a	31.5	26.9	14.0	26.6–27.2	13.8–14.2	A
				SRT	n/a	n/a	26.9	14.0	26.6–27.2	13.8–14.2	A
				Minifrac	n/a	31.5	26.3	13.7	25.9–26.6	13.5–13.8	B
C	1936–1938	IV-2	Sandstone	DFIT	n/a	40.8	29.7	15.3	28.3–32.4	14.6–16.7	C
				SRT	n/a	n/a	30.3	15.6	29.7–31.0	15.3–16.0	A
				Minifrac	n/a	43.7	31.7	16.4	31.7–37.9	16.4–19.6	B
D	1926	IV-2	silt/sand	RCI	30.9	n/a	n/a	16.0	n/a	n/a	C
D	1941	IV-1	Sandstone	RCI	n/a	n/a	29.1	15.0	29.0–31.0	14.9–16.0	B
				RCI	n/a	n/a	30.0	15.5	29.7–31.0	15.3–16.0	B
D	1946–1947	IV-1	Sandstone	SRT	n/a	n/a	29.7	15.3	29.0–30.3	14.9–15.6	A
				Minifrac	n/a	n/a	30.0	15.4	29.7–30.3	15.2–15.6	A
D	2071	III-3	Sandstone	RCI	n/a	n/a	32.2	15.5	30.0–33.4	14.5–16.2	B
				RCI	n/a	n/a	31.0	15.0	28.6–31.0	13.8–15.0	B
D	2086–2087	III-3	Sandstone	DFIT	n/a	33.2	32.1	15.4	31.6–32.4	15.1–15.5	A
				SRT	n/a	n/a	29.0	13.9	28.6–30.0	13.7–14.4	B
				Minifrac	31.7	n/a	31.6	15.1	31.4–31.7	15.0–15.2	B
Ez	1924–1933	IV-1	Sandstone	SRT	36.9	35.8	29.7	15.4	28.8–30.3	14.9–15.7	A
				Minifrac	36.6	36.0	28.6	14.8	28.3–29.3	14.7–15.2	A
Ez	1935–1954	IV-1	Sandstone	SRT	n/a	27.2	26.6	13.7	26.0–27.6	13.4–14.2	B
				Minifrac	n/a	30.6	26.9	13.8	26.6–27.2	13.7–14.0	A
Ez	1955–1980	IV-1	Sandstone	SRT	n/a	27.9	26.7	13.6	26.2–27.2	13.3–13.8	B
				Minifrac	n/a	29.2	28.1	14.3	28.0–28.3	14.2–14.4	B
F	1973–1974	IV-2	Sandstone	DFIT	57.9	53.8	46.2	23.4	44.8–47.6	22.7–24.1	A
				Minifrac	64.8	55.6	35.9	18.2	35.2–45.5	17.8–23.1	B
G	1931–1958	IV-3	Sandstone	DFIT2	n/a	67.0	32.1	16.5	31.9–32.2	16.4–16.6	A
G	1958–1975	IV-3	Sandstone	DFIT3	37.1	35.8	31.0	15.8	30.3–31.7	15.5–16.2	A
				SRT	n/a	n/a	31.7	16.1	31.4–32.4	16.0–16.5	A
				Minifrac	42.2	38.0	n/a	n/a	n/a	n/a	C
H	1950–1971	IV-3	Sandstone	SRT	n/a	n/a	25.2	12.8	24.5–26.2	12.5–13.3	A
				Minifrac	n/a	n/a	25.0	12.7	23.7–25.3	12.1–12.9	B

DFIT, diagnostic fracture injection test; FCP, fracture closure pressure; FPP, fracture propagation pressure; ISIP, instantaneous shut-in pressure; RCI Reservoir Characterization Instrument; SRT, step-rate test; TVD, true vertical depth.

Quality criteria: A, pressure points clearly interpretable; B, pressure points less clear; C, no pressure points interpretable.

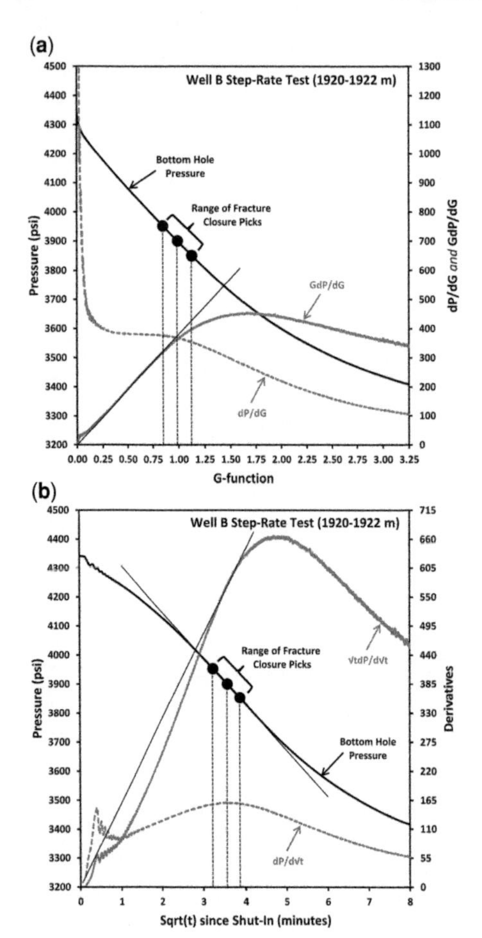

Fig. 7. Example of fracture closure pressure (FCP) interpretation from leak-off test in Well B. The FCP was interpreted in each test using both Nolte-G and square root (time) derivative analysis (see Table 3). A detailed summary of these approaches is provided by Barree *et al.* (2007). (**a**) On the G-function plot, fracture closure is identified at the departure of the semi-log derivative ($G\,dP/dG$) from the straight line through the origin. During normal leak-off prior to closure, the first derivative (dP/dG) should be constant and the primary pressure v. G-function curve should be a straight line. (**b**) On the square root (time) plot, fracture closure occurs at the inflection point on the pressure v. square root (time) curve; this is most readily identified by locating the maximum amplitude on the first derivative ($dP/d\sqrt{t}$) curve. The slope of the primary pressure curve starts low, increases to a maximum at the inflection point where fracture closure occurs, and then starts to decrease again. Fracture closure can also be identified at the departure of the semi-log derivative ($\sqrt{t}\,dP/d\sqrt{t}$) from the straight line through the origin. A high confidence fracture closure pressure interpretation (quality rank A as per Table 3) should be consistent on both the G-function and square root (time) plots – as is the case in the example shown.

(Table 3). This range for S_{hmin} partially overlaps with the theoretical frictional limits for the normal faulting stress regime assuming that S_v is the maximum principal stress and that μ ranges from 0.6 to 1.0. The highest gradients were interpreted in Well F (18.2 and 23.4 kPa m^{-1}), with the highest value being within 6% of the average overburden gradient at 2 km depth. The origin of the higher stress in Well F is poorly understood. The highest FCP pick in Well F comes from the lower injection volume DFIT, whereas the subsequent higher injection volume minifrac data display some evidence for multiple closure events in the fall-off data (see wide range of FCP estimates in Table 3). Therefore one possible explanation for the anomalously high FCP is that the initial DFIT was detecting the closure of a natural fracture that is not oriented parallel to S_{Hmax} and hence this FCP does not give a representative estimate of S_{hmin}. All the other wells indicate minimum principal horizontal stresses that are appreciably lower, with gradients generally in the range 13.5–16.5 kPa m^{-1} (Fig. 5).

The overall clustering of the S_{hmin} gradients in the range 13.5–16.5 kPa m^{-1} is in line with observations from other Ordovician fields in the Illizi Basin. Patton *et al.* (2003) reported a S_{hmin}/S_v ratio of 0.59 for the Tiguentourine field, which would have the typical S_{hmin} gradients in this field in the range 14.5–15.0 kPa m^{-1}. Based on 20 different tests in the Tin Fouyé Tabankort field, the fracture gradient is reported to range from 12.2 to 31.7 kPa m^{-1} (Baylocq *et al.* 1998); excluding the two highest and anomalous measurements, the average of the remaining 18 measurements is 16.0 kPa m^{-1}.

Maximum horizontal stress magnitude

The DITFs observed in the image logs from the vertical wells are mostly (sub-) axial (Fig. 8) and hence we infer that the overburden is a principal stress axis in the Ordovician reservoir of the Illizi Basin. From this, it follows that the *in situ* stress tensor consists of four independent parameters: one vertical stress magnitude; two orthogonal horizontal stress magnitudes; and one angle. The constraints on the magnitudes of S_v and S_{hmin} and the orientation of S_{Hmax} have been discussed earlier; in this section we discuss the available constraints on the magnitude of S_{Hmax}.

To constrain the magnitude of S_{Hmax} using the stress polygon approach, the distribution of stress-induced failures as a function of depth is required. DITFs are very common across the Ordovician section in the vertical wells studied (Fig. 8), while there are very few of these in the Silurian section. Where DITFs are present, a minimum constraint

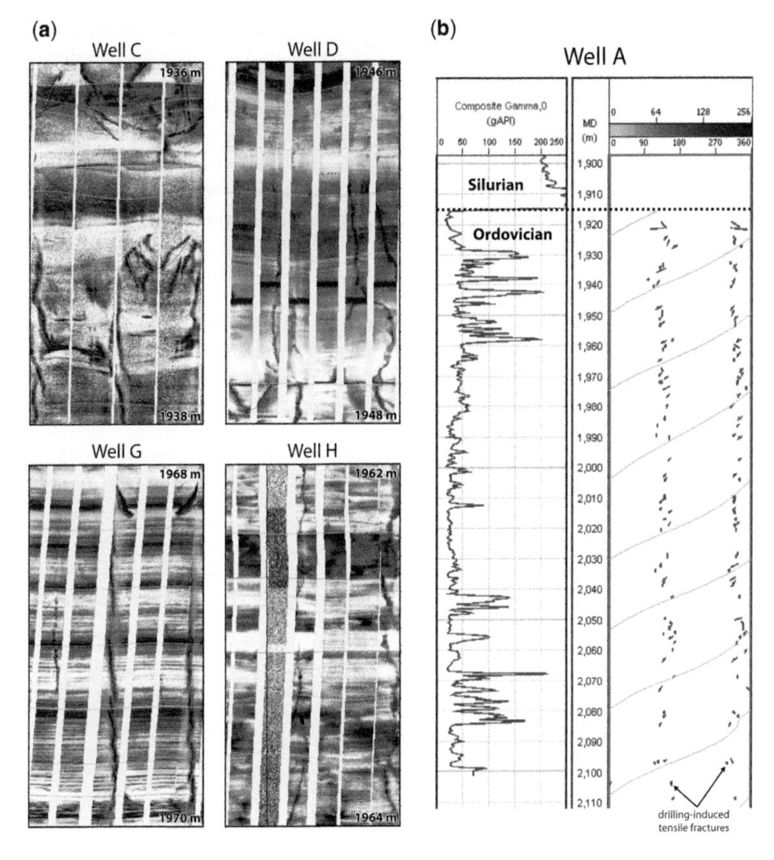

Fig. 8. (**a**) Unwrapped resistivity images showing drilling-induced tensile fractures as commonly observed in the Ordovician section of the vertical wells in the study area. All the examples shown here correspond to depth intervals where the fracture closure pressure was measured (Table 3). (**b**) Gamma ray (left track) and observed drilling-induced tensile fractures (blue ticks) as a function of depth in Well A across the entire imaged interval (right track). The green line denotes the Pad-1 azimuth, indicating that the imaging tool was rotating. This plot demonstrates how ubiquitous drilling-induced fractures are in the Ordovician sandstone.

on the magnitude of S_{Hmax} can be derived using the following equation (Zoback 2007):

$$S_{\mathrm{Hmax}} = 3S_{\mathrm{hmin}} - 2P_{\mathrm{p}} - \Delta P - T_0 - \sigma^{\Delta T} \quad (4)$$

where ΔP is the difference between the mud weight pressure and the formation pore pressure, T_0 is the tensile strength of the rock and $\sigma^{\Delta T}$ is the thermal stress. In terms of sign convention, T_0 is a negative quantity. The thermal stress is given by:

$$\sigma^{\Delta T} = \frac{\alpha_{\mathrm{T}} E \Delta T}{1 - \nu} \quad (5)$$

where α_{T} is the coefficient of linear thermal expansion, E is Young's modulus, ν is Poisson's ratio and ΔT is the temperature differential between the mud and the formation (this is a positive term when the mud temperature is lower than the formation temperature).

In contrast with the DITFs, only very few breakouts are interpreted within the wells studied. The exception is Well C, where breakouts are more common, and this also coincides with an anomalous S_{Hmax} azimuth (Table 2; Fig. 6). Based on the general absence of breakouts in the Ordovician sections, a maximum constraint on the magnitude of S_{Hmax} can be derived using the following equation (Barton *et al.* 1988; Zoback *et al.* 2003):

$$S_{\mathrm{Hmax}} = \frac{(C_{\mathrm{eff}} + 2P_{\mathrm{p}} + \Delta P + \sigma^{\Delta T}) - S_{\mathrm{hmin}}(1 + 2\cos 2\theta_{\mathrm{b}})}{1 - 2\cos 2\theta_{\mathrm{b}}} \quad (6)$$

where C_{eff} is the effective compressive strength of the rock and the angle $2\theta_{\mathrm{b}}$ is related to the width of the breakout (w_{bo}) and is given by:

$$2\theta_b = \pi - w_{bo} \quad (7)$$

Applying the linearized Mohr–Coulomb criterion, the effective compressive strength is given by:

$$C_{eff} = \Delta P[(\mu_i^2 + 1)^{1/2} + \mu_i]^2 + C_0 \quad (8)$$

where μ_i is the internal friction coefficient and C_0 is the unconfined compressive strength.

As most of the pore pressure and S_{hmin} data come from a narrow depth range, we constructed a single representative stress diagram for the Ordovician reservoir in the study area (Fig. 9). The various parameters assumed in the construction of the stress diagram are given in Figure 9. The input range of S_{hmin} is based on the typical gradients of 13.5–16.5 kPa m^{-1} (Fig. 5). The fact that the DITFs coincide with the interval where minifracs, step-rate tests, DFITs or Reservoir Characterization Instrument tests were conducted (Fig. 8) provides robust constraints for the stress polygon analysis. It is important to note that the image logs were run prior to conducting these tests.

The resulting range of possible S_{Hmax} magnitudes in the Ordovician sandstones is highlighted by shading and is bounded by (1) the onset of tensile failure for rock with zero tensile strength and (2) the onset of shear failure (breakout) for rock with a compressive strength of 90 MPa. This highlighted field corresponds to a set of conditions where DITFs are predicted in the Ordovician section, but with no expected breakout in rock with a compressive strength >90 MPa (a representative minimum

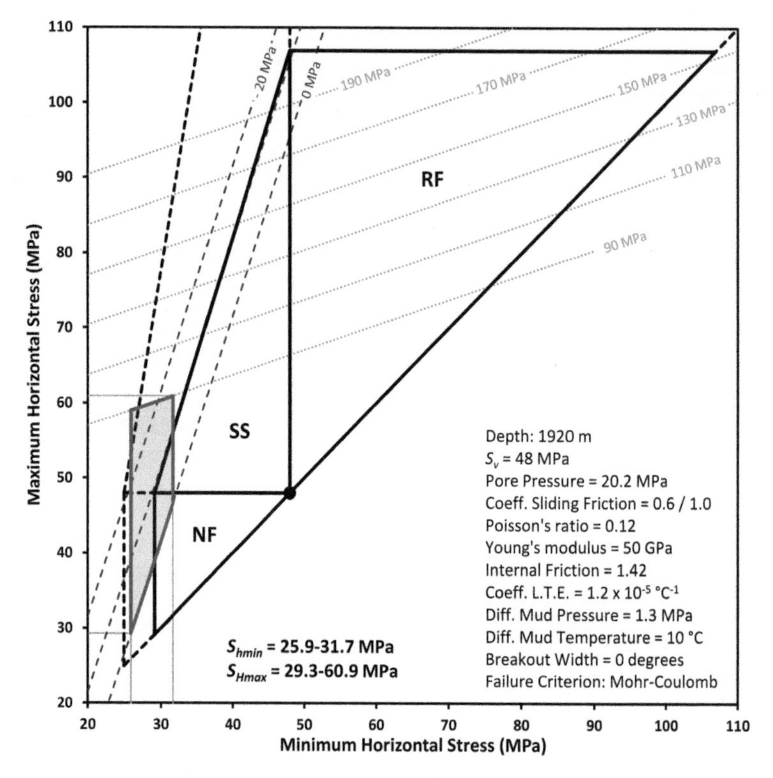

Fig. 9. Two stress polygons for the Ordovician reservoir in the study area. The polygon with the solid black lines is constructed assuming a sliding friction coefficient $\mu = 0.6$ (equivalent to a fault normal angle with respect to S_{Hmax} azimuth of $\beta = 62°$), whereas the one with the dashed black lines corresponds to $\mu = 1.0$, equivalent to $\beta = 67°$. The range of possible horizontal stress magnitudes is highlighted by the shaded polygon and is bound by (1) a representative S_{hmin} range of 13.5–16.5 kPa m^{-1}, (2) the observed tensile failure with an assumed tensile strength (purple dashed lines) of as low as zero and (3) the absence of stress-induced wellbore wall shear failure (i.e. breakouts) for rock with a compressive strength (green dotted lines) as low as 90 MPa (as per the rock mechanics test results in Table 1). Hence the highlighted field corresponds to an *in situ* stress state consistent with the occurrence of drilling-induced tensile fractures, but no breakouts, and the set of parameter conditions listed on the lower right. NF, normal faulting; RF, reverse faulting; SS, strike-slip faulting.

value for the Ordovician sandstones based on Table 1). Based on this analysis, S_{Hmax} magnitudes can range from 25.9 (equal to S_{hmin}) to 62.0 MPa (Fig. 9), corresponding to S_{Hmax} gradients of 13.5–32.3 kPa m^{-1}. The corresponding S_{Hmax}/S_v ratios are 0.54 and 1.29 and hence the allowable stress conditions straddle the boundary between the normal faulting and strike-slip faulting regimes (Fig. 9).

Critical stress in the exhumed Illizi Basin

Based on the data presented in this study, the stress state for this basin can be described as either strike-slip faulting ($S_{Hmax} \geq S_v > S_{hmin}$) or normal faulting ($S_v \geq S_{Hmax} > S_{hmin}$). The stress state is close to the perimeter of the polygon for sliding friction coefficients (μ) between 0.6 and 1.0, equivalent to β angles between *c.* 60 and 70° (Fig. 10). These β angles correspond to fault traces at an angle of 30–35° to S_{Hmax}. Taking the typical S_{Hmax} azimuth as 320–335°, critically stressed vertical (strike-slip) faults and fractures strike at either 285–305 or 350–010°. Comparing these results with local fault traces and their angular relationship to the mapped S_{Hmax} azimuths (Fig. 6), we conclude that some of the major vertical (strike-slip) faults in the area may be critically stressed under the present day stress field in the Illizi Basin. Similarly oriented faults are also present elsewhere in the basin (Fig. 1) and the ESRs described from the Tiguentourine field (S_{hmin}/S_v ratio 0.59 and S_{Hmax}/S_v ratio 1.04) (Patton *et al.* 2003) are consistent with those in our study (Fig. 10). Hence we conclude that some of the north–south- and NW–SE-trending vertical (strike-slip) faults in the basin are likely to be in a state of

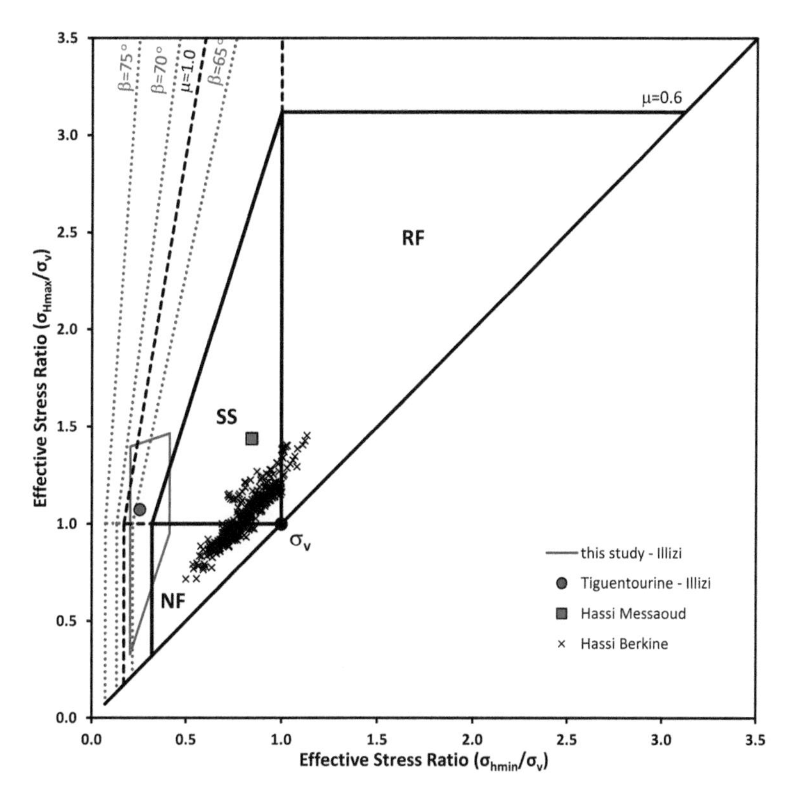

Fig. 10. Effective stress ratio diagram illustrating the current state of stress in the study area of the Illizi Basin. The plot shows several different polygon perimeters: the black solid and dashed lines show two different coefficients of sliding friction, $\mu = 0.6$ and $\mu = 1.0$ respectively; the brown dashed lines show different β angles (measured between the normal of a fault plane and the maximum principal stress direction). Based on the angular relationship between the local fault traces and the average S_{Hmax} azimuth (Fig. 6), we conclude that the present day stress field in the Illizi Basin is likely to be limited by the frictional strength of the major fault trends in the area. The effective stress ratios described from the Tiguentourine field in the Illizi Basin are consistent with those in our study. By contrast, the *in situ* stresses in the overpressured and unexhumed Berkine and Hassi Messaoud fields plot within the effective stress polygon and therefore even optimally oriented faults in these areas would not be in a critical state of incipient mechanical shear failure. NF, normal faulting; RF, reverse faulting; SS, strike-slip faulting.

incipient shear failure and hence limit the differential stress (i.e. the difference between S_{Hmax} or S_v and S_{hmin}). Unfortunately, direct seismic evidence for these faults being critically stressed is not available from existing earthquake catalogues (e.g. the United States Geological Survey and European–Mediterranean Seismological Centre). One possible reason is that an adequate seismic monitoring network to reliably detect (minor) earthquakes is unavailable in this part of the world. In addition, faults and fractures of these orientations are the most likely to be conductive and permeable under the present day *in situ* stress conditions.

Comparison of stress conditions with the unexhumed Berkine Basin and Hassi Messaoud Ridge

The east–west buried axis of the long-lived Ahara High separates the Illizi Basin from the Berkine Basin to the north (Galeazzi *et al.* 2010). Although *c.* 1.0–1.4 km of exhumation is estimated to have occurred in the study area in the Illizi Basin during the Eocene–Miocene (K.L. English *et al.* 2016*c*), the magnitude of this exhumation event decreases northwards into the Berkine Basin (Fig. 2) and is minimal in its main depocentre, where a largely complete Mesozoic–Cenozoic section is preserved (Yahi 1999; Yahi *et al.* 2001; Underdown & Redfern 2008). Previous sonic compaction analyses in the region have documented the difference in burial history across the region and described undercompacted sediments in the Berkine Basin and overcompacted sediments in the Illizi Basin (Dixon *et al.* 2010).

Potentiometric maps indicate the presence of an overpressured system beneath the Triassic salts of the unexhumed Berkine Basin to the north (Chiarelli 1978). The abnormally high pressures in the present day Hassi Messaoud region and the Oued Mya and Berkine basins originate in undercompacted shales sealed by the impermeable Triassic salt (Chiarelli 1978; Yahi *et al.* 2001). In parts of the basin characterized by deep Hercynian incision, Mesozoic salt-bearing strata were deposited directly on Cambro-Ordovician rocks (Fig. 1; Zeroug *et al.* 2007; Galeazzi *et al.* 2010). By contrast, the exhumed Illizi Basin is generally characterized by close to hydrostatic conditions (Chiarelli 1978; this study), suggesting that significant overpressure either never existed here or has dissipated since uplift from maximum burial conditions.

Hassi Messaoud field

The Hassi Messaoud field is located on a Hercynian structure located between the Berkine and Oued Mya basins in Algeria (Fig. 1). The reservoirs consist of Cambrian and Lower Ordovician sandstones at depths >3000 m, with oil reserves initially in excess of 25 billion barrels (Balducchi & Pommier 1970; Bacheller & Peterson 1991). On the structure, Triassic clastic sediments and salt unconformably overlie the Palaeozoic reservoir sequence. The field is compartmentalized by regional (sub-) vertical fault systems trending NE–SW and NNE–SSW, which are interpreted to be impermeable to the circulation of fluids (Zeroug *et al.* 2007). The maximum horizontal stress azimuth is reported to be 315–320° in the Hassi Messaoud area (Koceir & Tiab 2000), similar to that observed in the Illizi Basin (Fig. 6), and hence the sealing faults that compartmentalize the Hassi Messaoud field are generally sub-perpendicular to S_{Hmax}.

The initial reservoir pressure is reported to be 47.4 MPa at a depth of 3200 m (Bacheller & Peterson 1991), although the field has been in production since the 1960s. The initial reservoir pressure corresponds to an overpressured gradient of 14.8 kPa m^{-1} and this area was clearly not in lateral pressure communication with the normally pressured Illizi Basin to the south. The formation salinity is also extremely high at 400 000 ppm (Bacheller & Peterson 1991) and is probably derived from the overlying evaporitic sequence (K.L. English *et al.* 2016*b*). Assuming an overburden gradient of 24.5 kPa m^{-1} gives an S_v magnitude of 78.4 MPa at a depth of 3200 m. The minimum horizontal stress magnitude varies considerably across the field due to coupling between the horizontal stress and the pore pressure, and the spatial and temporal variability in levels of depletion over the production history of the field (McGowen *et al.* 1996; Zeroug *et al.* 2007). McGowen *et al.* (1996) presented a dataset of fracture gradient and pore pressure data from the Hassi Messaoud field, which can be used to evaluate the minimum horizontal stress gradient under virgin conditions (Fig. 11). Unfortunately, specific depth intervals were not provided in the dataset, so some of the scatter may be due to variations in the true vertical depth in each test. However, the dataset is still instructive and indicates that the fracture gradient generally decreases with decreasing pore pressure and that the fracture gradients can approach magnitudes close to the vertical stress gradient (Fig. 11). Using a best-fit trendline through the data, we estimate an initial S_{hmin} gradient of 23 kPa m^{-1}, which corresponds to a magnitude of 73.6 MPa at 3200 m depth. We acknowledge, however, that there is significant uncertainty with respect to this value and some of the data points indicate even higher S_{hmin} gradients.

The magnitude of S_{Hmax} is even less well constrained. Differential strain curve analysis on core measurements has indicated an S_{Hmax}/S_{hmin} ratio

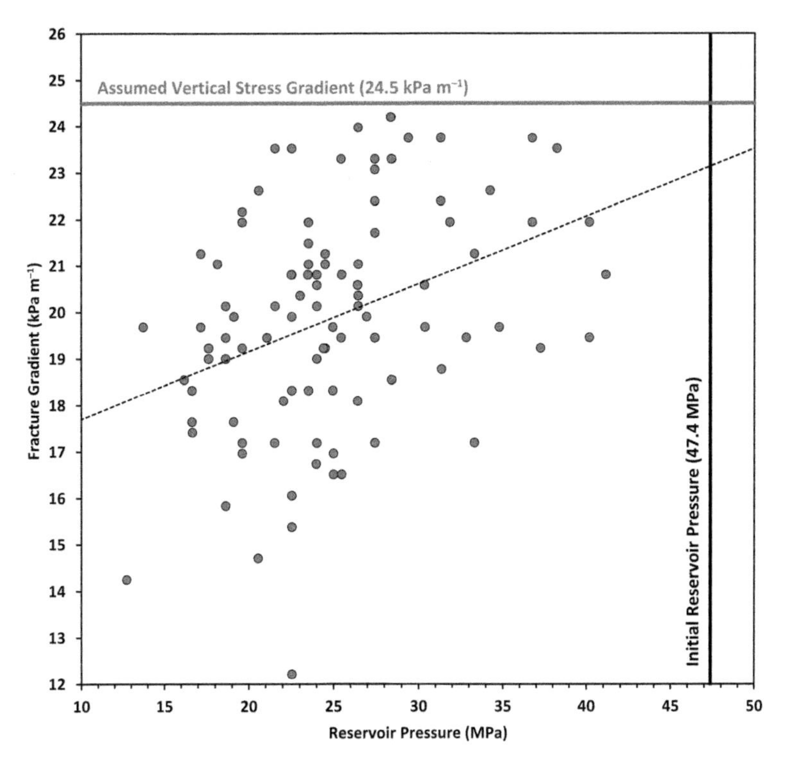

Fig. 11. Fracture gradient v. reservoir pressure data from the Hassi Messaoud field (data from McGowen *et al.* 1996). The Hassi Messaoud field has been in production since the 1960s and the minimum horizontal stress magnitude varies considerably across the field due to the coupling between horizontal stress and pore pressure, and the spatial and temporal variability in levels of depletion over the production history of the field. Although there is an appreciable amount of scatter and an unknown degree of uncertainty in these data, they nonetheless indicate that the fracture gradient generally decreases with decreasing pore pressure and that the fracture gradients in an undepleted state can approach magnitudes close to the vertical stress.

of 1.25 (Zeroug *et al.* 2007). Taking this ratio, we find that our estimate for the initial stress state of the Hassi Messaoud field plots inside the stress polygon (Fig. 10) and therefore even optimally oriented faults in this area would not be in a critical state of incipient mechanical shear failure. In spite of all the uncertainty, these data indicate that the Hassi Messaoud field is markedly different from the Illizi Basin in terms of *in situ* stresses; it is characterized by significant overpressure, a relatively benign stress field within the strike-slip faulting stress regime (i.e. $S_{Hmax} > S_v > S_{hmin}$) and an absence of critically stressed faults.

Berkine field

Zeroug *et al.* (2007) presented a case study looking at the sanding tendency in the Devonian Strunian reservoir of one of the central Berkine fields. As part of the workflow, the vertical stress was estimated from integrated bulk density logs, rock mechanical properties were calibrated to core data, and the least principal stress (S_{hmin}) was calibrated to hydraulic fracture data. All of these data were utilized to construct a mechanical Earth model and the S_{Hmax} magnitude was calibrated to match the borehole breakouts observed on image logs. Although none of the raw data were presented in this case study (Zeroug *et al.* 2007), we have digitized the resulting stress and pore pressure curves for inclusion in this study. The results from the Zeroug *et al.* (2007) study plot inside the stress polygon and mostly within the strike-slip faulting and normal faulting fields (Fig. 10). The average values based on these data at a depth of 3400 m are: S_v 80.5 MPa (23.7 kPa m^{-1}), S_{hmin} 73.0 MPa (21.5 kPa m^{-1}), S_{Hmax} 82.2 MPa (24.2 kPa m^{-1}) and pore pressure 40.8 MPa (12.0 kPa m^{-1}). As with the Hassi Messaoud example, even optimally oriented faults in this area would not be in a critical state of incipient mechanical shear failure. Once again, these data indicate that the unexhumed

Berkine Basin (Fig. 2) is characterized by a different, relatively benign, *in situ* stress state relative to the Illizi Basin and that the basin is characterized instead by overpressure and the absence of critically stressed faults.

Conceptual evolution of *in situ* stress during exhumation

Assuming that a sedimentary sequence is considered to consist of a series of linear-elastic, homogeneous and isotropic layers, we can utilize poroelastic theory to predict how the horizontal stresses change during exhumation (English 2012 and references cited therein). If we assume that the system is perfectly sealed (i.e. no pore pressures changes) and that there is no lateral strain, we can model the changes in both the horizontal principal stresses (ΔS_h) as a function of changing vertical stress (ΔS_v) and changing temperature (ΔT) using the following equation:

$$\Delta S_h = \frac{\nu}{1-\nu}\Delta S_v + \frac{E}{1-\nu}\alpha_T \Delta T \quad (9)$$

where ν is Poisson's ratio, E is Young's modulus and α_T is the coefficient of linear thermal expansion.

For the purposes of the conceptual models, we assume that the starting *in situ* stress state is equivalent to the average values from the Berkine field (Fig. 12) and we assume a constant pressure (i.e. sealed compartment), a constant Poisson's ratio ($\nu = 0.2$) and a constant temperature gradient (35°C km^{-1}) during exhumation from a depth of 3400 m. Four different stress path models are presented in Figure 12 with successively decreasing

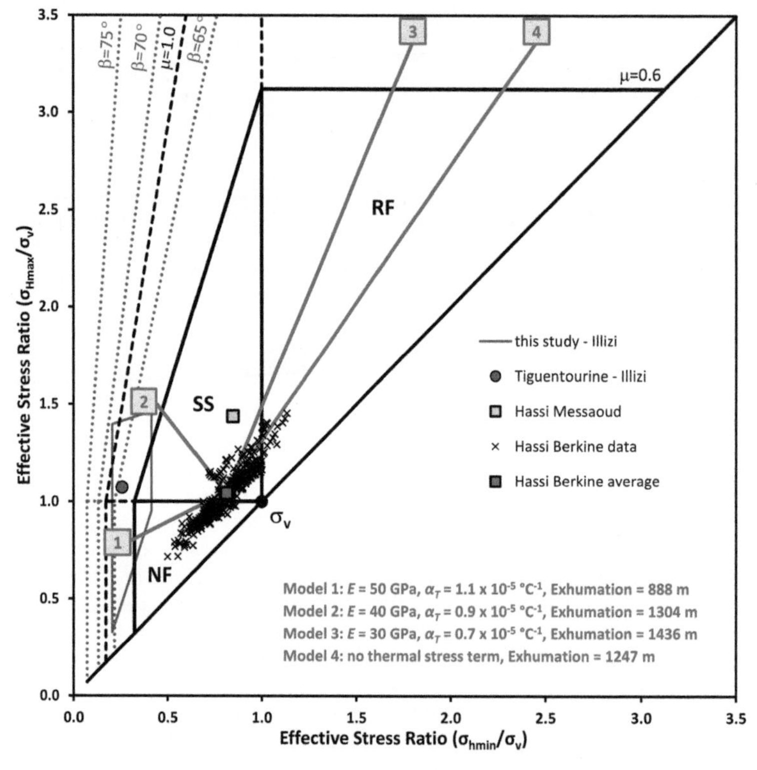

Fig. 12. Effective stress ratio diagram illustrating the current state of stress in the Illizi and Berkine basins and some conceptual stress paths during exhumation. The conceptual exhumation models adopt the present day stresses from the Berkine field for their initial conditions and utilize linear elasticity and poroelastic theory to predict how the horizontal stresses change during exhumation. The models also assume a constant pressure (i.e. sealed compartment), a constant Poisson's ratio ($\nu = 0.2$) and a constant temperature gradient (35°C km^{-1}) during exhumation from a depth of 3400 m. Four different stress path models are presented with successively decreasing importance of the thermal term (as per equation 9); Model 4 is an end-member model with no thermal effect. NF, normal faulting; RF, reverse faulting; SS, strike-slip faulting.

importance of the thermal term: (1) $E = 50$ GPa, $\alpha_T = 1.1 \times 10^{-5}$ °C^{-1}; (2) $E = 40$ GPa, $\alpha_T = 0.9 \times 10^{-5}$ °C^{-1}; (3) $E = 30$ GPa, $\alpha_T = 0.7 \times 10^{-5}$ °C^{-1}; and (4) no thermal term. The range of α_T values is representative for sedimentary rocks (English & Laubach 2017), whereas the range of E values is representative of the measured mechanical rock properties for the Ordovician sandstone (Table 1). These assumed parameter variations result in markedly different stress paths depending on the importance of the thermal term in equation (9). Note also from equation (9) that a specified fractional change in either E or α_T has the same effect on the horizontal stress calculations – that is, a 20% reduction in E will have the same impact on the calculations as a 20% reduction in α_T.

Model 1 moves into the normal faulting field, Model 2 moves out to the edge of the strike-slip faulting field, while models 3 and 4 move into the reverse faulting field (Fig. 12). When the exhumation stress path reaches the perimeter of the polygon, which is a function of the predominant fault orientation present in a given basin, the associated *in situ* principal stress magnitudes have caused the fault systems to reach the point of incipient critical shear failure (Fig. 12). In other words, the final stress environment is characterized by a critical differential stress that is markedly higher than at pre-exhumation conditions. Hence any faults optimally oriented for shear failure are reactivated and would enable fluids to flow and facilitate the dissipation of excess pore pressure. Note, however, that the stress paths depicted in Figure 12 only apply until fault reactivation occurs. After this point, the assumption of zero lateral strain used to derive equation (9) no longer applies and the stress paths are expected to deviate from those shown in the figure. The magnitudes of exhumation required in each model to reach critical stress conditions on optimally oriented faults (where $\mu \approx 0.65$) are 888, 1304, 1436 and 1247 m, respectively. Although the four models presented here are conceptual, we note that this range of exhumation magnitude is comparable with that documented in the study area (*c.* 1.0–1.4 km; K.L. English *et al.* 2016*b*). The amount of exhumation required for the reactivation of optimally oriented faults will depend on the trajectory of the stress path and on the initial stress conditions. Faults in basins with benign stress states (i.e. far from frictional shear failure and plotting in the interior of the stress polygon) will require greater magnitudes of exhumation, or have to be hosted in much stiffer rocks (i.e. very high E values), to reach a critical state even if the faults are optimally oriented. In some situations, they may, in fact, never be reactivated. On the contrary, if the initial stress conditions are not far from the critical state, a lower magnitude of exhumation is required to result in the predominant fault system becoming critically stressed, even if they are not optimally oriented for failure.

Discussion and implications

The comprehensive new dataset presented in this study has characterized the *in situ* stresses of the Illizi Basin in Algeria and established that the primary vertical fault systems in this exhumed basin are likely to be critically stressed (Fig. 6). By contrast, the *in situ* stresses in the overpressured Berkine Basin and Hassi Messaoud areas to the north indicate that these areas, which are currently at, or close to, their maximum burial depth (Fig. 2), are not characterized by critical stress conditions (Fig. 10). Although the stress data from the Hassi Messaoud and Berkine fields are limited, and we did not have the possibility of constructing our own independent geomechanical models, the significant overpressure and reservoir compartmentalization observed in the Hassi Messaoud field (Zeroug *et al.* 2007) is consistent with the interpreted absence of critical stress conditions here. In summary, we tentatively conclude from this Algerian study that the overpressured Berkine Basin at close to maximum burial is not characterized by critical stress conditions, whereas parts of the adjacent exhumed Illizi Basin appear to be at, or close to, critical stress conditions along pre-existing optimally oriented fault systems.

Exhumation: fault reactivation and fluid flow

Fluid inclusion and present day water chemistry data from Ordovician sandstones in the study area have indicated that lateral long-distance brine migration and mixing have occurred within the Illizi Basin (K.L. English *et al.* 2016*b*). The chemistry of the formation water in the Ordovician reservoir is consistent with derivation from a Triassic–Liassic halite-bearing sequence in the Berkine Basin further north (Fig. 1). The fluid inclusion data indicate that this high-salinity brine displaced a lower salinity formation water during the *c.* 1.0–1.4 km of exhumation that occurred in the study area during Eocene–Miocene time (K.L. English *et al.* 2016*b*, *d*). It has been proposed that breaching of a closed, overpressured system occurred during exhumation of the Illizi Basin and that this may have been a driving mechanism for the regional updip flow of high-salinity formation water (K.L. English *et al.* 2016*b*). Our conceptual stress path models have indicated that a sealed overpressure compartment may be driven towards critical stress conditions if significant cooling accompanies uplift and if optimally oriented faults are present (Fig. 12). More specifically, given the magnitude of exhumation in the Illizi Basin and its present day stress state, we can conclude that the Ordovician section may have followed a stress path akin to models 1 and 2. Therefore it is possible that Eocene–Miocene exhumation of the Illizi Basin resulted in the reactivation of optimally oriented (north–south and NW–SE) vertical faults that facilitated the southwards lateral migration of high-salinity brine into the study area.

Exhumation: implications for fault-bounded traps

The petroleum prospectivity of an exhumed basin is largely dependent on the ability of pre-existing traps to retain oil and gas volumes during and after the exhumation event. Based on our conceptual stress evolution modelling, a sedimentary basin with benign *in situ* stresses at maximum burial may change to being characterized by critical stress conditions on the predominant fault system during exhumation. This can occur even if the faults are not optimally oriented for shear failure, but the stress path during exhumation reaches a state where sufficient differential stress is eventually resolved on these fault surfaces. Therefore faulted hydrocarbon-bearing traps may become susceptible to breaching during exhumation if the fault systems are within the favourable orientation range with respect to S_{Hmax} and the respective level of critical differential stress is reached to reactivate these. The cooling of petroleum source rocks in exhumed basins generally leads to the cessation of active petroleum generation within the basin. Hence it is unlikely that leaky hydrocarbon traps will be recharged unless the exhumed basin is situated along a migration route from an adjacent normally subsiding basin. One possible implication of this is that fault-bounded traps with strike-slip or normal faults oriented at a high angle (or perpendicular) to S_{Hmax} may be more likely to seal and retain hydrocarbons in exhuming basins than fault-bounded traps with optimally oriented, and potentially critically stressed, faults.

Conclusions

A new geomechanical dataset from the exhumed Illizi Basin in Algeria has demonstrated that the primary north–south and NW–SE vertical (strike-slip) fault systems in this region are likely to be at, or close to, critical stress conditions. By contrast, the overpressured Berkine Basin and Hassi Messaoud areas to the north, which are currently at, or close to, their maximum burial depth, do not appear to be characterized by critical stress conditions. Conceptual models of stress evolution during exhumation have demonstrated that a sedimentary basin with benign *in situ* stresses at maximum burial may change to being characterized by critical stress conditions on existing fault systems if they are within the range of optimum orientations and reach critical differential stresses during exhumation. Therefore the exhumation of an overpressured sedimentary basin is likely to facilitate basin-scale fluid flow during depressurization if the pre-existing fault systems are reactivated. These models are supportive of the idea that the breaching of a closed, overpressured system during exhumation of the Illizi Basin may have been a driving mechanism for the regional updip flow of formation water within the Ordovician reservoirs during Eocene–Miocene time. Although this study focused primarily on a specific case study from Algeria, this methodology may be applied in other basins around the world where fault stability is an important factor for assessing the risk of seal failure. One possible implication of this work for petroleum exploration is that fault-bounded traps with strike-slip or normal faults oriented at a high angle to S_{Hmax} are more likely to have retained hydrocarbons in exhumed basins than fault-bounded traps with faults that are more optimally oriented for shear failure and have a greater propensity to become critically stressed during exhumation.

We thank Petroceltic, Sonatrach and Enel for sponsoring this study and granting permission for publication. We thank Jonathan Hunter, John Naismith, Ciaran Nolan, Siya Gancheva, Fabrice Toussaint and Tim Wynn for various discussions over the course of this project, and also Richard Plumb and one anonymous reviewer for constructive feedback that helped to improve the quality of this paper.

References

ALLAN, U.S. 1989. Model for hydrocarbon migration and entrapment within faulted structures. *American Association of Petroleum Geologists Bulletin*, **73**, 803–811, https://doi.org/10.1306/44B4A271-170A-11D7-8645000102C1865D

ANDERSON, E.M. 1951. *The Dynamics of Faulting*. Oliver and Boyd, Edinburgh.

BACHELLER, W.D. & PETERSON, R.M. 1991. Hassi Messaoud field – Algeria Trias Basin, eastern Sahara Desert. *In*: FOSTER, N.H. & BEAUMONT, E.A. (eds) *Structural Traps V*. AAPG, Treatise of Petroleum Geology, Atlas of Oil and Gas Fields, **5**, 211–225.

BALDUCCHI, A.M. & POMMIER, G. 1970. Cambrian oil field of Hassi Messaoud, Algeria. *In*: HALBOUTY, M.T. (ed.) *Geology of Giant Petroleum Fields*. AAPG Memoirs, **14**, 477–488.

BARREE, R.D., BARREE, V.L. & CRAIG, D. 2007. Holistic fracture diagnostics. *In*: *Rocky Mountain Oil & Gas Technology Symposium*. Denver, CO, USA. Society of Petroleum Engineers, https://doi.org/10.2118/107877-MS

Barton, C.A., Zoback, M.D. & Burns, K.L. 1988. In-situ stress orientation and magnitude at the Fenton Geothermal Site, New Mexico, determined from wellbore breakouts. *Geophysical Research Letters*, **15**, 467–470, https://doi.org/10.1029/GL015i005p00467

Barton, C.A., Zoback, M.D. & Moos, D. 1995. Fluid flow along potentially active faults in crystalline rock. *Geology*, **23**, 683–686, https://doi.org/10.1130/0091-7613(1995)023<0683:FFAPAF>2.3.CO;2

Baylocq, P., Sahnoune, A., Martin, A. & Sims, M. 1998. Tin Fouye Tabankort gas field – how understanding of abnormal fracture propagation and reservoir uncertainties led to successful hydraulic fracturing operations. Paper SPE-50610-MS, presented at the European Petroleum Conference, The Hague, the Netherlands, 20–22 October 1998, https://doi.org/10.2118/50610-MS

Bouvier, J.D., Kaars-Sijpesteijn, C.H., Kluesner, D.F., Onyejekwe, C.C. & Van Der Pal, R.C. 1989. Three-dimensional seismic interpretation and fault sealing investigations, Nun River Field, Nigeria. *American Association of Petroleum Geologists Bulletin*, **73**, 1397–1414, https://doi.org/10.1306/44B4AA5A-170A-11D7-8645000102C1865D

Byerlee, J.D. 1978. The fracture strength and frictional strength of Weber sandstone. *International Journal of Rock Mechanics and Mining Sciences & Geomechanics Abstracts*, **12**, 1–4, https://doi.org/10.1016/0148-9062%2875%2990736-6

Chiarelli, A. 1978. Hydrodynamic framework of eastern Algerian Sahara; influence on hydrocarbon occurrence. *American Association of Petroleum Geologists Bulletin*, **62**, 667–685.

Corcoran, D.V. & Doré, A.G. 2002. Depressurization of hydrocarbon-bearing reservoirs in exhumed basin settings: evidence from Atlantic margin and borderland basins. *In*: Doré, A.G., Cartwright, J.A., Stoker, M.S., Turner, J.P. & White, N. (eds) *Exhumation of the North Atlantic Margin: Timing, Mechanisms and Implications for Petroleum Exploration*. Geological Society, London, Special Publications, **196**, 457–483, https://doi.org/10.1144/GSL.SP.2002.196.01.25

Corcoran, D.V. & Mecklenburgh, R. 2005. Exhumation of the Corrib Gas Field, Slyne Basin, offshore Ireland. *Petroleum Geoscience*, **11**, 239–256, https://doi.org/10.1144/1354-079304-637

Dixon, R.J., Moore, J.K.S. *et al*. 2010. Integrated petroleum systems and play fairway analysis in a complex Palaeozoic basin: Ghadames-Illizi Basin, North Africa. *In*: Vining, B.A. & Pickering, S.C. (eds) *Petroleum Geology: From Mature Basins to New Frontiers – Proceedings of the 7th Petroleum Geology Conference*. Geological Society, London, 735–760, https://doi.org/10.1144/0070735

Donati, M., Piazza, J.L. *et al*. 2016. 3D-3C multicomponent seismic – a successful fracture characterization case study in Algeria. *First Break*, **34**, 35–47.

Doré, A.G., Corcoran, D.V. & Scotchman, I.C. 2002. Prediction of the hydrocarbon system in exhumed basins, and application to the NW European margin. *In*: Doré, A.G., Cartwright, J.A., Stoker, M.S., Turner, J.P. & White, N. (eds) *Exhumation of the North Atlantic Margin: Timing, Mechanisms and Implications for Petroleum Exploration*. Geological Society, London, Special Publications, **196**, 401–429, https://doi.org/10.1144/GSL.SP.2002.196.01.21

Echikh, K. 1998. Geology and hydrocarbon occurrences in the Ghadames Basin, Algeria, Tunisia, Libya. *In*: MacGregor, D.S., Moody, R.T.J. & Clark-Lowes, D.D. (eds) *Petroleum Geology of North Africa*. Geological Society, London, Special Publications, **132**, 109–129, https://doi.org/10.1144/GSL.SP.1998.132.01.06

England, W.A., Mackenzie, A.S., Mann, D.M. & Quigley, T.M. 1987. The movement and entrapment of petroleum fluids in the subsurface. *Journal of the Geological Society, London*, **144**, 327–347, https://doi.org/10.1144/gsjgs.144.2.0327

English, J.M. 2012. Thermomechanical origin of regional fracture systems. *American Association of Petroleum Geologists Bulletin*, **96**, 1597–1625, https://doi.org/10.1306/01021211018

English, J.M. & Laubach, S.E. 2017. Opening-mode fracture systems – insights from recent fluid inclusion microthermometry studies of crack-seal fracture cements. *In*: Turner, J.P., Healy, D., Hillis, R.R. & Welch, M.J. (eds) *Geomechanics and Geology*. Geological Society, London, Special Publications, **458**. First published online May 24, 2017, https://doi.org/10.1144/SP458.1

English, J.M., English, K.L., Corcoran, D.V. & Toussaint, F. 2016. Exhumation charge: the last gasp of a petroleum source rock and implications for unconventional shale resources. *American Association of Petroleum Geologists Bulletin*, **100**, 1–16, https://doi.org/10.1306/07271514224

English, K.L., English, J.M. *et al*. 2016*a*. Controls on reservoir quality in exhumed basins – an example from the Ordovician sandstones, Illizi basin, Algeria. *Marine and Petroleum Geology*, **80**, 203–227, https://doi.org/10.1016/j.marpetgeo.2016.11.011

English, K.L., English, J.M., Redfern, J., Hollis, C., Corcoran, D.V., Oxtoby, N. & Yahia Cherif, R. 2016*b*. Remobilization of deep basin brine during exhumation of the Illizi Basin, Algeria. *Marine and Petroleum Geology*, **78**, 679–689, https://doi.org/10.1016/j.marpetgeo.2016.08.016

English, K.L., Redfern, J., Bertotti, G., English, J.M. & Yahia Cherif, R. 2016*c*. Intraplate uplift: new constraints on the Hoggar dome from the Illizi basin (Algeria). *Basin Research*, **2**, https://doi.org/10.1111/bre.12182

English, K.L., Redfern, J., Corcoran, D.V., English, J.M. & Yahia Cherif, R. 2016*d*. Constraining burial history and petroleum charge in exhumed basins: new insights from the Illizi Basin, Algeria. *American Association of Petroleum Geologists Bulletin*, **100**, 623–655, https://doi.org/10.1306/12171515067

Finkbeiner, T., Zoback, M.D., Flemings, P. & Stump, B. 2001. Stress, pore pressure, and dynamically constrained hydrocarbon columns in the South Eugene Island 330 field, northern Gulf of Mexico. *American Association of Petroleum Geologists Bulletin*,

85, 1007–1031, https://doi.org/10.1306/8626CA55-173B-11D7-8645000102C1865D

Fjær, E., Holt, R.M., Horsrud, P., Raaen, A.M. & Risnes, R. 2008. *Petroleum Related Rock Mechanics*. Elsevier, Amsterdam.

Galeazzi, S., Point, O., Haddadi, N., Mather, J. & Druesne, D. 2010. Regional geology and petroleum systems of the Illizi–Berkine area of the Algerian Saharan Platform: an overview. *Marine and Petroleum Geology*, **27**, 143–178, https://doi.org/10.1016/j.marpetgeo.2008.10.002

Green, P.F. & Duddy, I.R. 2010. Synchronous exhumation events around the Arctic including examples from Barents Sea and Alaska North Slope. *In*: Vining, B.A. & Pickering, S.C. (eds) *Petroleum Geology: From Mature Basins to New Frontiers – Proceedings of the 7th Petroleum Geology Conference*. Geological Society, London, 633–644, https://doi.org/10.1144/0070633

Heidbach, O., Tingay, M., Barth, A., Reinecker, J., Kurfess, D. & Müller, B. 2009. *The World Stress Map Based on the Database Release 2008, Equatorial Scale 1:46,000,000*. Commission for the Geological Map of the World, Paris, https://doi.org/10.1594/GFZ.WSM.Map2009

Hillis, R.R., Holford, S.P. *et al.* 2008. Cenozoic exhumation of the southern British Isles. *Geology*, **36**, 371–374, https://doi.org/10.1130/G24699A.1

Issler, D.R., Beaumont, C., Willett, S.D., Donelick, R.A. & Grist, A.M. 1999. Paleotemperature history of two transects across the Western Canada sedimentary basin; constraints from apatite fission track analysis. *Bulletin of Canadian Petroleum Geology*, **47**, 475–486.

Jaeger, J.C., Cook, N.G.W. & Zimmerman, R. 2007. *Fundamentals of Rock Mechanics*. Wiley-Blackwell, New York.

Japsen, P. 1998. Regional velocity–depth anomalies, North Sea Chalk: a record of overpressure and Neogene uplift and erosion. *American Association of Petroleum Geologists Bulletin*, **82**, 2031–2074, https://doi.org/10.1306/00AA7BDA-1730-11D7-8645000102C1865D

Japsen, P., Green, P.F., Bonow, J.M., Rasmussen, E.S., Chalmers, J.A. & Kjennerud, T. 2010. Episodic uplift and exhumation along North Atlantic passive margins: implications for hydrocarbon prospectivity. *In*: Vining, B.A. & Pickering, S.C. (eds) *Petroleum Geology: From Mature Basins to New Frontiers – Proceedings of the 7th Petroleum Geology Conference*. Geological Society, London, 979–1004, https://doi.org/10.1144/0070979

Japsen, P., Chalmers, J.A., Green, P.F. & Bonow, J.M. 2012. Elevated, passive continental margins: not rift shoulders, but expressions of episodic, post-rift burial and exhumation. *Global and Planetary Change*, **90–91**, 73–86, https://doi.org/10.1016/j.gloplacha.2011.05.004

Jones, R.M. & Hillis, R.R. 2003. An integrated, quantitative approach to assessing fault-seal risk. *American Association of Petroleum Geologists Bulletin*, **87**, 507–524.

Koceir, M. & Tiab, D. 2000. Influence of stress and lithology on hydraulic fracturing in Hassi Messaoud reservoir, Algeria. SPE-62608-MS, presented at the SPE/AAPG Western Regional Meeting, Long Beach, CA, USA, 19–22 June 2000, https://doi.org/10.2118/62608-MS

Magoon, L.B. & Dow, W.G. 1994. The petroleum system. *In*: Magoon, L.B. & Dow, W.G. (eds) *The Petroleum System – From Source to Trap*. AAPG Memoirs, **60**, 3–24.

McGowen, J.M., Benani, A. & Ziada, A. 1996. Increasing oil production by hydraulic fracturing in the Hassi Messaoud Cambrian Formation, Algeria. Paper SPE-36904-MS, presented at the European Petroleum Conference, Milan, Italy, 22–24 October 1996, https://doi.org/10.2118/36904-MS

Moos, D. & Zoback, M.D. 1990. Utilization of observations of well bore failure to constrain the orientation and magnitude of crustal stresses: application to continental, Deep Sea Drilling Project, and Ocean Drilling Program boreholes. *Journal of Geophysical Research*, **95**, 9305–9325, https://doi.org/10.1029/JB095iB06p09305

Patton, T.L., Batchelor, A.S., Foxford, K.A., Hellman, T.J. & Maache, N. 2003. In situ stress state – Tiguentourine Field, southeastern Algeria. Paper presented at the AAPG Hedberg Conference, Paleozoic and Triassic Petroleum Systems in North Africa, Algiers, Algeria, 18–20 February 2003.

Rogers, S.F. 2003. Critical stress-related permeability in fractured rocks. *In*: Ameen, M. (ed.) *Fracture and In-Situ Stress Characterization of Hydrocarbon Reservoirs*. Geological Society, London, Special Publications, **209**, 7–16, https://doi.org/10.1144/GSL.SP.2003.209.01.02

Schowalter, T.T. 1979. Mechanics of secondary hydrocarbon migration and entrapment. *American Association of Petroleum Geologists Bulletin*, **63**, 723–760, https://doi.org/10.1306/2F9182CA-16CE-11D7-8645000102C1865D

Sibson, R.H. 1994. Crustal stress, faulting and fluid flow. *In*: Parnell, J. (ed.) *Geofluids: Origin, Migration and Evolution of Fluids in Sedimentary Basins*. Geological Society, London, Special Publications, **78**, 69–84, https://doi.org/10.1144/GSL.SP.1994.078.01.07

Sibson, R.H. 2000. Tectonic controls on maximum sustainable overpressure: fluid redistribution from stress transitions. *Journal of Geochemical Exploration*, **69–70**, 471–475, https://doi.org/10.1016/S0375-6742%2800%2900090-X

Terzaghi, K. 1943. *Theoretical Soil Mechanics*. Wiley, New York.

Tissot, B. & Espitalié, J. 1975. L'évolution thermique de la matière organique des sédiments: applications d'une simulation mathématique. *Revue de l'Institut Francais du Petrole*, **30**, 743–777, https://doi.org/10.2516/ogst:1975026

Townend, J. & Zoback, M.D. 2000. How faulting keeps the crust strong. *Geology*, **28**, 399–402, https://doi.org/10.1130/0091-7613(2000)28<399:HFKTCS>2.0.CO;2

Underdown, R. & Redfern, J. 2007. The importance of constraining regional exhumation in basin modelling: a hydrocarbon maturation history of the Ghadames Basin, North Africa. *Petroleum Geoscience*, **13**, 253–270, https://doi.org/10.1144/1354-079306-714

UNDERDOWN, R. & REDFERN, J. 2008. Petroleum generation and migration in the Ghadames Basin, north Africa: a two-dimensional basin-modeling study. *American Association of Petroleum Geologists Bulletin*, **92**, 53–76, https://doi.org/10.1306/08130706032

WIPRUT, D. & ZOBACK, M.D. 2002. Fault reactivation, leakage potential, and hydrocarbon column heights in the northern North Sea. *In*: KOESTLER, A.G. & HUNSDALE, R. (eds) *Hydrocarbon Seal Quantification*. Norwegian Petroleum Society, Special Publications, **11**, 203–219, https://doi.org/10.1016/S0928-8937 (02)80016-9

YAHI, N. 1999. *Petroleum generation and migration in the Berkine (Ghadames) Basin, eastern Algeria: an organic geochemical and basin modelling study*. PhD thesis, Aachen University of Technology.

YAHI, N., SCHAEFER, R.G. & LITTKE, R. 2001. Petroleum generation and accumulation in the Berkine basin, eastern Algeria. *American Association of Petroleum Geologists Bulletin*, **85**, 1439–1467, https://doi.org/10.1306/8626CAD7-173B-11D7-8645000102C1865D

YIELDING, G., FREEMAN, B. & NEEDHAM, D.T. 1997. Quantitative fault seal prediction. *American Association of Petroleum Geologists Bulletin*, **81**, 897–917, https://doi.org/10.1306/522B498D-1727-11D7-8645000102C1865D

ZEROUG, S., BOUNOUA, N. & LOUNISSI, R. 2007. *Well Evaluation Conference (WEC), Algérie 2007*. Sonatrach-Schlumberger/Wetmore Printing, Houston, TX.

ZOBACK, M.D. 2007. *Reservoir Geomechanics*. Cambridge University Press, Cambridge.

ZOBACK, M.D. & HEALY, J.H. 1984. Friction, faulting and 'in situ' stress. *Annales Geophysicae*, **2**, 689–698.

ZOBACK, M.D., MASTIN, L. & BARTON, C.A. 1986. In-situ stress measurements in deep boreholes using hydraulic fracturing, wellbore breakouts, and stonely wave polarization. *In*: ZOBACK, M.D., MASTIN, L. & BARTON, C. *Proceedings of the International Symposium on Rock Stress and Rock Stress Measurement*, 1–3 September 1986, Stockholm. Publ Lulea, Centek, 289–299.

ZOBACK, M.D., BARTON, C.A. *ET AL*. 2003. Determination of stress orientation and magnitude in deep wells. *International Journal of Rock Mechanics and Mining Sciences*, **40**, 1049–1076, https://doi.org/10.1016/j.ijrmms.2003.07.001

Chalk reservoir of the Ockley accumulation, North Sea: *in situ* stresses, geology and implications for stimulation

T. J. WYNN[1]*, R. KUMAR[2], R. JONES[2], K. HOWELL[1], D. MAXWELL[3] & P. BAILEY[1,4]

[1]*AGR TRACS International, Union Plaza, Aberdeen AB10 1SL, UK*

[2]*Maersk Oil North Sea UK Ltd, Maersk House, Crawpeel Road, Altens, Aberdeen AB12 3LG, UK*

[3]*Axis Well Technology, Units C, D&E, Kettock Lodge, Campus 2, Aberdeen Innovation Park, Balgownie Drive, Bridge of Don, Aberdeen AB22 8GU, UK*

[4]*PetroStars UK Ltd, Office 1, 23 Rubislaw Den North, Aberdeen AB15 4AL, UK*

**Correspondence: tim.wynn@agr.com*

Abstract: The Ockley discovery is a gas condensate accumulation contained within tight chalks of the Hod Formation. The observed faults and fractures are a combination of features radial to the main periclinal structure and parallel to the local structural grain 100° from north. Most natural fractures appear to be healed or cemented. The pore pressure gradient at Ockley is *c.* 0.199 bar/m. The *in situ* stresses are estimated to all be within *c.* 46 bar of each other, indicating a near-isotropic *in situ* stress system. Therefore the orientation of S_{Hmax} is hard to define and drilling optimally oriented wells to create transverse hydraulic fractures is difficult. The estimated intact rock fracture initiation pressure in a horizontal well would exceed all the *in situ* stress gradients. Therefore, even if a vertical planar induced fracture were created at the wellbore wall, it would probably exploit natural fractures and bedding planes, leading to complex fracture geometries. Acid fracture stimulations are plugged by the insoluble clay residue within the clay-rich Hod chalk, so this is not an optimal strategy. Proppant fracturing has more merit, but complex fracture geometries present significant challenges for successful treatment design while trying to avoid or minimize extremely costly early screen-outs.

It is well known that the optimization of hydraulic or acid fracture stimulations requires a knowledge of the magnitudes and orientations of the *in situ* stress system. This allows the prediction of the direction and geometry of the induced fractures. (e.g. Anderson 1951; Hubbert & Willis 1957; Williams & Nierode 1972). Although this design process is well established in systems with a normal or slightly elevated pore pressure and appreciable differences in magnitude between at least two *in situ* stresses (Economides & Nolte 2000; van Batenburg & Hellman 2002), it is less well understood in chalk reservoirs, where all three *in situ* stresses and the pore pressure approach similar values. However, some tight gas case histories are available (e.g. Holditch & Tschirhart 2005). This paper presents a case study from chalk where near-isotropic *in situ* stresses and pore pressures close to the minimum stress are shown to occur. We also discuss the implications for hydraulic fracture stimulations.

Geology

Ockley is situated in North Quad 30 in the East Central Graben of the North Sea (Fig. 1). The Ockley accumulation is a *c.* 60 m thick chalk sequence occurring within the Hod Formation of Campanian age chalk (Fig. 2) and occurs directly above a grounded Jurassic–Triassic tilted fault block (Fig. 3) The crest of the major tilted fault block at the Jurassic–Triassic level and the deep-seated structure have strongly controlled the structure in the chalk level above (Helgeson 1999; Davison *et al.* 2000). The crest of the Ockley structure is coincident with the deeper Jurassic crest because the lower Cretaceous sediments were deposited during the post-extensional sag phase and exploited pre-existing lows.

Continuing salt movement and salt withdrawal from beneath the Jurassic–Triassic fault block into the salt diapir to the north has caused rotation of the Ockley structure about an east–west axis towards the south since the original deposition. In addition, there has probably been some fault reactivation of the deep-seated pre-Base Cretaceous Unconformity fault system within the Jurassic–Triassic, which has resulted in an imprint and trace of the deep-seated faults at the Hod level on the Ockley structure (Helgeson 1999). During the deposition of the chalk package, seismic evidence shows that

From: TURNER, J. P., HEALY, D., HILLIS, R. R. & WELCH, M. J. (eds) 2017. *Geomechanics and Geology*. Geological Society, London, Special Publications, **458**, 113–129.
First published online May 30, 2017, https://doi.org/10.1144/SP458.3

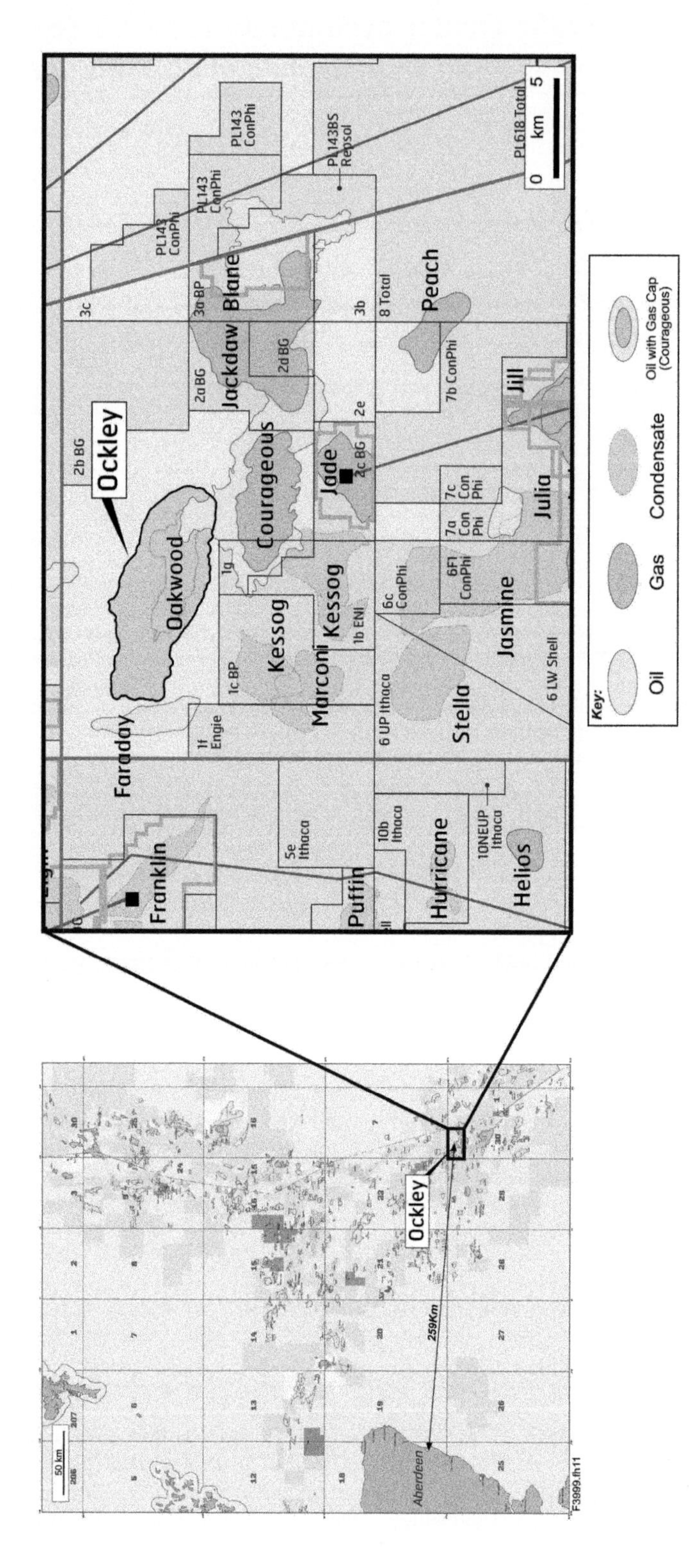

Fig. 1. Location map of Ockley Field, North Sea.

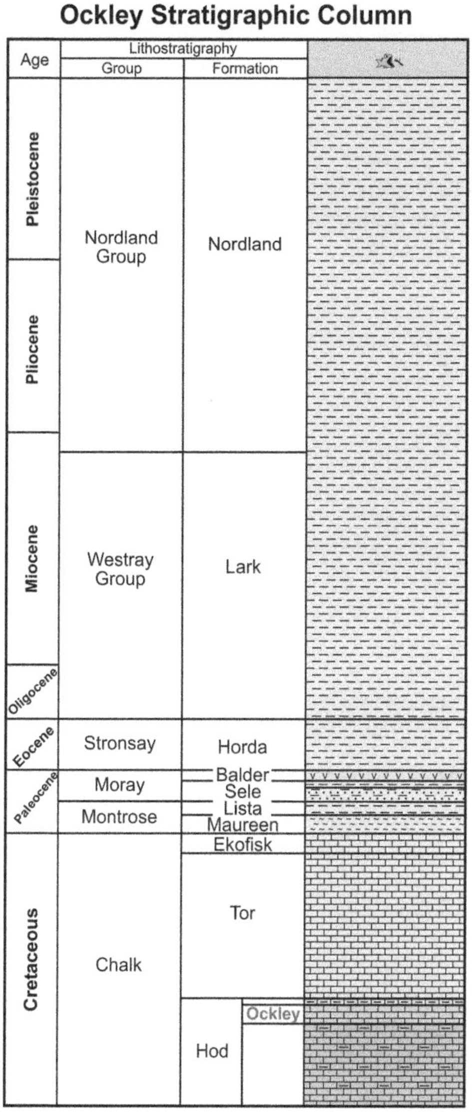

Ockley Stratigraphic Column

Age	Lithostratigraphy: Group	Lithostratigraphy: Formation
Pleistocene	Nordland Group	Nordland
Pliocene		
Miocene	Westray Group	Lark
Oligocene		
Eocene	Stronsay	Horda
Paleocene	Moray	Balder
		Sele
	Montrose	Lista
		Maureen
Cretaceous	Chalk	Ekofisk
		Tor
		Hod / Ockley

Fig. 2. Stratigraphic column of Ockley Field, North Sea.

there is very little variation in the depositional thickness of the Ockley interval, which suggests that it was a period of structural quiescence. This is supported by the consistent penetrated thicknesses of the Ockley interval in the surrounding wells and the correlatable porosity cycles over the 10–15 km length scale (Fig. 4).

Analysis of the slabbed core in well 30/1d-12 suggests that the Ockley chalk interval was deposited in a relatively calm and stable marine slope setting. The core shows very little textural variation between the individual beds. The argillaceous content is variable and detrital clay is observed to be interspersed with the carbonate chalk, suggesting that the chalks accumulated by suspension fallout of terrigenous clays. The succession displays cyclicity between the variably argillaceous chalk beds and the shale-rich layers. The presence of horizontal trace fossils on the bed tops indicates that the depositional setting was oxygenated and nutrient-rich. No large-scale soft sediment deformation or mass transport complexes are identified. The bedding remains undeformed throughout the cored interval.

There are nine distinct porosity cycles within the Ockley unit that can be correlated across the field with some variation in relative thickness. Each cycle in this area consists of regular interbeds of clay-rich, low-porosity (5–10%), very low permeability chalk and slightly less clay-rich, more porous (10–20%) chalk with a low permeability. Overprinted on this framework of variable clay content and porosity cyclicity, seismic amplitude maps show a strong link to elevated porosity, which appears to represent the preserved porosity from early hydrocarbon emplacement (Megson & Hardman 2001). The offset of the amplitude anomaly and structural contours seen in Figure 4 is the result of post-hydrocarbon emplacement tilting. The long-range correlation of cycles indicates a relatively simple reservoir architecture.

The origin of the high pore pressures in the Ockley unit is probably related to either disequilibrium compaction or the hydrocarbon charge–porosity preservation process (Swarbrick *et al.* 2000, 2010; Megson & Hardman 2001). This could occur from continued burial of the isolated hydrocarbon-bearing interval, where the very tight permeabilities in the reservoir section and non-hydrocarbon-bearing intervals would not allow the pore pressures to equilibrate to a normal pressure trend (Swarbrick *et al.* 2010).

Despite the structural history, involving nearby halokinesis and structural tilting, the structural elements within the Ockley accumulation are relatively small in terms of length and/or displacement. Features have been picked from seismic semblance (Kington 2015) at the Base Ockley (Fig. 5), although the origin of these structural elements is not clear from seismic data alone as they could be faults or clusters of fractures. These features indicate a variety of trends that vary spatially. The main strike trends appear to be WNW–ESE, NNW–ESE to north–south and east–west. It is possible that at least some features are accommodation structures formed from the reactivation of deeper basement faults at the Triassic level because these appear to have east–west and NNW–ESE trends.

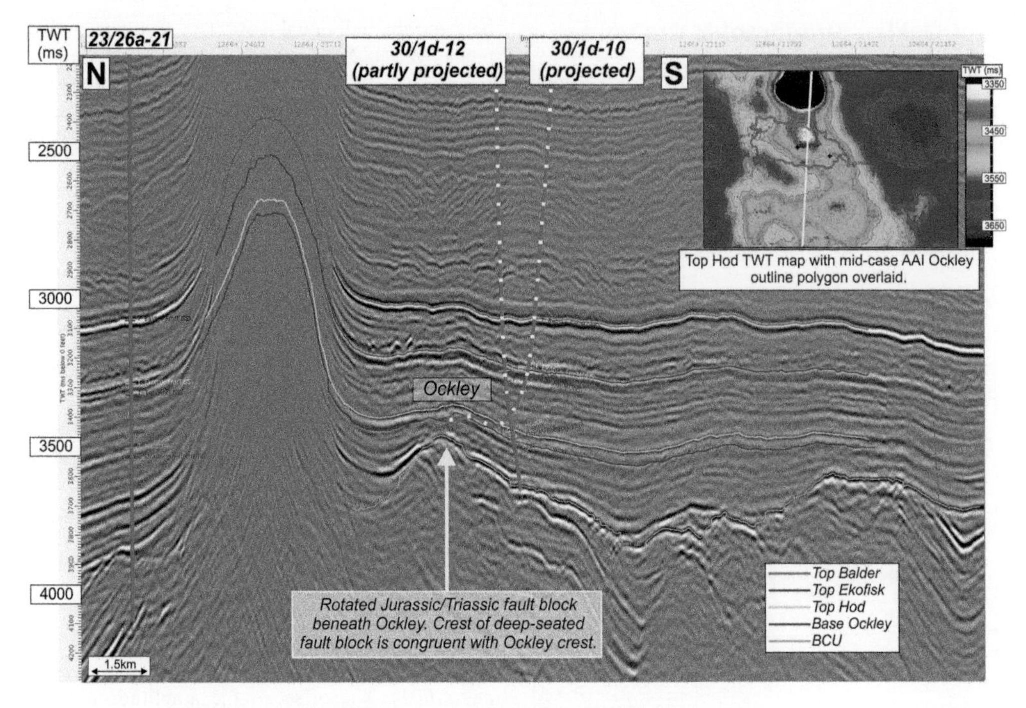

Fig. 3. Seismic section through the Ockley accumulation, NQuad 30, Central North Sea.

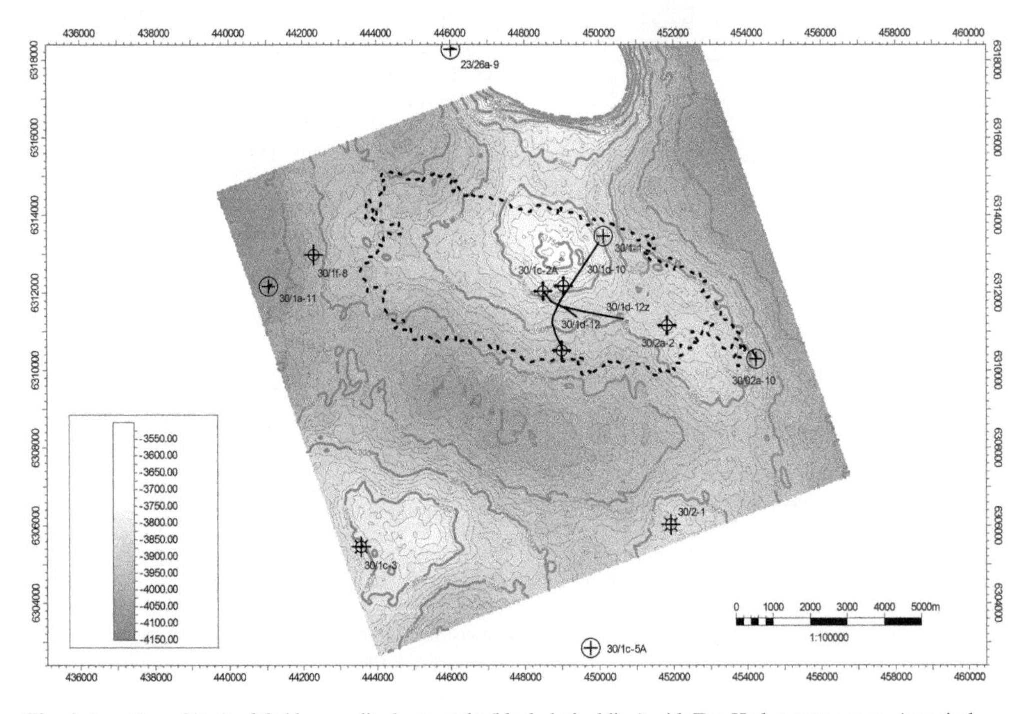

Fig. 4. Location of limit of Ockley amplitude anomaly (black dashed line) with Top Hod structure map. Appraisal wells shown as black lines. Regional exploration wells shown as symbols. Depth contours in metres.

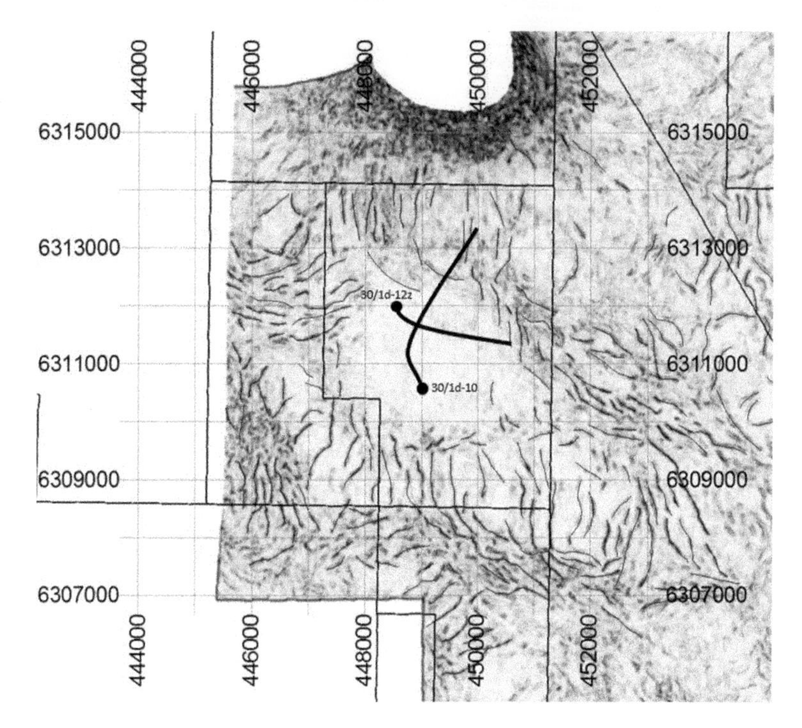

Fig. 5. Seismic semblance map at Base Hod level. Manual interpretation of structural elements shown as black lines. Horizontal appraisal well trajectories shown as thick black lines.

Appraisal data

The Ockley appraisal drilling and testing programme focused on the following elements: (1) the determination of the petrophysical properties of the matrix; (2) determination of the reservoir pressure and temperature; (3) identifying and characterizing natural fractures; (4) flow testing the hydrocarbon fluid to determine the reservoir permeability and fluid type; and (5) testing different stimulation options, including understanding the *in situ* stress system.

These objectives were largely met with three drilled appraisal wells: one deviated (30/1d-12) and two horizontal (20/1d-10 and 30/1d-12Z) (Fig. 6). The following sections describe various aspects of the appraisal data that are related to the *in situ* stress system, geomechanical parameters and relevant geology. The map in Figure 6 shows the distribution of the appraisal wells, plus some of the fault and fracture interpretations. The orientation sampling is excellent, with the three wells drilled in three different directions approaching orthogonal sampling. However, different image log suites were used for each well by different interpreters, meaning that systematic comparisons of image log interpretations are difficult. In particular, the 30/1d-12Z dataset was picked as part of a quick-look evaluation, meaning that the interpreted fracture intensity is much lower than in the other two wells.

Stimulation and testing

Well testing in the vertical exploration and appraisal wells generally consisted of short flow periods from perforated casing, occasionally with some acid wash stimulation that either yielded no fluid or very low rates of gas condensate or water depending on the well's location with respect to the hydrocarbon column. There is no evidence from any of these Hod chalk tests for natural fractures contributing to flow. However, the Ockley natural fractures are near-vertical and these are vertical wells testing a gas condensate over short periods. Therefore the chance of intersecting natural fractures is low and the radii of investigation of the well tests are very short.

Partly to address the shortcomings of the vertical sampling and well tests, two horizontal appraisal wells were drilled that improved the sampling of natural fractures. These were stimulated utilizing different stimulation techniques and longer well tests were feasible. Note a datum depth of 3897 mTVDss is assumed here when referencing all the pressures discussed in this paper.

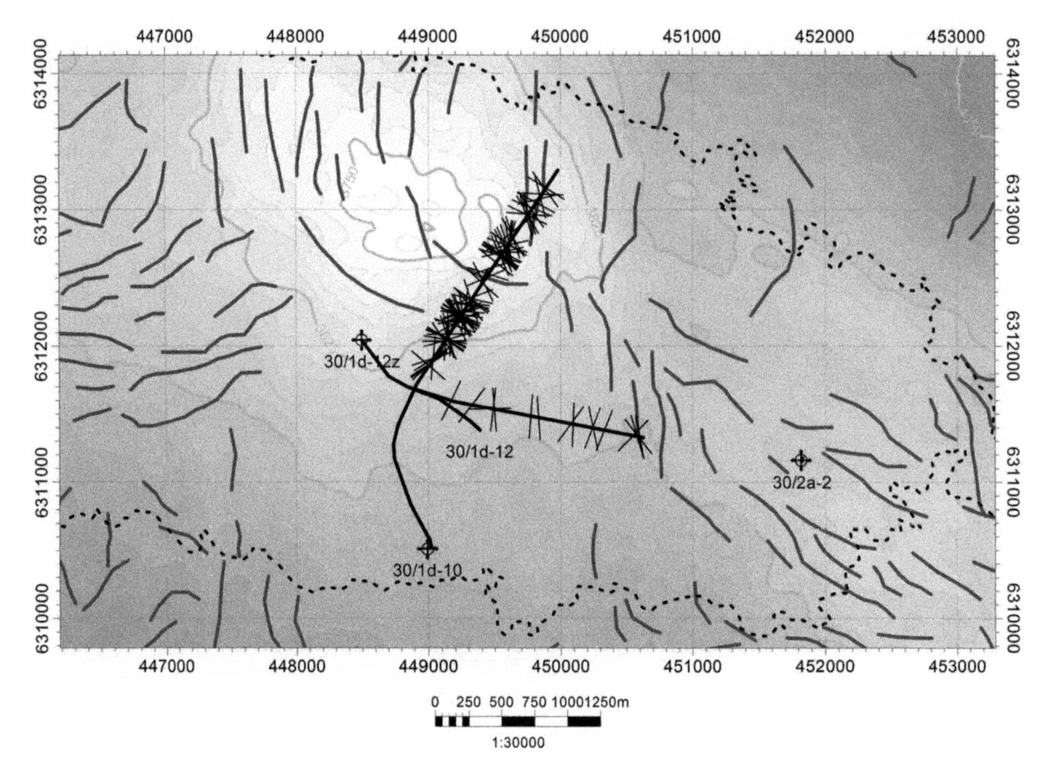

Fig. 6. Location map for Ockley appraisal wells. Grid is 1 km spacing. Map is top Ockley (Hod). Dashed outline is limit of seismic amplitude anomaly. Dark grey lines are fault/fracture clusters manually picked from seismic semblance at Base Ockley. Thin black lines at wells are strike directions of image log-derived fracture and fault picks.

Well 30/1d-10

This well was drilled in a NNE direction and was stimulated using a controlled acid jetting liner. This is a limited entry technique generally used for long horizontal wells in Danish chalks. Acid is bull-headed from the surface into a pre-drilled, uncemented liner, displacing the drilling mud. The design of the pre-drilled perforations facilitates the even distribution of the acid across the full length of the lower completion.

30/1d-10 produced for short periods over six days (Fig. 7). Flow was generally unstable and the pressure build-ups did not reach equilibrium even after the final three-day build-up period. This well test behaviour was best matched by a very low permeability matrix system with no clear evidence of contributions from natural fractures and it is consistent with the core poroperm data (Table 1).

It is likely that the matrix acid stimulation was largely unsuccessful due to the displacement of drilling mud into the tight carbonate and the high clay content of the matrix. This clay, having been released from the matrix by the acid, may have bridged and plugged any developed wormholes, natural fractures or possibly the liner annulus, flow along which the limited entry technique depends upon. Displaced mud may have plugged any natural fractures that were open. With very low flow rates from the formation, insufficient gas velocity was generated to fully remove the stimulation fluids from the well. The well therefore never fully cleaned up and the test results were severely affected, leaving considerable uncertainty in most key parameters.

Wells 30/1d-12 and 12Z

During the most recent appraisal campaign, a full thickness core, BakerAtlas Reservoir Characterization Explorer (RCX) pressures, samples and wireline logs, including image logs, were acquired in the pilot hole 30-1d-12 (Conti & Kumar 2013).

A side-track well, 30/1d-12Z, was drilled in a sub-horizontal direction to the ESE, which allowed the optimal sampling of different fractures to those sampled in 30/1d-10. From the seismic data

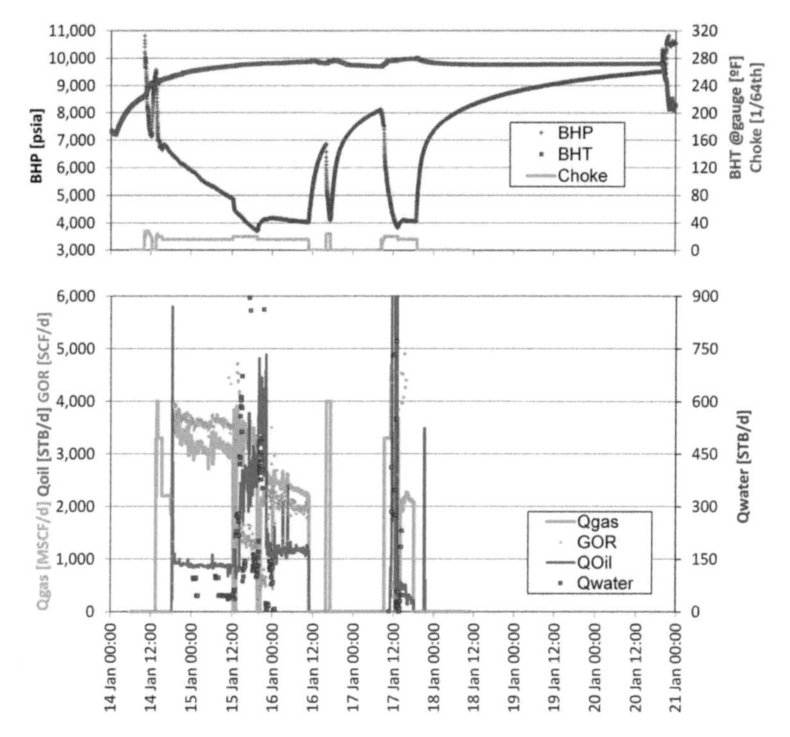

Fig. 7. Ockley well 30/1d-10 test results. The upper chart shows the downhole gauge pressure (BHP) and temperature (BHT), together with the production choke setting. The lower chart shows the data recorded at the surface during the 2008 well test: gas rate (Q_{gas}), oil rate (Q_{oil}), water rate (Q_{water}) and gas–oil ratio (GOR).

(Fig. 5), there appear to be fewer seismic-scale features in the area drilled, but several smaller faults and fractures were encountered.

Hydraulic propped fracture stimulation was selected as the most viable stimulation method for this well and a multistage stimulation completion was installed to facilitate efficient hydraulic fracture placement. However, issues with the completion and placement of the first two propped fractures led to the decision to stimulate the last four stages with acid fractures. Table 2 summarizes the details of the stimulations attempted in 30/1d-12Z and their locations are shown in Figure 8. Port #1 is at the toe of the well and port #8 is at the heel of the well. The propped hydraulic fractures attempted in ports #1 and #2 screened out early and, although several acid fractures were completed successfully,

Table 1. *Ockley well 30/1d-12Z stimulation summary*

Port No.	Fracture type	Comments
Port #1	Data and main fracture	Main fracture screened out at the 4 ppa stage
Port #2	Data fracture	Data fracture screened out; no attempt for main fracture
Port #3	Data and acid fracture	Data fracture unsatisfactory; moved to acid fracture
Port #4	Data and acid fracture	Completed
Port #5	Acid fracture	Completed
Port #6	Acid fracture	Completed
Port #7	Not activated	
Port #8	Not activated	

Table 2. *Ockley well test derived permeability and pressure interpretation summary*

Datum 3897 mTVDss	Permeability (mD)	Pressure (bar)
30/1d-12 pilot hole	0.0003–0.01 (RCX)	754–767 (RCX)
30/1d-12z WTA	0.035	767–778
30/1d-10 WTA	0.002–0.017	756
30/1C-2A	0.0078–0.02	708–724
30/1C-3	0.0032	730

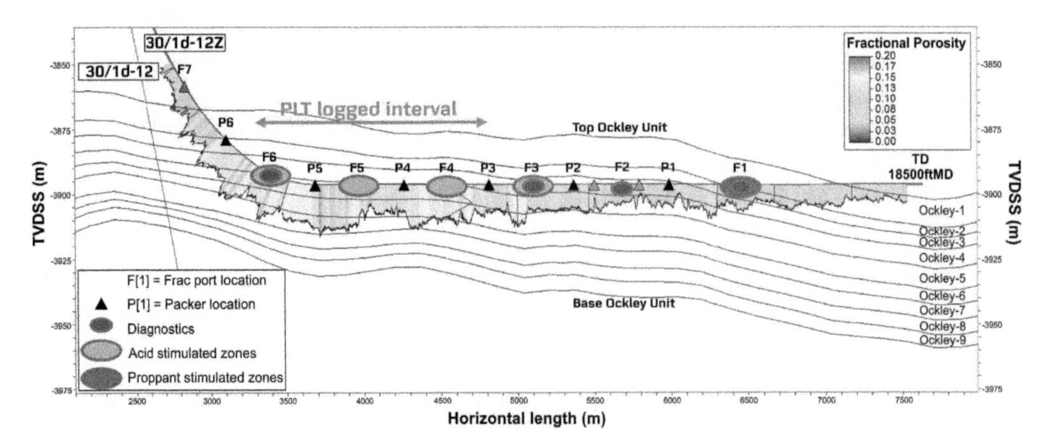

Fig. 8. Well section diagram from well 30/1d-12Z indicating location and type of stimulation jobs. Direction of well is ESE.

subsequent production logging tool (PLT) analysis indicated that they did not contribute very much to the total well flow. However, despite the issues with the propped hydraulic fractures, the PLT data indicated that 70% of the flow coming from below port #3 and 30% from ports #4–6. The evaluation of memory PLT data, the performance of acid fractures, numerical history match work, analytical well test analysis and consideration of the stimulation operation suggest a likelihood of 50% of the production being attributed to the toe fracture (port #1) and the rest evenly distributed from ports #2–6. Ports #7 and #8 were not activated because the lower completion had been set high and these ports sat above the Ockley Formation. It should be noted that this left an openhole interval of *c.* 300 m below the bottom packer and this whole zone was open to flow through port #1.

From Figure 9 it can be seen the well flowed for several short intervals over a 26-day period. A stable

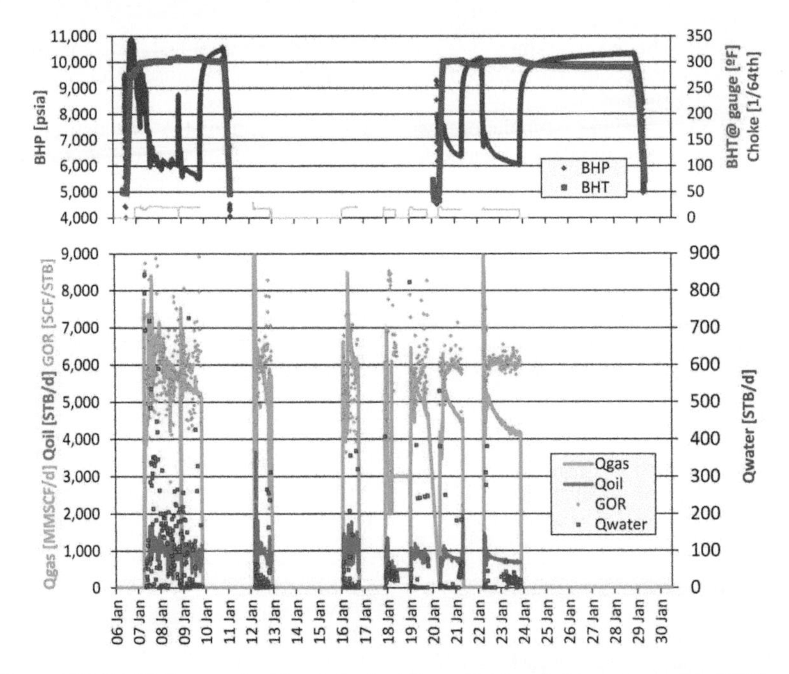

Fig. 9. Ockley well 30/1d-12Z test results. The upper chart shows the downhole gauge pressure (BHP) and temperature (BHT), together with the production choke setting. The lower chart shows the data recorded at surface during the 2013 well test: gas rate (Q_{gas}), oil rate (Q_{oil}), water rate (Q_{water}) and gas–oil ratio (GOR).

flow regime was achieved and multiple build-ups acquired, including a final build-up of 120 h. The best well test match achieved with the data was that of a horizontal well in a very low permeability matrix with induced fracture geometries as modelled from the fracture stimulation data. Again, no natural fracture contribution was required for the match. Note that this is at the limit of using transient well test analysis to interpret data on horizontal well multiple fractures where the fractures are of different geometries and natures; the results of the transient well test analysis were therefore further validated with numerical history match analysis.

Matrix properties

The variability in key logs within well 30/1d-12 is shown in Figure 10. Based on X-ray diffraction analysis from 30/1d-12 core data, the Ockley matrix consists of *c.* 80% calcite and the rest is quartz and clay. The quartz morphology is unclear, despite scanning electron microscopy analysis, and is believed to be a combination of authigenic and detrital components. The quartz component is relatively constant as a proportion of the total rock (based on cuttings data from the horizontal wells) and it has a weak trend with the clay content. Purely authigenic quartz might be expected to be more variably distributed. The clays consist of illite, chlorite and kaolinite.

Matrix porosity has been characterized from cores and logs in the three new appraisal wells and the older vertical exploration and appraisal wells. In general, the wells show a porosity range of 8–18% in the hydrocarbon-bearing interval and 2–12% in the water leg. The average porosity in the Ockley hydrocarbon-bearing interval is 12%. The bulk of the porosity is micro-porosity, but some of it is associated with vugs within millimetre-scale forams.

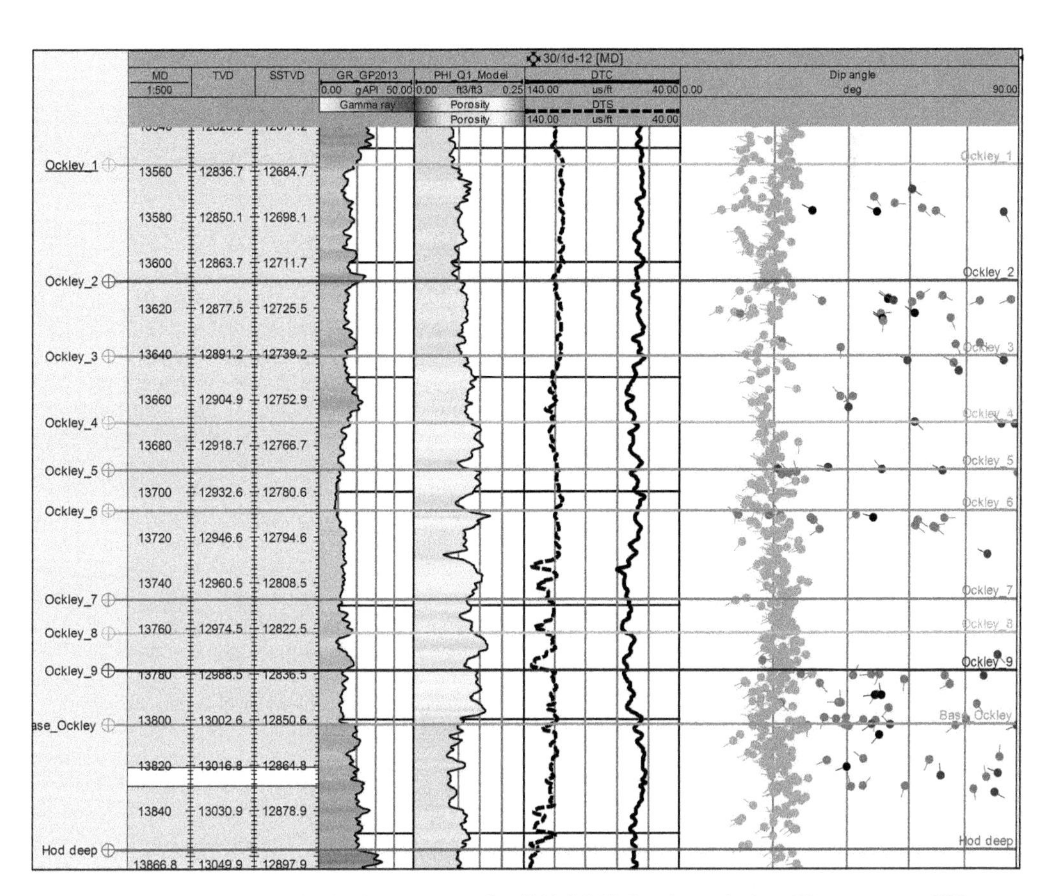

Fig. 10. Log variability within the Ockley sequence of well 30/1d-12. Depths are in feet. GR, gamma ray; PHI, effective porosity; DTC, compressional sonic; DTS, shear sonic. Dip angle: circle position represents dip magnitude, tick mark is dip azimuth. Pale points at 15–25° dips are bedding and darker steeper points are fractures. Note fracture clustering and variable orientations.

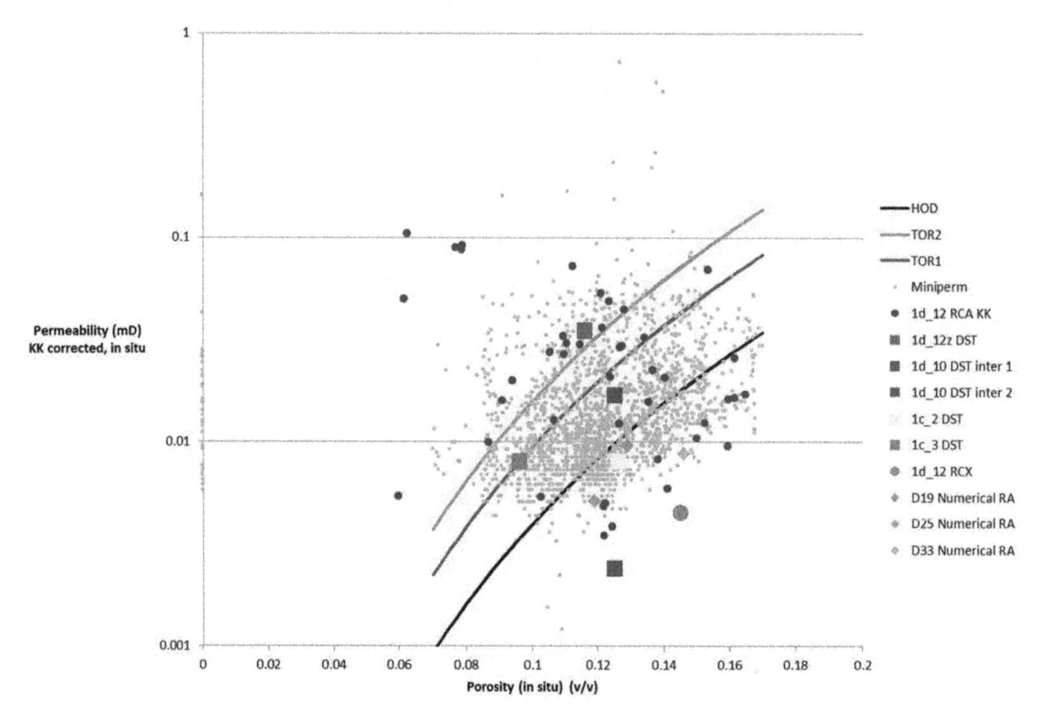

Fig. 11. Ockley core-derived poroperm data, regional data trend lines and well test derived average permeabilities.

The matrix permeability has been characterized from a variety of sources in the 30/1d-12 and 30/1d-12Z wells: core, well logs, well test data and formation fluid sampling (RCX tool). The key permeability data are summarized in Table 1 and Figure 11. These show that the very low average matrix permeabilities of 0.003–0.07 mD are characteristic of the core plug scale up to the well test scale.

Faults and natural fractures

From core available in well 30/1d-12, a number of different types of fractures have been identified over a cored interval of 240 m MD. Details of the feature classes identified are given in Figure 12. The cored natural features are generally small, irregular, healed or discontinuous. They are sparsely distributed within the core with an average of 0.1 fractures per metre, including all types of features and including multiple fractures in clusters.

Ockley faults and fractures are reviewed here briefly to establish the framework of structural elements that may be relevant to stimulation activities. Figure 13 shows a map of the distribution of image log picks in the three recent appraisal wells, together with the seismic semblance derived fault/fracture clusters. The rose diagrams in Figure 14 show open fracture strikes on the left and wellbore image log fault picks on the right. In the few places where the seismic-scale features coincide with the wells, there are generally picks in the wells with similar orientations. The pick types also generally confirm that the seismic-scale features are faults or clusters of fractures, possibly associated with minor faults.

In terms of orientation groupings, the natural fracture systems identified from image logs show a degree of scatter. Some of this may be due to undersampling of features in wells of different orientations. The image log fault pick orientations agree well with the seismic semblance feature orientations with a slightly more dominant WNW–NNW strike direction (Fig. 13). The image log 'open' (conductive) fractures generally show more features that are oriented NE, open east–west, compared to the total dataset (Fig. 14), but there are less obvious orientation clusters within these features. The filtering of features by dip $>50^\circ$ on the lower rose diagrams do not significantly change these observations (see Fig. 14).

Based on the various observations outlined here, the natural fractures with at least some evidence of openness occur in sparse clusters and do not appear to have well-developed open apertures. In summary:

(1) No mud loss was observed in wells 30/1d-10, 30/1d-12 or 30/1d-12z. However, wellbore breathing was observed in the Ockley interval during the drilling of well 30/1d-10.

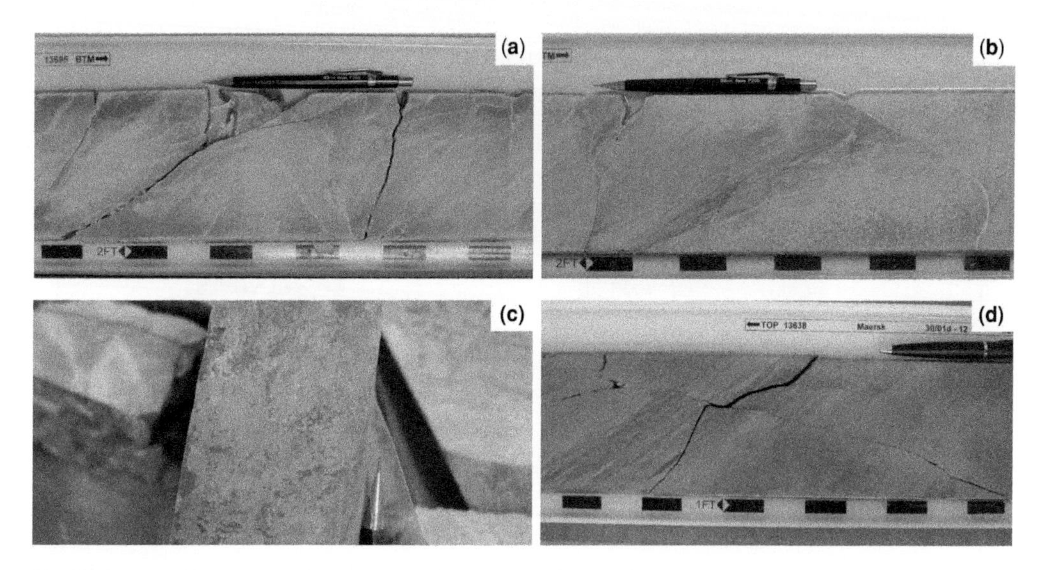

Fig. 12. Structural features seen in core. (**a**) Centimetre-scale clusters of irregular intersecting fractures in the cleaner chalk intervals. Possibly natural microfractures or possibly induced fractures created during coring/stress unloading/desiccation. (**b**) Partially or fully mineralized veins. These can be irregular in places and could be open in the subsurface. (**c**) Planar or irregular features with evidence of shear. Probably post-lithification faults or reactivated veins. Could be open in the subsurface. (**d**) Small-scale irregular healed features with evidence of shear. Early or synsedimentary faults.

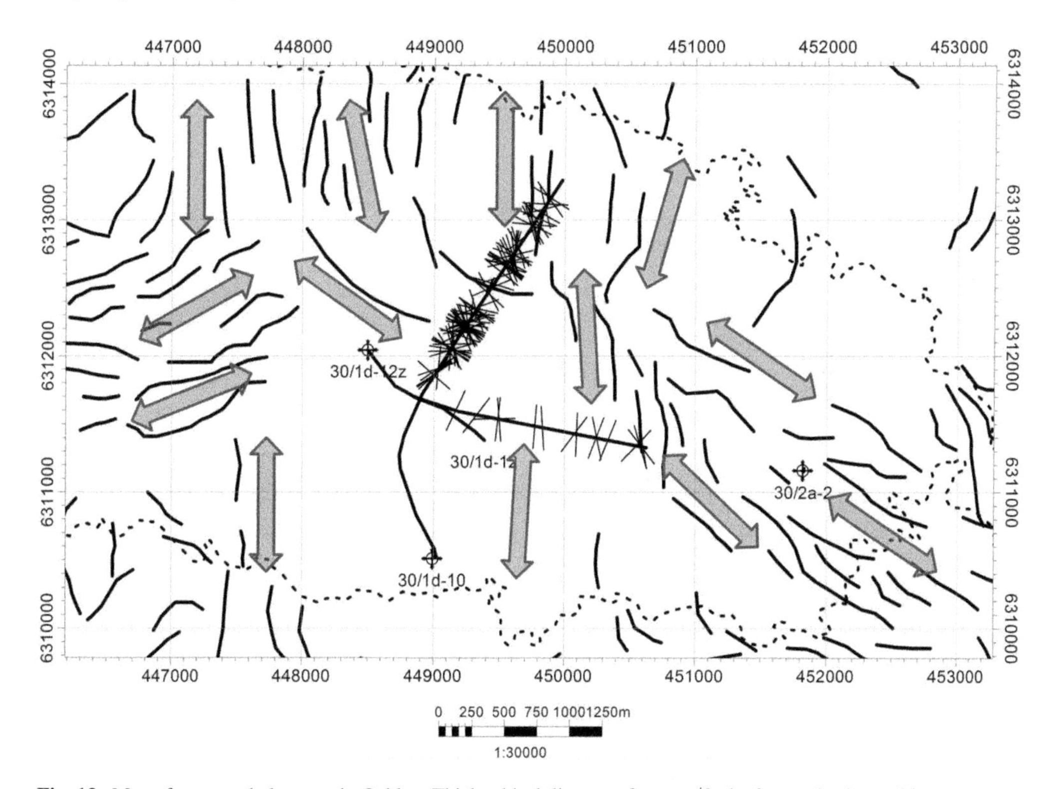

Fig. 13. Map of structural elements in Ockley. Thicker black lines are features/faults from seismic semblance attribute. Thin black lines at wells are strike directions of image log-derived fracture and fault picks. Double headed grey arrows are potential S_{Hmax} directions in option 4.

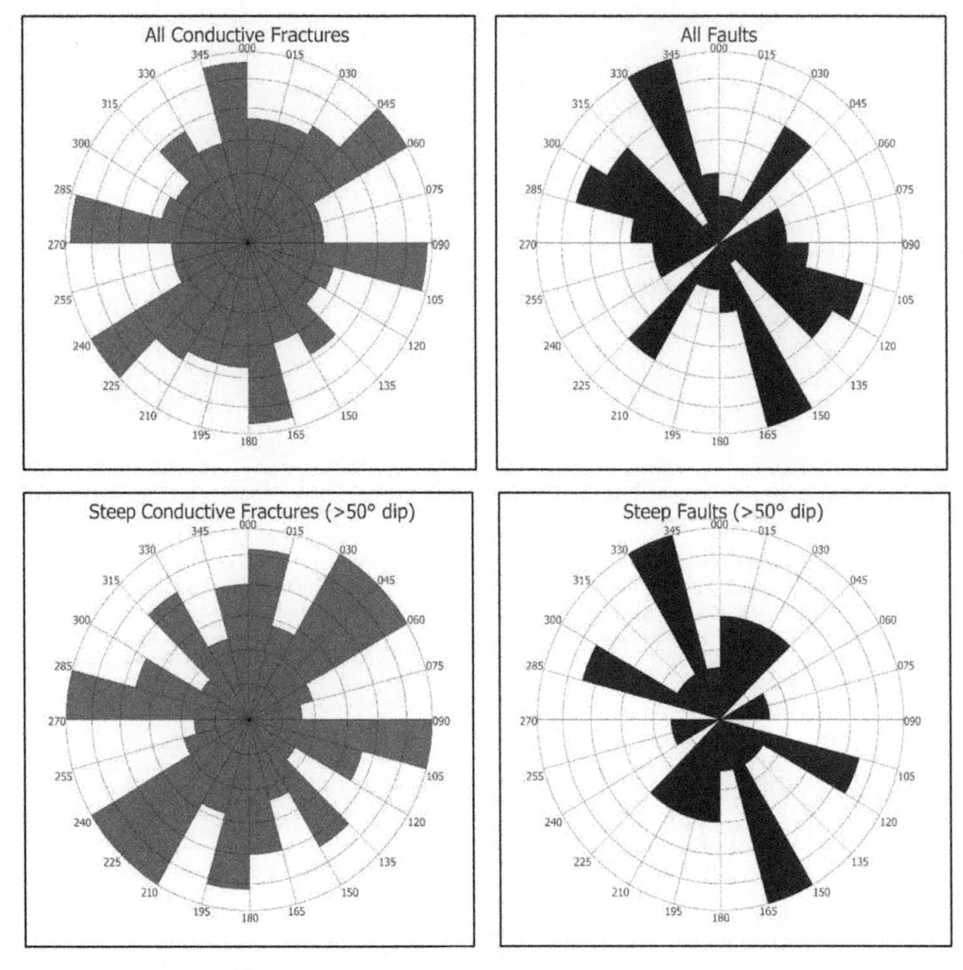

Fig. 14. Rose diagrams of fracture and fault strike directions from image log pick data in all three appraisal wells. Azimuth represents fracture strike direction in 15° orientation classes. Radius represents fracture frequency within each 15° orientation class.

(2) No clear fracture response was seen on azimuthal anisotropy, Stoneley reflectance or acoustic amplitude analysis in well 30/1d-12.

(3) Slickensides and partial mineralization were seen on some fracture faces observed in core. These were few in number and not very open.

(4) Production data from wells 30/1d-10 and 30/1d-12Z were not indicative of an extensive fracture network and were consistent with a contribution from the very low permeability Hod chalk matrix as defined by core plug data.

(5) The interpretation of flow from the partially successful port #1 propped fracture in well 30/1d-12Z indicated that multiple hydraulic fractures with full proppant loads could provide significant flow rates irrespective of the presence of natural fractures.

(6) Some evidence for natural fractures/faults at or beyond the well 30/1d-12z toe area was seen on seismic attribute maps, but this is speculative and their openness is unknown. If open fractures do occur, they are more likely in the toe area of well 30/1d-12z and some of the other peripheral areas with seismic semblance features.

Geomechanics

Pore pressure

There are a number of direct samples of pore pressure in Ockley and these were summarized in the previous section in Table 2. Pressures in very tight rocks such as the Hod chalk are difficult to determine due to the slow response time of the fluids via the very low permeability rock. The interpretation of the 30/1d–12Z extended well test were regarded as the most reliable and an estimated reservoir pressure of 765 bar is used here. This translates to a datum (3897 mTVDss) pressure gradient of 0.196 bar/m. However, the other well data indicate a valid reservoir pressure gradient range of 0.186 bar/m to 0.200 bar/m at this depth.

In situ *stress magnitudes*

The vertical principal total stress (S_V) was assessed from the integration of density logs in well 30/1d-12, plus additional calculated density logs in shallower intervals from Gardner's relation applied to the sonic data (Gardner *et al.* 1974). The vertical total stress was estimated as 841 bar at the datum depth (3897 mTVDss) or 0.216 bar/m. The horizontal minimum principal total stress (S_{hmin}) was assessed from the fracture closure pressures obtained from good quality 'data fracs', mini-fracs and step-rate tests in the 30/1d-12Z Zone #1 fracture stimulation. The interpreted values were clustered around 824–826 bar or 0.212 bar/m. These values are very close to the calculated vertical stress. The horizontal maximum principal total stress (S_{Hmax}) is hard to determine directly, although strong chalks such as Ockley can sustain high stresses, including strike-slip regimes. S_{Hmax} was estimated from regional offset data to be in the region of 1.05–1.07 times the magnitude of S_{hmin} (e.g. Hillis & Nelson 2005), which yields values of 862–881 bar or 0.221–0.226 bar/m at the datum depth.

Biot coefficient and effective stress

The Biot coefficient is an important component of poroelasticity and is defined by the following relation (Nur & Byerlee 1971, Fjaer *et al.* 2008):

$$\alpha = 1 - K_b/K_g$$

where α = Biot coefficient, K_b = the bulk modulus of rock and K_g = the bulk modulus of solid grains. The Biot coefficient limits are defined by $\phi < \alpha < 1$, where ϕ is the effective porosity, and it is a measure of the change in pore volume relative to the change in bulk volume at a constant pore pressure. For natural rocks under confinement, the gravitational and tectonic loads affecting the solid rock frame are, to some extent, counteracted by the hydrostatic pressure exerted by the pore fluid. The Biot coefficient is required to define the effective stress value in rocks that contain a connected fluid-filled matrix framework pore system (Nur & Byerlee 1971):

$$\sigma_n = S_n - \alpha P_p$$

where S_n = total normal or principal stress, σ_n = effective normal or principal stress and P_p = pore pressure.

In most porous reservoir rocks, the Biot coefficient is usually estimated or measured in the range 0.8–1. However, Fabricius (2007) has shown that for low-porosity North Sea chalks, the Biot coefficient can be in the range 0.5–0.85 for the low levels of porosity seen in Ockley. Estimates of the Biot coefficient based on Ockley core data yield a similar range, but this is poorly constrained because an accurate matrix grain modulus was not obtained. The implication of this range is that the pore pressure has less impact on the total stresses than is the case in many reservoir rocks. Therefore depletion during development would not be expected to reduce the fracture gradient by a significant amount within the matrix and any increase in pore pressure would not reduce the load on the matrix by a significant amount.

It should be noted that the effect of the Biot coefficient is only relevant in a matrix with no connected, open or compliant natural fractures. If natural fractures occur that can respond to pressure changes such that the effective compressibility of the aggregate is sufficiently greater than the grains, then it is more appropriate to use the Terzaghi law (Nur & Byerlee 1971):

$$\sigma_n = S_n - P_p$$

The Terzaghi law is also valid during any post-failure deformation because the matrix integrity will usually be reduced during any brittle deformation expected at these depths. It is an interesting issue as to which law is valid for defining the initial effective stress conditions of overpressured rocks containing sealed or cemented fractures that are weaker than matrix. If the Biot coefficient is assumed to be 0.68 or 1.00 (i.e. the Terzaghi law is valid), then the Ockley effective principal stress magnitudes at the initial conditions are as given in Table 3.

Table 3. *Ockley effective principal stress magnitudes at initial conditions for different Biot coefficients*

Stress	Biot 0.68 (bar)	Biot 1.00 (bar)
σ_v	320.47	75.57
σ_{hmin}	304.89	59.98
σ_{Hmax}	351.08	106.18

In both cases, this is a relatively narrow range with only 46.19 bar difference between the minimum and maximum principal effective stress values. At these low differential stresses, any increase in pore pressure from, say, a pressure increase created prior to hydraulic fracturing is likely to result in tensile failure rather than shear failure.

S_{Hmax} *direction*

The horizontal development wells need to be drilled parallel to the S_{hmin} direction such that hydraulic fractures can be created transverse to wells to maximize the fracture area per well, reduce tortuosity during the fracture stimulation and to maximize well rates. Given the similarity of the principal stress magnitudes it is practically very difficult to determine the S_{Hmax} direction and, even if it is possible to determine, it may vary significantly away from the measurement point. For Ockley, a direct measurement of an S_{Hmax} direction of *c.* 100° (ESE) was obtained in well 30/1d-12 from induced fractures in the basal part of the Tor chalk overlying the Hod (Tony Batchelor pers. comm. 2014).

A measurement in the Tor may not be relevant for the underlying Hod, so there is considerable uncertainty in the Hod S_{Hmax} direction. To fully assess the impact of a highly uncertain S_{Hmax} direction on future Ockley hydraulic fracture operations, a number of S_{Hmax} orientation scenarios were conceived. These are listed in the following and option 4 is shown in Figure 13

(1) All NE–SW. Some local *in situ* stress directions have this trend.
(2) All NW–SE. This is the dominant S_{Hmax} trend in the North Sea.
(3) Randomly switching NE to NW. This can occur when S_{Hmax} and S_{hmin} are close in magnitude.
(4) Follow local structure. This is a scaled-down implementation of observations across the North Sea that S_{Hmax} follows the dominant fault strike direction.

This variability in S_{Hmax} direction is not just restricted to Central Graben overpressured chalks; highly variable S_{Hmax} directions are observed throughout the Central Graben above the Zechstein salt, although S_{Hmax} is often parallel to the local structure (Hillis & Nelson 2005). Although option 4 (Fig. 13) may be the most likely based on regional experience, it is not directly confirmed from local Hod measurements. Therefore, because horizontal wells at this depth are time consuming and expensive to drill and stimulate, even unlikely alternative stress orientation scenarios need to be considered to mitigate risks.

Discussion

Fracture propagation

The tensile strength of intact, isotropic, low-porosity Hod chalk at scales relevant to the wellbore wall is not well known. Tensile strength correlations based on unconfined compressive strength tests are not reliable because there have been no rock mechanics core tests in Ockley. The tensile strength of intact Ockley Hod chalk is likely to be in the range 20–140 bar based on limestone tensile strengths (Perras & Diederichs 2014) and Griffith criteria estimates from the cohesion of 18–19% porosity chalk reported in Hickman (2004). However, the tensile strength of chalk with bedding planes or existing faults/fractures is likely to be significantly lower than intact Hod chalk and in the range 0–14 bar.

The fracture initiation pressure in a horizontal wellbore with an impermeable wellbore wall parallel to S_{hmin}, assuming an intact Hod chalk tensile strength of 52 bar or zero tensile strength (at 3897 mTVDss) and a Biot coefficient of 0.68 can be calculated using the following equations (Fjaer *et al.* 2008):

$$3S_v - S_{Hmax} - \alpha P_p + T_0$$
$$2522.65 - 871.50 - 0.68 \times 765.32 + 51.71 = 1182.45\,\text{bar}$$
$$2522.65 - 871.50 - 0.68 \times 765.32 + 0 = 1130.74\,\text{bar}$$

where α = Biot coefficient, P_p = pore pressure (bar) and T_0 = tensile strength.

If the Terzaghi law is assumed to be valid due to the presence of weak natural fractures (albeit poorly connected fractures) the relationship becomes:

$$3S_v - S_{Hmax} - P_p + T_0$$
$$2522.65 - 871.50 - 765.32 + 51.71 = 937.55\,\text{bar}$$
$$2522.65 - 871.50 - 765.32 + 0 = 885.84\,\text{bar}$$

These fracture initiation pressures are all greater than S_1 (S_{Hmax}) irrespective of whether the Biot or Terzaghi effective stress laws are used.

For an idealized horizontal wellbore in the Ockley stress system that is parallel to S_{hmin}, the hydraulic fracture will initiate at the wellbore wall in a horizontal direction parallel to S_1 (i.e. S_{Hmax}) and then reorient to exploit the direction with the lowest *in situ* stress plus tensile strength combination. In the case of isotropic intact Hod chalk, this direction is a vertical fracture perpendicular to the wellbore, opening perpendicular to S_{hmin} and propagating parallel to S_{Hmax}. However, given the low differential effective stresses and the potential for high variability in tensile strength in Ockley, other *in situ* stress plus strength combinations are possible for the

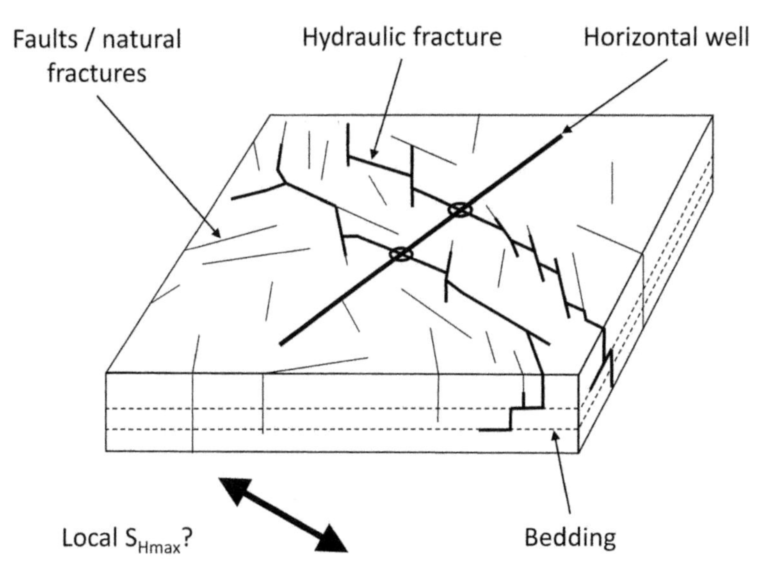

Fig. 15. Schematic diagram of potential hydraulic fractures in a horizontal Ockley well parallel to S_{hmin}.

optimum direction seen by a propagating fracture tip. For example, a fracture opening against S_V or S_{Hmax} plus a low or zero tensile strength may require a lower propagation pressure than a fracture opening against S_{hmin} plus the intact Hod chalk tensile strength. Given that induced fractures in a horizontal Ockley well would probably initiate in the horizontal direction at pressures $>S_1$ (S_{Hmax}), this could mean that fractures parallel to the bedding plane propagate for a significant distance away from the wellbore.

The diagram in Figure 15 graphically summarizes the scenario of a notional hydraulic fracture in a horizontal Ockley well parallel to S_{hmin}. This diagram ignores the near-wellbore sub-horizontal fracture initiation direction. The fracture starts growing in isotropic intact chalk in a transverse direction away from a horizontal wellbore. As it propagates, the relatively isotropic *in situ* stresses mean it could exploit a variety of pre-existing faults, fractures or bedding planes in any orientation. This will obviously have implications for the design of any hydraulic fractures.

Fracture designs

From the previous discussion it is apparent that hydraulic fractures in Ockley are likely to exploit a variety of pre-existing natural weakness planes and could propagate in any orientation (Taleghani & Olson 2014). This situation has implications for fracture designs because the propagating fracture is highly likely to have a complex geometry with a highly variable aperture (Jeffrey *et al.* 2009).

Given that stimulation is required for the Ockley Formation to produce commercial gas rates, and that acid stimulation is not a viable option due to the Hod clay content, multistage propped hydraulic fractures are seen as a requirement for development. However, due to the high cost of operating in a moderately deep, high-pressure, high-temperature environment, a low-risk approach to well and hydraulic fracture design is imperative. A cased hole (cemented liner) lower completion is the lowest risk well option, but design option contingencies will then be required to reduce the effects of stress concentration at the wellbore and ensure a good connection between the wellbore and the formation. This may be abrasive jetting, oriented (stimulation) perforating, cased hole stimulation sleeves and/or acid jetting or proppant slugs.

Tortuosity is defined as the pressure effects due to a near-wellbore convoluted hydraulic fracture geometry caused by the fracture orientation changing from the near-wellbore stress system to the far-field *in situ* stress system (Cleary *et al.* 1993, Economides & Nolte 2000). Significant tortuosity can be expected from any hydraulic fracture on Ockley, so screen-out during proppant placement becomes the most significant risk to successful fracture placement. To minimize this risk initially, proppant size should be minimized (30/60, for example) and proppant loading kept low at *c.* 4 pounds proppant added or lower. One option that may potentially achieve this is channel fractures (non-continuous proppant loading).

With so many unknowns associated with the placement of fractures in a near-isotropic *in situ* stress state such as Ockley, opportunities for continuous learning should be factored into the development programme. This may be the placement of 'test' fractures in the toe of the well, followed by long-term production tests to optimally adapt the treatment design (increased proppant size/loading or an alternate wellbore to formation connection strategy, for example) prior to full-scale implementation, or may simply be the acquisition of detailed geomechanical data, including micro-seismic (surface or well-to-well) for mapping the hydraulic fracture geometries and defining subsequent optimum well orientations. Laboratory test work would also help to define the optimum fracture width/proppant size for maximum clean-up and conductivity, proppant embedment quantification and formation damage mitigation with respect to fracture fluids. Learning from the shale gas industry may be more appropriate for the Ockley tight gas development than North Sea analogues.

Conclusions

Ockley is a highly overpressured Hod chalk gas condensate accumulation situated in the Central North Sea. The matrix properties display low porosities of 5–10% and very low (microdarcy) permeabilities. This low-porosity chalk is very strong, with intact chalk tensile strengths around 52 bar. Natural fractures, small faults and bedding plane discontinuities occur within the Ockley unit, but do not appear to be open and well connected in terms of fluid flow behaviour. These features may have a very low tensile strength (0–14 bar) and provide important paths of weakness that could influence hydraulic fracture propagation.

Appraisal well data indicate that Ockley has near-isotropic *in situ* stresses and a very high fluid overpressure. This situation reduces the effective differential stresses to a maximum of *c.* 46 bar. At these low differential stresses it is difficult to accurately define an S_{Hmax} direction to optimize well trajectories and fracture stimulation orientations.

To improve the flow performance, acid and propped hydraulic fracture stimulations have been attempted in a horizontal appraisal well. The acid fractures were not successful due the high clay content of the Hod chalk causing plugging the fracture. The propped hydraulic fracture stimulations showed some flow improvements, but were complicated by the unknown Ockley S_{Hmax} direction and complications from the near-isotropic *in situ* stresses affecting the fracture propagation direction and treating pressures.

Any future developments of Ockley need to account for the uncertainty in the S_{Hmax} direction and the likelihood of complex propped hydraulic fracture geometries. Design criteria obtained from shale gas experience and appraisal during development may help this process.

The authors thank Maersk Oil (UK) Ltd and AGR TRACS International Ltd for permission to publish this paper. The views expressed here are solely those of the authors. We thank the two anonymous reviewers for improving the manuscript. We also thank our many colleagues for technical input to our work on Ockley, which has helped the ideas presented here, for help and guidance during the writing of this paper, and for many discussions. In particular, we thank Stuart Pegg, Tony Batchelor, Karyn Hossack, Tracey Flynn and Nick Hocking.

References

Anderson, E.M. 1951. *The Dynamics of Faulting and Dyke Formation with Application to Britain*. Oliver & Boyd, Edinburgh.

Cleary, M.P., Johnson, D.E. *et al.* 1993. Field implementation of proppant slugs to avoid premature screen-out of hydraulic fractures with adequate proppant concentration. Paper SPE 25892, presented at the Low Permeability Reservoirs Symposium, 26–28 April, Denver, Colorado, https://doi.org/10.2118/25892-MS

Conti, F. & Kumar, R. 2013. Hydrocarbon sampling in tight chalk. Paper presented at the SPWLA 54th Annual Logging Symposium, 22–26 June 2013, New Orleans, Louisiana, SPWLA-2013-EE.

Davison, I., Alsop, I. *et al.* 2000. Geometry and late-stage structural evolution of Central Graben salt diapirs, North Sea. *Marine and Petroleum Geology*, **17**, 499–522.

Economides, M.J. & Nolte, K.G. 2000. *Reservoir Stimulation*. 3rd edn. Wiley, Chichester.

Fabricius, I.L. 2007. Chalk: composition, diagenesis and physical properties. *Bulletin of the Geological Society of Denmark*, **55**, 97–128.

Fjaer, E., Holt, R.M., Horsrud, P., Raaen, A.M. & Risnes, R. 2008. *Petroleum Related Rock Mechanics*. 2nd edn. Developments in Petroleum Geoscience, **53**. Elsevier, Amsterdam.

Gardner, G.H.F., Gardner, L.W. & Gregory, A.R. 1974. Formation velocity and density – the diagnostic basics for stratigraphic traps. *Geophysics*, **39**, 770–780.

Helgeson, D.E. 1999. Structural development and trap formation in the Central North Sea HP/HT play. *In*: Fleet, A.J. & Boldy, S.A.R. (eds) *Petroleum Geology of Northwest Europe – Proceedings of the 5th Petroleum Geology Conference*. Geological Society, London, 1029–1034, https://doi.org/10.1144/0051029

Hickman, R.J. 2004. *Formulation and implementation of a constitutive model for soft rock*. PhD thesis, Virginia Polytechnic Institute.

Hillis, R.R. & Nelson, E.J. 2005. In situ stresses in the North Sea and their applications: petroleum geomechanics from exploration to development. *In*: Doré, A.G. & Vining, B.A. (eds) *Petroleum Geology: North-West Europe and Global Perspectives – Proceedings of the 6th Petroleum Geology*

Conference. Geological Society, London, 551–564, https://doi.org/10.1144/0060551

Holditch, S.A. & Tschirhart, N.R. 2005. Optimal stimulation treatments in tight gas sands. Paper SPE 96104, presented at the SPE Annual Technical Conference and Exhibition, 9–12 October, Dallas, Texas, https://doi.org/10.2118/96104-MS

Hubbert, M.K. & Willis, D.G. 1957. Mechanics of hydraulic fracturing. *Transactions of the Society of Petroleum Engineers of AIME*, **210**, 153–168.

Jeffrey, R.G., Bunger, A.P. *et al.* 2009. Measuring hydraulic fracture growth in naturally fractured rock. Paper SPE 124919, presented at the SPE Annual Technical Conference and Exhibition, 4–7 October, New Orleans, Louisiana, https://doi.org/10.2118/124919-MS

Kington, J. 2015. Semblance, coherence, and other discontinuity attributes. *The Leading Edge*, **34**, 1510.

Megson, J. & Hardman, R. 2001. Exploration for and development of hydrocarbons in the Chalk of the North Sea: a low permeability system. *Petroleum Geoscience*, **7**, 3–12, https://doi.org/10.1144/petgeo.7.1.3

Nur, A. & Byerlee, J.D. 1971. An exact effective stress law for elastic deformation of rock with fluids. *Journal of Geophysical Research*, **76**, 6414–6419.

Perras, M.A. & Diederichs, M.S. 2014. A review of the tensile strength of rock: concepts and testing. *Geotechnical and Geological Engineering*, **32**, 525–546.

Swarbrick, R.E., Osborne, M.J. *et al.* 2000. Integrated study of the Judy Field (Block 30/7a) – an overpressured Central North Sea oil/gas field. *Marine and Petroleum Geology*, **17**, 993–1010.

Swarbrick, R.E., Lahann, R.W., O'Connor, S.A. & Mallon, A.J. 2010. Role of the Chalk in development of deep overpressure in the Central North Sea. *In*: Vining, B.A. & Pickering, S.C. (eds) *Petroleum Geology: From Mature Basins to New Frontiers – Proceedings of the 7th Petroleum Geology Conference*, 493–507, https://doi.org/10.1144/0070493

Taleghani, A.D. & Olson, J.E. 2014. How natural fractures could affect hydraulic-fracture geometry. *SPE Journal*, **19**, https://doi.org/10.2118/167608-PA

van Batenburg, D.W. & Hellman, T.J. 2002. Guidelines for the design of fracturing treatments for naturally fractured formations. Paper SPE 78320, presented at the European Petroleum Conference, 29–31 October, Aberdeen, UK, https://doi.org/10.2118/78320-MS

Williams, B.B. & Nierode, D.E. 1972. Design of acid fracturing treatments. *Journal of Petroleum Technology*, **24**, 849–860, https://doi.org/10.2118/3720-PA

The edge of failure: critical stress overpressure states in different tectonic regimes

RICHARD H. SIBSON

60 Brabant Drive, Ruby Bay, Mapua 7005, New Zealand

Department of Geology, University of Otago, PO Box 56, Dunedin 9054, New Zealand

rick.sibson@otago.ac.nz

Abstract: Earth's seismogenic crust is partitioned into the three Andersonian stress domains critically organized to the edge of failure both along plate boundaries and within plate interiors. Brittle/frictional failure in rocks (the formation and reactivation of faults and fractures) may be induced by two principal drivers: increasing differential stress ($\sigma_1-\sigma_3$) and/or pore fluid pressure, P_f, defined relative to vertical stress by $\lambda_v = P_f/\sigma_v$. Borehole measurements suggest the presence of hydrostatic-Byerlee conditions (the stress governed by the frictional strength of optimally oriented faults with Byerlee friction ($0.6 < \mu_s < 0.85$) under hydrostatic fluid pressure), sometimes postulated as the standard state for fractured seismogenic crust with a bulk permeability too high to allow fluid overpressuring. However, especially in areas of crust undergoing shortening and fluid release under compression, pore fluids are likely to be overpressured above the hydrostatic pressure (i.e. $\lambda_v > 0.4$) in the lower seismogenic zone (*c.* $T > 200$°C), where hydrothermal circulation and cementation reduces fracture permeability. In such regions, critical stress overpressure states prevail with differential stress inversely related to the degree of fluid overpressuring. Failure criteria on plots of ($\sigma_1-\sigma_3$) v. λ_v constructed for particular depths can be used to explore critical stress overpressure states, loading paths to failure and potential mineralizing scenarios in different settings.

Our interest's on the dangerous edge of things ...
(Robert Browning)

The stress state at any location in the Earth's crust is defined by the magnitude and orientation of three principal compressive stresses, $\sigma_1 > \sigma_2 > \sigma_3$, with one of these principal stresses approximately vertical because of the boundary condition imposed by the Earth's free surface. Differential stress ($\sigma_1-\sigma_3$) is pretty much ubiquitous, both along plate boundaries and within plate interiors. A compilation of known stress states into a global stress map (Zoback 1992) shows the crust to be divided into stress domains corresponding to the three basic stress/fault regimes originally identified by Anderson (1905): normal fault regimes in areas of crustal extension where $\sigma_v = \sigma_1$; strike-slip regimes where $\sigma_v = \sigma_2$; and thrust fault regimes in areas of crustal compression where $\sigma_v = \sigma_3$. Compilations of focal mechanisms (Célérier 2008), together with palaeostress analyses (Lisle *et al.* 2006), provide additional support for the existence of such stress domains both today and in the geological past.

The seismically active upper continental crust (generally bounded by isotherms in the 350–450°C range (e.g. Ito 1999) and commonly 10–20 km thick) consists of sedimentary basins (usually <5 km, but occasionally >10 km deep) overlying crystalline basement. Much of this potentially seismogenic crust appears to be critically stressed to the edge of failure, not only along seismically active plate boundary zones, but also in intraplate regions. This has become evident in recent years through the accumulation of evidence for fluid-driven failure from earthquakes induced by fluid injection down boreholes (e.g. Denver, Healy *et al.* 1968; Basel, Deichmann & Giardini 2009; Oklahoma, Ellsworth 2013), from reservoir-induced seismicity (Simpson *et al.* 1988) and through the modelling of Coulomb stress redistribution following large crustal earthquakes (King *et al.* 1984). McGarr (2014) argued that critically stressed crust generally lies within one earthquake shear stress drop, $\Delta\tau$, of failure. Stress drops for crustal earthquakes lie within an overall range $0.1 < \Delta\tau < 100$ MPa, but for larger earthquakes most are restricted to $\Delta\tau < 10$ MPa with a median value of *c.* 4 MPa (Kanamori & Anderson 1975; Allmann & Shearer 2009). Outlier stress drops >10 MPa may arise from comparatively short wavelength stress heterogeneity in the crust.

All modes of brittle failure (initial formation plus the subsequent reactivation of faults and extension fractures) depend on the pore fluid pressure, P_f (Table 1), suggesting that there are two principal drivers to the brittle failure of a rock mass: the accumulation of differential stress ($\sigma_1-\sigma_3$) and/or the overpressuring of fluids ($P_f >$ hydrostatic) in pore and/or fracture spaces.

From: Turner, J. P., Healy, D., Hillis, R. R. & Welch, M. J. (eds) 2017. *Geomechanics and Geology*. Geological Society, London, Special Publications, **458**, 131–141.
First published online May 24, 2017, https://doi.org/10.1144/SP458.5

Table 1. *Modes of brittle/frictional failure and associated failure criteria*

Failure mode (field of application)	τ/σ'_n space (Mohr diagram) $(\sigma_1-\sigma_3)/P_f$ space (BFM plot)	Attitude with respect to stress axes
Extensional $(\sigma_1-\sigma_3) < 4T_o$	$\tau^2 = 4T_o(\sigma_n-P_f) + 4T_o^2$ $P_f = \sigma_3 + T_o$ when $(\sigma_1-\sigma_3) < 4T_o$	Extension fractures forming perpendicular to σ_3
Extensional shear $4T_o < (\sigma_1-\sigma_3) < 5.66T_o$ (Griffith criterion)	$\tau^2 = 4T_o(\sigma_n-P_f) + 4T_o^2$ $P_f = \sigma_3 + [8T_o(\sigma_1-\sigma_3)]/3$	Extensional shear fractures along planes containing σ_2, lying at $\pm\theta < \theta_i$ to σ_1
Compressional shear $(\sigma_1-\sigma_3) > 5.66T_o$ (Coulomb criterion)	$\tau = C + \mu_i(\sigma_n-P_f)$ $P_f = \sigma_3 + [8T_o(\sigma_1-\sigma_3)-(\sigma_1-\sigma_3)^2]/16T_o$ for $\mu_i = 0.75$	Shear fractures developing on planes containing σ_2, lying at $\pm\theta_i = 0.5\tan^{-1}(1/\mu_i)$ to σ_1
Cohesionless fault re-shear (Amontons law)	$\tau = \mu_s(\sigma_n-P_f)$ $P_f = \sigma_3 + [(\sigma_1-\sigma_3)(1-0.6\tan\theta_r)]/[0.6(\cot\theta_r + \tan\theta_r)]$ for $\mu_s = 0.6$	Re-shear along fractures containing σ_2, lying at $\pm\theta_r$ to σ_1

Stress states driving faulting

The level of shear and/or differential stress driving faulting and other forms of crustal deformation remains one of the outstanding questions in tectonophysics, crucial to the proper understanding of earthquake mechanics and tectonic deformation in general (Brune & Thatcher 2002). A common presumption is the simple law of effective stress espoused by Hubbert & Rubey (1959), whereby all normal stresses are counteracted by the fluid pressure in pore and/or fracture space (P_f) so that the effective normal stress

$$\sigma'_n = (\sigma_n - P_f) \quad (1)$$

prevails throughout seismogenic crust. Note that there are some indications that this may break down towards the base of the seismogenic zone where temperature-dependent viscous deformation in rocks become increasingly important (Hirth & Beeler 2015). However, on the assumption that the simple principle of effective stress holds, the frictional shear strength of existing faults in seismogenic crust may be represented by a criterion of Coulomb form

$$\tau_{fr} = C_f + \mu_s\sigma'_n = C_f + \mu_s(\sigma_n - P_f) \quad (2)$$

where C_f is the fault cohesion (generally low compared with the cohesive strength of intact rocks), μ_s is the coefficient of sliding friction and P_f is the fluid pressure in the pore/fracture space. Faults may, however, reacquire a degree of cohesive strength through hydrothermal cementation, although this is unlikely to be uniform over an entire fault. Re-shear of cohesionless faults with $C_f = 0$ is considered as a limiting end-member case.

Pore fluid pressure is usefully defined relative to the vertical stress at a depth, z, by the pore fluid factor

$$\lambda_v = \frac{P_f}{\sigma_v} = \frac{P_f}{\rho_r g z} \quad (3)$$

where ρ_r is the average rock density (hereafter assumed to be 2500 kg m^{-3}) and g is gravitational acceleration. Hydrostatic fluid pressures prevail for fluids with a density, ρ_w, in pore/fracture space interconnected to a water table at the Earth's surface, in which case

$$\lambda_v = \frac{P_f}{\sigma_v} = \frac{\rho_w g z}{\rho_r g z} \approx 0.4 \quad (4)$$

Fluids overpressured above the hydrostatic pressure have $\lambda_v > 0.4$. Extreme overpressuring occurs when fluid pressures approach or even exceed the lithostatic load ($\lambda_v \approx 1.0$).

Hydrostatic Byerlee stress states

Compilation of borehole stress measurements in crystalline continental crust led Townend & Zoback (2000) to a general inference of critical hydrostatic Byerlee stress states. Under these circumstances, the crustal stress state is governed by the frictional strength of optimally oriented faults with Byerlee friction ($\mu_s \approx 0.6$) and under hydrostatic pore fluid pressure in the prevailing stress regime. Justification for the adoption of $\mu_s = 0.6$, at the lower end of the experimental range determined by Byerlee (1978), as a representative value for sliding friction comes from borehole stress measurements (Townend & Zoback 2000). Further support for a value of μ_s of *c.* 0.6 comes from the observed dip range for normal and reverse dip-slip faults, where frictional lock-up occurs at *c.* 60° to the inferred σ_1 (Collettini & Sibson 2002). Seismically active fractured crust is also inferred to be too permeable to allow fluid overpressuring. Aside from the 9 km deep KTB borehole, however, the stress measurements come from comparatively shallow boreholes (mostly <4 km) and, with one exception, are restricted to extensional and strike-slip settings where subsidiary fractures are likely to be subvertical, facilitating the easy drainage of overpressure.

Profiles of frictional shear strength for optimally oriented thrust, strike-slip and normal faults with $\mu_s = 0.6$ under hydrostatic levels of fluid pressure are illustrated in Figure 1. At 10–15 km depth, where large ruptures commonly nucleate, the anticipated shear strengths range from 50 to 70 MPa for normal faults in extensional regimes to 150–220 MPa for thrust faults in compressional regimes. These are an order of magnitude or more greater than typical shear stress drops accompanying earthquake rupturing (Allmann & Shearer 2009).

Critical stress overpressure states

By contrast, large areas of seismically active crust in areas of crustal shortening under horizontal compression, including sedimentary basins and their crystalline substrates (e.g. the Western Transverse Ranges and the Great Valley of California), appear to have pore fluids overpressured above hydrostatic pressure (i.e. $\lambda_v > 0.4$) at depths greater than a few kilometres (Yerkes *et al.* 1990; McPherson & Garven 1999; Suppe 2014). Overpressuring is increasingly likely in the lower seismogenic zone where hydrothermal activity becomes widespread at *c.* $T > 200°C$, with a consequent reduction in permeability, as demonstrated experimentally (Morrow *et al.* 2001; Tenthorey *et al.* 2003). Issues arise as to whether overpressuring is uniform or heterogeneous, particularly in relation to active fault structures. Hydrothermal circulation leads to alteration (e.g. feldspar → clay minerals) and to dissolution, especially in reactive, fine-grained cataclastic detritus in fault cores and damage zones, with the re-precipitation of silica and carbonates in the fracture space. Also important is the superior ability of a compressional stress field to contain overpressure (see Figs 2 & 4). In fluid-overpressured crust, differential stress and the degree of overpressuring become interdependent and inversely related. The maximum differential stress achievable before brittle failure depends on the level of fluid overpressuring and, conversely, the maximum sustainable overpressure before fluid loss through brittle failure depends on the level of differential stress. High fluid overpressures are incompatible with high differential stress and vice versa (see Fig. 4). Relevant here is the recent demonstration of an inverse relationship between the magnitude of induced earthquake stress drops and fluid pressure level as a consequence of fluid injected into the wall rock at *c.* 4.5 km depth from the Basel geothermal well (Goertz-Allmann *et al.* 2011).

How can strongly fluid-overpressured active faults be recognized? One expected characteristic is the rupturing of faults that appear to be badly

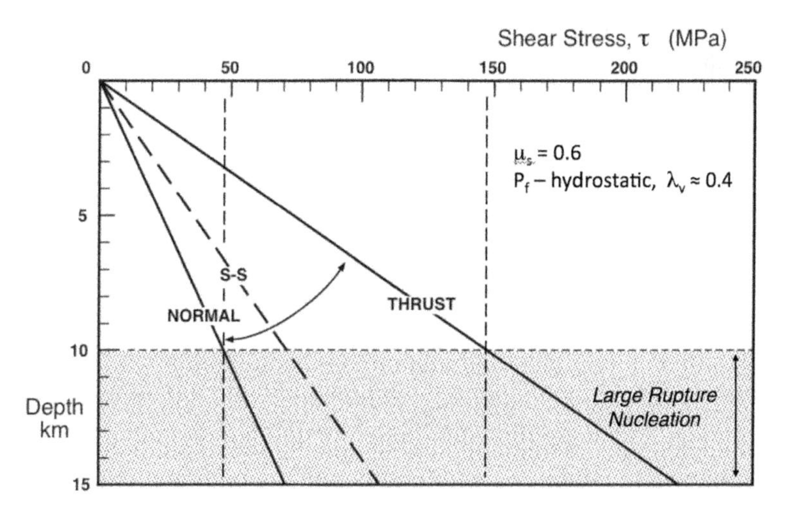

Fig. 1. Profiles of frictional shear strength for optimally oriented thrust, strike-slip (s-s) and normal faults with friction, $\mu_s = 0.6$, at hydrostatic pore fluid pressure (P_f) (after Sibson 1974).

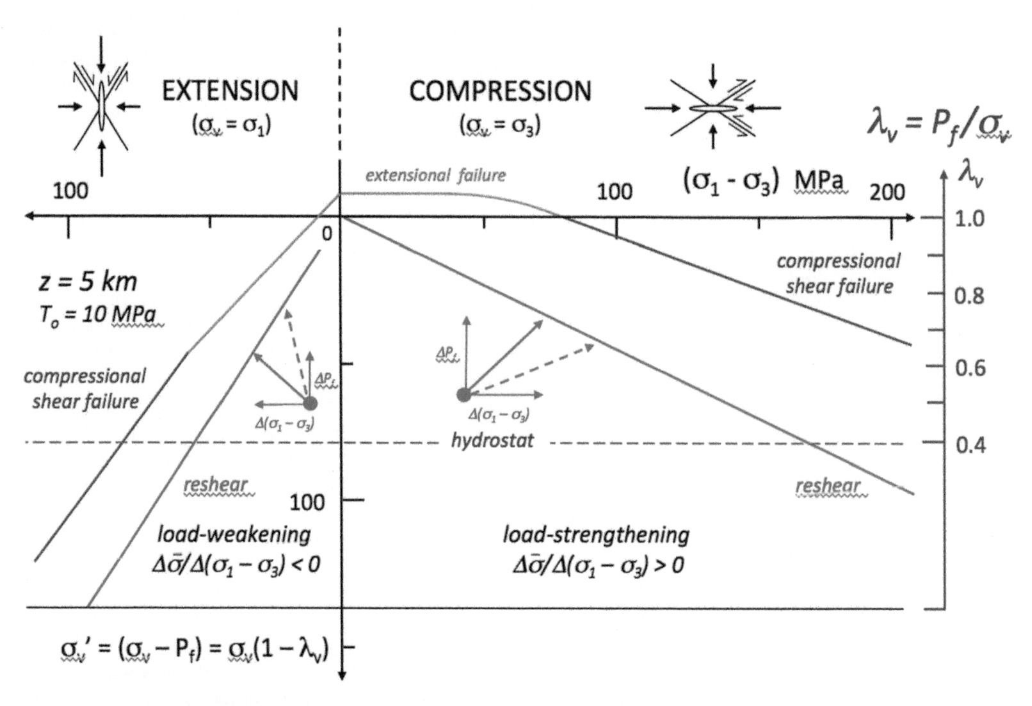

Fig. 2. BFM plots of differential stress v. effective vertical stress (corresponding to values of the pore fluid factor, λ_v) for intact rock ($T_o = 10$ MPa; $\mu_i = 0.75$) under compressional and extensional stress regimes at 5 km depth (green = extensional failure; dark red = compressional shear failure). Re-shear conditions (red lines) also shown for optimally oriented cohesionless faults with $\mu_s = 0.6$. Uniform rock properties ($T_o = 10$ MPa, $\rho = 2500$ kg m^{-3}) assumed together with Andersonian stress trajectories. Red arrows illustrate potential loading paths to failure from an initial point in $(\sigma_1 - \sigma_3)/\ \lambda_v$ space in terms of $\Delta(\sigma_1 - \sigma_3)$ and ΔP_f components. Expected orientations of newly forming extension fractures and faults in the two regimes as shown.

oriented for frictional reactivation in the prevailing stress field, for example the rupturing on steep reverse faults during compressional inversion. Many of these inversion structures are associated with geophysical anomalies (e.g. bright-spot reflectors, low-velocity zones, anomalous V_p/V_s ratios and high electrical conductivity) diagnostic of fluid overpressuring in the lower seismogenic zone and mid-crust (Sibson 2009). Another diagnostic feature may be ruptures followed by near-field aftershocks with diverse focal mechanisms. Where the stress drop is only partial ($\Delta\tau < \tau_{fail}$), the tectonic stress field is not greatly perturbed and the aftershock focal mechanisms tend to share the characteristics of the main shock mechanism. However, there are instances where near-field aftershock mechanisms are extremely diverse and inconsistent with the main shock rupture. In such cases it has been argued that the main shock stress drop is comparable with the shear stress at failure (i.e. the near-total stress drop with $\Delta\tau \approx \tau_{fail}$) so that the post-failure stress field is highly heterogeneous. Given that in most cases $\Delta\tau < 10$ MPa, it follows from equation (2) that either the frictional coefficient has abnormally low values, $\mu_s < 0.1$, or that the effective normal stress across the fault, $\sigma_n' = (\sigma_n - P_f) \rightarrow 0$, requiring $P_f \rightarrow \sigma_n$ and $\lambda_v \rightarrow 1.0$.

Examples of such ruptures include the 1952 M_w 7.5 Kern County earthquake involving left-reverse slip on ruptures dipping 50–75° (Castillo & Zoback 1995), the 1989 M_w 6.9 Loma Prieta earthquake involving right-reverse slip on a *c.* 70° dipping rupture (Michael *et al.* 1990; Zoback & Beroza 1993), the 2003 M_w 7.9 Denali earthquake involving dextral strike-slip on a subvertical fault for >300 km (Wesson & Boyd 2007) and the 2008 M_w 7.2 Iwate-Miyagi-Nairiku earthquake, which involved conjugate reverse faulting in northern Honshu (Yoshida *et al.* 2014*a*). Notably, three of these four events involved strong components of reverse faulting on moderate to steeply dipping planes. For the last event, the stress heterogeneity revealed by aftershock focal mechanisms has been inverted to infer near-lithostatic overpressures in the focal region (Yoshida *et al.* 2014*b*). On a far greater scale, Hasegawa *et al.* (2012) inferred a near-total stress drop for the 2011 M_w 9.0 Tohoku-Oki megathrust rupture, which appears to have been nearly lithostatically overpressured prefailure over the full seismogenic depth range (Sibson 2014).

Occasional rupturing of steep reverse faults also occurs in intraplate settings – for example the 1982 Miramachi, New Brunswick earthquake sequence (m_b 5.7, 5.1, 5.4 and 5.0) and the 1983 m_b 5.1 Goodnow, Adirondacks earthquake in eastern North America (Sibson 1990). Rupturing appears to involve localized compressional reactivation of normal faults inherited from Mesozoic rifting, possibly attributable to mantle degassing of CO_2 (Irwin & Barnes 1980).

Brittle failure: extensional v. compressional regimes

The behaviour of rocks close to failure has to be considered in terms of the two principal drivers: differential stress ($\sigma_1 - \sigma_3$) and pore fluid pressure P_f (or λ_v). In brittle failure mode (BFM) plots, both factors are taken into account on plots of differential stress against effective vertical stress at a depth, z, given by:

$$\sigma_v' = (\sigma_v - P_f) = \sigma_v(1 - \lambda_v) = \rho g z(1 - \lambda_v) \quad (5)$$

where ρ is the average rock density and g is the gravitational acceleration (Sibson 1998, 2000; Cox 2010). The vertical axis can then be calibrated to values of P_f or λ_v for a particular depth. Comparative BFM plots are constructed for the Andersonian compressional ($\sigma_v = \sigma_3$) and extensional ($\sigma_v = \sigma_1$) stress regimes representing end-member tectonic settings. Intermediary BFM plots can also be constructed for strike-slip regimes ($\sigma_v = \sigma_2$), but these vary depending where σ_2 lies between σ_1 and σ_3. The plots use the simplest two-dimensional stress-based failure criteria from classical rock mechanics (e.g. Jaeger & Cook 1979) listed in Table 1. Brittle failure then depends on the material properties and on the variables P_f, σ_1 and σ_3 alone, with the poles to all fault and fracture planes lying in the σ_1/σ_3 plane. Compositional and structural heterogeneity within the crust are the grounds for adopting these simple generic criteria.

Following the procedures outlined by Brace (1960) and Secor (1965) and using generic values for material properties (internal friction, $\mu_i = 0.75$; sliding friction, $\mu_s = 0.6$), a composite failure envelope for intact rock normalized to the rock tensile strength, T_o, is constructed by merging the Coulomb criterion for compressional shear failure with the macroscopic Griffith criterion for extensional and extensional shear failure. Assuming simple Andersonian stress trajectories with one of the principal stresses vertical, the failure curves are then recast for the BFM plots of differential stress v. effective vertical stress; see Sibson (2000) for the full procedure. Rock material properties are assumed not to vary with depth. It is important to keep the limitations of these BFM plots in mind – the two-dimensional character of the failure criteria, the unvarying material properties and the requirement that the vertical stress is one of the principal stresses throughout the brittle regime – nonetheless, they provide valuable insights into the factors governing brittle/frictional failure in different tectonic settings.

Comparative failure conditions

Failure curves for intact rock with $T_o = 10$ MPa and for the end-member case involving the re-shear of cohesionless, optimally oriented faults (i.e. faults oriented to contain σ_2 and lying at $\theta_r = 0.5 \tan^{-1}(1/\mu_s)$ to σ_1) have been constructed for compressional and extensional stress regimes at a depth of 5 km (Fig. 2). Note that for an existing fault with re-acquired cohesion, C_f, the re-shear lines for varying θ_r move to a common intercept: C_f/μ_s on the σ_v' axis of the failure mode plot (Sibson 2009).

Equivalent λ_v values to the effective vertical stress are calculated for rock and aqueous pore fluid densities of 2500 and 1000 kg m^{-3}, respectively, so that $\lambda_v = 0.4$ corresponds to the hydrostatic pore fluid pressure. Different failure modes (extensional, extensional shear and compressional shear failure) are indicated on the failure curves for intact rock; the expected orientation for newly forming extension fractures and faults are shown. The presence of extension fractures (often manifested as extension veins retaining crack-seal growth increments; Ramsay 1980) is especially significant because their formation requires $(\sigma_1 - \sigma_3) < 4T_o$ (Secor 1965), constraining the magnitude of differential stress during their formation (Etheridge 1983). For $T_o = 10$ and 30 MPa, characterizing many sedimentary rocks and competent crystalline rocks, respectively (Lockner 1995), the corresponding differential stress must be <40 and <120 MPa, respectively.

It is also apparent that the existence of optimally oriented existing faults lacking cohesive strength inhibits the formation of new faults and fractures within intact rock. A common field observation is for arrays of parallel extension veins formed by hydraulic extension fracturing to be cut by throughgoing faults that appear to have developed in the same stress field. Once developed, the presence of such faults favourably oriented for reactivation inhibits further hydraulic extension fracturing and vein formation. The continued formation of extension veins around existing faults is only likely if they retain cohesive strength through hydrothermal cementation or are severely misoriented for reactivation. These observations imply that the initial failure of intact rock under low differential stress driven by accumulating fluid overpressure, leading first to distributed hydraulic extension fracturing and then to the development of thoroughgoing faults, may be more widespread than is generally supposed.

Possible loading paths to failure range from purely stress-driven ($\Delta P_f = 0$) to purely fluid pressure driven ($\Delta(\sigma_1 - \sigma_3) = 0$), but there is an infinite range of potential loading paths incorporating changes in both differential stress and fluid pressure. In well-drained crust where poroelastic effects may be neglected, there are profound differences between load-strengthening compressional loading and load-weakening extensional loading (Sibson 1993). In a compressional regime with $\sigma_v = \sigma_3$ staying constant while horizontal σ_1 increases, the mean stress, $\sigma_m = \frac{1}{3}(\sigma_1 + \sigma_2 + \sigma_3)$ and fault frictional strength both increase as the differential stress and fault shear stress increase, that is, $\Delta\sigma_m/[\Delta(\sigma_1 - \sigma_3)] > 0$. By contrast, for an extensional regime with $\sigma_v = \sigma_1$ staying constant during decreasing horizontal σ_3, $\Delta\sigma_m/[\Delta(\sigma_1 - \sigma_3)] < 0$ with the fault frictional strength decreasing as the differential stress and fault shear stress increase. In undrained crust, poroelastic interactions need to be considered and may drastically modify some of these effects (Rice & Cleary 1976; Simpson 2001).

Contrasts in failure sensitivity

For simplicity, we restrict consideration here to the conditions for frictional re-shear in compressional and extensional regimes (Fig. 3). The slopes of the failure lines for optimal re-shear with $\mu_s = 0.6$ determine the relative sensitivity of failure to $\Delta(\sigma_1 - \sigma_3)$ and ΔP_f with a pronounced contrast in failure sensitivity between the two regimes. For an optimally oriented thrust fault dipping *c.* 30° in a compressional regime, $\Delta(\sigma_1 - \sigma_3) = -2.12\Delta P_f$, whereas for an optimum normal fault dipping 60° in an extensional regime, $\Delta(\sigma_1 - \sigma_3) = -0.68\Delta P_f$. Significant decreases in thrust fault strength may thus be induced by comparatively small changes in pore fluid pressure, but changes in the normal fault strength are lower than the change in pore fluid pressure.

The effects become more pronounced if faults are unfavourably oriented for reactivation. Thus for reverse faults dipping 50 or 9° (at equivalent misorientation to the horizontal (σ_1), $\Delta(\sigma_1 - \sigma_3) = -4.28\Delta P_f$, heightening the failure sensitivity to changes in pore fluid pressure. As the reverse fault dip increases towards frictional lock-up (dip = 59° for $\mu_s = 0.6$), $\lambda_v \rightarrow 1.0$ becomes necessary for reactivation, making this a particularly good tectonic setting for holding in overpressure. In the case of a normal fault dipping 40° (at the same reactivation angle of 50°), the reactivation condition becomes $\Delta(\sigma_1 - \sigma_3) = -0.81\Delta P_f$.

Maximum sustainable overpressure

Near-lithostatic overpressures may develop in compacting sedimentary basins where sealing horizons, such as shale units, may have permeabilities approaching $10^{-20}\,m^{-2}$ or less. Overpressuring may be enhanced by active tectonic shortening through poroelastic effects (Simpson 2001). For tectonically active continental crust undergoing

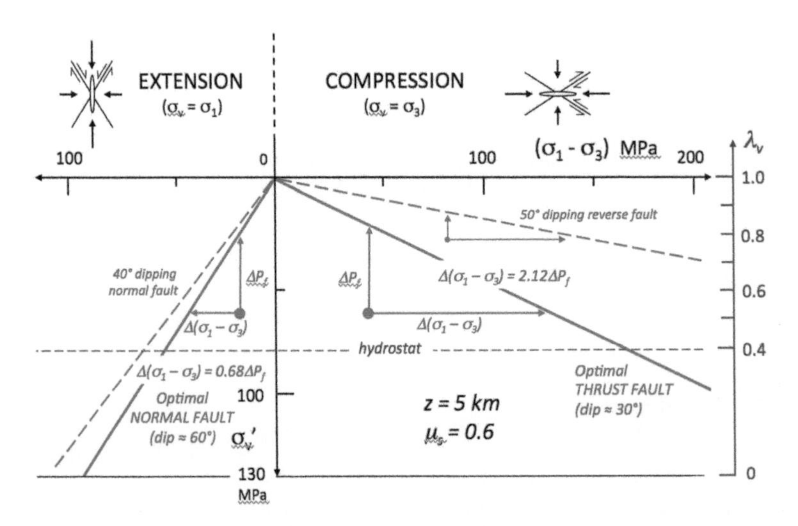

Fig. 3. Contrasts in failure sensitivity between compressional and extensional stress regimes at 5 km depth for the re-shear of optimally oriented thrust and normal faults. Rock properties as in Figure 2.

prograde metamorphism at depth, Ingebritsen & Manning (2010) suggest a mean bulk permeability $<10^{-18}\ m^{-2}$ at depths equivalent to the lower seismogenic zone (*c.* 12 km) to account for thermal models and metamorphic fluxes that allow the accumulation and maintenance of near-lithostatic overpressure in the ductile regime. The activation of faults and fractures across low-permeability sealing horizons allows drainage from overpressured portions of the crust, buffering overpressure development. This is particularly the case for extensional (and strike-slip) stress regimes, where newly forming faults and fractures tend to be subvertical or steep. Failure curves for intact rock and for the re-shear of optimally oriented faults are shown in Figure 4. The vertical blue arrows define the degree of overpressuring above hydrostatic pressure ($\lambda_v > 0.4$) at different values of differential stress before brittle failure or re-shear takes place. It is evident, again, that the presence of optimally oriented faults lacking cohesion lowers the sustainable overpressure and prevents other forms of brittle failure, such as hydraulic extension fracturing. Values of differential stress and fluid overpressure are inversely related; the maximum sustainable overpressure before brittle failure and fluid loss depends on the level of differential stress and, conversely, the maximum achievable differential stress before brittle failure depends on the degree of fluid overpressuring.

It is therefore evident that compressional stress regimes are inherently better at holding in fluid overpressure, partly from the geometry of newly forming faults and fractures, but also from the mechanical constraints illustrated in Figure 4 (Sibson 2003).

The edge of failure

Evidence of a crust critically stressed to the edge of failure comes from increasing occurrences of induced seismicity from borehole injection (e.g. the Oklahoma oilfield operations; Ellsworth 2013) and reservoir impoundment in intraplate settings (Nicholson & Wesson 1990), together with the implication from Coulomb stress modelling that very small stress changes ($\Delta\tau \approx 0.1$ MPa or less) can switch aftershock activity on or off (King *et al.* 1984; Harris 1998). These circumstances led McGarr (2014) to hypothesize that, as a general rule, the crust is mostly stressed to within one earthquake shear stress drop of failure – that is, $\Delta\tau < 10$ MPa. For a fault at the optimum orientation for re-shear with $\mu_s = 0.6$, this corresponds to $\Delta(\sigma_1-\sigma_3) < 23$ MPa. In Figure 5, the re-shear lines for compressional and extensional stress regimes are shown on a BFM plot constructed for a depth of 5 km; the shaded areas are the loci of critical stress/fluid pressure states satisfying this near-failure condition.

Loading paths to failure

Consider a point corresponding to a particular stress/fluid pressure condition (with $P_f >$ hydrostatic) close to a failure curve on a BFM plot (Fig. 6). The rock could be brought to failure either by increasing differential stress, $\Delta(\sigma_1-\sigma_3)$ (purely stress-driven

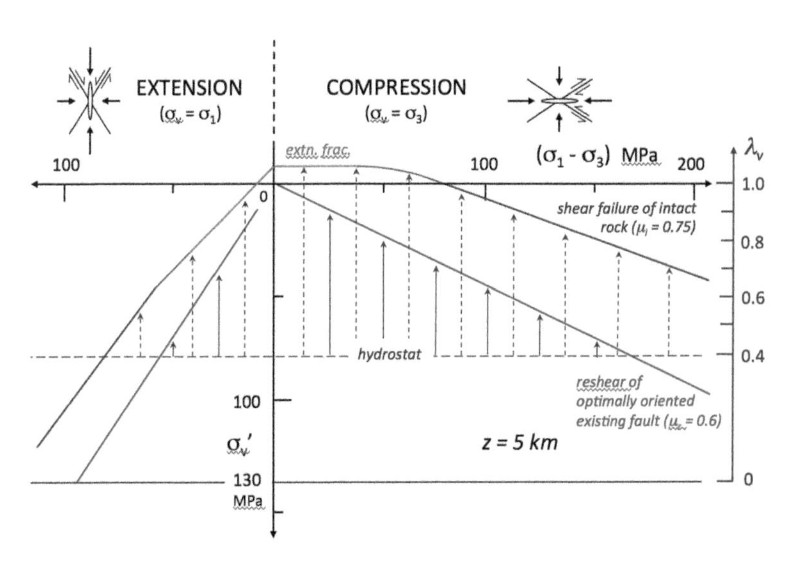

Fig. 4. Maximum sustainable overpressure at 5 km depth in compressional and extensional stress regimes for intact rock and for optimally oriented cohesionless faults. Vertical blue arrows represent sustainable overpressure above hydrostatic pressure before brittle failure as a function of differential stress. Colour codes as in Figure 2.

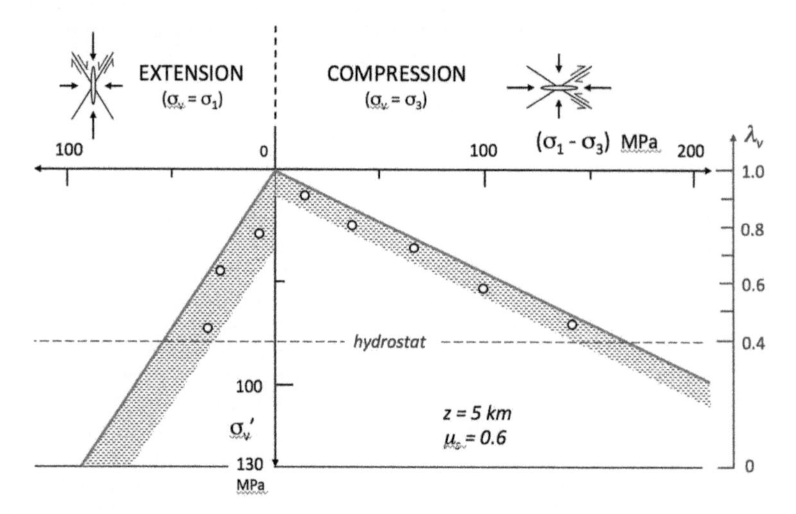

Fig. 5. Edge of failure represented by critical stress overpressure states (circles) within one earthquake stress drop of failure ($\Delta\tau < 10$ MPa) at 5 km depth for compressional and extensional stress regimes.

failure under constant P_f), or by increasing fluid pressure, ΔP_f (purely fluid-driven failure under constant $(\sigma_1 - \sigma_3)$). However, these are just two possibilities in a broad spectrum of brittle failure scenarios (cf. Cox 2010). In field A, differential stress and fluid pressure are both increasing along the path to failure. Possible settings for such behaviour include: areas of progressive crustal shortening, thickening and dewatering, such as accretionary prisms in subduction zones (Saffer & Tobin 2011; Suppe 2014); progressive shortening of fluid-rich sedimentary basins in areas of crustal convergence (e.g. the Sacramento Basin of California; McPherson & Garven 1999); and areas where forceful igneous intrusion into fluid-rich host rocks has generated an overpressured brittle carapace (Fournier 1999). In field B, the fluid pressure decreases while the differential stress increases. This might, for example, represent a situation where the development of microcrack dilatancy at progressively higher differential stress increases the bulk permeability, allowing fluid loss. In field C, the differential stress decreases (perhaps from incipient plastic yielding) while the fluid pressure continues to increase.

In a single loading cycle to failure, it is also necessary to consider potential poroelastic coupling between ΔP_f and $\Delta\sigma_m$, depending whether the regime is load-strengthening ($\Delta\sigma_m > 0 \rightarrow \Delta P_f > 0$) or load-weakening ($\Delta\sigma_m < 0 \rightarrow \Delta P_f < 0$) (Rice & Cleary 1976). The magnitude of such changes depends on the drainage behaviour of the rock mass over the loading cycle. For example, Simpson (2001) and Micklethwaite (2008) showed that the shortening of fluid-saturated crust under undrained conditions can create a change from load-strengthening to load-weakening behaviour.

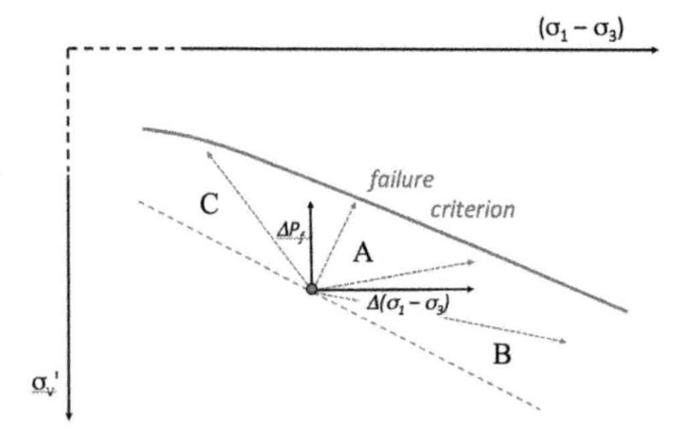

Fig. 6. Range of potential loading paths to failure on a BFM plot. Dashed arrows illustrate loading paths to failure from an initial point in $(\sigma_1 - \sigma_3)/P_f$ space.

Discussion

Simple stress-based failure criteria have been used to construct BFM plots exploring the relative roles of the two principal drivers to brittle rock failure in the upper crust: differential stress $(\sigma_1 - \sigma_3)$ and pore fluid pressure (P_f or λ_v). These plots could be further elaborated by using more comprehensive three-dimensional failure criteria (e.g. Healy *et al.* 2006) and by exploring poroelastic interactions during loading to a far greater extent. Much has been learned about poroelastic interactions – for instance, from industrial applications of fluid-induced fracturing and the ensuing seismicity (e.g. Shapiro 2015). In addition, the structural consequences of fluid overpressuring could alternatively be analysed in terms of seepage forces arising from suprahydrostatic gradients in fluid pressure (Cobbold & Rodrigues 2007; Mourgues & Cobbold 2003).

Loading paths to the formation and reactivation of brittle/frictional structures have been compared between compressional and extensional tectonic regimes. There is an infinite range of possible loading paths to failure or reactivation from changes in differential stress and pore fluid pressure. Broadly speaking, stress-driven failure is likely to dominate when $\Delta(\sigma_1-\sigma_3)/\Delta t \gg \Delta P_f/\Delta t$ and fluid-driven failure may dominate when $\Delta P_f/\Delta t \gg \Delta(\sigma_1-\sigma_3)/\Delta t$. The latter becomes likely in regions of intense fluid release – for example, in compacting sedimentary basins undergoing fast sedimentation, in areas where magma is actively intruding fluid-rich rocks and in subduction forearcs. The permeability structure of the seismogenic crust ($T < 350–450°C$) and, in particular, the relative permeabilities of sedimentary cover sequences and crystalline basement assemblages, are clearly of huge importance. A low-permeability cover overlying a high-permeability fractured basement will have a very different mechanical response from a high-permeability cover overlying a low-permeability basement.

The hydrostatic Byerlee condition as demonstrated by borehole measurements (Townend & Zoback 2000) is likely to prevail at shallow depths (<5–7 km), especially where the basement assemblages are close to the Earth's surface, and in extensional or strike-slip settings. The attainment of critical stress overpressure states is more probable in the lower half of seismogenic crust, where hydrothermal activity (*c.* $T > 200°C$) reduces both fracture and bulk permeability. As a general rule, fluid overpressuring leading to critical stress overpressure conditions is more likely to develop and be sustained in compressional regimes, where ongoing thrusting and crustal thickening affect both the sedimentary cover and the underlying basement, leading over time to prograde metamorphism and accompanying fluid loss. It has been shown, for instance, how the progressive emplacement of foreland thrust sheets self-generates fluid overpressures in the overridden substrate (Cello & Nur 1988). Such processes reach their extreme in subduction settings, where the supply of fluid is particularly abundant (Saffer & Tobin 2011), and in areas of active compressional inversion (Sibson 2009), where the reactivation of steep reverse faults leading to fluid loss is only possible at extreme fluid overpressures ($\lambda_v \rightarrow 1.0$).

The general presumption that failure is stress-driven arises from geodetic measurements of strain–stress accumulation around major plate boundary faults. By contrast, strain rates associated with intraplate earthquakes tend to be rather low (failure under approximately near-constant stress conditions) and, at least for the failure of poorly oriented faults, is plausibly fluid-driven (Sibson 1990). The containment of overpressures in critical stress overpressure states is precarious and highly dependent on the stress state: abrupt changes in stress accompanying earthquake rupture must invariably lead to fluid redistribution. The formation and/or reactivation of faults and fractures causes fluid loss and frictional strengthening through some form of fault/fracture valving action, which is either localized on a single dominant structure or may be dispersed across arrays of brittle faults and fractures (Sibson 1992; Cox 2016). In such settings, earthquake recurrence from fault rupture may be governed just as much by the cycling of frictional strength through changes in fluid pressure as by the cycling of stress acting on the fault.

Packages of overpressured fluid migrating through the crust are likely to self-generate fault–fracture permeability through microearthquake swarm activity with accompanying mineral deposition (Sibson 1996; Cox 2016): critical stress overpressure states appear to be intrinsic to several varieties of hydrothermal mineralization. Specific examples include fault-hosted orogenic gold–quartz vein systems derived from valving action in the lower seismogenic zone (Sibson *et al.* 1988) and porphyry copper systems associated with shallow calc-alkaline intrusive magmatism (Fournier 1999). It remains to be established how many other forms of hydrothermal mineralization (e.g. iron oxide–copper–gold ore deposits; Williams *et al.* 2005) necessarily involve critical stress overpressure states.

Sincere thanks to the organizers for making it possible for me to participate in *The Geology of Geomechanics*. Neville Price first drew my attention to the role of pore fluids in brittle rock failure, with additional valued input over many years, conferences and field excursions from Stephen Cox, Rob Knipe, Francois Robert, Howard Poulsen and many others. Special thanks also to the reviewers for their constructive criticism and for drawing my attention to relevant recent publications. This paper is dedicated to my late colleague, Richard J. Norris.

References

Allmann, B.P. & Shearer, P.M. 2009. Global variations of stress drop for moderate to large earthquakes. *Journal of Geophysical Research*, **114**, B01310, https://doi.org/10.1029/2008JB005821

Anderson, E.M. 1905. The dynamics of faulting. *Transactions of the Edinburgh Geological Society*, **8**, 387–402, https://doi.org/10.1144/transed.8.3.387

Brace, W.F. 1960. An extension of the Griffith theory of fracture to rocks. *Journal of Geophysical Research*, **65**, 3477–3480.

BRUNE, J.N. & THATCHER, W. 2002. Strength and energetics of active fault zones. *In*: LEE, W.H.K., KANAMORI, H., JENNINGS, P.C. & KISSLINGER, C. (eds) *International Handbook of Earthquake and Engineering Seismology*. Part A. International Geophysics Series, **81A**. Academic Press, Amsterdam, 69–588.

BYERLEE, J.D. 1978. Friction of rocks. *Pure & Applied Geophysics*, **116**, 615–626.

CASTILLO, D.A. & ZOBACK, M.D. 1995. Systematic stress variation in the southern San Joaquin Valley and along the White Wolf fault: implications for the rupture mechanics of the 1952 M_s 7.8 Kern County earthquake. *Journal of Geophysical Research*, **100**, 6249–6264.

CÉLÉRIER, B. 2008. Seeking Anderson's faulting in seismicity: a centennial celebration. *Reviews of Geophysics*, **46**, RG4001, https://doi.org/10.1029/2007RG000240

CELLO, G. & NUR, A. 1988. Emplacement of foreland thrust systems. *Tectonics*, **7**, 261–271.

COBBOLD, P.R. & RODRIGUES, N. 2007. Seepage forces, important factors in the formation of horizontal hydraulic fractures and bedding-parallel fibrous veins ('beef' and 'cone-in-cone'). *Geofluids*, **7**, 313–322.

COLLETTINI, C. & SIBSON, R.H. 2002. Normal faults, normal friction? *Geology*, **29**, 927–930.

COX, S.F. 2010. The application of failure mode diagrams for exploring the role of fluid pressure and stress states in controlling styles of fracture-controlled permeability enhancement in faults and shear zones. *Geofluids*, **10**, 217–233.

COX, S.F. 2016. Injection-driven swarm seismicity and permeability enhancement: implications for the dynamics of hydrothermal ore systems in high fluid-flux overpressured faulting regimes – an invited paper. *Economic Geology*, **111**, 559–587.

DEICHMANN, N. & GIARDINI, D. 2009. Earthquakes induced by the stimulation of an enhanced geothermal system beneath Basel (Switzerland). *Seismological Research Letters*, **80**, 784–798.

ELLSWORTH, W.L. 2013. Injection-induced earthquakes. *Science*, **341**, https://doi.org/10.1126/science.1225942

ETHERIDGE, M.A. 1983. Differential stress magnitudes during regional deformation and metamorphism: upper bound imposed by tensile fracturing. *Geology*, **11**, 231–234.

FOURNIER, R.O. 1999. Hydrothermal processes related to movement of fluid from plastic into brittle rock in the magmatic-epithermal environment. *Economic Geology*, **94**, 1193–1211.

GOERTZ-ALLMANN, B.P., GOERTZ, A. & WIEMER, S. 2011. Stress drop variations of induced earthquakes at the Basel geothermal site. *Geophysical Research Letters*, **38**, L09308, https://doi.org/10.1029/2011GL047498

HARRIS, R.A. 1998. Introduction to special section: stress triggers, stress shadows, and implications for seismic hazard. *Journal of Geophysical Research*, **103**, 24 347–24 358.

HASEGAWA, A., YOSHIDA, K., ASANO, Y., OKADA, T., IINUMA, T. & ITO, Y. 2012. Change in stress field after the 2011 great Tohoku-Oki earthquake. *Earth & Planetary Science Letters*, **355–356**, 231–243.

HEALY, D., JONES, R. & HOLSWORTH, R. 2006. Three-dimensional brittle shear fracturing by tensile crack interaction. *Nature*, **439**, 64–67.

HEALY, J.H., RUBEY, W.W., GRIGGS, D.T. & RALEIGH, C.B. 1968. The Denver earthquakes. *Science*, **161**, 1301–1310.

HIRTH, G. & BEELER, N.M. 2015. The role of fluid pressure on frictional behaviour at the base of the seismogenic zone. *Geology*, **43**, 223–226.

HUBBERT, M.K. & RUBEY, W.L. 1959. Role of fluid pressure in the mechanics of overthrust faulting. *Geological Society of America Bulletin*, **70**, 115–166.

INGEBRITSEN, S.E. & MANNING, C.E. 2010. Permeability of the continental crust: dynamic variations inferred from seismicity and metamorphism. *Geofluids*, **10**, 193–205.

IRWIN, W.P. & BARNES, I. 1980. Tectonic relations of carbon dioxide discharges and earthquakes. *Journal of Geophysical Research*, **85**, 3115–3121.

ITO, K. 1999. Seismogenic layer, reflective lower crust, surface heat flow and large inland earthquakes. *Tectonophysics*, **306**, 423–433.

JAEGER, J.C. & COOK, N.G.W. 1979. *Fundamentals of Rock Mechanics*. 3rd edn. Chapman & Hall, London.

KANAMORI, H. & ANDERSON, D. 1975. Theoretical basis of some empirical relations in seismology. *Bulletin of the Seismological Society of America*, **84**, 1073–1095.

KING, G.C.P., STEIN, R.S. & LIN, J. 1984. Static stress changes and the triggering of earthquakes. *Bulletin of the Seismological Society of America*, **84**, 935–953.

LISLE, R.J., ORIFE, T.O., ARLEGUI, L., LIESA, C. & SRIVASTAVA, D.C. 2006. Favoured states of stress in the Earth's crust: evidence from fault-slip data. *Journal of Structural Geology*, **28**, 1051–1066.

LOCKNER, D.A. 1995. Rock failure. *In*: AHRENS, T.J. (ed.) *Rock Physics and Phase Relations: A Handbook of Physical Constants*. American Geophysical Union Reference Shelf, **3**, 127–147.

MCGARR, A. 2014. Maximum magnitude earthquakes induced by fluid injection. *Journal of Geophysical Research: Solid Earth*, **119**, 1008–1019, https://doi.org/10.1002/2013JB010597

MCPHERSON, B.J.O.L. & GARVEN, G. 1999. Hydrodynamics and overpressure mechanisms in the Sacramento basin, California. *American Journal of Science*, **299**, 429–466.

MICHAEL, A.J., ELLSWORTH, W.L. & OPPENHEIMER, D.H. 1990. Coseismic stress changes induced by the 1989 Loma Prieta, California, earthquake. *Geophysical Research Letters*, **17**, 1441–1444.

MICKLETHWAITE, S. 2008. Optimally oriented 'fault-valve' thrusts: Evidence for aftershock-related fluid pressure pulses? *Geochemistry, Geophysics, Geosystems*, **9**, Q04012, https://doi.org/10.1029/2007GC001916

MORROW, C.A., MOORE, D.E. & LOCKNER, D.A. 2001. Permeability reduction in granite under hydrothermal conditions. *Journal of Geophysical Research*, **106**, 30551–30560.

MOURGUES, R. & COBBOLD, P.R. 2003. Some tectonic consequences of fluid overpressures and seepage forces as demonstrated by sandbox modelling. *Tectonophysics*, **376**, 75–97.

NICHOLSON, C. & WESSON, R. 1990. *Earthquake Hazard Associated with Deep Well Injection – a Report to the U.S. Environmental Protection Agency*. United States Geological Survey Bulletin, **1951**.

Ramsay, J.G. 1980. The crack-seal mechanism of rock deformation. *Nature*, **284**, 135–139.

Rice, J.R. & Cleary, M.P. 1976. Some basic stress-diffusion solutions for fluid-saturated elastic porous media with compressible constituents. *Reviews of Geophysics & Space Physics*, **14**, 227–241.

Saffer, D.M. & Tobin, H.J. 2011. Hydrogeology and mechanics of subduction zone forearcs: fluid flow and pore pressure. *Annual Review of Earth and Planetary Sciences*, **39**, 157–186.

Secor, D.T. 1965. Role of fluid pressure in jointing. *American Journal of Science*, **263**, 633–646.

Shapiro, S.A. 2015. *Fluid-Induced Seismicity*. Cambridge University Press, Cambridge.

Sibson, R.H. 1974. Frictional constraints on thrust, wrench, and normal faults. *Nature*, **249**, 542–544.

Sibson, R.H. 1990. Rupture nucleation on unfavourably oriented faults. *Bulletin of the Seismological Society of America*, **80**, 1580–1604.

Sibson, R.H. 1992. Implications of fault-valve behaviour for rupture nucleation and recurrence. *Tectonophysics*, **211**, 283–293.

Sibson, R.H. 1993. Load-strengthening v. load-weakening faulting. *Journal of Structural Geology*, **15**, 123–128.

Sibson, R.H. 1996. Structural permeability of fluid-driven fault-fracture meshes. *Journal of Structural Geology*, **18**, 1031–1042.

Sibson, R.H. 1998. Brittle failure mode plots for compressional and extensional tectonic regimes. *Journal of Structural Geology*, **20**, 655–660.

Sibson, R.H. 2000. A brittle failure mode plot defining conditions for high-flux flow. *Economic Geology*, **95**, 41–48.

Sibson, R.H. 2003. Brittle failure controls on maximum sustainable overpressure in different tectonic regimes. *American Association of Petroleum Geologists Bulletin*, **87**, 901–908.

Sibson, R.H. 2009. Rupturing in overpressured crust during compressional inversion—the case from NE Honshu, Japan. *Tectonophysics*, **473**, 404–416.

Sibson, R.H. 2014. Earthquake rupturing in fluid-overpressured crust: how common? *Pure & Applied Geophysics*, **171**, 2867–2885.

Sibson, R.H., Robert, F. & Poulsen, K.H. 1988. High-angle reverse faults, fluid pressure cycling, and mesothermal gold-quartz deposits. *Geology*, **16**, 551–555.

Simpson, D.W., Leith, W.S. & Scholz, C.H. 1988. Two types of reservoir-induced seismicity. *Bulletin of the Seismological Society of America*, **78**, 2025–2040.

Simpson, G. 2001. Influence of compression-induced fluid pressures on rock strength in the brittle crust. *Journal of Geophysical Research*, **106**, 19465–19478.

Suppe, J. 2014. Fluid overpressures and strength of the sedimentary upper crust. *Journal of Structural Geology*, **69**, 481–492.

Tenthorey, E., Cox, S.F. & Todd, H.F. 2003. Evolution of strength recovery and permeability during fluid-rock reaction in experimental fault zones. *Earth & Planetary Science Letters*, **206**, 161–172.

Townend, J. & Zoback, M.D. 2000. How faulting keeps the crust strong. *Geology*, **28**, 399–402.

Wesson, R.L. & Boyd, O.S. 2007. Stress before and after the 2002 Denali fault earthquake. *Geophysical Research Letters*, **34**, L07303, https://doi.org/10.1029/2007GL029189

Williams, P.J., Barton, M.D. *et al.* 2005. Iron oxide copper gold deposits: geology, space-time distribution, and possible modes of origin. *Economic Geology*, **100**, 371–405.

Yerkes, R.F., Levine, P. & Wentworth, C.M. 1990. *Abnormally High Fluid Pressures in the Region of the Coalinga Earthquake Sequence and their Significance*. United States Geological Survey, Professional Papers, **1487**, 235–257.

Yoshida, K., Hasegawa, A., Okada, T. & Iinuma, T. 2014*a*. Changes in the stress field after the 2008 M7.2 Iwate-Miyagi-Nairiku earthquake in northeastern Japan. *Journal of Geophysical Research*, **119**, https://doi.org/10.1002/2014JB011291

Yoshida, K., Hasegawa, A. *et al.* 2014*b*. Pore pressure distribution in the focal region of the 2008 M7.2 Iwate-Miyagi-Nairiku earthquake. *Earth, Planets and Space*, **63**, 703–707.

Zoback, M.D. & Beroza, G.C. 1993. Evidence for near-frictionless faulting in the 1989 (M6.9) Loma Prieta, California, earthquake and its aftershocks. *Geology*, **21**, 181–185.

Zoback, M.L. 1992. First- and second-order patterns of stress in the lithosphere: the World Stress Map project. *Journal of Geophysical Research*, **97**, 11 703–11 728.

Active low-angle normal faults in the deep water Santos Basin, offshore Brazil: a geomechanical analogy between salt tectonics and crustal deformation

MARCOS FETTER[1]*, ANDERSON MORAES[2] & ANDRE MULLER[3]

[1]*Petrobras/E&P, Av. Republica do Chile, 330; 13th floor, Rio de Janeiro 20031-170, RJ Brazil*

[2]*Petrobras/R&D-CENPES, Av. Republica do Chile, 330; 13th floor, Rio de Janeiro 20031-170, RJ Brazil*

[3]*Catholic University of Rio de Janeiro – PUC-RIO, Av. Republica do Chile, 330; 13th floor, Rio de Janeiro 20031-170, RJ Brazil*

**Correspondence: fetter@petrobras.com.br*

Abstract: We provide a structural analysis and propose a geomechanical model for an extensional fault system observed in seismic data from the deep water Santos Basin, offshore Brazil. The system includes low-angle normal faults (LANFs) associated with underlying listric faults and salt rollers. The LANFs cut through shallow sedimentary successions up to the seafloor, reaching displacements of 1 km during the last 5 myr, at least one order of magnitude larger than the displacement of the coeval Andersonian high-angle normal faults. We suggest that the LANFs were produced by clockwise rotation of the stress field caused by the presence of salt intruded along the footwall of the main listric normal faults of the extensional system. Numerical models indicate that the proposed geomechanical model can be attained with strain weakening and widespread plastic yielding along the main normal faults. We suggest that the studied LANFs are related to salt flow in the same way that the LANFs described in regional extensional provinces are related to flow of either the lower crust or the mantle. The rheological analogy is remarkable and therefore we suggest that this work will contribute to the solution of the LANFs paradox.

When loaded to their critical limits by gravitational and tectonic forces, rock bodies in the upper crust usually deform by a process of localized shear failure to produce fault structures. The geomechanics of rock faulting has been approached with a number mechanical concepts, namely: (1) continuum mechanics, by means of Cauchy's stress tensor and the Mohr–Coulomb yielding criteria (Anderson 1951); (2) poroelasticity, by means of Terzaghi's effective stress concept (Hubbert & Rubey 1959); and (3) Amonton's law of friction (Sibson 1974; Byerlee 1978).

At almost the same time as fault mechanics were well established at scales ranging from plug samples in laboratory tests to outcrops, earthquakes and first-order plate kinematics, an enigmatic class of faults called detachments or low-angle normal faults (LANFs) was reported in regions of active crustal extension (Wernicke 2009; Collettini 2011), where the deformation was supposed to be accommodated by high-angle Andersonian normal faults (Anderson 1951). Crustal-scale LANFs are mechanically intriguing as they make a high angle with the vertical stress component, the maximum principal stress for Andersonian normal faults in extensional tectonic environments. According to continuum mechanics, any fault close to a Cauchy's shearless principal plane, as is the case for LANFs, should be locked. Surprisingly, individual LANFs are characterized by shear displacements of up to tens of kilometres (Snow & Wernicke 2000; Howard 2003), at least one order of magnitude larger than the displacement of coeval individual normal faults (Forsyth 1992). LANF solutions are also scarce in the seismological record. Therefore LANFs still represent a mechanical paradox (Collettini 2011) and the geomechanical model for LANFs is still debated. Two kinds of solution have been suggested for the LANF puzzle: (1) mechanical solutions, based on poroelastic effects (including anisotropy), weakening and stress rotation; and (2) geometric solutions, based either on the rotation of planar normal faults or on the down-dip coalescence of listric normal faults.

This work contributes to the study of LANFs and is based on a well-imaged fault system observed in three-dimensional seismic data from the Santos Basin along the southeastern continental margin of Brazil. A geomechanical model is proposed that fully supports the mechanical solutions because the observed LANFs both nucleate and propagate at a low angle. We compare this fault system, with its underlying salt layer, with crustal-scale LANFs with an underlying viscoelastic lower crust or upper mantle.

From: TURNER, J. P., HEALY, D., HILLIS, R. R. & WELCH, M. J. (eds) 2017. *Geomechanics and Geology*. Geological Society, London, Special Publications, **458**, 143–154.
First published online May 26, 2017, https://doi.org/10.1144/SP458.11

Geological setting

The Santos Basin, on the southeastern coast of Brazil (Fig. 1), was formed by rifting, which ended up with the break-up of the Gondwana supercontinent and the opening of the South Atlantic Ocean by the end of the Early Cretaceous. During the late stages of the extensional process, in the Aptian, rift shale successions were capped by a lacustrine carbonate system (LK1a; Fig. 2) and by a thick salt layer (LK1b; Fig. 2). The Lower Cretaceous stratigraphy of the Santos Basin consists of a petroleum system that is related to the most extensive oil discoveries of the last decade.

The Santos Basin then evolved to a typical intraplate continental margin and this evolution from the Cretaceous to the Quaternary was strongly controlled by the interaction between basement reactivations and the downdip flow of the Aptian salt layer (Fig. 2). The LANFs studied in this work are not related to the Cretaceous rift system; instead, they cut through shallow sedimentary successions up to the seafloor (2 km water depth) and are clearly related to the underlying normal faults and salt rollers (Figs 1 & 2). In agreement with the model proposed by Forsyth (1992), one single LANF accommodates a finite displacement at least one order of magnitude larger than the coeval set of Andersonian normal faults.

Structural analysis

The studied fault system is located at 2 km water depth in the deep water Santos Basin (Figs 1 & 2). The mapped faults accommodate the extensional deformation of the post-salt sedimentary successions in an area of *c.* 100 km^2 and they cut through post-salt Lower Cretaceous marine carbonates (LK2; Fig. 2) and through deep water siliciclastic sediments from the Upper Cretaceous to the Quaternary (UK, PALEOG, NEOG and QUAT; Fig. 2).

The structural style mixes both high-angle extensional faults (65° average dip) and low-angle extensional faults (or LANFs; 10° average dip), both trending east–west (stereonet plot; Fig. 2). The LANFs affect most the younger sedimentary successions from the Neogene to the Quaternary (blue faults; Fig. 2). The normal fault system consists of conjugate synthetic (green faults; Fig. 2) and antithetic faults (orange faults; Fig. 2). The main listric normal faults are antithetic (orange faults; Fig. 2), dipping towards the basin margin to the north (Fig. 1), and are associated with thick salt rollers along the footwall blocks (salt/LK1b; Fig. 2) and with growth strata, particularly in the Upper Cretaceous (UK; Fig. 2) and Palaeogene (PALEOG; Fig. 2) strata. Synthetic normal faults downthrown towards the basin (green faults; Fig. 2) are subordinate structures accommodating the deformation of the hanging wall of the main antithetic normal faults. The LANFs (blue faults; Fig. 2) are also antithetic, dipping towards the basin margin to the north, and are located along the upper portion of the footwall of the main listric normal faults (Fig. 2).

The kinematics of the studied area is defined by an overall extension that continued from the Late Cretaceous to the Quaternary. The average fault displacement is *c.* 50 m for the synthetic normal faults (green faults; Fig. 2), *c.* 250 m for the antithetic normal faults (orange faults; Fig. 2) and >750 m for the LANFs (blue faults; Fig. 2). The displacement of the LANFs was developed during the last 5 myr, whereas the normal fault system was active during the last 80 myr. Therefore the displacement data for the studied fault system agree with the finite extension model proposed by Forsyth (1992), which states that only low-angle faults can accommodate large amounts of extension.

Regarding the dynamics of the fault system, it is commonplace to attribute all the extensional deformation at salt-bearing intraplate margins to gravitational salt gliding. However, in this study it appears that the salt layer is reactive to tectonic extension because the salt rollers represent the flow of the salt to the low pressure zones at the bottom of the footwalls of the main antithetic normal faults (Figs 2 & 3). Therefore it is also possible that the observed extension has been caused by intraplate tectonics.

Geomechanical model

We suggest a conceptual geomechanical model for the studied fault system based on three main features: (1) stress rotation; (2) low effective stress; and (3) interaction between elastic and viscous rheologies. The relief of the viscous layer is also a key feature of the model.

Stress rotation is supported by the Coulomb angle observed between the LANFs and the upper portion of the main listric normal fault. It is suggested that the upwards flow of the salt constrained along the footwall of the main listric normal fault causes the clockwise rotation of the principal stress from the vertical to a position parallel with the fault plane (Fig. 3). Jackson & Talbot (1986) suggested a similar flow pattern for salt structures caused by extension. If we admit some shear weakening of the rock in the damage zone of the main listric fault, then this fault plane will act as a free surface and hence must form a principal plane of stress. The other rotated principal plane of stress is the shearless top of the SSE flank of the salt roller, which is normal to the fault plane (Fig. 3).

As the mapped LANFs cut through shallow successions at 2 km water depth (blue faults; Fig. 2), the

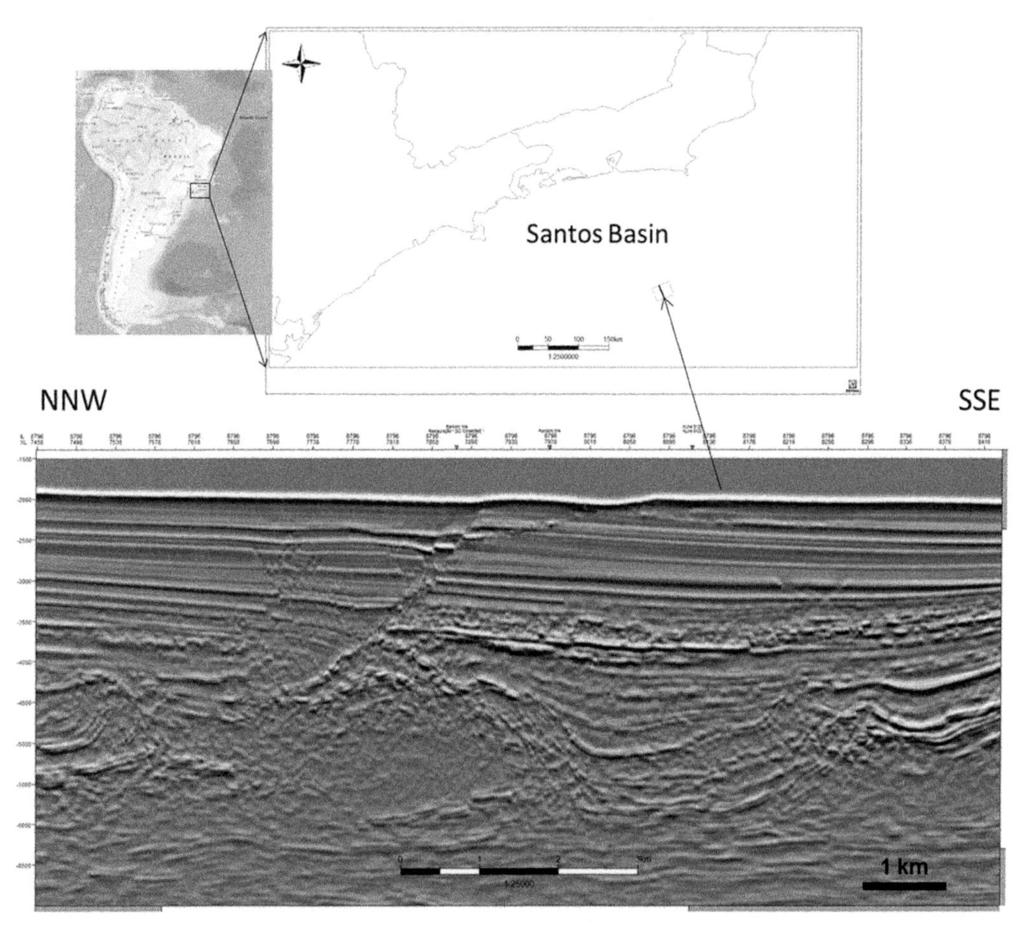

Fig. 1. Seismic section from the Santos Basin showing the studied fault system (no vertical exaggeration). Sea bottom is at 2 km water depth. The LANF can be seen at the centre of the section cutting through shallow sedimentary successions with 500 m of displacement at the sea bottom and with about 1 km maximum displacement at 500 m depth.

ratio between the pore pressure and the average stress is high (close to unity). This situation configures a very low effective stress environment characterized by high fluid pressures.

The rheology of the proposed geomechanical model is complex, with interaction between an underlying viscous salt layer and an elastic cohesive overburden. Mathematically, this viscoelastic coupling can be addressed by Biot's correspondence principle (Biot 1965), which states that elastic constants can be interchanged with viscoelastic operators in the Navier equations in continuum mechanics models.

It is important to point out the mechanical analogy of the studied area with the rheological configuration of most of the regional extensional provinces in which crustal-scale LANFs have been described (Wernicke 1981; Buck 1991; Johnson 2006; Lavier & Manatschal 2006; Collettini 2011; Morley 2014; Pinvidic & Osmundsen 2016). The studied LANFs are related to salt flow rheology in the same way as metamorphic core complexes are related to quartzo-feldspathic flow rheology in the lower crust (Davis & Coney 1979) and exhumed mantle detachments are related to olivine flow in the mantle (Manatschal 2004). Figure 4 compares the power law flow rheology of salt in scale with the studied seismic section with the flow rheology equations defined for the lower crust and mantle (Jackson & Talbot 1986; Van Keken *et al.* 1993; Van Der Pluijm & Marshak 2004; Fossen 2010; Gerya 2010; Turcotte & Schubert 2014). The relief of the viscous salt layer in the studied area is also proportional to the relief reported for the viscous layers in the lower crust and mantle related to crustal-scale LANFs in regional extensional

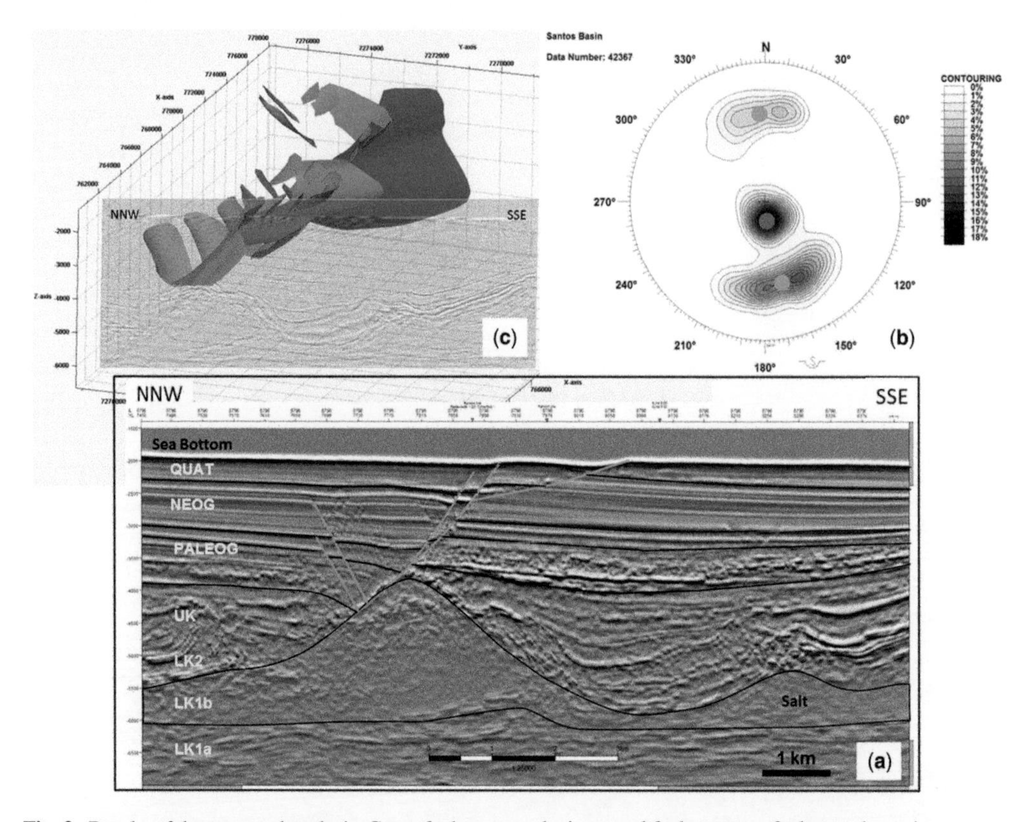

Fig. 2. Results of the structural analysis. Green faults are synthetic normal faults; orange faults are the main antithetic normal faults; and blue faults are the LANFs. (**a**) Interpreted seismic section showing the Aptian salt layer and the studied fault system with the normal fault conjugated set and one of the LANFs. Stratigraphy: LK1a, Aptian (carbonates); LK1b, Aptian (salt); LK2, Albian (carbonates); UK, Upper Cretaceous (shales and sandstones); PALEOG, Palaeogene (shales and sandstones); NEOG, Neogene (shales); and QUAT, Quaternary (shales). (**b**) Stereonet plot (Daisy3 package/uniroma3) with fault pole density showing the average dips: 65° for normal faults and 10° for LANFs. (**c**) Three-dimensional view of the studied fault system.

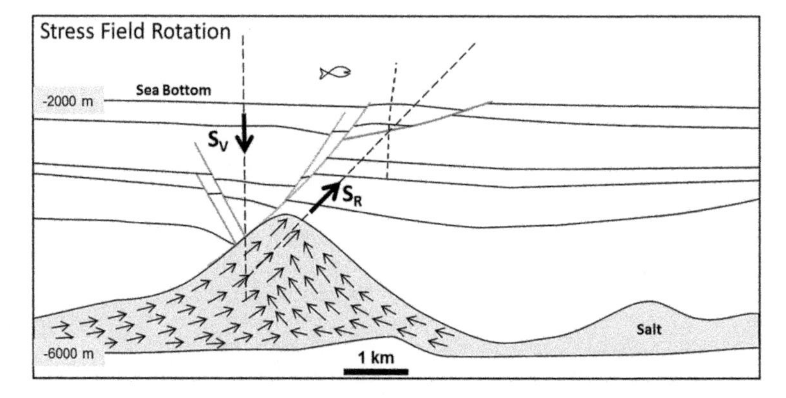

Fig. 3. Sketch of the proposed conceptual geomechanical model for the studied LANFs. We suggest that the extension of the section caused both the conjugated normal fault system and the flow of the salt constrained by the fault plane along the footwall of the main listric normal fault. The flow of salt, in turn, caused rotation of the principal stress (S_V) from the vertical to a position parallel with the upper portion of the main normal fault producing nucleation of the observed LANF as a typical Coulomb shear fracture. Jackson & Talbot (1986) suggested a similar flow pattern for salt structures caused by extension.

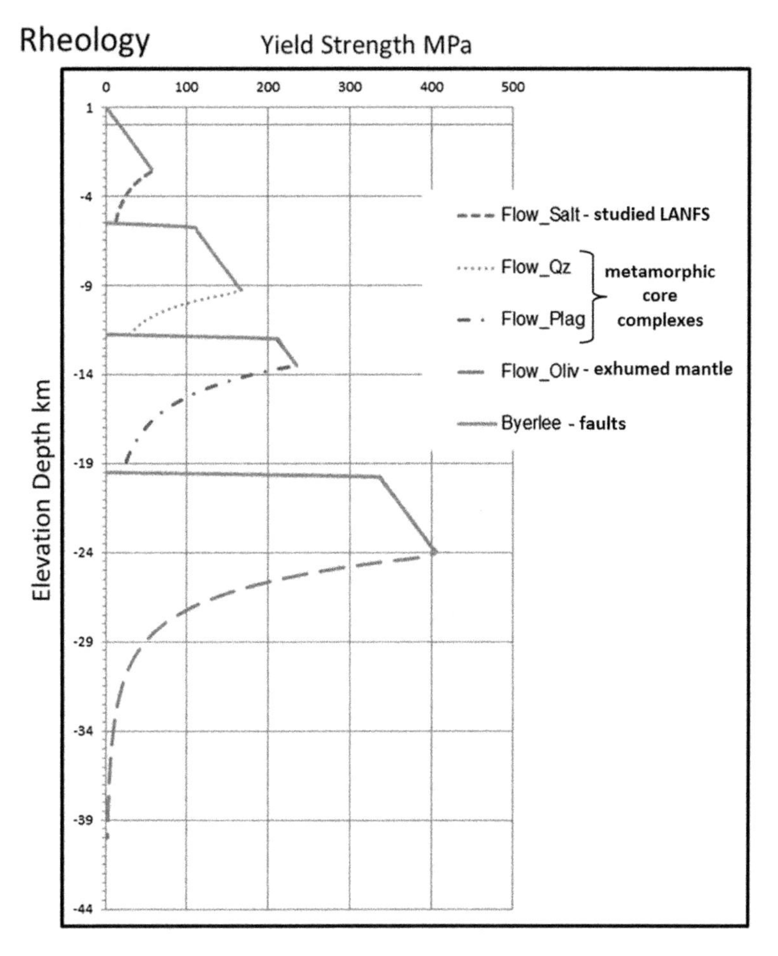

Fig. 4. Yield strength profile showing the analogy between the mechanical stratigraphy of the studied area and some rheological structures at crustal and mantle scale. The studied LANFs are related to a salt flow law in much the same way as metamorphic core complexes are related to a quartzo-feldspathic flow law into the lower crust (Davis & Coney 1979) and exhumed mantle detachments are related to an olivine flow law into the mantle (Manatschal 2004). The salt strength profile is in scale with the seismic section in Figures 1 and 2 and represents the power law equation for halite flow in the depth–temperature interval of the studied example.

provinces (Johnson 2006; Lavier & Manatschal 2006; Pinvidic & Osmundsen 2016).

Model parameters

The key parameters of the geomechanical model are well constrained. As the observed LANFs nucleated at 500 m depth, under a 2 km water table (Fig. 2), the hydrostatic pore pressure can be set at 25 MPa. The shallow sediments have densities between 1900 and 2100 kg m^{-3}, so the ratio between the pore pressure and the lithostatic stress is high (>0.8), indicating an environment of low effective stress where poroelastic effects must be strong. The studied environment perfectly fits the concept of effective stress because the water table acts on the sediments both as a load from above and as a pore pressure from inside. Therefore the net effect of the water column on the loading of the shallow sediments is null.

According to the Petrobras geohazards database, which is consistent with published data (Jaeger *et al.* 2007; Zoback 2007; Fjaer *et al.* 2008), it is also possible to constrain the mechanical properties of the shallow sediments to evaluate the model with the Mohr diagram (Fig. 5). The tensile strength can be set at 0.4 MPa and the angle of internal friction at 22° (0.4 friction coefficient). For an average density of 1900 kg m^{-3} the stress difference for fault nucleation according to the Griffith–Coulomb yield criteria is *c.* 3.5 MPa (Fig. 5a). In turn, the stress difference for fault propagation considering Byerlee's

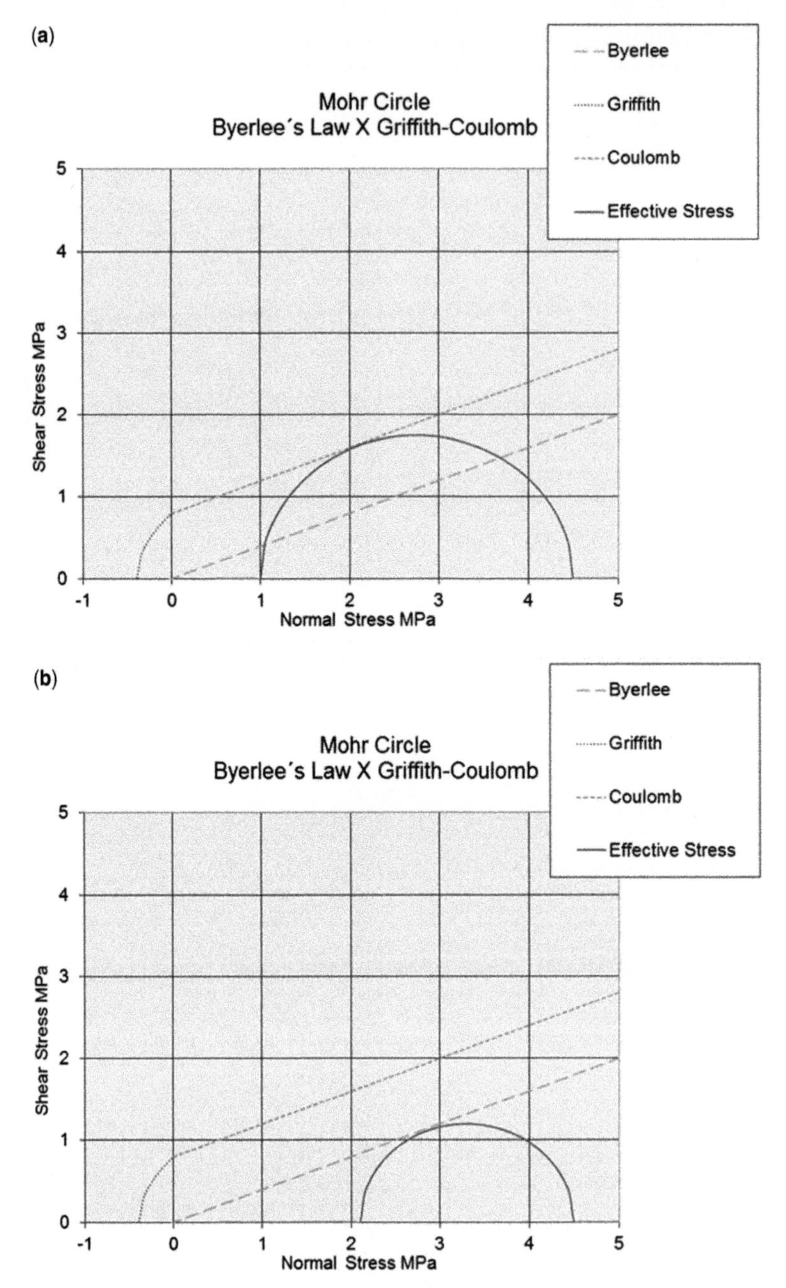

Fig. 5. Mohr diagrams showing the effective stress state for a LANF at 500 m depth under a 2 km water table. Pore pressure $p = 25$ MPa. Shallow sediment properties: density $\rho = 1900$ kg m^{-3}; tensile strength $T_0 = 0.4$ MPa; friction angle $\phi = 22°$ (friction coefficient $u = 0.4$). (**a**) Stress difference for fault nucleation (Griffith–Coulomb criteria) = 3.5 MPa. (**b**) Stress difference for fault propagation (Byerlee's law) = 2.4 MPa. Both values are reasonable and can be attained in extensional environments in sedimentary basins.

law is *c.* 2.4 MPa (Fig. 5b). Both are reasonable values of stress difference that can be attained in extensional environments in sedimentary basins. In this way, the geomechanical model proposed for the studied LANFs is coherent from the point of view of well-established fault mechanics.

We set the salt properties according to published data (Jackson & Talbot 1986; Van Keken *et al.*

1993) with flow rates between 10^{-7} and 10^{-13} s^{-1} and a salt viscosity of 10^{18} Pa s. The thermal effects on salt viscosity are negligible for the depth range of the studied section (Jackson & Talbot 1986; Van Keken *et al.* 1993). The yield strength curve for salt in Figure 4 was computed for a 10^{-7} s^{-1} flow rate.

Numerical model

We numerically simulated the proposed geomechanical model with the TECTOS finite element package (a partnership between Petrobras and the Catholic University of Rio de Janeiro-PUC-RIO). The TECTOS package has the capability to run coupled elastic–viscous models in accordance with Biot's correspondence principle (Biot 1965). We defined a two-dimensional finite element mesh based on the seismic section (NNW–SSE) orthogonal to the trend of the studied fault system (Figs 1 & 2). The TECTOS package contemplates continuum mechanics in solid, porous and incompressible media with pore pressure and fluid mechanical coupling. The analysis program implements algorithms for solving contact problems, large deformations and large displacements. TECTOS is also able to represent different constitutive elastic–plastic and viscous models such as the Maxwell, double creep and power law creep models. The TECTOS package simulates multi-scale evolutionary models in time.

The finite element model has four elastic layers with a Mohr–Coulomb constitutive model above the viscous salt layer (Fig. 6). We used a Maxwell constitutive model for salt flow with a viscosity of 10^{18} Pa s. The main listric normal fault and the LANF in the section were represented in the mesh by interface elements. The model was fixed along the boundary to the right (SSE) and along the bottom. Gravity loading was set as a body force and we prescribed a 0.8% extension towards the left boundary (NNW). Using the described loading and boundary conditions (Fig. 6), we tested two cases, without and with strain softening (Fig. 7). During the strain softening stage, the strength parameters decrease, particularly along the interface elements of the faults. The strain softening parameter is described in terms of a norm of deviatoric volumetric strains as bilinear functions. The strength parameter dilation angle (Lecomte *et al.* 2012) decreases linearly from its initial value (30°) to a residual value (−10°), obtained when the strain softening parameter reaches a critical value (10^{-3}). The numerical simulation reproduced the weakening along the main listric normal fault and along the LANF according to the model of the thick fault theory described by Lecomte *et al.* (2012).

In the first case tested without strain softening, the plastic yield along the main normal fault was restricted (Fig. 7a) and there was neither nucleation of the LANF nor significant stress rotation. In the second case with strain softening, plastic yielding along the main normal fault is widespread (Fig. 7b) and there was both nucleation of the LANF and clockwise rotation of the stress field in such a way that the maximum principal stress (S_R) became

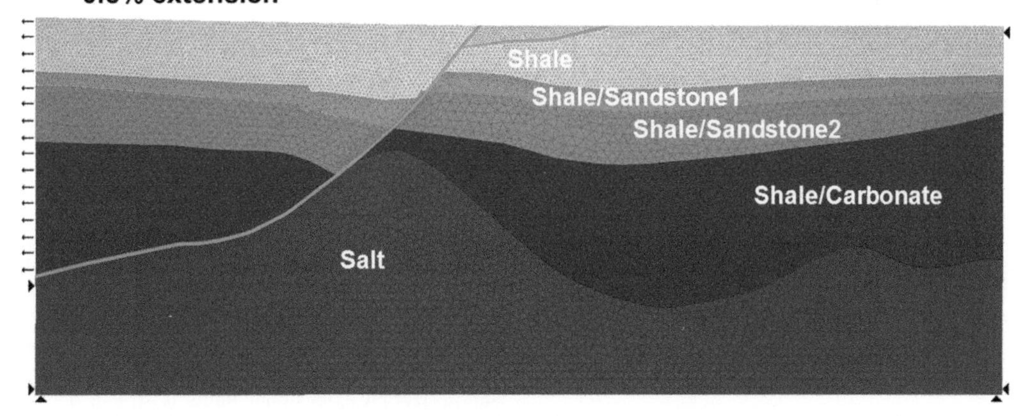

Fig. 6. Finite element mesh with five layers defined based on the seismic section in Figures 1 and 2. Model properties: shale (yellow), Young's modulus $E = 5$ GPa, Poisson's ratio $\nu = 0.28$, $\phi = 30°$, $\rho = 2100$ kg m^{-3}; shale/sandstone 1 (orange), $E = 15$ GPa, $\nu = 0.28$, $\phi = 28°$, $\rho = 2325$ kg m^{-3}; shale/sandstone 2 (green), $E = 30$ GPa, $\nu = 0.18$, $\phi = 24°$, $\rho = 2400$ kg m^{-3}; shale/carbonate (dark blue), $E = 45$ GPa, $\nu = 0.20$, $\phi = 26°$, $\rho = 2500$ kg m^{-3}; salt layer (purple), $E = 5$ GPa, $\nu = 0.48$, $\rho = 2200$ kg m^{-3}, viscosity $\mu = 10^{18}$ Pa s. The main listric normal fault and the LANF were represented with interface elements (light blue): $E = 5$ GPa, $\nu = 0.25$, $\rho = 2100$ kg m^{-3}. The model was set fixed along the lateral limit to the right (SSE) and also along the bottom limit. A 0.8% extension was prescribed towards the left limit of the model (NNW).

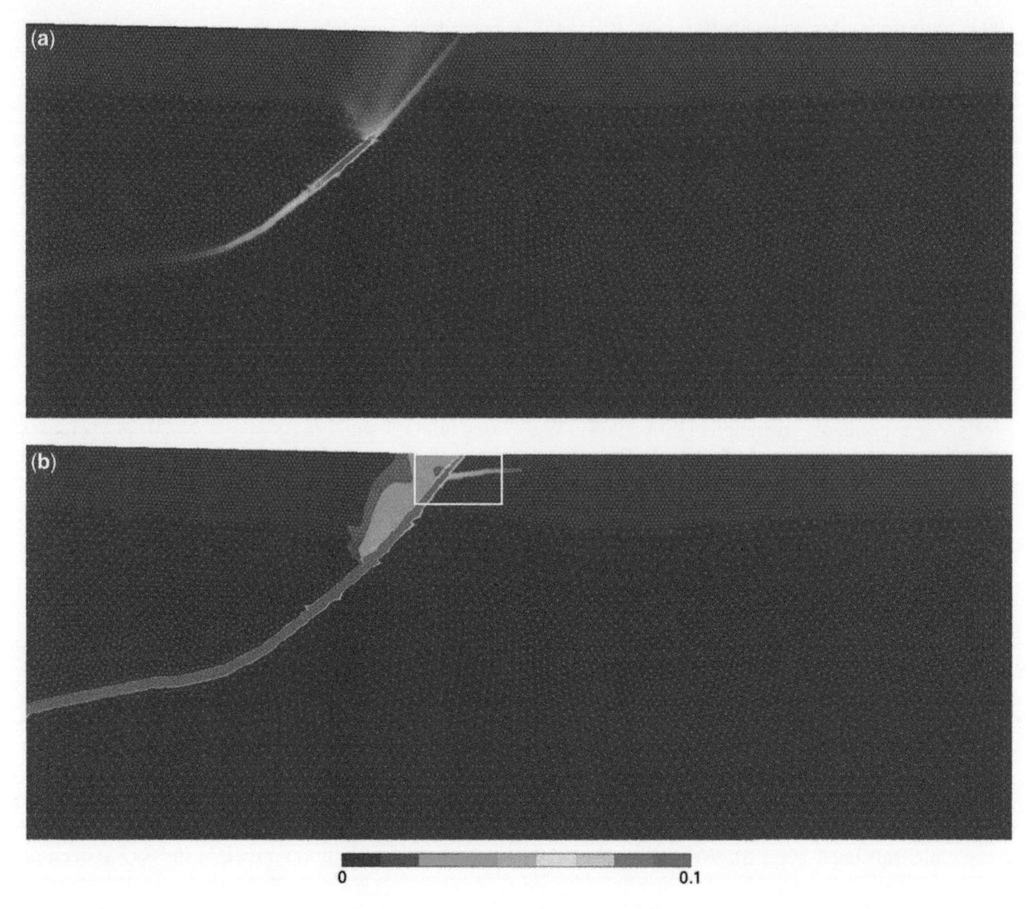

Fig. 7. Results of the numerical simulation; the colour bar represents a plastic yielding index (higher values in red). (**a**) First case tested without strain softening: plastic yielding is restricted to the central portion of the main listric normal fault; there is no nucleation of the LANF. (**b**) Second case simulated with strain softening: widespread plastic yielding of the main normal fault; observe the nucleation of the LANF; the white rectangle is the area detailed in Figure 8.

sub-parallel with the main normal fault and normal to the flank of the salt roller (Figs 3, 6 & 8). The finite element simulation of the proposed model indicates that it is possible to produce clockwise rotation of the principal stress and nucleation of the studied LANF by assuming strain weakening.

Discussion and conclusions

Wernicke (1995) reviewed the mechanical paradox of LANFs, originally described in the Basin and Range extensional province of the western USA (Wernicke 1981; Allmendinger *et al.* 1983). The paradox refers to the fact that LANFs are both precluded by classic Andersonian fault mechanics (Anderson 1951) and are scarce in the seismological record (Jackson & McKenzie 1983; Jackson & White 1989). Since then, crustal-scale faults characterized by sub-horizontal geometry and extensional kinematics with large displacements have been reported by Lavier *et al.* (1999), Hayman *et al.* (2003), Miller & Pavlis (2005), Westaway (2005), Johnson (2006), Lavier & Manatschal (2006), Wernicke *et al.* (2008), Collettini (2011), Morley (2014) and, more recently, by Pinvidic & Osmundsen (2016). Nevertheless, the geomechanical model for sub-horizontal extensional faults is still in debate (Collettini 2011). Axen (1992) indicated that high pore pressures and fault weakening are the key elements of the dynamic model. Wdowinski & Axen (1992) proposed a model with viscous flow in the lower crust. Forsyth (1992) relied more on the role of finite strain, suggesting that only LANFs could accommodate the large displacements observed in regional extensional provinces.

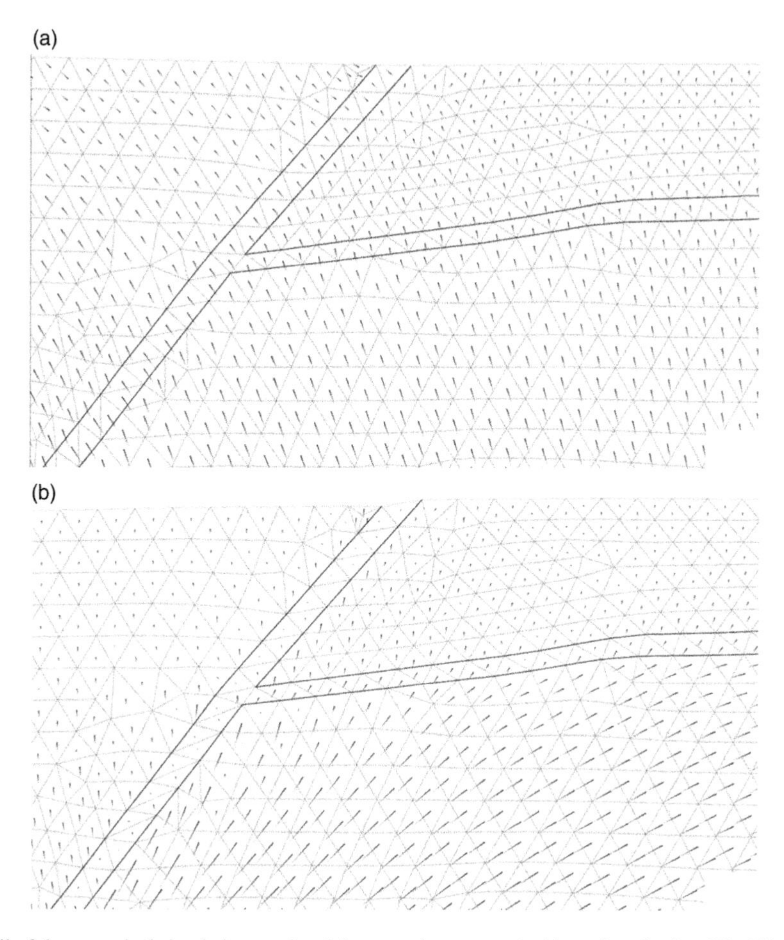

Fig. 8. Detail of the numerical simulation results of the second case tested with strain softening (Fig. 7b). (**a**) Maximum stress bars before loading: observe that the principal stress is sub-vertical or slightly orthogonal to the main normal fault. (**b**) Maximum stress bars after loading: observe the clockwise rotation of the stress field and the intensification of the principal stress caused by the flow of the salt along the footwall of the main normal fault.

Wernicke (1995) also suggested that LANFs actually nucleate and propagate at a low angle, probably because of the rotation of the principal stress axes, and that their activity has a recurrence time one order of magnitude longer than the seismicity of high-angle normal faults. Other researchers have proposed that LANFs represent either the rotation of normal faults (Proffett 1977; Jackson & McKenzie 1983; Buck 1988; Jackson & White 1989; Lavier *et al.* 1999) or the downdip coalescence of listric normal faults (Hayman *et al.* 2003; Miller & Pavlis 2005). Healy (2009) proposed that LANFs could be related to anisotropic poroelastic models.

We interpret the studied extensional faults as salt-related active LANFs formed under very low effective stress at shallow sedimentary successions close to the seafloor, where poroelastic effects, including poroelastic anisotropy, are strong. Following the idea of Wdowinski & Axen (1992), we suggest that the rotation of the maximum principal stress from the vertical is caused by channelized viscous flow of the underlying salt constrained along the footwall of a major Andersonian listric normal fault (Fig. 3). According to the kinematics of this geomechanical approach, the observed LANFs (blue faults; Fig. 2) could have been nucleated as sub-horizontal Mohr–Coulomb shear fractures. Following the energy balance approach of Forsyth (1992), after nucleation they easily propagate at a low angle according to Byerlee's law of friction (Byerlee 1978), accumulating the observed 1 km displacement, which is one order of magnitude larger than the displacement of the coeval normal faults (orange and green faults; Fig. 2). Forsyth (1992) observed that Anderson's theory of faulting was developed for infinitesimal strain only and

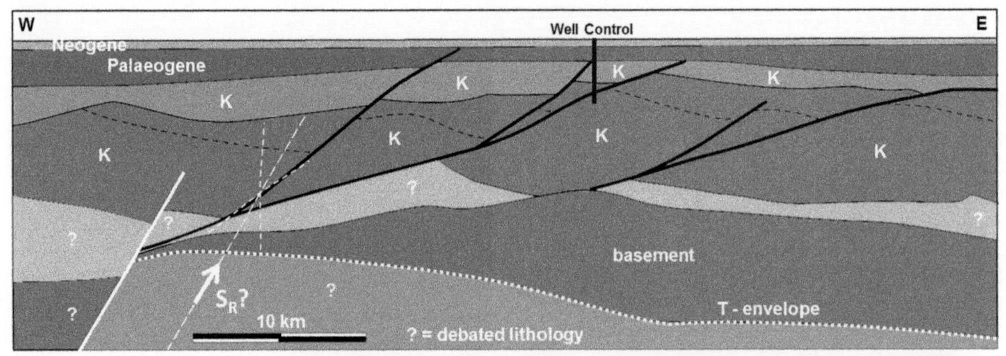

Fig. 9. Sketch of the east–west section through the Gjallar Ridge (Voring Basin, offshore Norway; modified from Pinvidic & Osmundsen 2016; K, Cretaceous; no vertical exaggeration). The T-envelope represents a series of high-amplitude seismic reflectors with dome shapes below the structural highs of the Gjallar Ridge. We interpreted the white normal fault in the western part of the section and the stress rotation assuming that the T-envelope is a rheological boundary at the top of a viscous lower crust or upper mantle.

only includes the energy necessary to overcome the friction resistance of the fault plane. He proposed a model that also included the energy necessary to overcome the flexural resistance and isostatic adjustment along the fault zone to evaluate the finite extension. He observed that the maximum accumulated displacement along a single fault for the same regional stress is inversely proportional to the dip-angle of the fault. According to this model, one single LANF accommodates a finite displacement at least one order of magnitude larger than the average displacement of the coeval set of Andersonian normal faults.

We suggest that the proposed geomechanical model is general enough to encompass crustal-scale LANFs because of the striking rheological analogy (Buck 1991, 1993; Wdowinski & Axen 1992; Lavier *et al.* 1999). The proposed model fully endorses the low effective stress hypothesis (Axen 1992; Nur & Walder 1992) and also the poroelastic anisotropic effects (Healy 2009) that can generate the differential stress required for fault nucleation. Compared with the lower plate rheology, the proposed hypothesis of viscous flow (Wdowinski & Axen 1992) is between the end-members of an inviscid fluid (Forsyth 1992; Lavier *et al.* 1999) and a material with a high shear traction (Yin 1989; Melosh 1990), so the Maxwell viscous flow approach for rheology is also reliable.

The geometric analogy of the relief of the salt layer in the studied area with the relief of viscous layers in the lower plate related to crustal-scale LANFs is also noteworthy. We suggest that the crustal-scale LANFs observed at Gjallar Ridge (Fig. 9) (modified from Pinvidic & Osmundsen 2016) could also be produced by stress rotation associated with the relief of the ridge structure, which is similar to the relief of the salt layer in the studied area (compare Figs 2 & 9).

LANFs characterize the structural style of regional extensional provinces (Wernicke 2009; Collettini 2011), which can evolve to the continental break-up phase of the Wilson cycle. Numerical experiments with large strains indicated that the lithospheric-scale strain localization that enables continental plate tectonics to create such weak plate boundaries is related to a rheological process of strain weakening (Gueydan *et al.* 2014). Correspondingly, our numerical experiment showed that we need to assume strain weakening to produce the stress rotation required to nucleate the observed LANFs. We therefore conclude that LANFs at any scale are related to viscoelastic rheological models with strain weakening behaviour that can produce rotation of the stress field and nucleation of Coulomb shear fractures with low-angle dips.

The studied area corresponds to a distal domain of the Santos Basin, where the stress state related to gravitational gliding above the salt layer is mostly compressive (Cobbold & Szatmari 1989). This is a further argument for the hypothesis that the studied fault system and the LANFs were produced by localized extension of the basement associated with intraplate tectonics. The role of basement reactivation during the evolution of the southeastern continental margin of Brazil has been widely reported and supports this hypothesis (Cobbold *et al.* 2001; Fetter 2009; Alves *et al.* 2017). It is important to point out that in the proximal domain of the Santos Basin, the structural style of the extensional environment related to salt gravitational gliding, with listric fault/rollover systems (Cobbold & Szatmari 1989), is completely different from the structural style of the fault system studied in this work.

It is important to observe that the proposed geomechanical model strictly follows the fundamental concepts of fault mechanics (continuum mechanics, poroelasticity and friction). The main physical process of the conceptual model is viscoelastic coupling, which can be addressed by Biot's correspondence principle (Biot 1965) and is supported by finite element models. In other words, the proposed model explains LANFs by means of general and well-established principles of rock mechanics.

We thank Petrobras for support to publish this work and the Tecgraf team (PUC-RIO) for help with the numerical model. Jonathan Turner, Mike Welch and two anonymous reviewers are acknowledged for their constructive contributions.

References

Allmendinger, R.W., Sharp, J.W., Von Tish, D., Serpa, L., Brown, L., Kaufman, S. & Oliver, J. 1983. Cenozoic and Mesozoic structure of the eastern Basin & Range province, Utah, from COCORP seismic-reflection data. *Geology*, **11**, 532–536.

Alves, T.M., Fetter, M. *et al.* 2017. An incomplete correlation between pre-salt topography, top reservoir erosion, and salt deformation in deep-water Santos Basin (SE Brazil). *Marine and Petroleum Geology*, **79**, 300–320.

Anderson, E.M. 1951. *The Dynamics of Faulting and Dyke Formation with Applications to Britain*. 2nd edn. Oliver & Boyd, Edinburgh.

Axen, G.J. 1992. Pore pressure, stress increase, and fault weakening in low-angle normal faulting. *Journal of Geophysical Research*, **97/B6**, 8979–8991.

Biot, M.A. 1965. *Mechanics of Incremental Deformations*. Wiley, Chichester, 349–365.

Buck, W.R. 1988. Flexural rotation of normal faults. *Tectonics*, **7**, 959–973.

Buck, W.R. 1991. Modes of continental lithospheric extension. *Journal of Geophysical Research*, **96/B12**, 20161–20178.

Buck, W.R. 1993. Effect of lithospheric thickness on the formation of high- and low-angle normal faults. *Geology*, **21**, 933–936.

Byerlee, J.D. 1978. Friction of rocks. *Pure & Applied Geophysics*, **116**, 615–626.

Cobbold, P.R. & Szatmari, P. 1989. Radial gravitational gliding on divergent margins. *Tectonophysics*, **188**, 249–289.

Cobbold, P.R., Meisling, K.E. & Mount, S.V. 2001. Reactivation of an obliquely-rifted margin: Campos and Santos basins, SE Brazil. *American Association of Petroleum Geologists Bulletin*, **85**, 1925–1944.

Collettini, C. 2011. The mechanical paradox of low-angle normal faults: current understanding and open questions. *Tectonophysics*, **510**, 253–268.

Davis, G.H. & Coney, P.J. 1979. Geologic development of Cordilleran metamorphic core complexes. *Geology*, **7**, 120–124.

Fetter, M. 2009. The role of basement tectonic reactivation on the structural evolution of Campos Basin, offshore Brazil: evidence from 3D seismic analysis and section restoration. *Marine and Petroleum Geology*, **26**, 873–886.

Fjaer, E., Holt, R.M., Horsrud, P., Raaen, A.M. & Risnes, R. 2008. *Petroleum Related Rock Mechanics*. Elsevier, Amsterdam.

Forsyth, D.W. 1992. Finite extension and low-angle normal faulting. *Geology*, **20**, 27–30.

Fossen, H. 2010. *Structural Geology*. Cambridge University Press, Cambridge.

Gerya, T.V. 2010. *Introduction to Numerical Geodynamic Modelling*. Cambridge University Press, Cambridge.

Gueydan, F., Précigout, J. & Montési, L.G.J. 2014. Strain weakening enables continental plate tectonics. *Tectonophysics*, **631**, 189–196.

Hayman, N.W., Knott, J.R., Cowan, D.S., Memser, E. & Sarna-Wojcicki, A.M. 2003. Quaternary low-angle slip on detachment faults in Death Valley, California. *Geology*, **31**, 343–346.

Healy, D. 2009. Anisotropy, pore fluid pressure and low angle normal faults. *Journal of Structural Geology*, **31**, 561–574.

Howard, K.A. 2003. Crustal structure in the Elko-Carlin region, Nevada, during Eocene gold mineralization: Ruby-East Humboldt metamorphic core complex as a guide to deep crust. *Economic Geology*, **98**, 249–258.

Hubbert, M.K. & Rubey, W.W. 1959. Role of fluid pressure in the mechanics of overthrust faulting. *Geological Society of America Bulletin*, **70**, 115–205.

Jackson, J.A. & McKenzie, D. 1983. The geometrical evolution of normal fault systems. *Journal of Structural Geology*, **5**, 471–482.

Jackson, J.A. & White, N.J. 1989. Normal faulting in the upper continental crust: observation from regions of active extension. *Journal of Structural Geology*, **11**, 15–36.

Jackson, M.P.A. & Talbot, C.J. 1986. External shapes, strain rates and dynamics of salt structures. *Geological Society of America Bulletin*, **97**, 305–323.

Jaeger, J.C., Cook, N.G. & Zimmerman, R.W. 2007. *Fundamentals of Rock Mechanics*. Blackwell, Oxford.

Johnson, B.J. 2006. Extensional shear zones, granitic melts, and linkage of overstepping normal faults bounding the Shuswap metamorphic core complex, British Columbia. *Geological Society of America Bulletin*, **118**, 366–382.

Lavier, L. & Manatschal, G. 2006. A mechanism to thin the continental lithosphere at magma-poor margins. *Nature*, **440**, 324–328.

Lavier, L., Buck, W.R. & Poliakov, A.N.B. 1999. Self-consistent rolling-hinge model for the evolution of large-offset low-angle normal faults. *Geology*, **27**, 1127–1130.

Lecomte, E., Le Pourhiet, L. & Lacombe, O. 2012. Mechanical basis for slip along low-angle normal faults. *Geophysical Research Letters*, **39**, L03307.

Manatschal, G. 2004. New models for evolution of magma-poor rifted margins based on a review of data and concepts from West Iberia and the Alps. *International Journal of Earth Sciences (Geologische Rundschau)*, **93**, 432–466.

Melosh, H.J. 1990. Mechanical basis for low-angle normal faulting in the Basin & Range province. *Nature*, **343**, 331–335.

Miller, M.B. & Pavlis, T.L. 2005. The Black Mountains turtlebacks: rosetta stones of Death Valley tectonics. *Earth Science Reviews*, **73**, 115–138.

Morley, C.K. 2014. The widespread occurrence of low-angle normal faults in a rift setting: review of examples from Thailand, and implications for their origin. *Earth Science Reviews*, **133**, 18–42.

Nur, A. & Walder, J. 1992. Hydraulic pulses in the Earth's crust. *In*: Evans, B. & Wong, T.F. (eds) *Fault Mechanics and Transport Properties of Rocks*. Academic Press, London, 461–473.

Pinvidic, G.P. & Osmundsen, P.T. 2016. Architecture of the distal and outer domains of the Mid-Norwegian rifted margin: insights from the Ran-Gjallar ridges system. *Marine and Petroleum Geology*, **77**, 280–299.

Proffett, J.M., Jr 1977. Cenozoic geology of the Yerington district, Nevada, and implications for the nature and origin of Basin & Range faulting. *Geological Society of America Bulletin*, **88**, 247–266.

Sibson, R.H. 1974. Frictional constraints on thrust, wrench and normal faults. *Nature*, **249**, 542–544.

Snow, J.K. & Wernicke, B.P. 2000. Cenozoic tectonism in the central Basin & Range: magnitude, rate, and distribution of upper crustal strain. *American Journal of Science*, **500**, 659–719.

Turcotte, D. & Schubert, G. 2014. *Geodynamics*. Cambridge University Press, Cambridge.

Van Der Pluijm, B.A. & Marshak, S. 2004. *Earth Structure*. W.W. Norton, London.

Van Keken, P.E., Spiers, C.J., Van Den Berg, A.P. & Muyzert, E.J. 1993. The effective viscosity of rock salt: implementation of steady-state creep laws in numerical models of salt diapirism. *Tectonophysics*, **225**, 457–476.

Wdowinski, S. & Axen, G.J. 1992. Isostatic rebound due to tectonic denudation: a viscous flow model of a layered lithosphere. *Tectonics*, **11**, 303–315.

Wernicke, B.P. 1981. Low angle normal faults in the Basin & Range province: nappe tectonics in an extending orogeny. *Nature*, **291**, 645–648.

Wernicke, B.P. 1995. Low angle normal faults and seismicity: a review. *Journal of Geophysical Research*, **100/B10**, 20159–20174.

Wernicke, B.P. 2009. The detachment era (1977-1982) and its role in revolutionizing continental tectonics. *In*: Ring, U. & Wernicke, B.P. (eds) *Extending a Continent: Architecture, Rheology and Heat Budget*. Geological Society, London, Special Publications, **321**, 1–8, https://doi.org/10.1144/SP321.1

Wernicke, B.P., Davis, J.L., Niemi, N.A., Luffi, P. & Bisnath, S. 2008. Active megadetachment beneath the western United States. *Journal of Geophysical Research*, **1113/B11409**, 1–26.

Westaway, R. 2005. Active low-angle normal faulting in the Woodlark extensional province, Papua New Guinea: a physical model. *Tectonics*, **24**, TC6003, 1–25.

Yin, A. 1989. Origin of regional, rooted low-angle normal faults: a mechanical model and its tectonics implications. *Tectonics*, **8**, 469–482.

Zoback, M.D. 2007. *Reservoir Geomechanics*. Cambridge University Press, Cambridge.

Estimating friction in normal fault systems of the Basin and Range province and examining its geological context

CARSON A. RICHARDSON* & ERIC SEEDORFF

Department of Geosciences and Lowell Institute for Mineral Resources, University of Arizona, 1040 East Fourth Street, Tucson, AZ 85721-0077, USA

**Correspondence: carichardson@email.arizona.edu*

Abstract: The life cycle of a fault following initiation is governed in part by the reshear criterion, of which rock surface friction is the critical factor limiting the dip of a fault at its death. Using structural restorations where the initial and final dips of faults can be ascertained, the coefficient of rock surface friction is calculated for well-characterized extended locales ($n = 20$) in the Basin and Range province, many with multiple fault generations ($n = 34$). The calculated values exhibit a considerably wider range (0.19–1.33) than previously reported. The amount of tilting associated with each fault generation is compared with eight characteristics (mean slip magnitude, tilting per unit of slip, fault spacing, percentage extension, absence or presence and composition of magmatism, duration of extension, timing of extension and strain rate). No statistically strong correlation was found with any of the examined characteristics, although tentative linkages were noted with percentage extension, strain rate and mean slip magnitude from weighted regression analysis. These results are consistent with normal faults behaving as non-linear systems, with friction being an emergent property.

Fault mechanics can help guide our thinking on the nature of continental extension (e.g. Axen 2004), providing insights regarding the geometric evolution of fault systems, the potential controls on fault activity, dormancy, death, and the role of fluid migration in actively extending regions. Following initiation, normal faults can repeatedly reuse a fault plane within a fault zone and undergo tilting concurrent with slip, as governed by the reshear criterion. Tilting is controlled by the coefficient of rock surface friction. The coefficient of rock surface friction is an experimentally determined value from rock deformation experiments (e.g. Byerlee 1978). Sibson (1994) developed a relationship from which to calculate such values from natural geological settings that have cross-cutting normal faults. Sibson (1994) found that many fault systems, regardless of the tectonic regime, had frictional values within the range 0.6–0.85 reported by Byerlee (1978).

Since the publication of the paper by Sibson (1994), studies in the Basin and Range province have continued to identify locales with multiple sets of superimposed normal faults that initiated at high angles, rotated to lower angles and were then cut by faults of younger generations (e.g. Emmons 1907; Morton & Black 1975; Proffett 1977). Much of the work in these areas also included detailed studies of the timing of extension and related magmatism (e.g. Gans *et al.* 1989; Hudson *et al.* 2000; Konstantinou *et al.* 2013) and this compilation draws on areas with superimposed generations of normal faults. The initial and final dips of the faults are constrained from geological data and cross-cutting relationships. The coefficient of rock surface friction can be estimated with the assumption that a new fault initiates because the frictional resistance of the fault could not be overcome, rather than because of a change in the local stress field.

There were two aims to this study. The first was to quantitatively estimate the coefficient of rock surface friction in natural normal fault systems following the approach of Sibson (1994), but with a much larger sample set from the Basin and Range province (Fig. 1), to assess the applicability of the coefficient of rock surface friction values of Byerlee (1978). The second was to examine possible factors and controls on tilting in this style of rotational planar normal fault systems. A number of alternative names are used to describe these extensional fault systems, including domino-style, bookshelf faulting and block tilting. For consistency, we utilize the term rotational planar normal faulting, as described by Wernicke & Burchfiel (1982). We examine eight factors that might play a part: mean slip magnitude; tilting per unit of slip; fault spacing; percentage extension; the absence/presence and composition of synextensional magmatism; the duration of extension; the timing of extension; and the strain rate. We make explicit several assumptions in this paper: (1) all work discussed here is restricted to the brittle regime with temperatures *c.* <300°C (Voll 1976; Kerrich *et al.* 1977; Crider & Peacock 2004); (2) the stress trajectories do not deviate with depth, although this has been invoked by some to suggest

From: Turner, J. P., Healy, D., Hillis, R. R. & Welch, M. J. (eds) 2017. *Geomechanics and Geology*. Geological Society, London, Special Publications, **458**, 155–179.
First published online May 25, 2017, https://doi.org/10.1144/SP458.8

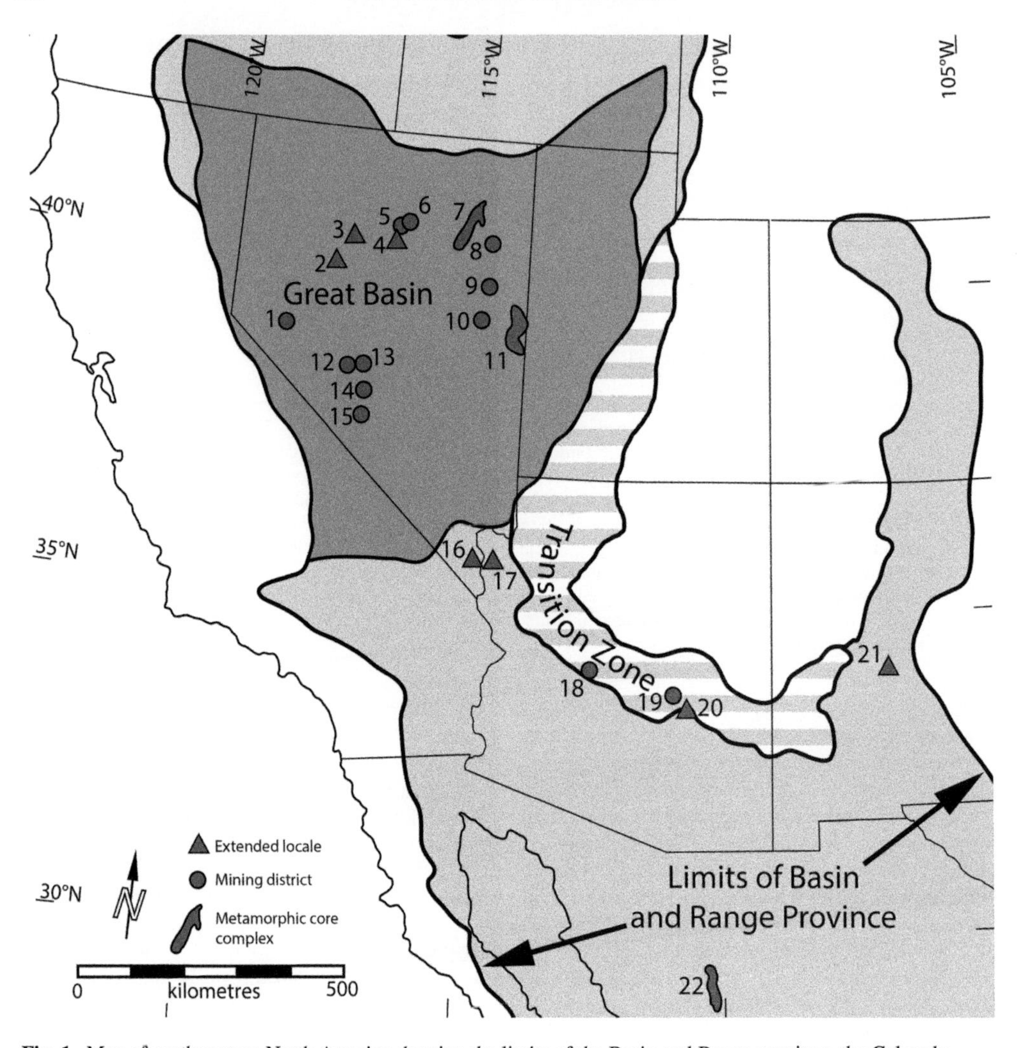

Fig. 1. Map of southwestern North America showing the limits of the Basin and Range province, the Colorado Plateau transition zone and the Great Basin, and the extended field areas discussed in this paper. Modified from Seedorff (1991*a*), Dickinson (2006) and Richardson & Seedorff (2015). Locales are: 1, Yerington district; 2, Stillwater Range; 3, Tobin Range; 4, Caetano Caldera; 5, Lewis district and Mill Creek area, northern Shoshone Range; 6, Hilltop district, northern Shoshone Range; 7, Ruby-East Humboldt metamorphic core complex; 8, Spruce Mountain; 9, Hunter district; 10, Robinson district; 11, Snake Range metamorphic core complex; 12, Royston prospect; 13, Hall deposit; 14, Tonopah district; 15, Goldfield district; 16, Eldorado Mountains; 17, Mt Perkins; 18, White Picacho and Sheep Mountains districts, Wickenburg; 19, Tea Cup system; 20, Romero Wash-Tecolote Ranch; 21, Lemitar Mountains; and 22, Sierra Mazatán metamorphic core complex.

the low-angle initiation of normal faults (e.g. Parsons & Thompson 1993; Axen *et al.* 2015); and (3) pore fluid pressure does not play a part in the initiation of normal faults, but may allow normal faults to continue to slip at a low angle because mineralization and associated hydrothermal fluid flow synchronous with faulting is widely documented in ore deposits and active geothermal systems (e.g. Dreier 1984; Sibson 1987; Micklethwaite 2009; Faulds *et al.* 2015; Rhys *et al.* 2015).

Traditional factors in extensional fault mechanics

Principles of Andersonian–Byerlee fault mechanics

Faults in the Earth's brittle upper crust are discrete discontinuities where offset is accomplished along compressional shear fractures. The classification of any one fault (normal, reverse, strike-slip, or a

Table 1. *Abbreviations used in text*

Notation	Definition
σ_1	Maximum principal stress
σ_2	Intermediate principal stress
σ_3	Minimum principal stress
τ	Shear stress
σ_n	Normal stress
C_0	Cohesive strength
ϕ	Angle of internal friction
μ_i	Coefficient of internal friction
Θ	Angle between maximum principal stress and fault plane
μ_s	Coefficient of rock surface friction
P_f	Pore fluid pressure
Θ_l	Angle of frictional lock-up between the maximum principal stress and the fault plane
Θ_r^*	Angle of optimum reactivation between the maximum principal stress and the fault plane
σ_1'	Effective maximum principal stress
σ_2'	Effective intermediate principal stress
σ_3'	Effective minimum principal stress
δ	Dip or angle from horizontal of a planar feature

combination thereof) is a function of the orientations of the three principal stress axes ($\sigma_1 > \sigma_2 > \sigma_3$ where compression is reckoned to be positive; all oriented in directions of zero shear stress) and the disrupted geometry across the resultant fault plane at the time of fault initiation (see Table 1 for definitions of abbreviations used). Anderson (1905, 1951) developed this dynamic classification with two major assumptions: (1) two of the principal compressive stress axes are oriented at 90° to one another along the Earth's surface and the third is orthogonal to the other two and perpendicular to the Earth's surface; and (2) faults form as planar surfaces containing σ_2 and inclined at an angle to σ_1, with the initial angle controlled by the coefficient of internal friction. Planes of maximum tangential stress will be oriented at 45° between σ_1 and σ_3 and normal to σ_2, but planes of shear failure are found to nucleate with an acute angle (22–32°) from σ_1 to σ_3 and containing σ_2 (Sibson 2002; Jaeger *et al.* 2007). These planes of shear failure are the initial fault planes and develop through repeated earthquake events.

Fault initiation is most commonly approximated in compressional Mohr circle space with the Coulomb failure criterion for brittle failure:

$$\tau = C_0 + \mu_i \times \sigma_n, \tag{1}$$

where τ is the shear stress, C_0 is initial cohesive strength of unfaulted rock, μ_i is the coefficient of internal friction and σ_n is the normal stress. Typical values of μ_i from deformation experiments range between 0.5 and 1.0 (Hoek 1965), but values greater than unity are reported for igneous and strong sedimentary rocks (Axen 2004). In Mohr circle space (Fig. 2), with σ_n as the x-axis and τ as the y-axis, the Coulomb failure criterion is controlled by C_0 (the y-intercept) and μ_i (the slope of the line). Note that μ_i can also be written as tan ϕ, with ϕ equal to 90° minus 2Θ; Θ is the acute angle between σ_1 and the fault plane at the initiation of faulting. The Mohr circle approximates the two-dimensional stress field between σ_1 and σ_3, with σ_2 in and out of the plane. A line representing the initiation of fault failure is plotted as the radius of the Mohr circle using the angle 2Θ, with the angle being measured from the σ_3 (left) side of the circle (Fig. 2).

Following initiation, faults are observed to repeatedly reutilize a fault plane and undergo tilting concurrent with slip, rather than initiating a new fault plane for every slip event (King *et al.* 1988; Stein *et al.* 1988). The continual reutilization of the fault plane is characterized by Amonton's law (hereafter referred to as the reshear criterion):

$$\tau = C_0 + \mu_s \times (\sigma_n - P_f), \tag{2}$$

where μ_s is the coefficient of rock surface friction (also known as sliding or Byerlee friction), and P_f is the pore fluid pressure. Experimentally determined values for μ_s lie in the range 0.6–0.85 (Byerlee 1978). Faults in the reshear criterion are commonly given a lower, if not zero, cohesive strength than intact rock, as a result of the difficulty in separating the relative cohesive strengths of a fault from its country rock and in determining the effective lithostatic pressure (Donath & Cranwell 1981; Etheridge 1986; Nur *et al.* 1986; Choi & Buck 2012). Axen (2004) points out an important component of the experimental work of Byerlee (1978). The 'ranges' commonly cited are in actuality two trends, (1) a low-pressure (normal stresses 5–200 MPa) trend with a fit of

$$\tau = 0.85 \times \sigma_n \tag{3}$$

and (2) a high-pressure (normal stresses 200–2000 MPa) trend with a fit of

$$\tau = 0.6 \times \sigma_n + 50\,\text{MPa}. \tag{4}$$

Axen (2004) determined the low-pressure trend characterized the crust to depths of *c.* 20.6 km by solving the low-pressure trend with the effective lithostatic stress, a maximum normal stress of 200 MPa and the maximum differential stress (i.e. $\sigma_3 = 0$). Thus if the low-pressure trend of Byerlee

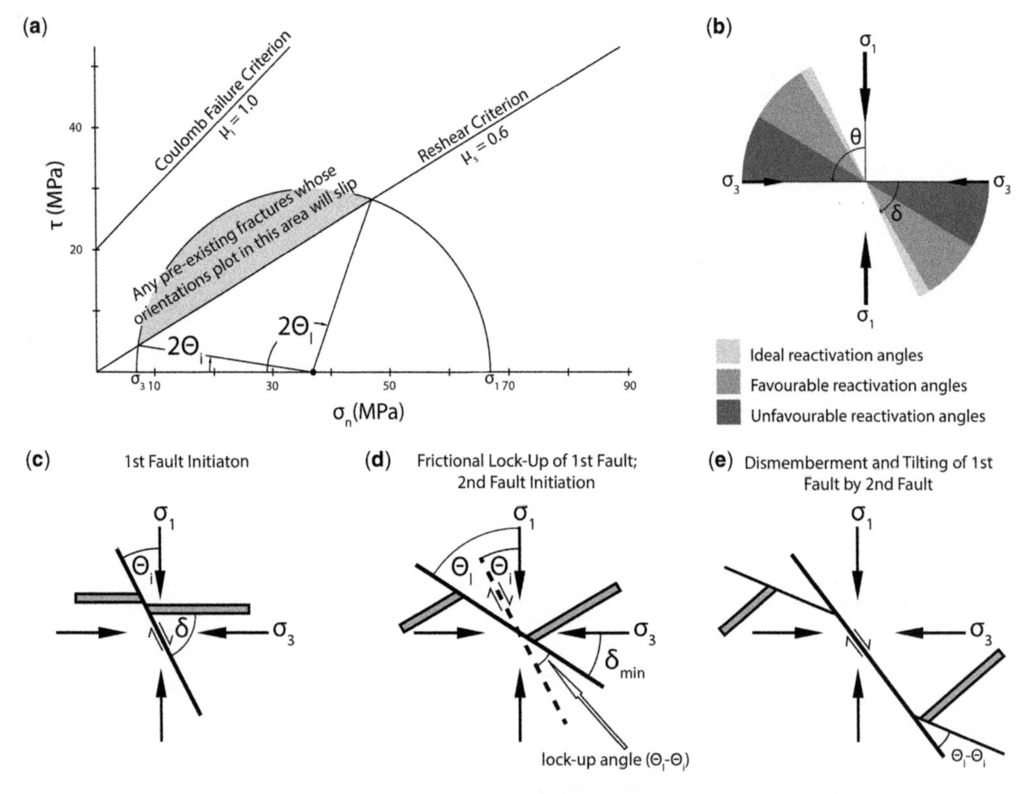

Fig. 2. Fault mechanics and geometrical relationships in normal fault systems. (**a**) Mohr diagram of shear stress against normal stress, with the Coulomb failure envelope and cohesionless reshear envelope illustrating the stress conditions required for fault failure to occur. The shaded area of the plotted Mohr circle represents the range of 2Θ orientations under the imposed stress field that would fail. Modified from Davis & Reynolds (1996, fig. 5.48) and Sibson (2000). (**b**) Range of permissible orientations for normal faults with respect to σ_1 under typical ranges of rock surface friction. Modified from Collettini & Sibson (2001). (**c**) Initiation of first normal fault in the upper crust with bedding shown to demonstrate how to determine Θ and dip (δ) from the initial cut-off angle. (**d**) Tilting of first normal fault to frictional lock-up with the initiation of a mechanically favourable high-angle second normal fault. (**e**) Continued tilting and dismemberment of first normal fault by second fault (c–e modified from Sibson 1994, fig. 4).

(1978) were to characterize the entirety of the brittle crust, the expected tilting in normal fault systems would be *c.* 20°, in stark contrast with the high amounts of tilting observed in many extended locales. In addition, fault cohesion plays a part in the high-pressure data, despite it commonly being dropped from equation (2), because it is observed to be negligible in relation to other values near the surface. Fault cohesion increases with depth and may increase during interseismic intervals but is commonly ignored in papers on fault mechanics (Jamison & Cook 1980; Angevine *et al.* 1982; Tenthorey & Cox 2006, fig. 10). Consequently, the reshear criterion commonly originates at the origin in the Mohr circle space, lies below the Coulomb failure criterion, and mechanically governs much of the life cycle of a fault (Fig. 2c–e). The slope of the reshear criterion is defined as μ_s, with lower values of μ_s corresponding to greater amounts of tilting prior to fault lock-up than higher μ_s values because lower values allow more potential orientations to serve as failure planes (Fig. 2a). Normal faults tilt as they slip, decreasing in dip, until it is energetically more favourable to initiate a new fault plane than to continue reshearing along the existing fault plane and to overcome the frictional resistance; this process is known as frictional lock-up (Thatcher & Hill 1991; Hill & Thatcher 1992). Frictional lock-up is represented by the increasing 2Θ angle (represented by a line originating from the radius in the Mohr circle and plotting a point on the circumference of the Mohr circle) until the point along the Mohr circle circumference plots below the reshear criterion, at which time slip along the fault is mechanically unfeasible. The final Θ value limits possible values for μ_s through the following equation from Sibson (1974, 1985, 1994), where Θ_r^* is the optimum reactivation angle and Θ_l is the angle of frictional lock-up:

$$\Theta_l = 2\Theta_r^* = \arctan\left(\frac{1}{\mu_s}\right). \quad (5)$$

Pore fluid pressure

A wealth of research on the role of pore fluid pressure on unfavourably oriented slip in fault systems has been conducted over the last half-century (Hubbert & Rubey 1959; Rubey & Hubbert 1959; Sibson 1973, 2000; Axen 1992, 2004; Healy 2009; Collettini 2011). In a fluid-saturated rock with a pore fluid pressure (P_f), the principal stresses are reduced and become effective principal stresses: $\sigma_2' = (\sigma_2 - P_f)$, $\sigma_2' = (\sigma_2 - P_f)$, $\sigma_3' = (\sigma_3 - P_f)$. The effective principal stresses shift the Mohr circle left towards the origin so that a greater range of dip values are mechanically feasible under the reshear criterion. Although elevated pore fluid pressure has been suggested as a satisfactory mechanism to explain slip on unfavourably oriented faults (e.g. Sibson 1990), the mere presence or absence of fluids in a fault zone is not evidence for low-angle slip. Many locales have evidence for fluid flow along low-angle fault zones (e.g. chloritic breccias in fault rocks in metamorphic core complexes; Davis 2013); however, other field sites (e.g. Carlin-type and epithermal-type deposits; Micklethwaite *et al.* 2010) show evidence for more than ample syntectonic fluid circulation, yet the faults do not tilt beyond dips that are mechanically unfavourable in fluid-free environments. Thus there must be other factors operating alongside pore fluid pressure that enhance or inhibit tilting and low-angle normal fault slip.

Principal stress trajectories

An implicit assumption behind Andersonian fault mechanics is that the principal stress trajectories do not deviate with depth but remain constant. One of the primary advantages of a rotated stress field is that it allows the initiation and slip of low-angle normal faults, such as the detachment faults associated with metamorphic core complexes. This has been suggested by previous workers (Yin 1989; Parsons & Thompson 1993), but field evidence for rotated palaeostress fields has been scant until recent work (Axen *et al.* 2015), which suggests this may be a permissible hypothesis. One line of evidence commonly cited to support Andersonian stress trajectories and standard fault mechanics is the preponderance of moderate to steep dips from active normal fault earthquakes, with no large-magnitude seismic events confidently resolved onto a normal fault that dips at low angles (Jackson & White 1989; Collettini & Sibson 2001; Hreinsdóttir & Bennett 2009).

Fault rock composition and variability

Fault zones, both the fault core and the fault damage/process zones, can be highly variable along-strike or downdip in lithology, deformation style and fault rock composition. The presence of materials, such as foliated phyllosilicates with relatively low values for μ_s, has also been used to account for low-angle slip in normal fault systems (Collettini & Holdsworth 2004; Collettini *et al.* 2005; Numelin *et al.* 2007; Behnsen & Faulkner 2012). Debate continues on the topics of the necessary proportion of phyllosilicates needed to weaken a fault, how this varies depending on the fault rock type and how this scales from discrete areas to an entire fault (Jefferies *et al.* 2006; Collettini *et al.* 2009*a*; Niemeijer *et al.* 2010; Collettini 2011). Thus the application of the frictional properties of phyllosilicates to characterize an entire fault is questionable if phyllosilicates are not a major constituent of the entire fault and compose a continuous network (Collettini 2011). In addition, the depth range in which these materials with low values of μ_s can form varies mineralogically but is typically viewed as forming within 8 km of the surface (Collettini *et al.* 2009*b*). This suggests that low μ_s values attributed to phyllosilicates are restricted to the upper parts of faults by virtue of the limited stability at depth. Nonetheless, the stability of phyllosilicates may extend to greater depths than previously predicted (Schmalholz & Podladchikov 2014; Wheeler 2014).

Fault interactions

The interaction, growth and rupture history of an individual fault play a significant part in the long-term evolution of a fault system. Fault interaction is commonly characterized by two main types: (1) hard linkage, where two faults become physically linked and the resultant fault plane has a new geometry; and (2) soft linkage, where faults interact through their overlapping stress fields (Gupta & Scholz 2000) or via ductile strain of the rock volume between two faults (Walsh & Watterson 1991). Soft linkage between faults is largely the result of the interaction of the stress fields of individual faults within a fault system, which evolves over time due to the stress changes in the surrounding region of a discrete fault following a slip event. Scholz (2007) describes three main styles of soft linkage in normal fault systems. The first style is pinning, where two faults propagate along-strike until they are overlapping and within the stress shadow of the other fault's tip, where further propagation is reduced and eventually ceases. Each fault's displacement profile is then shifted towards the overlapping tips from the idealized bell curve of an isolated normal fault as the faults continue to accumulate displacement without elongating along-strike (Peacock & Sanderson 1994; Contreras *et al.* 2000). The second style is coalescence, where the net result of displacement along multiple faults within a fault system resembles that of a single fault (Contreras *et al.* 2000). The third style is nucleation inhibition, where the stress shadows of larger, slightly older (but not necessarily of a different generation) faults will prevent new faults from forming (Ackermann & Schlische 1997).

The timescale of fault interactions is crucial in the short-term (tens to 10 000s of years) evolution of a fault system, as they can trigger or inhibit slip within the fault system (e.g. King *et al.* 1994; Hubert *et al.* 1996; Beeler *et al.* 2000). Wallace (1984, 1987) suggested that this variation in activity is also displayed at the wider scale of entire extensional provinces, where zones of concentrated, active deformation (i.e. the Central Nevada Seismic Zone) spatially migrate over time in the Basin and Range province. The continual modification of the local stress field by hundreds to thousands of slip events in a fault system, as well as the regional migration in activity, would lead to a highly complex stress pattern. The above factors clearly play a significant part in the local evolution of fault systems, but questions remain as to whether stress interactions can permanently deactivate faults and how far stress interactions can impact the activity of faults beyond the immediate vicinity of the fault system (Jackson 1999). Teasing out the influence of fault interactions in the evolution of a fault system is clearly a difficult enterprise – particularly in dead fault systems.

Natural fault systems are inherently complex, emergent systems with a large number of interacting elements. Systems can be fundamentally viewed as two distinct classes: linear and non-linear. In linear systems, the products of the system are proportional to one or more of the inputs; however, most natural systems are fundamentally non-linear and may appear chaotic or unpredictable due to the non-linearity of the interactions of the inputs (Holland 1998). Non-linear systems have outputs that are not directly proportional to the inputs, but some can be solved with linear approximations. Emergent systems are a subset of non-linear systems where larger entities arise through interactions of smaller or simpler entities. Those entities, although affecting the outputs, generally fail the statistical tests of correlation applicable to linear systems. The chief characteristic of complex systems is the self-organization (e.g. emergence) of an overall behaviour that is not determined solely by the individual interactions and properties of the local entities. Fault systems display diverse evolutionary patterns in terms of structural style and deformation that are recognized at a variety of scales, yet are unpredictable based on local interactions of inputs (Navarro 2002). This unpredictability produces order, but with elements of chaos due to the non-linearity of the local interactions of the inputs.

It is thus apparent that there are a multitude of questions in fault mechanics that are still highly debated. How important are fluids and to what degree can their presence account for variable tilt? How consistent are stress trajectories with depth? How widespread and consistent are low-friction phyllosilicates? These are all crucial questions that warrant further attention and highlight the debate that continues in the fault mechanics community. However, there are a number of other factors that are commonly invoked by field geologists working in highly extended regions, to which we now turn our attention.

Methods

This study examines the geological conditions that can influence the mechanical evolution of highly extended, tilted areas. Using previously published palinspastic structural restorations and observations from extensionally faulted areas within the Basin and Range province (Fig. 2c–e), we estimate μ_s values, which are calculated for each generation of faults at each extended locale using equation (5) (from Sibson 1985, 1994). Table 2 compiles the values of μ_s with the associated geological characteristics of faulting, such as geochronology (both timing and duration), the percentage extension, the strain rate, the presence/absence of synextensional magmatism and composition, the mean amount of slip per fault set, the fault spacing, the amount of tilting and the amount of tilting per unit slip. The calculated values of μ_s capture the mechanical and tilting history of a fault assuming Andersonian stress trajectories. Synextensional magmatism is defined as igneous activity that occurs nearly contemporaneously with extensional tectonism. This is typically in the form of volcanism, but intrusions such as dykes are also represented in the data table. A key assumption in this work is that, at each location, a new generation of faults develops because it is mechanically unfavourable to reutilize the older generation of faults, not because there was a change in the regional/local stress regime.

Palinspastic structural restorations

Palinspastic structural restorations have been used for over a century to test structural hypotheses and interpretations in deformed terrains (Ransome *et al.* 1910; Wernicke & Axen 1988; Smith *et al.* 1991; Karlstrom *et al.* 2010). Although this technique is not unique, some previously studied localities with palinspastic structural restorations were not included due to outstanding questions around the initial dips of faults and/or the absence of a well-defined pre-extensional datum to use in restorations (e.g. the Eureka district, Nevada; Di Fiori *et al.* 2015; Hoge *et al.* 2015).

Most palinspastic restorations are two-dimensional, cross-sectional reconstructions, oriented parallel to the slip direction, and aim to maintain the material balance (i.e. conserve volume) between the deformed state and the restored (initial) state sections (Dahlstrom 1969). The palinspastic restorations compiled in this study primarily utilized rigid-body translations and rotations of fault blocks (sometimes called jigsaw restorations) involving little to no distortion of the rock units along the fault by folding to approximate the observed deformation (Gibbs 1983; Lingrey & Vidal-Royo 2015). Material balance is represented in two-dimensional restorations by area balance and line-length balance (Woodward *et al.* 1989) between the deformed and restored state sections. Line-length balance is typically utilized in restorations through contractional settings, where the thrust faults predominantly involve sedimentary cover rocks (e.g. the Sevier thrust belt; DeCelles 1994). In settings such as the eastern Great Basin, polyphase structural deformation complicates palinspastic restorations due to the pre-extensional geometry already being in a deformed state, which increases the number of potentially valid retrodeformations, except where flat-lying, pre-extensional strata constrain the initial fault orientations (e.g. Gans & Miller 1983; Miller *et al.* 1999; Pape *et al.* 2016). The pre-extensional strata need to be observed to: (1) mantle the stratigraphy and contractional structures and thus define the pre-extensional (or initial state) architecture; and (2) be offset by extensional structures. This is highlighted in areas such as Spruce Mountain, where the primary structural markers for restorations were previously deformed Palaeozoic strata and an iterative approach progressively modifying the pre-extensional architecture was used to produce a valid reconstruction that satisfied the existing geological constraints (Elliott 1983; Pape *et al.* 2016). Many of the locales in this study lack a 'layer cake' stratigraphy and instead are dominated by plutonic rocks; as such, line-length balance is poorly suited to address palinspastic restorations in these locales, and instead area balance is utilized. Area balance strives to maintain the area of each rock unit in cross-section between the deformed and initial state sections and to minimize overlaps and gaps between fault blocks because these represent material gains and losses between the different section states (Lingrey & Vidal-Royo 2015).

In areas where the extensional direction changes between superimposed fault generations, cross-sectional restorations are not recommended because material is moving in and out of section and thus material balance fails. In these situations, the alternative approach of fault surface maps (also referred to as Allan diagrams after Allan 1989), where geological maps of the hanging wall and footwall of a fault are constructed and the hanging wall map is then restored over the footwall (see Seedorff *et al.* 2015 for a detailed description of the methodology). Fault surface maps are particularly useful where there are abundant subsurface piercements of faults, such as in densely drilled mining districts (e.g. the Yerington district; Richardson & Seedorff 2015). Some of the palinspastic restorations were not constructed with material balance fully maintained between the deformed and initial state sections and were instead schematic restorations that captured many of the geological constraints (e.g. amount of slip, tilting of pre/synextensional units, fault spacing) and conveyed the structural evolution of the locales, but were not rigorous structural restorations in terms of maintaining area balance (e.g. Mount Perkins, Sierra Mazatán metamorphic core complex; Faulds *et al.* 1995; Wong & Gans 2008). The geological evidence and documentation of fault orientations in these locales and their relationships with strata that can determine their initial configuration are nonetheless still valid and were integrated into this study. The palinspastic restoration methods used for each locale are listed in Table 2. The number of localities utilized in this study prevents even a cursory discussion of each of their geological and structural histories; however, the highly extended Tea Cup prospect is selected and discussed in detail to illustrate the how the data in Table 2 were generated.

Locality example: Tea Cup porphyry system, Arizona

The southwestern North America porphyry copper province, one of the world's major sources of copper, molybdenum and rhenium, is predominantly a result of mineralization related to magmatism of the Laramide arc from *c.* 80–45 Ma; the portion in southern Arizona hosts *c.* 60% of the known Laramide porphyry copper deposits (Leveille & Stegen 2012). Mineral exploration was heavily influenced by the Lowell (1968) and Lowell & Guilbert (1970) model of alteration-mineralization zonation in porphyry systems; they were among the first to recognize the significance of extensional dismemberment and tilting in the Basin and Range province (Wilkins & Heidrick 1995). Despite over a hundred

Table 2. *Geochronological, magmatic and structural data*

Map No.	Locality	Fault generation	δ_i	δ_f	$\Delta\delta$	θ_f	μ_s	Average fault slip (km)	Tilting per unit slip (° km^{-1})	Fault spacing (km)*
1	Yerington district, Nevada	Second	70	35	35	55	0.70	3.2	10.9	0.7
		Third	60	42	18	48	0.90	1.4	12.9	0.6
3	Tobin Range, Nevada	First	66	41.5	25	48.5	0.88	0.5	49.0	0.7
4	Caetano Caldera, Nevada	First	65	20	45	70	0.36	4.0	11.3	1.5
5	Lewis district and Mill Creek area, northern Shoshone Range, Nevada	First	60	20	40	70	0.36	1.0	40.0	0.5
6	Hilltop district, northern Shoshone Range, Nevada	First	55	25	30	65	0.47	0.4	75.0	0.3
7	Ruby-Humboldt metamorphic core complex, Nevada	First	53	11	42	79	0.19	14.0	3.0	2.7
8	Spruce Mountain district, Nevada	First	47	15	32	75	0.27	1.8	18.3	
		Second	65	42	23	48	0.90	1.0	23.0	1.9
		Fourth	77	30	47	60	0.58	1.5	31.3	1.8
9	Hunter district, Nevada	First	63	25	38	65	0.47	3.0	12.7	0.4
		Second	71	42	29	48	0.90	1.0	29.0	7.9
11	Snake Range metamorphic core complex, Nevada/Utah	First	60	20	40	70	0.36	1.5	26.7	1.2
		Second	60	20	40	70	0.36	1.1	36.4	0.2
12	Royston district, Nevada	First	60	35	25	55	0.70	3.2	7.8	0.9
		Second	60	35	25	55	0.70	1.2	21.4	0.4
13	Hall district, Nevada	First	65	35	30	55	0.70	1.8	16.7	
		Second	68	31	37	59	0.60	2.8	13.2	0.6
14	Tonopah district, Nevada	First	70	35	35	55	0.70	0.5	70.0	0.1
15	Goldfield district, Nevada	First	45	28	17	62	0.53	0.4	42.5	0.2
16	Eldorado Mountains, Nevada	First	90	20	70	70	0.36			
17	Mount Perkins, Arizona	First	70	18	52	72	0.32	5.5	9.5	
18	White Picacho and Sheep Mountains districts, Wickenburg, Arizona	First	65	19	46	71	0.34	5.2	8.8	0.8
		Third	64	47	17	43	1.07	0.5	34.0	2.4
19	Tea Cup porphyry system, Kelvin and Riverside mining districts, Arizona	First	75	45	30	45	1.00	3.0	10.0	1.9
		Second	60	30	30	60	0.58	4.5	6.7	1.6
		Third	67	46	21	44	1.04	2.0	10.5	4.2
20	Romero Wash-Tecolote Ranch, Arizona	First	72	42	30	48	0.90	1.3	23.1	2.3
		Second	60	40	20	50	0.84	2.7	7.4	3.7
		Third	59	39	20	51	0.81	1.7	11.8	3.0
		Fourth	63	53	10	37	1.33	1.4	7.1	3.1
21	Lemitar Mountains, New Mexico	First	72	44	28	46	0.97	1.3	21.5	0.5
		Second	55	42	13	48	0.90	0.5	26.0	0.4
22	Sierra Mazatán metamorphic core complex, Sonora, Mexico	First	60	15	45	75	0.27	15.0	3.0	

*Fault block thickness based on thickness normal to fault planes, not horizontal separation at modern/palaeosurface.
†Magmatism codes are as follows: A, andesite; B, basalt; BA, basaltic andesite; D, dacite; N, no magmatism; R, rhyolite; RD, rhyodacite.

years of significant exploration and mining in the region, continued exploration has led to the discovery of new deposits (e.g. Resolution), better understanding and the refinement of the porphyry system 'model' from the increased number of observations (Seedorff *et al.* 2008; Hehnke *et al.* 2012). Tea Cup is one such system that has benefited from decades of activity in east-central Arizona,

Percentage extension	Synextensional magmatism†	Timing of extension (Ma)	Duration of extension (Ma)	Strain rate (s^{-1})	Palinspastic restoration method‡	Source
169.0	D–A	13.8–12.6	1.2	4.48×10^{-12}	JS, FSM, AB	Proffett (1977), Proffett & Dilles (1984), Dilles & Gans (1995), Stockli *et al.* (2002), Richardson & Seedorff (2015)
5.3	BA	11–8	3	5.56×10^{-14}		
36	R–(B?)	24.72–14.0	12.11	8.58×10^{-14}	JS, AB	Gonsior & Dilles (2008)
114	B–BA–R?	16–(10–12)	5	7.23×10^{-13}	JS, AB	Colgan *et al.* (2008)
189	D	<36			JS, AB	Colgan *et al.* (2014)
50	D	<36			JS, AB	Colgan *et al.* (2014), Richardson *et al.* (2016)
190	R–RD?	(17–15)–(12–10)	5		JS, AB	Colgan *et al.* (2010)
42	N	*c.* 38			JS, AB	Pape *et al.* (2016)
18	N	*c.* 38				
15	N	<38				
53	R	36–27	9	1.87×10^{-13}	JS	Gans & Miller (1983), Gans *et al.* (1989), Seedorff (1991*a*)
12	N					
66	R–D	39–35	4	5.23×10^{-13}	JS	Gans & Miller (1983), Gans *et al.* (1985), Miller *et al.* (1983, 1999)
38	N	17–14	3	4.02×10^{-13}		
200	R–RD	29–26.7	2.3	2.76×10^{-12}	JS, AB	Seedorff (1991*b*)
24	R–RD	26.7–20?	6.7	1.14×10^{-13}		
30	D-A	(29–18)–15?	3–14		JS	Shaver & McWilliams (1987)
76	N	<15?				
50	R	21–(20.4–18.7)	1.5	1.06×10^{-12}	JS	Seedorff (1991*a*)
15	R–RD	21.8–21.2	0.6	7.93×10^{-13}	JS	Seedorff (1991*a*)
100	R–B	15.1–14.1	1			Gans & Landau (1993), Gans *et al.* (1994), Gans & Bohrson (1998), Miller & Miller (2002)
	R–B	15.7–14	1.7		JS	Faulds *et al.* (1995)
60	N				JS, AB	Nickerson & Seedorff (2016)
2	N					
18	N	25–15?			JS, AB	Nickerson *et al.* (2010), Wong *et al.* (2015)
127	N	25–15?				
13	N	25–15?				
122	N	29–22.5	6.5	5.95×10^{-13}	JS, AB	Dickinson (1991), Favorito (2016), Favorito & Seedorff (2017)
37	N	22.5–21	1.5	7.82×10^{-13}		
16	N	21–16	5	1.01×10^{-13}		
7	N					
132	R–D–A	28.5–27	1.5	2.79×10^{-12}	JS	Chamberlin (1982, 1983), Cather *et al.* (1994)
42	R	12–7	5	2.66×10^{-13}		
175	D–B	21–16	5	1.11×10^{-12}	JS	Wong & Gans (2008)

‡Codes for methods used in the palinspastic structural restorations are as follows: AB, area balance; FSM, fault surface map; JS, jigsaw.

with some of the earliest work in the area dating back to more than a century ago (Ransome 1903).

The Tea Cup porphyry system is hosted in the *c.* 72–70 Ma Tea Cup composite pluton and surrounding Proterozoic Ruin Granite. It is currently in fault contact with Oligo-Miocene sedimentary and volcanic rocks (Barton *et al.* 2005; Nickerson *et al.* 2010; Nickerson 2012). The timing of

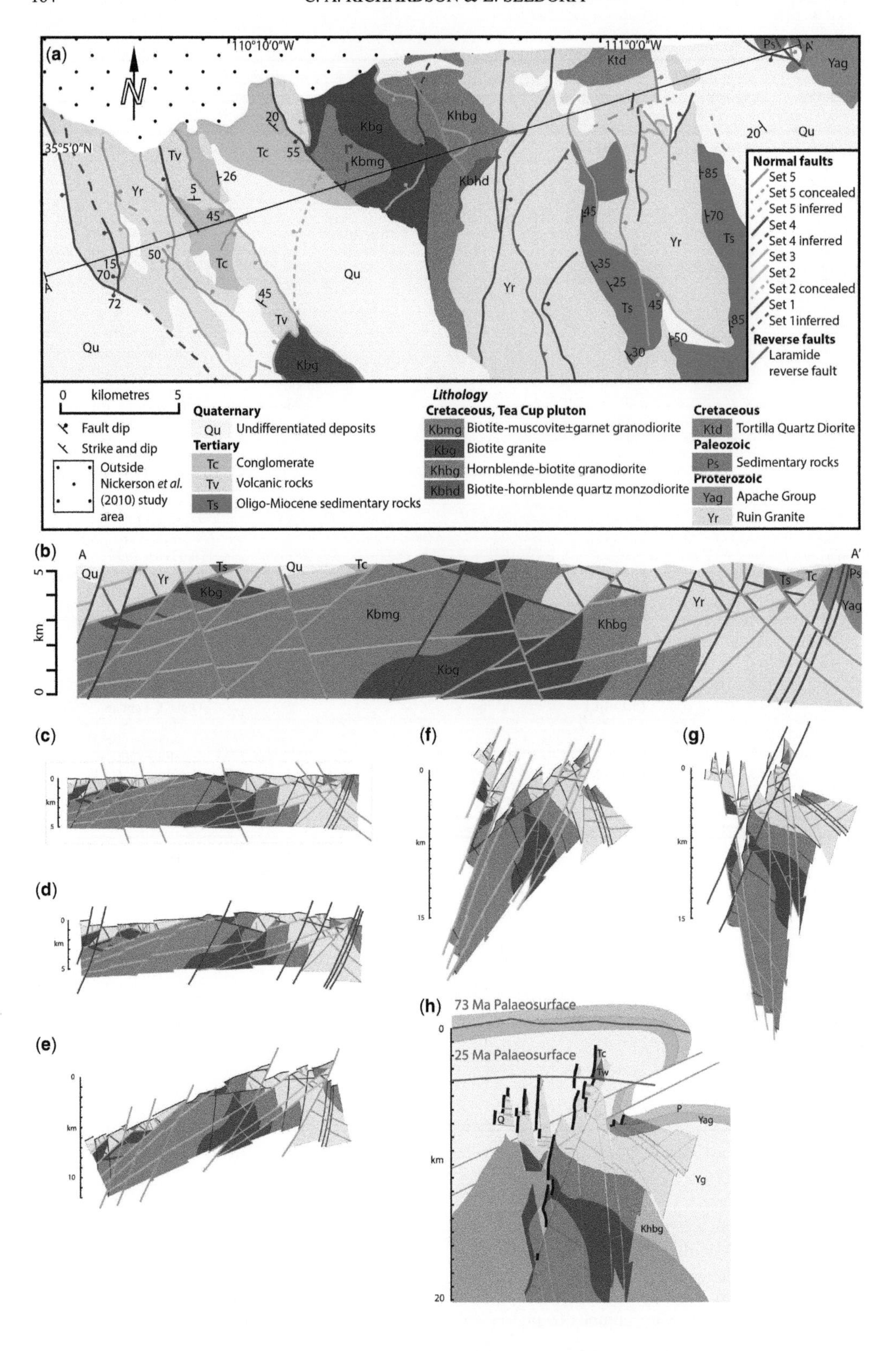
(a)
110°10'0"W
111°0'0"W
35°5'0"N
N
A
A'
Normal faults
Set 5
Set 5 concealed
Set 5 inferred
Set 4
Set 4 inferred
Set 3
Set 2
Set 2 concealed
Set 1
Set 1 inferred
Reverse faults
Laramide reverse fault
0 kilometres 5
Fault dip
Strike and dip
Outside Nickerson et al. (2010) study area
Lithology
Quaternary
Qu Undifferentiated deposits
Tertiary
Tc Conglomerate
Tv Volcanic rocks
Ts Oligo-Miocene sedimentary rocks
Cretaceous, Tea Cup pluton
Kbmg Biotite-muscovite±garnet granodiorite
Kbg Biotite granite
Khbg Hornblende-biotite granodiorite
Kbhd Biotite-hornblende quartz monzodiorite
Cretaceous
Ktd Tortilla Quartz Diorite
Paleozoic
Ps Sedimentary rocks
Proterozoic
Yag Apache Group
Yr Ruin Granite
(b)
(c)
(d)
(e)
(f)
(g)
(h)
73 Ma Palaeosurface
25 Ma Palaeosurface
km

extension is constrained by the steep dips of Oligocene strata. Five generations of normal faults were identified by exposures of clay gouge and breccias, the offset of porphyry dykes and, locally, of (syn-) sedimentary rocks, and within the pluton by the juxtaposition of hydrothermal alteration types belonging to different levels of the palaeohydrothermal system (Nickerson *et al.* 2010). Faults of the oldest generation strike northwards, dip *c.* 15° E, have offsets of ≤3 km and are tilted *c.* 30° E. Faults of the second generation strike southwards, dip *c.* 10–15° W, have offsets between 2 and 7 km and are tilted *c.* 30° E. Faults of the third generation strike southwards, dip *c.* 45° W, have offsets of *c.* 2 km and are tilted *c.* 21° E. Faults of the fourth generation strike *c.* 190°, dip 70° W, have offsets of *c.* 1 km and are tilted *c.* 5° E. Faults of the fifth generation strike *c.* 350°, dip *c.* 70° E, have offsets ≤1 km and are tilted *c.* 5° W. The cross-cutting relationships observed in map pattern are shown in Figure 3a; a detailed discussion of the geological relationships between the faults and the constraints used in this structural restoration can be found in Nickerson *et al.* (2010).

The timing of extensional deformation by the first three generations of faults is constrained to *c.* 25–15 Ma by the offset of variably tilted volcanic and synsedimentary cover rocks. A structural restoration of the Tea Cup system (Fig. 3b–h) shows *c.* 210% extension between the five fault generations. However, only data from the first three generations of faults are included in Table 2; the latter two fault generations were excluded due to their high-angle initial dip and minimal associated tilting (Nickerson *et al.* 2010).

Following restoration of the fourth and fifth sets, faults of the third fault generation would have dips of *c.* 45° W (see Table 2). Some faults display some curvature along their dip profile and, in these situations, the fault dip was measured off its upper, steeper portion. The cut-off angle (the acute angle between the older and younger faults) between the third and second fault generations is *c.* 21°. These cut-off angles provide the amount of tilting that occurred between two fault sets related to the younger of the two fault sets. The third fault generation would restore to *c.* 61° W, and restore the faults that presently dip *c.* 10–15° W to *c.* 30–35° W. The cut-off angle between the second and first fault generations is *c.* 30°, which would require restoring the second generation faults to dips of *c.* 60–65° W, and the first fault generation faults to dips of *c.* 30° W. Evidence for the total amount of tilting in the Tea Cup also come from Oligo-Miocene sedimentary units and Proterozoic diabase sill-like sheets in the Ruin Granite (not shown in Fig. 3a). Two Oligo-Miocene sedimentary units are present in the Tea Cup area: (1) an upper conglomerate (Tc in Fig. 3a) with dips of *c.* 20–25° E; and (2) a lower synsedimentary sequence (Ts in Fig. 3a) with dips of *c.* 25–45° in its upper half and dips of 70–85° at the base of the unit. These units are interpreted to be deposited into the hanging walls of the first, second and third fault generations and then later cut by the fourth and fifth fault generations (Fig. 3b). The diabase sheets were mapped by Nickerson *et al.* (2010) at near-vertical dips and increase in abundance in the upper 1 km of the Ruin Granite and overlying Apache Group. These near-vertical dips of the diabase sheets and the lowermost Ts would be *c.* 35–40° E following restoration of the first four fault generations (Fig. 3c–f). Restoring them to horizontal would then restore the initial fault generation to dips of 75° W or steeper (Fig. 3g, h). No magmatism accompanied the faulting episodes, but Wong *et al.* (2015) documented the presently gently dipping Oligocene basaltic dykes in the area that were emplaced vertically prior to extensional deformation.

Results

All the characteristics were plotted against the tilting for each fault generation to analyse which factors might control tilting (and thus μ_s). Tilting was used instead of μ_s because 14 of the fault generations have non-Andersonian initial dips (outside the range of 58–68° dips). The use of tilting avoids the issue created by faults with non-Andersonian initial dips and undergoing the same amount of tilting producing different calculated friction values, thus implying different amounts of tilting. For example, assume that three faults (initial dips 75, 60 and 45°) all underwent 30° of tilting, which yields final dips of 45, 30 and 15°. Using equation (5) from Sibson (1985, 1994), the calculated μ_s would be 1.00, 0.58 and 0.27, respectively, which imply three different tilting

Fig. 3. Map, cross-section and palinspastic structural restoration of the Tea Cup system, Arizona. (**a**) Geological map of the Tea Cup area. (**b**) Modern cross-section. (**c**) Restoration of deformation associated with fifth generation faults. (**d**) Restoration of deformation associated with fourth generation faults. (**e**) Restoration of deformation associated with third generation faults. (**f**) Restoration of deformation associated with second generation faults. (**g**) Restoration of deformation associated with first-generation faults. (**h**) Hypothetical Laramide cross-section showing the composite Tea Cup pluton intrude a basement-cored uplift. The black and maroon lines represent the modern surface and palaeosurfaces, respectively. Modified from Nickerson *et al.* (2010).

histories (15, 30 and 45°) rather than the actual 30° of tilting.

Both unweighted and weighted linear regressions were conducted. The weighted regressions used the bisquare weighting function (robustfit.m in the statistics toolbox of MATLAB). This method weights each data point based on its distance from the unweighted line, then iteratively reweights the regression matrix to reduce the influence of strong outliers within datasets that may not be normally distributed. Calculated μ_s values are sorted into 0.05 bins and are shown in Figure 4.

Percentage extension

The percentage extension was calculated by the change in length between the undeformed and deformed states from the structural restorations for each generation of faults in each extended locale. No overall statistical correlation was found in Figure 5a, but the weighted regression resulted in an r^2 value of *c.* 0.7. This is due to the down-weighting of localities with high-magnitude extension (>120%).

Duration of extension

This dataset was limited relative to the others due to the paucity of geochronological data to constrain the timing of movement of faults. No statistical correlation was found in Figure 5b.

Strain rate

The corresponding percentage extension and duration of extension data were used to calculate the strain rate. The rate was converted to the conventional strain rate unit of s^{-1}. Figure 5c shows the data with no observed overall pattern, although the weighted regression resulted in an r^2 value of *c.* 0.69. The three data points strongly down-weighted by the weighted regression were the early fault generations associated with large-magnitude extension over relatively brief time intervals at Yerington, Royston and the Lemitar Mountains.

Mean slip magnitude

The mean slip magnitude is defined as the mean of the maximum cumulative slip on all constituent faults belonging to a single fault generation. Although the unweighted regression shows no statistical correlation, the weighted regression that reduces the impact of two data points in the upper left (the Sierra Mazatán and Ruby-East Humboldt metamorphic core complexes) has an r^2 value of *c.* 0.78. More data from other Cordilleran metamorphic core complexes with faults exhibiting similar high magnitudes of slip could determine whether or not there is a fundamental difference in behaviour between extensional faulting within and without metamorphic core complexes.

Amount of tilting per unit slip

The amount of tilting per unit slip ($° km^{-1}$) was determined by dividing the amount of tilting for each fault set by the average amount of slip (Fig. 5e). The scatter indicates no clear correlation.

Fault spacing

Fault spacing is defined here as the distance between two faults as measured orthogonal to their dips because that distance will not change with tilting, whereas the apparent spacing measured at the

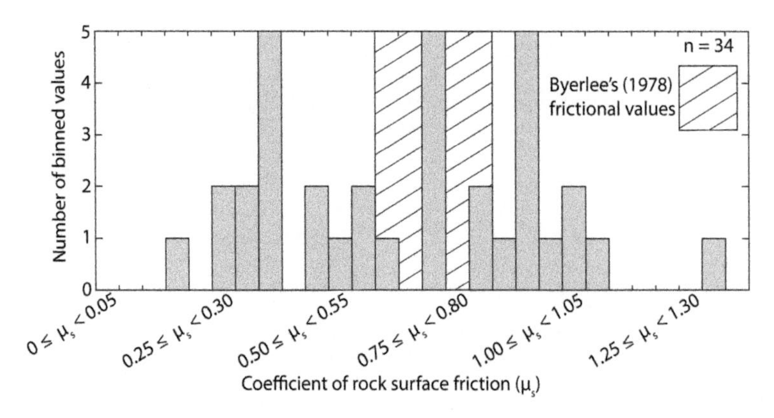

Fig. 4. Histogram showing distribution of calculated values of the coefficient of rock surface friction (μ_s). Values are sorted into 0.05 bins, each of which includes the lower value and goes up to the following next 0.05 bin. The value range Byerlee (1978) of 0.60–0.85 is in the diagonally ruled box.

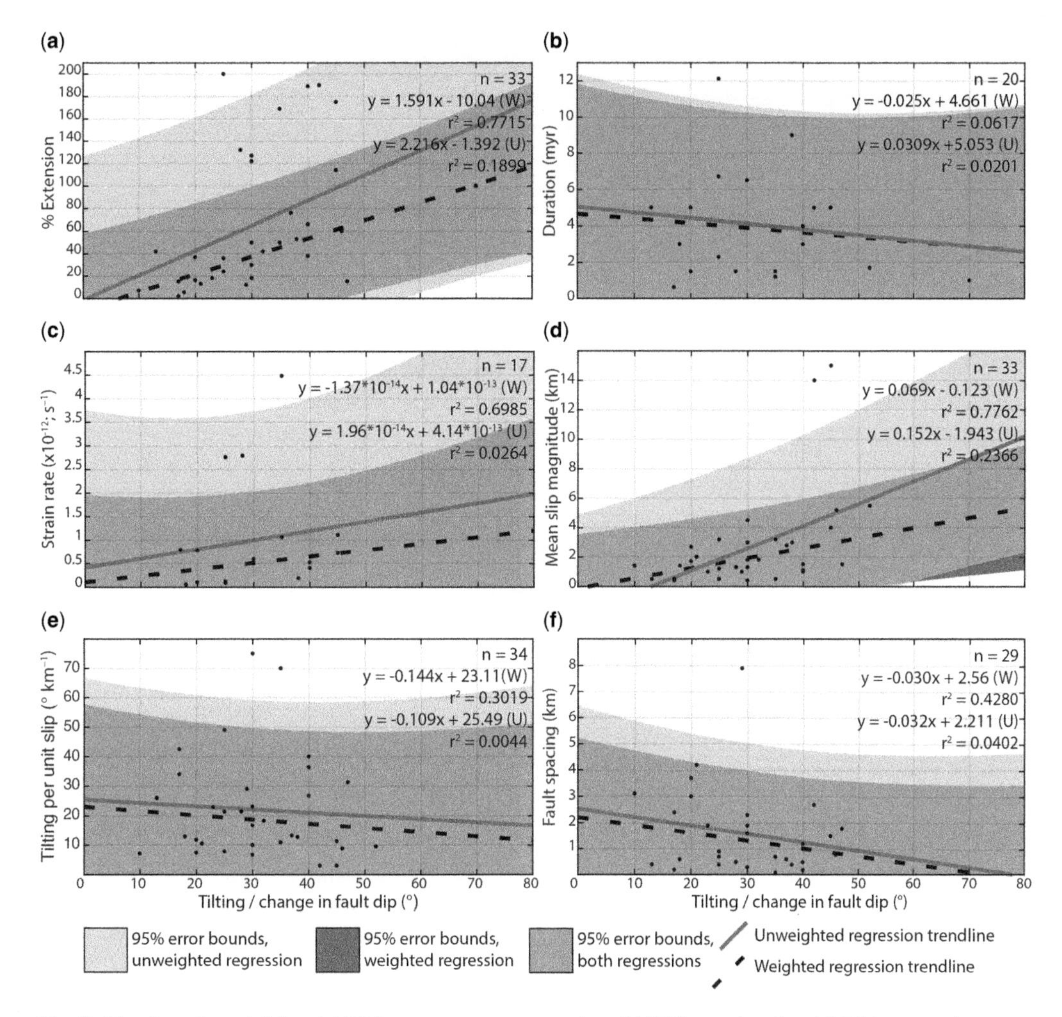

Fig. 5. Bivariate plots of tilting. (**a**) Tilting v. percentage extension. (**b**) Tilting v. duration. (**c**) Tilting v. strain rate. (**d**) Tilting v. mean slip magnitude. (**e**) Tilting v. tilting per unit slip. (**f**) Tilting v. fault spacing.

Earth's surface between fault traces would change with tilting. Fault spacing (km) was measured from each cross-sectional restoration. The weighted regression results in an r^2 value of *c.* 0.43, but the scatter indicates no clear correlation (Fig. 5f).

Timing of extension

Figure 6 shows the timing of extension of each locale (where geochronological constraints exist) to examine whether certain geological epochs were more prone to extension and tilting, which would be expressed by lower values of μ_s. There is a small cluster of Miocene extended locales between 0.2 and 0.4; at values >0.4 there is broadly an inverse relationship between the timing of extension and μ_s.

Synextensional magmatism

Figure 7 shows the composition (or absence) of synextensional magmas plotted against tilting. A wide spectrum of compositions is observed with no clear relationship. The scatter indicates no clear correlation.

Interpretations

The calculated μ_s values are much wider than the reported values of Byerlee (1978) and Sibson (1994). The distribution of values (Fig. 3) shows that only eight of the 34 values fall within the friction window of Byerlee (1978) of 0.60–0.85, with the other 27 values falling outside the window. This demonstrates that there is not only a much wider range of

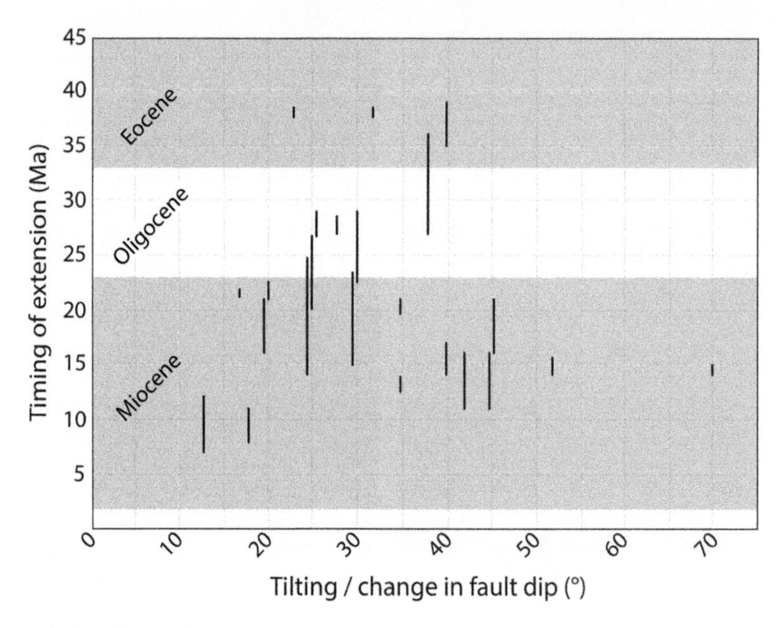

Fig. 6. Tilting v. timing of extension.

μ_s values in natural fault systems than suggested by the earlier experimental study of Byerlee (1978), but also that many of the calculated values have significantly higher frictional values.

None of the imposed trend lines on plots of the normal fault characteristics v. tilting exhibits a strong bivariate correlation. The percentage extension, strain rate and mean slip magnitude show a better fit for weighted trend line than the other factors examined (Fig. 5a, c, d), but the statistical correlation remains poor. All of the remaining factors – tilting per unit slip, fault spacing, duration of extension, timing of extension and synextensional magmatism – exhibit poor correlations. Therefore no single factor is recognized as the driving cause of highly tilted fault systems.

The failure of the analysis to isolate a single dominating factor suggests several possible broader interpretations. There may be a single factor that controls normal fault behaviour, but such a

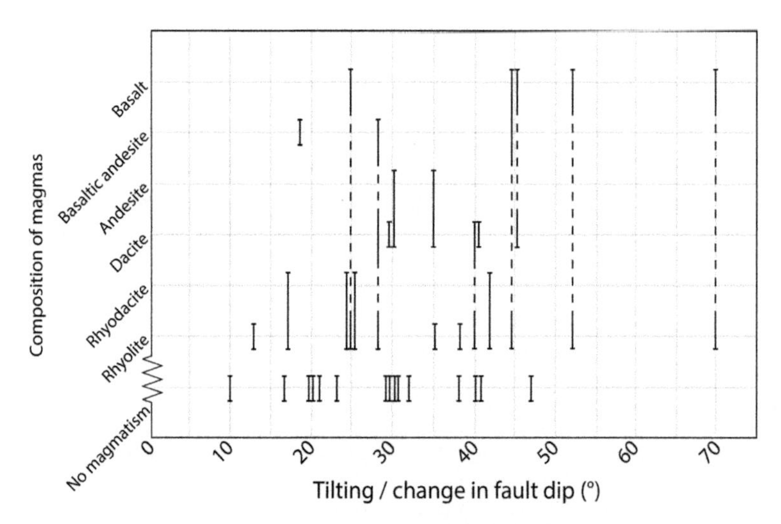

Fig. 7. Tilting v. composition of magmatism. For locales where a spectrum of magmatic compositions were observed, a solid vertical line was drawn through the observed compositions.

geological factor is not one of the eight factors considered here. Alternatively, some combination of the eight factors considered here controls fault behaviour, but this relatively simple analysis using bivariate plots does not reveal these combinations. Yet another possibility is that combinations of factors control behaviour, but the group of controlling factors differs depending on the local geological settings of extension, which would not be revealed in this analysis. Another possibility is that a single factor largely controls behaviour, but the explicit simplifying assumptions made here preclude identification of the controlling factor. Assumptions were made regarding thermal regime, stress trajectories and pore fluid pressure, as well as the operating assumptions here and in Sibson (1994) that a new generation of faults develops because it is mechanically unfavourable to reutilize the older generation of faults, not because there was a change in the regional/local stress regime. Our preferred interpretation is that tilting (and the calculated μ_s values) is an emergent property in fault systems, rather than an input, and thus would have no statistical correlation with any one of the other factors associated with the fault systems.

It is not possible to discuss tilting in fault systems without also considering the implications of friction, bearing in mind their interrelationship. Friction is an experimentally determined value measured with respect to σ_1 from deformation experiments where σ_1 is known or (as in this study) determined from palinspastic restorations where the palaeohorizontal can be inferred from geological constraints (Byerlee 1978; Sibson 1994). Friction has been shown to vary with the composition of the mineralogy of the fault gouge (e.g. Collettini 2011), which suggests that friction is a fundamental physical property of the rock/mineral type. However, as discussed earlier, this does not always equate with the expected tilting for the associated friction.

Fault systems display clear degrees of complexity and chaos and can be categorized as emergent systems. Erickson *et al.* (2011) examined emergence in the numerical modelling of earthquakes and suggested that the non-linear rate-and-state friction law is scale-dependent. Thus the emergent behaviour exhibited by natural earthquake ruptures is lost when looking at laboratory results because the full geological system with all of its inputs is not properly represented. The lack of statistically significant correlations between tilting and any of the examined variables suggests that tilting and friction in fault systems are emergent physical properties.

The overall implication is that values of rock surface friction exhibit a much wider range in nature than in the canonical study of Byerlee (1978). Careful consideration must be taken when applying the reshear criterion to an understanding of the evolution of natural normal fault systems because there are many implicit and explicit assumptions made by its utilization, which likely deviate across time and space in geological settings.

Discussion

Synextensional magmatism and its role in faulting

The relationship between extensional deformation and magmatism has been a continued source of debate over the last three decades in both continental and oceanic tectonic settings (Franke 2013; Whitney *et al.* 2013). In the Basin and Range province, hypothesized relationships between the two processes include: magmatism localizing the development of metamorphic core complexes by thermally weakening the lower crust (Armstrong & Ward 1991); rotation of the stress field by the rate of magmatism exceeding the extension rate, causing an increase in horizontal stresses (Parsons & Thompson 1993); the release of volatiles during crystallization to weaken the crust and promote extension (Whitney *et al.* 2013); extension unrelated to magmatism (Best & Christiansen 1991); extension concentrated in two north–south belts in the Great Basin with southwards sweeping magmatism and only locally concurrent (Axen *et al.* 1993); and magmatism immediately precedeing extension and potentially preconditioning the crust for extension (Gans *et al.* 1989; Gans 1990; Seedorff 1991*a*; Parsons *et al.* 1994; Gans & Bohrson 1998; Hudson *et al.* 2000).

In the areas of rapid, large-magnitude extension with suggested relationships with synextensional magmatism, a temporal pattern has been observed of: (1) pre-extensional, large-volume calc-alkaline dacitic–rhyolitic ash-flow tuffs; (2) 'bimodal' eruption of dacite–andesite and lesser volumes of rhyolite to high-silica rhyolite immediately preceding and during the onset of extension; (3) the suppression of volcanism during rapid crustal extension (but possible voluminous plutonism during this time); and (4) bimodal, basalt-dominated magmatism with peralkaline silicic rocks and topaz rhyolites associated with the waning stages of extension (Gans *et al.* 1989; Seedorff 1991*a*; Gans & Bohrson 1998).

No relationship between the presence/absence or composition of magmas and tilting was observed (Fig. 7). This does not preclude a relationship between magmatism and tilting because the majority of locations contained abundant magmatism, but rather suggests there is no compositional control on tilting. Although no broad generalization can be

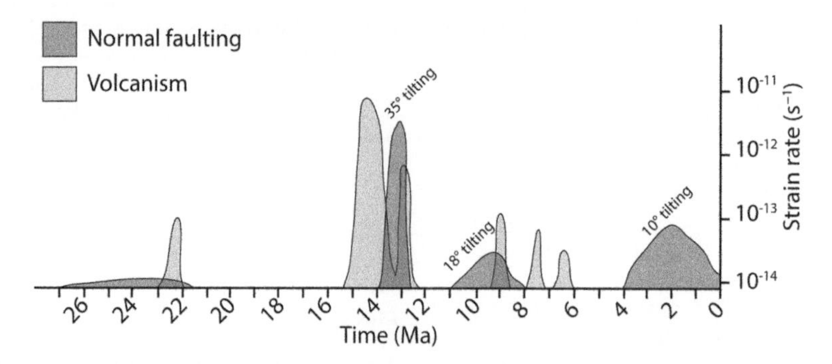

Fig. 8. Time–strain–qualitative eruptive volume diagram of the Yerington district, Nevada. Modified from Gans & Bohrson (1998) with information from Richardson & Seedorff (2015).

made about the relationship between magmatism and extension, there are extended areas with clear temporal relationships with synextensional magmatism, such as the Yerington district, western Nevada (Fig. 8).

There are four episodes of extensional faulting in the Yerington district. The earliest generation of faults, oriented *c.* 60° to the three successive east-dipping fault generations, were originally south-dipping normal faults that accommodated very low amounts of extension and may be analogous to similarly oriented faults that preceded high-magnitude extension in the east-central Nevada (Gans 1990; Richardson & Seedorff 2015). These faults are hypothesized to have been active during or following the deposition of large volumes of ash-flow tuff (*c.* 27–21 Ma) sourced from calderas *c.* 200 km east, but only coincide with local minor magmatism in the form of pyroxene andesitic dykes that intruded strike-slip and normal faults in the Wassuk Range (Dilles 1993; Garside *et al.* 2002). The second fault generation, which was responsible for the majority of the extensional deformation, was active between 13.8 and 12.6 Ma following the eruption of the majority of the Lincoln Flat andesite and dacite (Dilles & Gans 1995). The third generation of faults was active between *c.* 11 and 8 Ma, as demonstrated by offsets of a 9–8 Ma basaltic andesite in the Wassuk Range and predating a 7.5 Ma basaltic andesite in the Singatse Range.

The modern range-bounding faults in the greater Yerington district are interpreted to have initiated at *c.* 4 Ma based on apatite fission track and (U–Th)/He thermochronology (Stockli *et al.* 2002). This continues to support the notion of an evolutionary trend in synextensional magmatism – in some areas – beginning with an initial magmatic event of dacite and andesite that could weaken or precondition the crust prior to the first major extensional episode, followed by lesser extensional episodes that broadly correlate in time and space with more mafic magmatism. Similar evolutionary trends have been suggested for east-central Nevada (e.g. Gans *et al.* 1989). Centres of rhyolite to high-silica rhyolite are present – for example, in the Hunter district in the northern Egan Range (Gans & Miller 1983) and in the Robinson district in the central Egan Range (Gans *et al.* 2001) – whereas dacite–andesite magma was erupted broadly contemporaneously from a centre that is spatially between those two locales, near Robinson Summit (Feeley & Grunder 1991; Feeley 1993).

Sibson *et al.* (1988) demonstrate that unfavourably oriented fault systems near magmatic/metamorphic environments that produced fluid could act as valves, where the faults would slip only when the fluid pressure exceeded the lithostatic pressure. The normal fault systems with μ_s values <0.6 (i.e. faults that slipped with dips $<30°$) could have been mechanically encouraged by degassing magmatic plumbing systems in the crust that were active during extension. In addition, Gans & Bohrson (1998) suggest that volcanism is suppressed during extension in favour of enhanced crystallization from the reduction of confining pressures on hydrous magmas, resulting in the exsolution of magmatic volatiles and crystallization of the magma body (e.g. Burnham 1997). Modelling has suggested that fluid release from crystallizing igneous intrusions has varying effects on faulting, with high-viscosity fluids limiting seismicity and low-viscosity fluids (e.g. aqueous, volatile-rich phases) encouraging seismicity (Anderson 1978). This has been supported by Muirhead *et al.* (2016), who demonstrate that the release and migration of magmatic volatiles in fluid flow systems helped to focus the loci of extensional faulting in the East African Rift. These scenarios, depending on the proximity of the magmatic reservoirs, would supply additional fluid to the fluid already present in pore spaces and fractures from groundwater and continental crustal water reservoirs (Bodnar *et al.* 2013), reducing the effective stresses and encouraging slip along unfavourable orientations.

Tilting and fault spacing

One often-cited control on the amount of tilting is the fault spacing, with more tightly spaced faults belonging to rotational planar normal fault systems capable of large amounts of tilting and more widely spaced faults belonging to horst and graben normal fault systems (Proffett 1977, p. 261; Gans & Landau 1993). However, the control on fault spacing in extended domains and the relationship of fault spacing and the mean slip magnitude remains unclear. Fault spacing is rarely defined in published papers. Previously suggested controls of fault spacing that are not studied here are: the thickness of the layer (i.e. the brittle crust) undergoing deformation (Vendeville *et al.* 1987; Jackson & White 1989; Ackermann *et al.* 2001; Soliva *et al.* 2006); the amount of strain a volume of rock can accommodate without faulting (Mandl 1987); and the aspect ratio of a fault and its overlap with neighbouring fault segments (Willemse 1997).

The lack of a bivariate correlation between tilting and fault spacing does not demonstrate a control for the tilting of normal faults. In addition, fault spacing v. percentage extension and mean slip magnitude (Fig. 9) do not show a correlation with fault spacing, although the weighted regression of fault spacing v. mean slip magnitude suggests that a positive correlation could exist between spacing and slip magnitude. The question then arises that if there is not a universal bivariate control on fault spacing, then what other characteristics and conditions of extensional tectonism might influence fault spacing?

It has been argued that the large-magnitude extension in the Eldorado Mountains, Nevada, characterized by closely spaced, highly tilted (up to 90°) normal faults that had initial and final non-Andersonian orientations, is due to rapid strain rates between 15.0 and 14.1 Ma that exceeded the readjustment of the local geothermal gradient

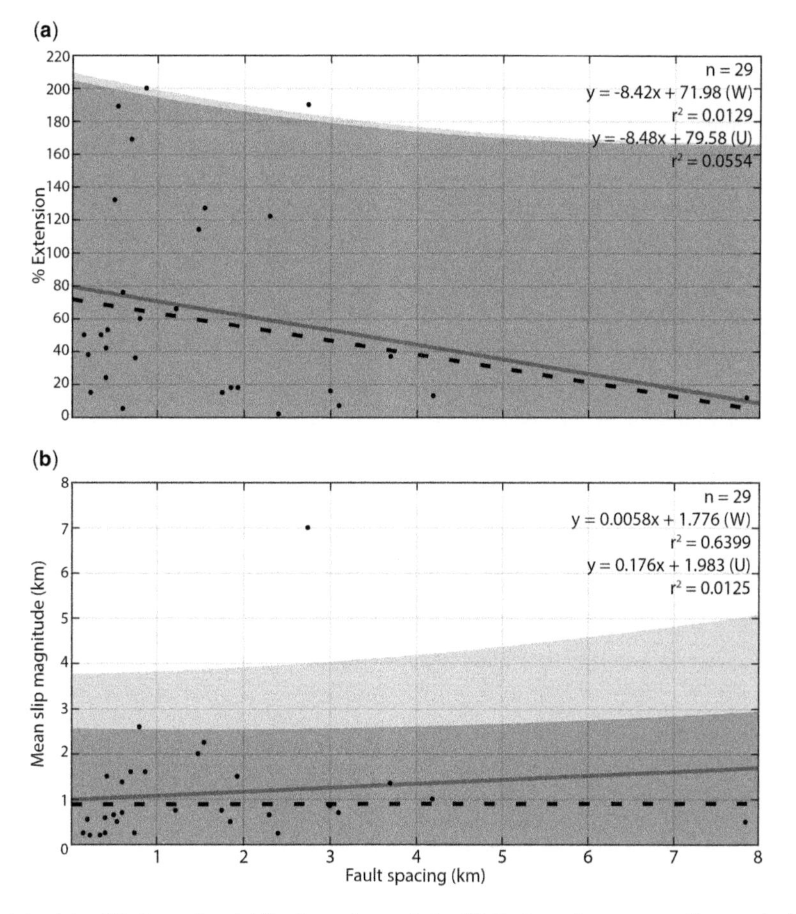

Fig. 9. Bivariate plots of fault spacing. (**a**) Fault spacing v. strain. (**b**) Fault spacing v. mean slip magnitude.

(Gans & Landau 1993). Since 14.1 Ma, minor extension and tilting have continued along widely spaced, high-angle normal faults (Gans *et al.* 1994; Gans & Bohrson 1998). In the Yerington district, the fault spacing broadly increases with time (0.7, 0.6 and 1.3 km) and tilting decreases (35, 18 and 10°), but the strain rate actually increases for the fourth fault generation. In the Snake Range metamorphic core complex, it is noted that as the faults become more widely spaced, the total amount of extension decreases (Miller *et al.* 1983, p. 243). Modelling by Gans *et al.* (1991) suggests that the initial geothermal gradient and strain rate during the thermal evolution of fault blocks may influence the fault spacing.

The nature of fault spacing and its relationship to tilting and other characteristics of normal fault systems is still unclear. A better understanding of the thermal regime present during extensional deformation, bracketing the timing of extension through geochronology and thermochronology to estimate strain rates, and multivariate analysis of these characteristics is needed to further explore these relationships.

Timing of extension

Controversy continues around the timing and relative magnitudes of extension in the northern Basin and Range province, with recent emphasis on Miocene extension (McQuarrie & Wernicke 2005; Colgan & Henry 2009; Cassel *et al.* 2014), despite the documentation of significant extension at various times in the Eocene and Oligocene (e.g. Wernicke *et al.* 1987; Seedorff 1991*a*; Cather *et al.* 1994; Hudson *et al.* 2000; Gans *et al.* 2001; Rahl *et al.* 2002; Pape *et al.* 2016). The driving mechanism for extensional deformation during the Cenozoic in the North American Cordillera is still debated, with arguments for the gravitational collapse of a thickened crust (Coney & Harms 1984), magmatism weakening the crust (Parsons *et al.* 1994; Camp *et al.* 2015) and far-field plate effects (Kreemer & Hammond 2007). Although extensional deformation was pervasive across much of the northern Basin and Range province, it was heterogeneous in its distribution in terms of magnitude, diachronous in its development and distributed into high-strain and low-strain zones (Gonsior & Dilles 2008, fig. 2; Colgan & Henry 2009, fig. 12; Van Buer *et al.* 2009, fig. 1C).

Figure 6 shows the timing of extension v. tilting. Although there was not a single controlling factor here, what is of interest is that for tilting amounts <40° there is a generally linear relationship, with lesser amounts of tilting as the faulting becomes younger. This could represent a reduction in the overall strain rate with time, with younger faults subject to lower strain rates and the locus of faulting migrating away from them before they can tilt as much (Wallace 1987). Alternatively, this could be viewed as a function of strain hardening, with the more deformed, faulted locales being more resistant to continued faulting (i.e. less tilted), with the graph potentially showing this for much of southwestern North America (Yamasaki & Stephenson 2009). This temporal evolution is seen in certain well-studied locales, such as the Yerington district. At Yerington, each successive generation of faults (ignoring the first-generation faults, for which the distribution is as yet poorly known; Richardson & Seedorff 2015) is seen to decrease in its amount of tilting (and its amount of slip). It is also worth noting that, although Yerington has multiple superimposed fault generations, the well-constrained structural evolution and geochronology of the district indicates that there are hiatuses between each episode of faulting (Dilles & Gans 1995). Could this be a signature of a change in the stress regime between each episode, even if only temporary, or that focused extensional deformation has migrated elsewhere for a period of time (e.g. Wallace 1987)?

Suggestions for future research

The results suggest that there is: (1) a wider spread in values for rock surface friction in natural fault systems than previously reported and (2) no statistical correlation between tilting and the examined variables. Although this study does not demonstrate any convincing trends to account for highly tilted normal fault systems in western North America, it does demonstrate that no single factor serves as the 'silver bullet' that can be attributed as the driving cause of highly tilted fault systems. This lack of a linear relationship argues that tilting and friction in fault systems – even when estimated from sound geometrical relationships (Sibson 1985, 1994) within the fault system – is an emergent property in relation to the examined variables. Instead, a more integrated, multicomponent approach is needed to characterize any one location before trying to characterize extensional faults more broadly. This study does, however, suggest several future avenues of research to address the questions presented here.

First, better documentation of the properties and internal structure of fault zones would allow a more nuanced look at how friction behaves in different fault systems. Many of the field studies incorporated here, although addressing questions in structural geology, were not in-depth, detailed structural studies and had their focus in other subdisciplines (e.g. petrology, ore deposits, regional geology). Those studies documented the large-scale structures, but they generally did not examine properties such as the thickness of the fault zone, the fault rock type/proportions, the number of slip surfaces and synthetic structures. As such, we could not compare the mineralogy of the fault gouge with the calculated friction for each fault generation. Considering the sizeable literature that has developed around the frictional properties of fault gouge and low-angle normal faults, better documentation of fault zone properties in locales in the Basin and Range province, noting the mineralogy and abundance of fault gouge, would provide for interesting future research and may reveal hitherto unseen correlations.

Second, assumptions of Andersonian stress trajectories at depth should be reconsidered. Multiple mechanisms, including basal traction forces in the lower crust (Westaway 1999) and the role of magmatism (Parsons & Thompson 1993), have been postulated to explain stress trajectories at depth. Detailed work, such as that of Axen *et al.* (2015) examining the orientation of structures within the fault zone to obtain palaeo-σ_1 orientations, presents the opportunity to construct a detailed understanding of palaeostress trajectories in fault systems. In this study, 14 of 34 fault generations have initial dips that fall outside the expected range of 58–68° for the initial dips of normal faults (Collettini & Sibson 2001). Three possible explanations could explain this: (1) non-Andersonian stress trajectories existed at the time of fault initiation; (2) the values for μ_s used to derive the expected initial dips of 58–68° do not capture the full range of natural variability; or (3) mechanical stratigraphy (e.g. Ferrill *et al.* 2014) plays a greater part in large-scale fault systems than previously documented. Better documentation of palaeo-σ_1 orientations in fault systems would allow these alternatives to be tested and would further refine our knowledge of crustal conditions at the time of fault initiation.

Third, tilting in normal fault systems is still a poorly understood phenomenon, with no clear control on the origin of tilting. Thompson & Parsons (2016) synthesized geodetic data from several historical normal fault earthquakes in the western USA and integrated field studies on major range-bounding normal faults. In common with similar models proposed by earlier workers (e.g. King *et al.* 1988; Kusznir *et al.* 1991; Westaway 1992), Thompson & Parsons (2016) reinterpreted tilting of the hanging wall and footwall blocks as crustal flexure or bending around a tilting normal fault over time. The rotation of fault blocks as competent, intact blocks are then inferred as only occurring between tightly spaced normal faults, with bending/flexure occurring on the exterior sides of the outermost normal faults and decreasing to zero *c.* 10 km away from the faults. The primary caveat with this mechanism is that the maximum amount of tilting a single fault should experience is 45° – not an unreasonable limit, as only three locales in this study reported greater amounts of tilting (Table 2). Incorporating flexure would alter some published structural restorations and may require new geological scenarios to be considered. Where evidence on strain rate and tilting is available, there is a cursory correlation with high strain rates resulting in higher amounts of tilting, as well as an inverse correlation between fault block thickness and tilting (Gans & Bohrson 1998; Seedorff & Richardson 2014). Additional work in fault systems on constraining the timing of faulting through geo/thermochronology (and thus allowing strain rates to be calculated) would allow this tentative link to be properly investigated.

Conclusions

Continental extension is a fundamental feature in the long-term evolution of orogenic systems and the Basin and Range province continues to serve as one of the type localities for studying extensional tectonic processes. Our understanding of the life cycle of normal faults, although improving, remains incomplete, as indicated by the areas of conflict between mechanical models and geological observations. This study is an attempt to reconcile some of these conflicts.

Our first aim in this study was to quantitatively estimate and assess the μ_s values of Byerlee (1978) compared with newly calculated μ_s values from natural normal fault systems. We document a much larger range of estimated μ_s values in natural extensional fault systems – not only values below the range of Byerlee (1978) that have been the focus of extensive, previous research, but also higher μ_s values than the range of Byerlee (1978) that are nonetheless still associated with significantly tilted normal fault systems with substantial amounts of slip.

Our second aim was to examine potential factors that could influence and control tilting in rotational planar normal fault systems. Statistical analysis revealed no compelling linear correlation, although weighted regressions suggest possible links with the percentage of extension, the strain rate and the mean slip magnitude. We attribute this lack of correlation to fault systems being inherently non-linear, with friction arising as an emergent property of these complexly interacting geosystems. Multivariate analysis may yield future insights, especially when considering long-standing, albeit sparse, observations of relationships between fault spacing, the percentage of extension and the strain rate with tilting.

Financial support was provided by the Lowell Institute for Mineral Resources at the University of Arizona. We thank Gary Axen, Rick Bennett, Paul Kapp and Phil Nickerson for stimulating conversations and helpful insights. This manuscript was improved by detailed feedback from two anonymous reviewers and editorial comments by David Healy. We acknowledge Rick Sibson for his fascinating papers that have inspired much of the work reported here.

References

Ackermann, R.V. & Schlische, R.W. 1997. Anticlustering of small normal faults around larger faults. *Geology*, **25**, 1127–1130.

Ackermann, R.V., Schlische, R.W. & Withjack, M.O. 2001. The geometric and statistical evolution of normal fault systems: an experimental study of the effects of mechanical layer thickness on scaling laws. *Journal of Structural Geology*, **23**, 1803–1819.

Allan, U.S. 1989. Model for hydrocarbon migration and entrapment within faulted structures. *American Association of Petroleum Geologists Bulletin*, **73**, 803–811.

Anderson, E.M. 1905. The dynamics of faulting. *Transactions of the Edinburgh Geological Society*, **8**, 387–402, https://doi.org/10.1144/transed.8.3.387

Anderson, E.M. 1951. *The Dynamics of Faulting and Dyke Formation with Applications to Britain*. 2nd edn. Oliver and Boyd, Edinburgh.

Anderson, O.L. 1978. The role of magma vapors in volcanic tremors and rapid eruptions. *Bulletin Volcanologique*, **41**, 341–353.

Angevine, C., Turcotte, D. & Furnish, M. 1982. Pressure solution lithification as a mechanism for the stick-slip behavior of faults. *Tectonics*, **1**, 151–160.

Armstrong, R.L. & Ward, P. 1991. Evolving geographic patterns of Cenozoic magmatism in the North American Cordillera: the temporal and spatial association of magmatism and metamorphic core complexes. *Journal of Geophysical Research*, **B96**, 13 201–13 224.

Axen, G.J. 1992. Pore pressure, stress increase, and fault weakening in low-angle normal faulting. *Journal of Geophysical Research*, **97**, 8979–8991.

Axen, G.J. 2004. Mechanics of low-angle normal faults. *In*: Karner, G.D., Taylor, B., Driscoll, N.W. & Kohlstedt, D.L. (eds) *Rheology and Deformation of the Lithosphere at Continental Margins*. Columbia University Press, New York, 46–91.

Axen, G.J., Taylor, W.J. & Bartley, J.M. 1993. Space-time patterns and tectonic controls of Tertiary extension and magmatism in the Great Basin of the western United States. *Geological Society of America Bulletin*, **105**, 56–76.

Axen, G.J., Luther, A. & Selverstone, J. 2015. Paleostress directions near two low-angle normal faults: testing mechanical models of weak faults and off-fault damage. *Geosphere*, **11**, 1996–2014.

Barton, M.D., Brown, J. et al. 2005. *Porphyry Copper Deposit Life Cycles Field Conference*, 21–22 May 2002, Southeastern Arizona. US Geological Survey, Scientific Investigations Reports, **2005–5020**.

Beeler, N.M., Simpson, R.W., Hickman, S.H. & Lockner, D.A. 2000. Pore fluid pressure, apparent friction, and Coulomb failure. *Journal of Geophysical Research: Solid Earth*, **105**, 25 533–25 542.

Behnsen, J. & Faulkner, D.R. 2012. The effect of mineralogy and effective normal stress on frictional strength of sheet silicates. *Journal of Structural Geology*, **42**, 49–61.

Best, M.G. & Christiansen, E.H. 1991. Limited extension during peak Tertiary volcanism, Great Basin of Nevada and Utah. *Journal of Geophysical Research: Solid Earth*, **96**, 13 509–13 528.

Bodnar, R.J., Azbej, T., Becker, S.P., Cannatelli, C., Fall, A. & Severs, M.J. 2013. *Whole Earth Geohydrologic Cycle, from the Clouds to the Core: the Distribution of Water in the Dynamic Earth System*. Geological Society of America, Special Papers, **500**, 431–461.

Burnham, C.W. 1997. Magmas and hydrothermal fluids. *In*: Barnes, H.L. (ed.) *Geochemistry of Hydrothermal Ore Deposits*. 3rd edn. Wiley, New York, 63–123.

Byerlee, J. 1978. Friction of rocks. *Pure and Applied Geophysics*, **116**, 615–626.

Camp, V.E., Pierce, K.L. & Morgan, L.A. 2015. Yellowstone plume trigger for Basin and Range extension, and coeval emplacement of the Nevada–Columbia Basin magmatic belt. *Geosphere*, **11**, 203–225.

Cassel, E.J., Breecker, D.O., Henry, C.D., Larson, T.E. & Stockli, D.F. 2014. Profile of a paleo-orogen: high topography across the present-day Basin and Range from 40 to 23 Ma. *Geology*, **42**, 1007–1010.

Cather, S.M., Chamberlin, R.M., Chapin, C.E. & McIntosh, W.C. 1994. *Stratigraphic Consequences of Episodic Extension in the Lemitar Mountains, Central Rio Grande Rift*. Geological Society of America, Special Papers, **291**, 157–170.

Chamberlin, R.M. 1982. *Geologic Map, Cross Sections, and Map Units of the Lemitar Mountains, Socorro County, New Mexico*. New Mexico Bureau of Geology and Mineral Resources Open-File Report 169, scale 1:12,000.

Chamberlin, R.M. 1983. Cenozoic domino-style crustal extension in the Lemitar Mountains, New Mexico: a summary. *In*: Chapin, C.E. (ed.) *Socorro Region II*. New Mexico Geological Society, 34th Annual Field Conference, 111–118.

Choi, E. & Buck, W.R. 2012. Constraints on the strength of faults from the geometry of rider blocks in continental and oceanic core complexes. *Journal of Geophysical Research: Solid Earth*, **117**, B04410.

Colgan, J.P. & Henry, C.D. 2009. Rapid middle Miocene collapse of the Mesozoic orogenic plateau in north-central Nevada. *International Geology Review*, **51**, 920–961.

Colgan, J.P., John, D.A., Henry, C.D. & Fleck, R.J. 2008. Large-magnitude Miocene extension of the Eocene Caetano caldera, Shoshone and Toiyabe Ranges, Nevada. *Geosphere*, **4**, 107–130.

Colgan, J.P., Howard, K.A., Fleck, R.J. & Wooden, J.L. 2010. Rapid middle Miocene extension and unroofing of the southern Ruby Mountains, Nevada. *Tectonics*, **29**, TC6022.

Colgan, J.P., Henry, C.D. & John, D.A. 2014. Evidence for large-magnitude, post-Eocene extension in the northern Shoshone Range, Nevada, and its implications for the structural setting of Carlin-type gold deposits in the lower plate of the Roberts Mountains allochthon. *Economic Geology*, **109**, 1843–1862.

Collettini, C. 2011. The mechanical paradox of low-angle normal faults: current understanding and open questions. *Tectonophysics*, **510**, 253–268.

Collettini, C. & Holdsworth, R. 2004. Fault zone weakening and character of slip along low-angle normal faults: insights from the Zuccale fault, Elba, Italy. *Journal of the Geological Society, London*, **161**, 1039–1051, https://doi.org/10.1144/0016-764903-179

Collettini, C. & Sibson, R.H. 2001. Normal faults, normal friction? *Geology*, **29**, 927–930.

Collettini, C., Chiaraluce, L., Pucci, S., Barchi, M.R. & Cocco, M. 2005. Looking at fault reactivation matching structural geology and seismological data. *Journal of Structural Geology*, **27**, 937–942.

Collettini, C., Niemeijer, A., Viti, C. & Marone, C. 2009*a*. Fault zone fabric and fault weakness. *Nature*, **462**, 907–910.

Collettini, C., Viti, C., Smith, S.A.F. & Holdsworth, R.E. 2009*b*. Development of interconnected talc networks and weakening of continental low-angle normal faults. *Geology*, **37**, 567–570.

Coney, P.J. & Harms, T.A. 1984. Cordilleran metamorphic core complexes: Cenozoic extensional relics of Mesozoic compression. *Geology*, **12**, 550–554.

Contreras, J., Anders, M.H. & Scholz, C.H. 2000. Growth of a normal fault system: observations from the Lake Malawi basin of the east African rift. *Journal of Structural Geology*, **22**, 159–168.

Crider, J.G. & Peacock, D.C.P. 2004. Initiation of brittle faults in the upper crust: a review of field observations. *Journal of Structural Geology*, **26**, 691–707.

Dahlstrom, C. 1969. Balanced cross sections. *Canadian Journal of Earth Sciences*, **6**, 743–757.

Davis, G.H. 2013. Localization control for chlorite breccia deformation beneath Catalina detachment fault, Rincon Mountains, Tucson, Arizona. *Journal of Structural Geology*, **50**, 237–253.

Davis, G.H. & Reynolds, S.J. 1996. *Structural Geology of Rocks and Regions*. Wiley, New York.

DeCelles, P.G. 1994. Late Cretaceous–Paleocene synorogenic sedimentation and kinematic history of the Sevier thrust belt, northeast Utah and southwest Wyoming. *Geological Society of America Bulletin*, **106**, 32–56.

Dickinson, W.R. 1991. *Tectonic Setting of Faulted Tertiary Strata Associated with the Catalina Core Complex in Southern Arizona*. Geological Society of America, Special Papers, **264**, scale 1:125,000.

Dickinson, W.R. 2006. Geotectonic evolution of the Great Basin. *Geosphere*, **2**, 353–368.

Di Fiori, R.V., Long, S.P., Muntean, J.L. & Edmondo, G.P. 2015. Structural analysis of gold mineralization in the southern Eureka mining district, Eureka county, Nevada: a predictive structural setting for Carlin-type mineralization. *In*: Pennell, W.M. & Garside, L.J. (eds) *New Concepts and Discoveries. Geological Society of Nevada 2015 Symposium*. Vol. 2. Geological Society of Nevada, Reno/Sparks, NV, 885–903.

Dilles, J.H. 1993. Cenozoic normal and strike-slip faults in the northern Wassuk Range, western Nevada. *In*: Craig, S.D. (ed.) *Structure, Tectonics, and Mineralization of the Walker Lane: Walker Lane Symposium, Proceedings Volume*, April 24 1992. Geological Society of Nevada, Reno, NV, 114–136.

Dilles, J.H. & Gans, P.B. 1995. The chronology of Cenozoic volcanism and deformation in the Yerington area, western Basin and Range and Walker Lane. *Geological Society of America Bulletin*, **107**, 474–486.

Donath, F.A. & Cranwell, R.M. 1981. Probabilistic treatment of faulting in geologic media. *In*: Carter, N.L., Friedman, M., Logan, J.M. & Stearns, D.W. (eds) *Mechanical Behavior of Crustal Rocks: the Handin Volume*. American Geophysical Union, Monographs, **24**, 231–241.

Dreier, J.E. 1984. Regional tectonic control of epithermal veins in the western United States and Mexico. *In*: Wilkins, J., Jr. (ed.) *Gold and Silver Deposits of the Basin and Range Province, Western U.S.A.* Arizona Geological Society Digest, **15**, 28–50.

Elliott, D. 1983. The construction of balanced cross-sections. *Journal of Structural Geology*, **5**, 101.

Emmons, W.H. 1907. Normal faulting in the Bullfrog district. *Science*, **26**, 221–222.

Erickson, B.A., Birnir, B. & Lavallée, D. 2011. Periodicity, chaos and localization in a Burridge–Knopoff model of an earthquake with rate-and-state friction. *Geophysical Journal International*, **187**, 178–198.

Etheridge, M.A. 1986. On the reactivation of extensional fault systems. *Philosophical Transactions of the Royal Society of London, Series A, Mathematical and Physical Sciences*, **317**, 179–194.

Faulds, J.E., Feuerbach, D.L., Reagan, M.K., Metcalf, R.V., Gans, P. & Walker, J.D. 1995. The Mount Perkins block, northwestern Arizona: an exposed cross section of an evolving, preextensional to synextensional magmatic system. *Journal of Geophysical Research: Solid Earth*, **100**, 15 249–15 266.

Faulds, J.E., Coolbaugh, M. & Hinz, N. 2015. Favourable structural settings of active geothermal and young epithermal systems in the Great Basin region, western USA: implications for exploration strategies [extended abstract]. *New Concepts and Discoveries, Geological Society of Nevada 2015 Symposium*, 18 May 2015, Sparks, NV, 52–53.

Favorito, D.A. 2016. *Characterization and reconstruction of Laramide deformation and porphyry-style alteration in the Romero Wash-Tecolote Ranch area, southeastern Arizona*. MS thesis, University of Arizona.

Favorito, D.A. & Seedorff, E. 2017. Characterization and reconstruction of Laramide shortening and superimposed Cenozoic extension, Romero Wash-Tecolote Ranch area, southeastern Arizona. *Geosphere*, **13**, 577–607.

Feeley, T.C. 1993. *Geologic Map of the Robinson Summit Quadrangle, Nevada*. Nevada Bureau of Mines and Geology, Field Studies Map 2, scale 1:24,000.

Feeley, T.C. & Grunder, A.L. 1991. Mantle contribution to the evolution of middle Tertiary silicic magmatism during early stages of extension: the Egan Range volcanic complex, east-central Nevada. *Contributions to Mineralogy and Petrology*, **106**, 154–169.

Ferrill, D.A., McGinnis, R.N. *et al.* 2014. Control of mechanical stratigraphy on bed-restricted jointing and normal faulting: Eagle Ford Formation, south-central

Texas. *American Association of Petroleum Geologists Bulletin*, **98**, 2477–2506.

Franke, D. 2013. Rifting, lithosphere breakup and volcanism: comparison of magma-poor and volcanic rifted margins. *Marine and Petroleum Geology*, **43**, 63–87.

Gans, P.B. 1990. Space-time patterns of Cenozoic N-S extension, N-S shortening, E-W extension and magmatism in the Basin and Range Province: evidence for active rifting. *Geological Society of America, Abstracts with Programs*, **22**, 24.

Gans, P.B. & Bohrson, W.A. 1998. Suppression of volcanism during rapid extension in the Basin and Range Province, United States. *Science*, **279**, 66–68.

Gans, P.B. & Landau, B. 1993. Initially vertical normal faults and runaway block rotations in the Basin and Range province: some geological and mechanical observations. *Geological Society of America, Abstracts with Programs*, **25**, 40.

Gans, P.B. & Miller, E.L. 1983. Styles of mid-Tertiary extension in east-central Nevada. *In*: Nash, W.P. & Gurgel, K.D. (eds) *Geological Excursions in the Overthrust Belt and Metamorphic Core Complexes of the Intermountain Region*. Utah Geological Mineral Survey, Special Studies, **59**, 107–139.

Gans, P.B., Miller, E., McCarthy, J. & Ouldcott, M. 1985. Tertiary extensional faulting and evolving ductile–brittle transition zones in the northern Snake Range and vicinity: new insights from seismic data. *Geology*, **13**, 189–193.

Gans, P.B., Mahood, G.A. & Schermer, E. 1989. *Synextensional Magmatism in the Basin and Range Province: a Case Study from the Eastern Great Basin*. Geological Society of America, Special Papers, **233**.

Gans, P.B., Miller, E.L., Houseman, G. & Lister, G.S. 1991. Assessing the amount, rate, and timing of tilting in normal fault blocks: a case study of tilted granites in the Kern-Deep Creek Mountains, Utah. *Geological Society of America, Abstracts with Programs*, **23**, 28.

Gans, P.B., Landau, B. & Darvall, P. 1994. Ashes, ashes, all fall down: caldera-forming eruptions and extensional collapse of the Eldorado Mountains, southern Nevada. *Geological Society of America, Abstracts with Programs*, **26**, 53.

Gans, P.B., Seedorff, E., Fahey, P.L., Hasler, R.W., Maher, D.J., Jeanne, R.A. & Shaver, S.A. 2001. Rapid Eocene extension in the Robinson district, White Pine County, Nevada: constraints from $^{40}Ar/^{39}Ar$ dating. *Geology*, **29**, 475–478.

Garside, L.J., Henry, C.D. & Boden, D.R. 2002. Far-flung ash-flow tuffs of Yerington, western Nevada erupted from calderas in the Toquima Range, central Nevada. *Geological Society of America, Abstracts with Programs*, **34**, 44.

Gibbs, A.D. 1983. Balanced cross-section construction from seismic sections in areas of extensional tectonics. *Journal of Structural Geology*, **5**, 153–160.

Gonsior, Z.J. & Dilles, J.H. 2008. Timing and evolution of Cenozoic extensional normal faulting and magmatism in the southern Tobin Range, Nevada. *Geosphere*, **4**, 687–712.

Gupta, A. & Scholz, C.H. 2000. A model of normal fault interaction based on observations and theory. *Journal of Structural Geology*, **22**, 865–879.

Healy, D. 2009. Anisotropy, pore fluid pressure and low angle normal faults. *Journal of Structural Geology*, **31**, 561–574.

Hehnke, C., Ballantyne, G., Martin, H., Hart, W., Schwarz, A. & Stein, H. 2012. Geology and exploration progress at the Resolution Cu–Mo deposit, Arizona. *In*: Hedenquist, J.W., Harris, M. & Camus, F. (eds) *Geology and Genesis of Major Copper Deposits and Districts of the World: a Tribute to Richard H. Sillitoe*. Society of Economic Geologists, Special Publications, **16**, 147–166.

Hill, D.P. & Thatcher, W. 1992. An energy constraint for frictional slip on misoriented faults. *Bulletin of the Seismological Society of America*, **82**, 883–897.

Hoek, E. 1965. *Rock Fracture Under Static Stress Conditions*. Council for Scientific and Industrial Research Report MEG **383**.

Hoge, A.K., Seedorff, E., Barton, M.D., Richardson, C.A. & Favorito, D.A. 2015. The Jackson-Lawton-Bowman normal fault system and its relationship to Carlin-type gold mineralization, Eureka district, Nevada. *In*: Pennell, W.M. & Garside, L.J. (eds) *New Concepts and Discoveries. Geological Society of Nevada 2015 Symposium*. Vol. 2. Geological Society of Nevada, Reno/Sparks, NV, 967–1000.

Holland, J.H. 1998. *Emergence: from Chaos to Order*. Basic Books, New York.

Hreinsdóttir, S. & Bennett, R.A. 2009. Active aseismic creep on the Alto Tiberina low-angle normal fault, Italy. *Geology*, **37**, 683–686.

Hubbert, M.K. & Rubey, W.W. 1959. Role of fluid pressure in mechanics of overthrust faulting I. Mechanics of fluid-filled porous solids and its application to overthrust faulting. *Geological Society of America Bulletin*, **70**, 115–166.

Hubert, A., King, G., Armijo, R., Meyer, B. & Papanastasiou, D. 1996. Fault re-activation, stress interaction and rupture propagation of the 1981 Corinth earthquake sequence. *Earth and Planetary Science Letters*, **142**, 573–585.

Hudson, M.R., John, D.A., Conrad, J.E. & McKee, E.H. 2000. Style and age of late Oligocene–early Miocene deformation in the southern Stillwater Range, west central Nevada: paleomagnetism, geochronology, and field relations. *Journal of Geophysical Research: Solid Earth*, **105**, 929–954.

Jackson, J.A. 1999. Fault death: a perspective from actively deforming regions. *Journal of Structural Geology*, **21**, 1003–1010.

Jackson, J.A. & White, N.J. 1989. Normal faulting in the upper continental crust: observations from regions of active extension. *Journal of Structural Geology*, **11**, 15–36.

Jaeger, J.C., Cook, N.G.W. & Zimmerman, R.W. 2007. *Fundamentals of Rock Mechanics*. Blackwell, Oxford.

Jamison, D.B. & Cook, N.G. 1980. Note on measured values for the state of stress in the Earth's crust. *Journal of Geophysical Research*, **B85**, 1833–1838.

Jefferies, S.P., Holdsworth, R.E., Wibberley, C.A.J., Shimamoto, T., Spiers, C.J., Niemeijer, A.R. & Lloyd, G.E. 2006. The nature and importance of phyllonite development in crustal-scale fault cores: an example from the Median Tectonic Line, Japan. *Journal of Structural Geology*, **28**, 220–235.

KARLSTROM, K.E., HEIZLER, M. & QUIGLEY, M.C. 2010. Structure and $^{40}Ar/^{39}Ar$ K-feldspar thermal history of the Gold Butte block: reevaluation of the tilted crustal section model. *In*: UMHOEFER, P.J., BEARD, L.S. & LAMB, M.A. (eds) *Miocene Tectonics of the Lake Mead Region, Central and Range*. Geological Society of America, Special Papers, **463**, 331–352.

KERRICH, R., BECKINSALE, R.D. & DURHAM, J.J. 1977. The transition between regimes dominated by intercrystalline diffusion and intracrystalline creep evaluated by oxygen isotope thermometry. *Tectonophysics*, **38**, 241–257.

KING, G.C.P., STEIN, R.S. & RUNDLE, J.B. 1988. The growth of geological structures by repeated earthquakes 1. Conceptual framework. *Journal of Geophysical Research*, **B93**, 13 307–13 318.

KING, G.C.P., STEIN, R.S. & LIN, J. 1994. Static stress changes and the triggering of earthquakes. *Bulletin of the Seismological Society of America*, **84**, 935–953.

KONSTANTINOU, A., STRICKLAND, A., MILLER, E., VERVOORT, J., FISHER, C.M., WOODEN, J. & VALLEY, J. 2013. Synextensional magmatism leading to crustal flow in the Albion–Raft River–Grouse Creek metamorphic core complex, northeastern Basin and Range. *Tectonics*, **32**, 1384–1403.

KREEMER, C. & HAMMOND, W.C. 2007. Geodetic constraints on areal changes in the Pacific–North America plate boundary zone: what controls Basin and Range extension? *Geology*, **35**, 943–946.

KUSZNIR, N.J., MARSDEN, G. & EGAN, S.S. 1991. A flexural-cantilever simple-shear/pure-shear model of continental lithosphere extension: applications to the Jeanne d'Arc basin, Grand Banks and Viking graben, North Sea. *In*: ROBERTS, A.M., YIELDING, G. & FREEMAN, B. (eds) *The Geometry of Normal Faults*. Geological Society, London, Special Publications, **56**, 41–60, https://doi.org/10.1144/GSL.SP.1991.056.01.04

LEVEILLE, R.A. & STEGEN, R.J. 2012. The southwestern North America porphyry copper province. *In*: HEDENQUIST, J.W., HARRIS, M. & CAMUS, F. (eds) *Geology and Genesis of Major Copper Deposits and Districts of the World: a Tribute to Richard H. Sillitoe*. Society of Economic Geologists, Special Publications, **16**, 361–401.

LINGREY, S. & VIDAL-ROYO, O. 2015. Evaluating the quality of bed length and area balance in 2D structural restorations. *Interpretation*, **3**, SAA133–SAA160.

LOWELL, J.D. 1968. Geology of the Kalamazoo orebody, San Manuel district, Arizona. *Economic Geology*, **63**, 645–654.

LOWELL, J.D. & GUILBERT, J.M. 1970. Lateral and vertical alteration-mineralization zoning in porphyry ore deposits. *Economic Geology*, **65**, 373–408.

MANDL, G. 1987. Tectonic deformation by rotating parallel faults: the 'bookshelf' mechanism. *Tectonophysics*, **141**, 277–316.

MCQUARRIE, N. & WERNICKE, B.P. 2005. An animated tectonic reconstruction of southwestern North America since 36 Ma. *Geosphere*, **1**, 147–172.

MICKLETHWAITE, S. 2009. Mechanisms of faulting and permeability enhancement during epithermal mineralisation: Cracow goldfield, Australia. *Journal of Structural Geology*, **31**, 288–300.

MICKLETHWAITE, S., SHELDON, H.A. & BAKER, T. 2010. Active fault and shear processes and their implications for mineral deposit formation and discovery. *Journal of Structural Geology*, **32**, 151–165.

MILLER, C.F. & MILLER, J.S. 2002. Contrasting stratified plutons exposed in tilt blocks, Eldorado Mountains, Colorado River Rift, NV, USA. *Lithos*, **61**, 209–224.

MILLER, E.L., GANS, P.B. & GARING, J. 1983. The Snake Range décollement: an exhumed mid-Tertiary ductile–brittle transition. *Tectonics*, **2**, 239–263.

MILLER, E.L., DUMITRU, T.A., BROWN, R.W. & GANS, P.B. 1999. Rapid Miocene slip on the Snake Range–Deep Creek Range fault system, east-central Nevada. *Geological Society of America Bulletin*, **111**, 886–905.

MORTON, W.H. & BLACK, R. 1975. Crustal attenuation in Afar. *In*: PILGER, A. & RÖSLER, A. (eds) *Afar Depression of Ethiopia: International Symposium on the Afar Region and Related Rift Problems*. Proceedings of the Inter-Union Commission on Geodynamics, Scientific Report 14. E. Schweizerbart'sche Verlagbuchhandlung, Stuttgart, 55–65.

MUIRHEAD, J.D., KATTENHORN, S.A. *ET AL*. 2016. Evolution of upper crustal faulting assisted by magmatic volatile release during early-stage continental rift development in the East African Rift. *Geosphere*, **12**, 1670–1700.

NAVARRO, M. 2002. *Fault roughness and fault complexity: field study, multi-scale analysis and numerical fault model*. Dr. rer. nat. dissertation, University of Bonn.

NICKERSON, P.A. 2012. *Post-mineral normal faulting in Arizona porphyry systems*. PhD thesis, University of Arizona.

NICKERSON, P.A. & SEEDORFF, E. 2016. Dismembered porphyry systems near Wickenburg, Arizona: district-scale reconstruction with an arc-scale context. *Economic Geology*, **111**, 447–466.

NICKERSON, P.A., BARTON, M.D. & SEEDORFF, E. 2010. Characterization and reconstruction of the multiple copper-bearing hydrothermal systems in the Tea Cup porphyry systems, Pinal County Arizona. *In*: GOLDFARB, R.J., MARSH, E.E. & MONECKE, T. (eds) *The Challenge of Finding New Mineral Resources: Global Metallogeny, Innovative Exploration, and New Discoveries*. Society of Economic Geologists, Special Publications, **15**, 299–316.

NIEMEIJER, A., MARONE, C. & ELSWORTH, D. 2010. Fabric induced weakness of tectonic faults. *Geophysical Research Letters*, **37**.

NUMELIN, T., MARONE, C. & KIRBY, E. 2007. Frictional properties of natural fault gouge from a low-angle normal fault, Panamint Valley, California. *Tectonics*, **26**, https://doi.org/10.1029/2005TC001916

NUR, A., RON, H. & SCOTTI, O. 1986. Fault mechanics and the kinematics of block rotations. *Geology*, **14**, 746–749.

PAPE, J.R., SEEDORFF, E., BARIL, T.C. & THOMPSON, T.B. 2016. Structural reconstruction and age of an extensionally faulted porphyry molybdenum system at Spruce Mountain, Elko County, Nevada. *Geosphere*, **12**, 237–263.

PARSONS, T. & THOMPSON, G.A. 1993. Does magmatism influence low-angle normal faulting? *Geology*, **21**, 247–250.

PARSONS, T., THOMPSON, G.A. & SLEEP, N.H. 1994. Mantle plume influence on the Neogene uplift and extension of the US western Cordillera? *Geology*, **22**, 83–86.

PEACOCK, D.C.P. & SANDERSON, D.J. 1994. Geometry and development of relay ramps in normal fault systems. *American Association of Petroleum Geologists Bulletin*, **78**, 147–167.

PROFFETT, J.M., JR. 1977. Cenozoic geology of the Yerington district, Nevada, and implications for the nature and origin of Basin and Range faulting. *Geological Society of America Bulletin*, **88**, 247–266.

PROFFETT, J.M., JR. & DILLES, J.H. 1984. *Geologic Map of the Yerington District, Nevada*. Nevada Bureau of Mines and Geology, Map 77, scale 1:24,000.

RAHL, J.M., MCGREW, A.J. & FOLAND, K.A. 2002. Transition from contraction to extension in the northeastern Basin and Range: new evidence from the Copper Mountains, Nevada. *Journal of Geology*, **110**, 179–194.

RANSOME, F.L. 1903. *Geology of the Globe Copper District*. US Geological Survey, Professional Papers, **12**.

RANSOME, F.L., EMMONS, W.H. & GARREY, G.H. 1910. Geology and ore deposits of the Bullfrog district, Nevada. *US Geological Survey Bulletin*, **407**.

RHYS, D., VALLI, F., BURGESS, R., HEITT, D., GRIESEL, G. & HART, K. 2015. Controls on fault and fold geometry on the distribution of gold mineralization on the Carlin trend. *In*: PRENNELL, W.M. & GARSIDE, L.J. (eds) *New Concepts and Discoveries. Geological Society of Nevada 2015 Symposium*. Vol. 1. Geological Society of Nevada, Reno/Sparks, NV, 333–389.

RICHARDSON, C.A. & SEEDORFF, E. 2015. Reconstruction of normal fault blocks in the Ann-Mason and Blue Hill areas, Yerington district, Lyon county, western Nevada. *In*: PENNELL, W.M. & GARSIDE, L.J. (eds) *New Concepts and Discoveries. Geological Society of Nevada 2015 Symposium*. Vol. 2. Geological Society of Nevada, Reno/Sparks, NV, 1153–1178.

RICHARDSON, C.A., SEEDORFF, E. & KING, C.A. 2016. Extensional dismemberment of the Hilltop mining district, northern Shoshone Range, north-central Nevada. *Geological Society of America, Abstracts with Programs*, **48**, https://doi.org/10.1130/abs/2016AM-278000

RUBEY, W.W. & HUBBERT, M.K. 1959. Role of fluid pressure in mechanics of overthrust faulting II. Overthrust belt in geosynclinal area of western Wyoming in light of fluid-pressure hypothesis. *Geological Society of America Bulletin*, **70**, 167–206.

SCHMALHOLZ, S.M. & PODLADCHIKOV, Y. 2014. Metamorphism under stress: the problem of relating minerals to depth. *Geology*, **42**, 733–734.

SCHOLZ, C.H. 2007. Fault mechanics. *In*: SCHUBERT, G. (ed.) *Treatise on Geophysics*. Elsevier, Amsterdam, 441–483.

SEEDORFF, E. 1991*a*. Magmatism, extension, and ore deposits of Eocene to Holocene age in the Great Basin: mutual effects and preliminary proposed genetic relationships. *In*: RAINES, G.L., LISLE, R.J., SCHAFER, R.W. & WILKINSON, W.H. (eds) *Geology and Ore Deposits of the Great Basin. Geological Society of Nevada, Symposium Proceedings*, Vol. 1. Geological Society of Nevada, Reno, NV, 133–178.

SEEDORFF, E. 1991*b*. Royston district, western Nevada – a Mesozoic porphyry copper system that was tilted and dismembered by Tertiary normal faults. *In*: RAINES, G.L., LISLE, R.E., SCHAFER, R.W. & WILKINSON, W.H. (eds) *Geology and Ore Deposits of the Great Basin. Geological Society of Nevada, Symposium Proceedings*, Vol. 1. Geological Society of Nevada, Reno, NV, 359–391.

SEEDORFF, E. & RICHARDSON, C.A. 2014. Diverse geometries and temporal relationships of normal faults and fault systems in continental extension: a perspective from the Basin and Range province and mineral deposits [extended abstract]. Geological Society of London Conference, 23–25 June 2014, London, UK, Programme and Abstract Volume, 20–22.

SEEDORFF, E., BARTON, M.D., STAVAST, W.J.A. & MAHER, D.J. 2008. Root zones of porphyry systems: extending the porphyry model to depth. *Economic Geology*, **103**, 939–956.

SEEDORFF, E., RICHARDSON, C.A. & MAHER, D.J. 2015. Fault surface maps: three-dimensional structural reconstructions and their utility in mining and exploration. *In*: PENNELL, W.M. & GARSIDE, L.J. (eds) *New Concepts and Discoveries. Geological Society of Nevada 2015 Symposium*. Vol. 2. Geological Society of Nevada, Reno/Sparks, NV, 1179–1206.

SHAVER, S.A. & MCWILLIAMS, M. 1987. Cenozoic extension and tilting recorded in Upper Cretaceous and Tertiary rocks at the Hall molybdenum deposit, northern San Antonio Mountains, Nevada. *Geological Society of America Bulletin*, **99**, 341–353.

SIBSON, R.H. 1973. Interactions between temperature and pore-fluid pressure during earthquake faulting and a mechanism for partial or total stress relief. *Nature*, **243**, 66–68.

SIBSON, R.H. 1974. Frictional constraints on thrust, wrench and normal faults. *Nature*, **249**, 542–544.

SIBSON, R.H. 1985. A note on fault reactivation. *Journal of Structural Geology*, **7**, 751–754.

SIBSON, R.H. 1987. Earthquake rupturing as a mineralizing agent in hydrothermal systems. *Geology*, **15**, 701–704.

SIBSON, R.H. 1990. Rupture nucleation on unfavorably oriented faults. *Bulletin of the Seismological Society of America*, **80**, 1580–1604.

SIBSON, R.H. 1994. An assessment of field evidence for 'Byerlee' friction. *Pure and Applied Geophysics*, **142**, 645–662.

SIBSON, R.H. 2000. Fluid involvement in normal faulting. *Journal of Geodynamics*, **29**, 469–499.

SIBSON, R.H. 2002. Geology of the crustal earthquake source. *In*: LEE, W.H.K., KANAMORI, H., JENNINGS, P.C. & KISSLINGER, C. (eds) *International Handbook of Earthquake and Engineering Seismology: International Geophysics, Part A*, **81**. Academic Press, 455–473.

SIBSON, R.H., ROBERT, F. & POULSEN, K.H. 1988. High-angle reverse faults, fluid-pressure cycling, and mesothermal gold–quartz deposits. *Geology*, **16**, 551–555.

SMITH, D.L., GANS, P.B. & MILLER, E.L. 1991. Palinspastic restoration of Cenozoic extension in the central and eastern Basin and Range province at latitude 39–40°N. *In*: RAINES, G.L., LISLE, R.E., SCHAFER, R.W. & WILKINSON, W.H. (eds) *Geology and Ore Deposits of the Great Basin. Geological Society of Nevada,*

Symposium Proceedings, Vol. 1. Geological Society of Nevada, Reno, NV, 75–86.

Soliva, R., Benedicto, A. & Maerten, L. 2006. Spacing and linkage of confined normal faults: importance of mechanical thickness. *Journal of Geophysical Research*, **111**, B01402, https://doi.org/10.1029/2004JB003507

Stein, R.S., King, G.C.P. & Rundle, J.B. 1988. The growth of geological structures by repeated earthquakes 2. Field examples of continental dip-slip faults. *Journal of Geophysical Research*, **93**, 13 319–13 331.

Stockli, D.F., Surpless, B.E., Dumitru, T.A. & Farley, K.A. 2002. Thermochronological constraints on the timing and magnitude of Miocene and Pliocene extension in the central Wassuk Range, western Nevada. *Tectonics*, **21**, https://doi.org/10.1029/2001TC 001295

Tenthorey, E. & Cox, S.F. 2006. Cohesive strengthening of fault zones during the interseismic period: an experimental study. *Journal of Geophysical Research*, **111**, B09202, https://doi.org/10.1029/2005JB00 4122

Thatcher, W. & Hill, D.P. 1991. Fault orientations in extension and conjugate strike-slip environments and their implications. *Geology*, **19**, 1116–1120.

Thompson, G.A. & Parsons, T. 2016. Vertical deformation associated with normal fault systems evolved over coseismic, postseismic, and multiseismic periods. *Journal of Geophysical Research: Solid Earth*, **121**, 2153–2173.

Van Buer, N.J., Miller, E.L. & Dumitru, T.A. 2009. Early Tertiary paleogeologic map of the northern Sierra Nevada batholith and the northwestern Basin and Range. *Geology*, **37**, 371–374.

Vendeville, B., Cobbold, P.R., Davy, P., Choukroune, P. & Brun, J.P. 1987. Physical models of extensional tectonics at various scales. *In*: Coward, M.P., Dewey, J.F. & Hancock, P.L. (eds) *Continental Extensional Tectonics*. Geological Society, London, Special Publications, **28**, 95–107, https://doi.org/10.1144/GSL.SP.1987.028.01.08

Voll, G. 1976. Recrystallization of quartz, biotite, and feldspars from Erstfeld to the Levantina Nappe, Swiss Alps, and its geological significance. *Schweizerische Mineralogische und Petrographische Mitteilungen*, **56**, 641–647.

Wallace, R.E. 1984. Patterns and timing of Late Quaternary faulting in the Great Basin Province and relation to some regional tectonic features. *Journal of Geophysical Research*, **89**, 5763–5769.

Wallace, R.E. 1987. Grouping and migration of surface faulting and variations in slip rates on faults in the Great Basin province. *Bulletin of the Seismological Society of America*, **77**, 868–876.

Walsh, J.J. & Watterson, J. 1991. Geometric and kinematic coherence and scale effects in normal fault systems. *In*: Roberts, A.M., Yielding, G. & Freeman, B. (eds) *The Geometry of Normal Faults*. Geological Society, London, Special Publications, **56**, 193–203, https://doi.org/10.1144/GSL.SP.1991.056.01.13

Wernicke, B. & Axen, G.J. 1988. On the role of isostasy in the evolution of normal fault systems. *Geology*, **16**, 848–851.

Wernicke, B. & Burchfiel, B. 1982. Modes of extensional tectonics. *Journal of Structural Geology*, **4**, 105–115.

Wernicke, B.P., England, P.C., Sonder, L.J. & Christiansen, R.L. 1987. Tectonomagmatic evolution of Cenozoic extension in the North American Cordillera. *In*: Coward, M.P., Dewey, J.F. & Hancock, P.L. (eds) *Continental Extensional Tectonics*. Geological Society, London, Special Publications, **28**, 203–221, https://doi.org/10.1144/GSL.SP.1987.028.01.15

Westaway, R. 1992. Analysis of tilting near normal faults using calculus of variations: implications for upper crustal stress and rheology. *Journal of Structural Geology*, **14**, 857–871.

Westaway, R. 1999. The mechanical feasibility of low-angle normal faulting. *Tectonophysics*, **308**, 407–443.

Wheeler, J. 2014. Dramatic effects of stress on metamorphic reactions. *Geology*, **42**, 647–650.

Whitney, D.L., Teyssier, C., Rey, P. & Buck, W.R. 2013. Continental and oceanic core complexes. *Geological Society of America Bulletin*, **125**, 273–298.

Wilkins, J., Jr. & Heidrick, T.L. 1995. Post-Laramide extension and rotation of porphyry copper deposits, southwestern United States. *In*: Pierce, F.W. & Bolm, J.G. (eds) *Porphyry Copper Deposits of the America Cordillera*. Arizona Geological Society Digest, **20**, 109–127.

Willemse, E.J. 1997. Segmented normal faults: correspondence between three-dimensional mechanical models and field data. *Journal of Geophysical Research: Solid Earth*, **102**, 675–692.

Wong, M.S. & Gans, P.B. 2008. Geologic, structural, and thermochronologic constraints on the tectonic evolution of the Sierra Mazatán core complex, Sonora, Mexico: new insights into metamorphic core complex formation. *Tectonics*, **27**, TC4103, https://doi.org/10.1029/2007TC002173

Wong, M.S., Gleason, D.M.B., O'Brien, H.P. & Idleman, B.D. 2015. Confirmation of a low pre-extensional geothermal gradient in the Grayback normal fault block, Arizona: structural and He thermochronologic evidence. *Geological Society of America Bulletin*, **127**, 200–210.

Woodward, N.B., Boyer, S.E. & Suppe, J. 1989. *Balanced Geological Cross Sections: an Essential Technique in Geological Research and Exploration*. American Geophysical Union, Short Course in Geology, **6**.

Yamasaki, T. & Stephenson, R. 2009. Change in tectonic force inferred from basin subsidence: implications for the dynamical aspects of back-arc rifting in the western Mediterranean. *Earth and Planetary Science Letters*, **277**, 174–183.

Yin, A. 1989. Origin of regional, rooted low-angle normal faults: a mechanical model and its tectonic implications. *Tectonics*, **8**, 469–482.

Natural CO_2 sites in Italy show the importance of overburden geopressure, fractures and faults for CO_2 storage performance and risk management

JENNIFER J. ROBERTS[1,2]*, MARK WILKINSON[1], MARK NAYLOR[1], ZOE K. SHIPTON[2], RACHEL A. WOOD[1] & R. STUART HASZELDINE[1]

[1]*Scottish Carbon Capture and Storage, School of GeoSciences, University of Edinburgh, Edinburgh EH9 3JW, UK*

[2]*Department of Civil and Environmental Engineering, University of Strathclyde, James Weir Building, Glasgow G1 1XJ, UK*

**Correspondence: jen.roberts@strath.ac.uk*

Abstract: The study of natural analogues can inform the long-term performance security of engineered CO_2 storage. There are natural CO_2 reservoirs and CO_2 seeps in Italy. Here, we study nine reservoirs and establish which are sealed or are leaking CO_2 to surface. Their characteristics are compared to elucidate which conditions control CO_2 leakage. All of the case studies would fail current CO_2 storage site selection criteria, although only two leak CO_2 to surface. The factors found to systematically affect seal performance are overburden geopressure and proximity to modern extensional faults. Amongst our case studies, the sealing reservoirs show elevated overburden geopressure whereas the leaking reservoirs do not. Since the leaking reservoirs are located within <10 km of modern extensional faults, pressure equilibration within the overburden may be facilitated by enhanced crustal permeability related to faulting. Modelling of the properties that could enable the observed CO_2 leakage rates finds that high-permeability pathways (such as transmissive faults or fractures) become increasingly necessary to sustain leak rates as CO_2 density decreases during ascent to surface, regardless of the leakage mechanism into the overburden. This work illustrates the value of characterizing the overburden geology during CO_2 storage site selection to inform screening criterion, risk assessment and monitoring strategy.

Carbon capture and storage (CCS) could significantly reduce anthropogenic CO_2 emissions from large industrial sources of CO_2. However, to be an effective climate change mitigation strategy, the injected CO_2 must remain in the subsurface for timescales of multiple thousands of years (Shaffer 2010). Leakage of CO_2 out of a reservoir could compromise the long-term emission reductions achieved by a CCS project (EU CSS Directive 2009; Zwaan & Gerlagh 2009), and if leaked CO_2 then migrates to the surface or into aquifers there may be local environmental and human health impacts (Jones *et al.* 2015). Unintended leakage to surface in the early phase of technology roll-out could compromise the public acceptability of future CCS, as well as the economic feasibility due to remediation expenditure and liability pay-out (Zwaan & Gerlagh 2009; Heptonstall *et al.* 2012), and, in the EU, possible fines for CO_2 emissions (Dixon *et al.* 2015). Thus, any incidence of leakage from engineered stores may have ramifications for the CCS industry on a global scale.

For these reasons, it is important that the storage site characterization and selection process maximizes the likelihood that injected CO_2 will be securely retained in the subsurface for the timescales intended (thousands of years). Characterization and selection criteria must be applicable in a range of geological settings, and without imposing excessive financial costs. In addition to geological, technical and economic considerations, siting is also constrained by the proximity of the CO_2 source, permitting procedures and public perception. Sites selected for storage do not, therefore, have to be the most geologically favourable (Hannon & Esposito 2015) but must comply with selection criteria. To ensure that CO_2 leakage is avoided, these criteria must be guided by a thorough understanding of the geological characteristics that are most relevant to site integrity. Table 1 summarizes the site selection criteria from guidelines published to date (Miocic *et al.* 2016), which are intended to maximize the likelihood of long-term CO_2 containment. The site selection process must characterize the risks of

From: TURNER, J. P., HEALY, D., HILLIS, R. R. & WELCH, M. J. (eds) 2017. *Geomechanics and Geology*. Geological Society, London, Special Publications, **458**, 181–211.
First published online June 19, 2017, updated June 23, 2017, https://doi.org/10.1144/SP458.14

Table 1. *A summary of published CO_2 storage site selection criteria*

	Feature	Criteria/Requirement	Source
CO_2 properties	CO_2 state	Dense phase	Chadwick *et al.* (2008)
Reservoir properties	Structure	No faults, or small faults. Low faulting frequency	Chadwick *et al.* (2008), IEA-GHG (2009), Smith *et al.* (2011)
		Multilayered system	IEA-GHG (2009)
	Depth (m)	Between 800 and 2500 m	Chadwick *et al.* (2008), IEA-GHG (2009), Smith *et al.* (2011)
		800 m minimum depth	IEA-GHG (2009)
	Temperature	>35°C	IEA-GHG (2009)
	Pressure (MPa)	>7.5	IEA-GHG (2009)
Cap-rock property	Thickness (m)	Between 10 and 100 m	Chadwick *et al.* (2008), IEA-GHG (2009), Smith *et al.* (2011)
	Continuity	Uniform	Chadwick *et al.* (2008)
		(laterally) Extensive	IEA-GHG (2009)

These criteria are recommended to minimize the risks of CO_2 leakage. Adapted from Miocic *et al.* (2016).

geological storage, and understanding how CO_2 could move out of a reservoir and potentially through the overburden and to the surface is critical for constraining and managing risk.

Natural CO_2 reservoirs cannot serve as direct analogues to engineered CO_2 storage sites. The latter are specifically selected for characteristics that minimize leakage, and are charged from a point source at rates and for timescales that are unlikely to mimic any natural process. However, instances of CO_2 migration to the surface from naturally occurring reservoirs provide an opportunity to assess the conditions required for leakage from the reservoirs and to understand the crustal fluid pathways for migration from depth (Annunziatellis *et al.* 2008; Wilkinson *et al.* 2009; Dockrill & Shipton 2010; Kampman *et al.* 2010). Similarly, instances where CO_2 has been successfully retained for geologically long time periods offer opportunities to assess the conditions that will enable effective storage and CO_2–water–rock interactions (Allis *et al.* 2001; Gilfillan *et al.* 2009). The most important controls on the security of CO_2 retention can be established by comparing the characteristics of reservoirs that leak with reservoirs that seal (Miocic *et al.* 2016).

Resource exploration drilling in Italy has revealed the presence of CO_2 accumulations at a range of depths below surface (Casero 2004; Collettini *et al.* 2008; Chiodini *et al.* 2010; Trippetta *et al.* 2013). Italy is also a region of widespread surface CO_2 degassing; over 308 CO_2 seeps have been catalogued at 270 locations in mainland Italy and Sicily (Chiodini & Valenza 2008). Here, we explore the geological conditions that govern whether reservoirs leak or retain CO_2 and establish the mechanisms of leakage. To do this, we identify CO_2-bearing reservoirs from borehole data and establish which boreholes are located geographically close to CO_2 seeps that may represent leakage from that reservoir to surface. We then examine and compare the structure and conditions of the CO_2 boreholes to investigate the controls on whether a reservoir leaks or retains CO_2. Finally, to inform the potential mechanisms of leakage, we assess the properties of possible pathways through the overburden that could enable CO_2 seepage at the rates and styles observed at the Earth surface. Understanding these natural processes over geological timescales is important for informing the long-term performance security of engineered CO_2 storage, since engineered storage sites will be selected and managed to minimize the risks of CO_2 migration. This work can therefore guide effective site assessment, injection strategy and remediation strategies in the case of leakage.

A summary of the geology and CO_2 fluids in Italy is presented in the next section. For completeness, the section that follows outlines CO_2 flow in rocks and the potential mechanisms of migration from a containing reservoir into the overburden (CO_2 leakage) and to the Earth surface (CO_2 seepage).

An overview of CO_2 geofluids in Italy

Hydrocarbon exploration drilling in Italy has encountered subsurface CO_2 accumulations, either as a component within hydrocarbon reservoirs (Casero 2005) or as the dominant gas (Collettini & Barchi 2002; Bicocchi *et al.* 2013). Such accumulations are mostly found in central Italy, which is also where CO_2 degassing is most intense (Chiodini *et al.* 2004). In this region, there is a strong regional NW–SE structural trend (Fig. 1) resulting from the westwards subduction of the Adria Plate beneath the European margin. Crustal shortening stacked several tectonic–stratigraphic units originally located on the Apulian Palaeozoic crystalline carbonate basement, with flysch and synorogenic foredeep sediments (Scrocca *et al.* 2005) (Fig. 1a). To the west, coeval extension opened marine and continental basins and the Tyrrhenian Sea (Ghisetti & Vezzani 2002), leading to high heat flow and volcanism. Seismogenic normal faulting is currently active in the Apennines where exposed fault scarps date from 12 to 18 ka (Roberts 2008). Here, CO_2 fluids trapped at depths >5 km play a critical role in the nucleation and evolution of seismogenesis, and therefore in the deformation style and geodynamics of the region (Miller *et al.* 2004; Malagnini *et al.* 2012).

CO_2 degassing is most active towards the Tyrrhenian, west of the region of active extension (Chiodini *et al.* 2004). CO_2 seeps are mostly low-temperature emissions that are manifested as vents (pressurized CO_2 release, some are referred to as CO_2-driven mud volcanoes: e.g. Bonini 2009*b*), diffuse soil degassing, springs and pools of bubbling water (Minissale 2004; Chiodini *et al.* 2008; Roberts *et al.* 2011). Geochemical studies find that the CO_2 has a number of sources, including shallow biogenic processes, carbonate hydrolysis, mechanical breakdown and thermometamorphism of carbonates, and mantle degassing (Italiano *et al.* 2008; Frezzotti *et al.* 2009). There are few studies of the origins of CO_2 trapped in the subsurface in Italy.

Previous work at CO_2 seeps in Italy has explored the factors affecting the human health risk they pose, and the geological and geomorphological controls on their distribution and characteristics, to inform risk assessment and monitoring design above storage sites (Roberts *et al.* 2011, 2014). This article extends this work to examine the subsurface geological attributes that control CO_2 leakage to surface. This is the first study to summarize the properties of CO_2 traps in Italy and learn from these analogues to minimize risk of leakage at engineered CO_2 stores.

CO_2 flow in geological formations

CO_2 fluids are retained in geological formations either as a free phase or dissolved in formation water. Depending on the subsurface conditions, free-phase CO_2 may be present in a dense or light form, where we define 'dense' as CO_2 with densities greater than the critical density ($\rho_c = 464$ kg m^{-3}) and 'light' as CO_2 with densities below the critical density. CO_2 phase behaviour is primarily controlled by subsurface temperature and pressure conditions, but is also affected by the presence of other fluids. Free-phase CO_2 will dissolve when it is in contact with undersaturated formation waters, which results in an increase in the water density (Spycher *et al.* 2003).

How free-phase or dissolved CO_2 flows through a rock formation is dependent on the properties of the fluid itself and of the rock. A rock volume will commonly exhibit a distribution of fluid pathway geometries due to heterogeneity intrinsic to geological units and the presence (and orientation) of fractures (Krevor *et al.* 2015). The permeability of the rock will vary according to the properties of the fluid flowing through it. This 'effective permeability' (K_E) is determined by the bulk permeability of the rock (K_{rock}) and the fraction of the total permeability accessible to each fluid phase (the relative permeability, K_r):

$$K_E = K_{rock} K_{r_{CO_2}} \quad (1)$$

In single-phase flow, all pores are saturated with a single fluid. For CO_2-saturated water flowing through a water-wet rock, K_E is equal to K_{rock}. However, for two-phase flow, such as free-phase CO_2 flowing through a water-wet rock, $K_{r_{CO_2}}$ is influenced by the saturation of formation water in the pores or fractures through which the CO_2 is flowing.

Rate of fluid flow per unit area, otherwise known as fluid flux, through a rock volume increases with effective permeability and fluid pressure gradients, and lower fluid viscosity. This is commonly approximated by capillary (or 'Darcy') flow:

$$\frac{Q}{A} = \frac{K_E}{\mu} \frac{\delta P}{\delta z} \quad (2)$$

where Q (flux) is the CO_2 flow rate (m^3 s^{-1}) over the seepage area, A (m^2), K_E is the effective permeability of the fluid (m^2), $\delta P/\delta z$ is the pressure gradient, (Pa m^{-1}), where P is pressure and z is depth (m), and μ is CO_2 viscosity (Pa s^{-1}).

Light-phase CO_2 is less viscous and more buoyant than dense phase, and experiments find that the effective permeability for light-phase CO_2 is higher than dense phase (Bachu & Bennion 2008). According to equations (1) and (2), light-phase CO_2 will flow more readily than its dense phase. Once flow is established in a rock, the relative permeability to CO_2 may increase due to drying effects, whereby formation fluids dissolve into the flowing CO_2 phase, decreasing the water saturation (Pruess 2008*b*).

Darcy's law does not characterize fluid flow in fractures, where permeability pathways are spatially focused rather than distributed throughout the rock volume. If the fracture spacing, orientation and aperture are known, then fracture flow

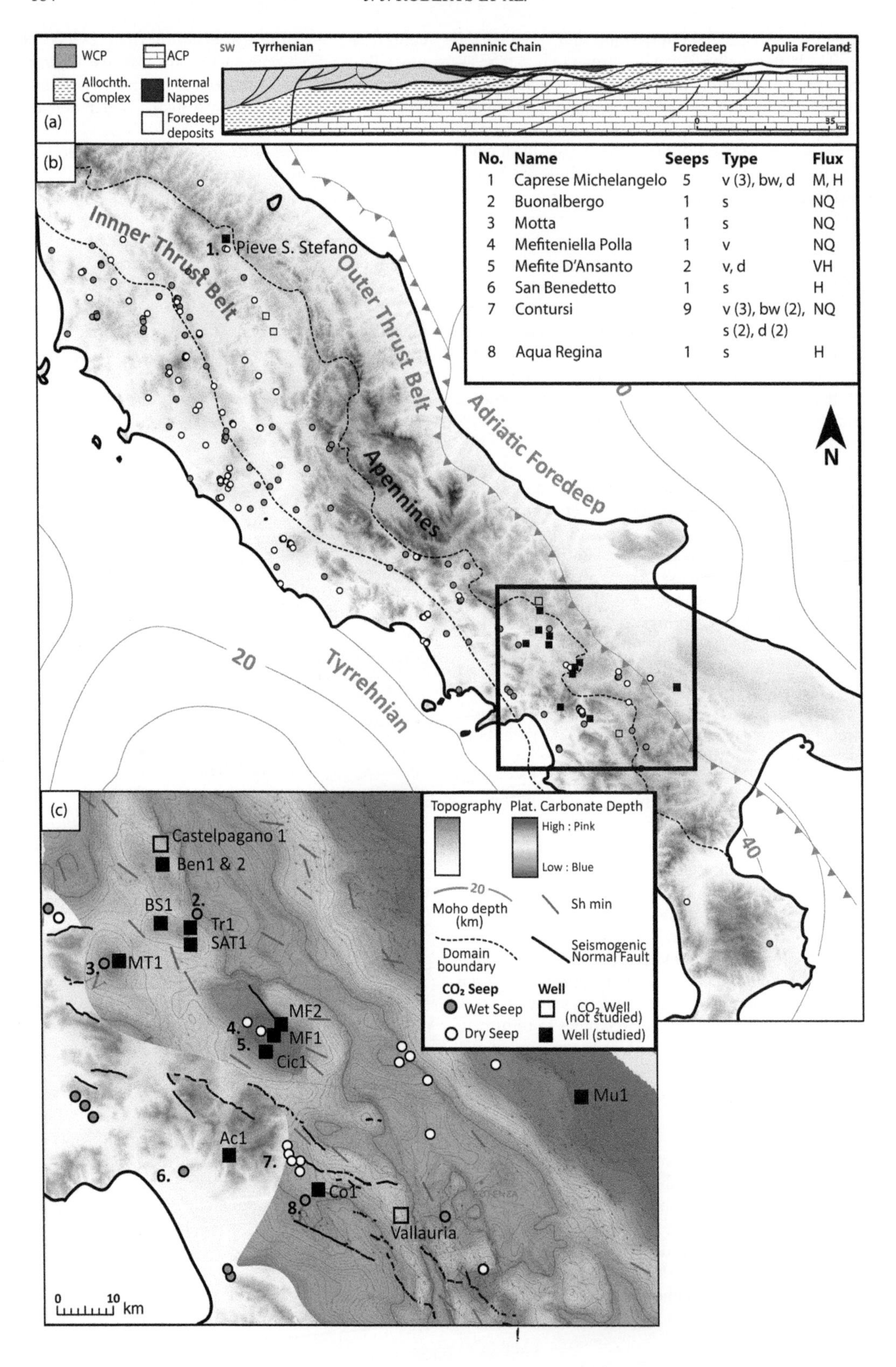

No.	Name	Seeps	Type	Flux
1	Caprese Michelangelo	5	v (3), bw, d	M, H
2	Buonalbergo	1	s	NQ
3	Motta	1	s	NQ
4	Mefiteniella Polla	1	v	NQ
5	Mefite D'Ansanto	2	v, d	VH
6	San Benedetto	1	s	H
7	Contursi	9	v (3), bw (2), s (2), d (2)	NQ
8	Aqua Regina	1	s	H

can be modelled discretely. However, where this information is not available, then fractured rocks can be approximated as porous media, where fracture permeability is upscaled to bulk permeability values that represent the flow properties of the rock volume (see Kuhlman *et al.* 2015 and references therein).

Mechanisms of CO_2 migration from its containing reservoir to surface

Some naturally occurring CO_2 reservoirs have been found to successfully retain CO_2 for millennia (Gilfillan *et al.* 2008) and in a range of geological environments (Lewicki *et al.* 2007). Other sites leak CO_2 to surface (Miocic *et al.* 2016), which can occur through a range of mechanisms. Free-phase CO_2 is less dense than surrounding porewaters and will rise buoyantly, becoming structurally trapped beneath a sealing unit if both a low-permeability rock and a containing structure are present. Dissolved and free-phase CO_2 can leak from the reservoir formation by diffusion, but this is an extremely slow process (Lu *et al.* 2009). However, high leak rates (e.g. tonnes CO_2 per day) through the overburden of natural reservoirs will most likely arise from a buoyant free phase of CO_2 which may leak by capillary transport through pores or microfractures in the overburden, or along unsealed faults and associated damage zones (Zweigel *et al.* 2004; Bachu 2008). Otherwise, in natural CO_2 systems, a low-permeability seal may be bypassed if free-phase CO_2 'spills' from a trapping structure. In this case, the overburden directly above the CO_2 reservoir has not been compromised, but the space for CO_2 in the reservoir-trap structure has simply been exceeded, although CO_2 storage reservoirs will be engineered such that the capacity will not be exceeded. Advective flow of CO_2-bearing waters could transport CO_2 from the reservoir more rapidly than diffusion. Such flow could occur through high-permeability pathways through the overburden, or by hydrodynamic flow within the aquifer.

The capillary entry pressure of free-phase CO_2 within a given cap rock determines the maximum column height of the free-phase CO_2 in the reservoir before it invades the seal by capillary transport. The capillary entry pressure is a function of the fluid properties (e.g. interfacial tension, contact angle) and rock properties (e.g. rock pore and pore throat geometry), or fracture aperture and roughness (Bachu & Bennion 2008; Naylor *et al.* 2011). An extremely high gas pressure will be necessary to overcome the capillary threshold pressure of a shale cap rock; however, following capillary breakthrough, CO_2 can then flow through the seal and overlying overburden at capillary pressures below the initial entry pressure, facilitating further leakage. Subsidiary fluid components can affect the fluid properties of the CO_2 phase, and so can enhance or retard capillary breakthrough.

Confining pressure, defined by overburden thickness and density, improves seal quality by increasing the mean stress, and therefore rock strength, until the yield stress of the rock is approached. Hence, at higher confining pressures, greater fluid pressures are required for the seal to mechanically fail (Osborne & Swarbrick 1997). Confining pressure also reduces rock permeability by closing pores. Additionally, as rocks are buried, diagenesis may occlude pore throats and fracture apertures by cementation (Nara *et al.* 2011). Elevated fluid pressures can lead to the hydraulic opening of fractures or shear on existing fractures due to a reduction in effective stress resulting in enhanced permeability (Gudmundsson *et al.* 2001; Yang & Manga 2009). The effect of pore fluid pressure (P_f) on rock strength can be represented by the pore fluid factor, λv, which is the ratio of pore fluid pressure (i.e. reservoir pressure, P_{res}) to lithostatic stress (Streit & Cox 2001) after Hubbert & Rubey (1959):

$$\lambda v = \frac{P_{res}}{\rho_{rock}\, g\, z} \quad (3)$$

where ρ_{rock} is rock density (typically assumed to be 2500 kg m^{-3}), g is acceleration due to gravity (m s^{-2}) and z is depth. For hydrostatic pressure, $\lambda v = 0.4$, and for lithostatic pressure, $\lambda v = 1$. The pore fluid factor indicates how close a coherent rock body is to failure, and will therefore be underestimated when applied to a fractured rock unit. Rock bulk permeability could therefore be increased by CO_2 buoyancy or fluid pressure, which in

Fig. 1. (**a**) Cross-section of Italy modified from (Improta *et al.* 2000). Most CO_2 reservoirs are hosted by the Apulian Carbonate Platform (ACP) in the Inner Thrust Belt, and the Allochthonous Complex forms the overburden. (**b**) Map of Italy modified from Patacca *et al.* (2008), detailing topography and the location of CO_2 seeps (dry and wet) and wells (including CO_2 wells not studied here), and geological terrane boundaries (Ghisetti & Vezzani 2002). (**c**) Map of the Southern Apennine region, showing the location of CO_2-bearing wells, stress field data (Barba *et al.* 2009; Sh min denotes minimum horizontal stress), mapped seismogenic normal faults (Roberts 2008) and deep platform carbonate structure (Nicolai & Gambini 2007). Table inset: detail of seeps shows in (c), including name, seep type (v, vent; bw, bubbling water; s, spring; d, diffuse) and flux information (M, medium; H, high; NQ, not quantified: see the text for definitions).

extreme cases can encourage rock failure (Collettini *et al.* 2008).

Should CO_2 leak from its primary reservoir, it may continue to migrate via available pathways through the overburden. During ascent, it is likely to encounter multiple reservoir and cap-rock units, and so CO_2 may accumulate in any overlying secondary reservoirs (Pruess 2008*a*) until these reservoirs, too, are breached or bypassed. Several mechanisms may attenuate the mass of migrating CO_2 during its ascent, including residual trapping or dissolution into unsaturated porewaters. Therefore the mass of CO_2 that reaches the near-surface it likely to be only a proportion of the mass that migrated from the deep rock formations, or none may reach the surface at all. Depending on the lithostatic pressure and geothermal gradient profile, CO_2 will typically become subcritical at depths shallower than 1 km. Subcritical CO_2 will pass through two hydrological zones: the phreatic (groundwater saturated) zone and the vadose (unsaturated) zone. In its light phase, CO_2 will be significantly more buoyant than groundwater in the phreatic zone. However, CO_2 at ambient temperatures will be denser than soil gas in the vadose zone, and so may disperse laterally in the shallow subsurface, perhaps above the water table (Annunziatellis *et al.* 2008; Kirk 2011). Depending on the soil properties, this could make the area of elevated soil CO_2 degassing substantially larger than the leak pathway from depth.

Observations at CO_2 and CH_4 seeps around the world find that gas release typically occurs over a discrete area (<0.01 km^2), although in some cases the region of CO_2 phenomena where seeps occur may be much larger (several km^2 or more) (Chiodini *et al.* 2004, 2010; Heinicke *et al.* 2006; McGinnis *et al.* 2011; Talukder 2012; Burnside *et al.* 2013; Elío *et al.* 2015; Nickschick *et al.* 2015).

Methods

CO_2 seep data was taken from Googas (Chiodini & Valenza 2008), a web-based catalogue of degassing sites in Italy and Sicily, which documents seep location and seep type, and, where available, rate of CO_2 degassing, gas composition and temperature. In the catalogue, the rate of CO_2 degassing is classified into low (<1 t CO_2/day), medium (1–10 t CO_2/day), high (10–100 t CO_2/day) and very high (>100 t CO_2/day).

Step 1: Selecting case studies

First, we identified CO_2 reservoirs, and established which are geographically close to CO_2 seeps, and therefore may be leaking.

Well logs and accompanying drilling notes for non-commercial boreholes in Italy are publicly available (www.videpi.com). By examining the VIDEPI dataset and in consultation with ENI and Independent Resources PLC, we selected non-commercial boreholes where test results document that the reservoir fluids are predominantly composed of CO_2. These boreholes, and boreholes nearby, were studied to constrain the subsurface structure and conditions.

A geographical information system (GIS) was populated with seep and well bore data. We included data on: 'active' faults, as defined by seismogenic fault scarps mapped (Roberts 2008); seismically capable faults (ISPRA 2007); the present-day stress field (Barba *et al.* 2009); elevation (SRTM 90 m: Jarvis *et al.* 2008); subsurface carbonate structure (Nicolai & Gambini 2007); and isotherms at 1 and 2 km depth (Geothopica 2010). The distance from the well bore to the nearest CO_2 seeps was calculated from this GIS.

Step 2: Case study geology

To explore the geological conditions that affect whether a CO_2 reservoir is sealed or is leaking, we determined the geology for each reservoir, the overburden thickness and properties, regional structure, and subsurface conditions, using the publicly available well logs and other published data.

Many of the selected boreholes were drilled in the 1960s and 1970s, and therefore lack downhole information available from more modern boreholes, including many well tests. Downhole rock formations and their properties were determined from the well log (from well cuttings and core) and accompanying lithological descriptions, and any data from formation tests. We define the CO_2 reservoir as the shallowest rock formation that has high CO_2 gas saturation, and 'seal thickness' was defined by the maximum and minimum thickness of impermeable rock units overlying the CO_2 reservoir. Deviated drilling was corrected to the true vertical depth (TVD) and assumed a standard depth error of approximately ±10 m and normally distributed to 1 SD.

To reconstruct downhole conditions, a number of assumptions had to be made. Formation pressure information was estimated from formation tests, where available, or from the density of drilling fluids (mud weight). Pore fluid pressures from mud weights usually exceed the actual formation pressure. To account for this, we adjust the formation pressures by 10%, following the methodology of Wilkinson *et al.* (2013). Formation pressure measurements were not sufficiently regularly spaced, nor were mud weight data sufficiently detailed, to distinguish the CO_2 and water legs from the pressure profiles. However, where formation tests documented a transition from CO_2 to the water leg, minimum CO_2 column heights were estimated.

A number of boreholes provided the information to calculate corrected geothermal gradient from downhole temperature measurements. Unusually low downhole temperatures are the expected result of circulation of cold drilling mud within the borehole. In these cases, and cases where downhole temperatures are unavailable, the geothermal gradient was interpolated from the geospatial dataset. Loss of drilling fluid circulation or significant mud absorption, which are sometimes recorded on the borehole logs, can be useful indicators of geological horizons with enhanced permeability, or where the rock fracture gradient has been exceeded. Overpressure was defined as measured pressure exceeding calculated hydrostatic pressure by 3 MPa, which allows for uncertainty in both measured depth and the mud weights.

Step 3: Modelling CO_2 properties

Downhole pressure (P) and temperature (T) conditions constructed from the well logs were used to model CO_2 properties, including density, viscosity, buoyancy and solubility at depth. The sensitivity of these fluid properties to the calculated $P-T$ conditions were tested at 10 m intervals, for which we assumed a standard error of ± 0.1 MPa and ± 5°C, to 2 SD. Surface temperature and pressure of 15°C and 1 atm, respectively, were assumed.

CO_2 density and viscosity was modelled using the Huang *et al.* (1985) and Span & Wagner (1996) equation of state. CO_2 solubility was calculated using the equation of Spycher *et al.* (2003) using values for freshwater rather than brines because formation tests in most of the wells show low salinities in the reservoir units. The viscosity of freshwater was calculated using the polynomial equation for variable temperature and pressure from Likhachev (2003). We neglect the effect of dissolved CO_2 on the viscosity of water, which is small at these temperatures (Islam & Carlson 2012), and also the effect of subsidiary gases such as CH_4 and H_2S which can affect CO_2 behaviour (Savary *et al.* 2012).

For column heights that could be estimated from the well log, the buoyancy pressure (B) of the CO_2 at the crest of the reservoir structure intersected by the well was calculated as:

$$B = (\rho_{H_2O} - \rho_{CO_2})\, g\, h_{CO_2} \tag{4}$$

where ρ_{H_2O} is the density of water (kg m^{-3}), ρ_{CO_2} is the density of the CO_2 (kg m^{-3}), h_{CO_2} is the CO_2 column height (m) and g is gravitational acceleration (m s^{-2}).

Step 4: Classifying the reservoirs

Each CO_2-bearing borehole was classified according to whether the corresponding reservoir is interpreted to be leaking CO_2 to surface or not, depending on its proximity to CO_2 seeps and the nature of the seep itself. Whether the distance between a borehole and a seep is considered to be 'near' or 'far' was determined by spatial analysis of the distance distribution between the boreholes, the CO_2 reservoir structure, CO_2 seeps and faults. If there are no CO_2 seeps located at the surface close to the CO_2-bearing borehole, the CO_2 reservoir is determined to be sealing. If there are high-flux or dry CO_2 seeps located at the surface near to the CO_2-bearing borehole, then the reservoir is inferred to be leaking. It is not uncommon for springs emerging from carbonate rocks to contain small quantities of CO_2 from the dissolution of carbonates, which is not related to CO_2 leakage from depth, and so if the seeps are CO_2 springs with small CO_2 content, or are located relatively far away from the borehole, then the leakage is classified as inconclusive.

Step 5: CO_2 leakage pathways

To evaluate the geological conditions that could enable the observed fluid leak rates from case studies inferred to be leaking CO_2 to surface, we examined the Darcy flow equation for CO_2 fluids at reservoir conditions. Reservoir fluid leakage into the overburden could occur by distributed migration through the overburden over a broad area (small K_E, large A) or focused migration via fault-related fracturing in the overburden (large K_E, small A). So, we calculate the combinations of overburden effective permeability (K_E) and area (A) necessary to sustain leakage from the reservoir at the observed rate of surface seepage.

Conservative mass transport is assumed: that is, no CO_2 attenuation/loss during ascent to surface. For leakage of free-phase CO_2, the minimum leak rate from the reservoir (m^3 s^{-1}) to deliver CO_2 to the surface at the measured rates (often reported in t/day) was calculated from CO_2 densities (ρ_{CO_2}) modelled at $P-T$ conditions in the reservoir. The minimum leak rate of CO_2-saturated formation water needed to supply CO_2 to the near-surface at the measured CO_2 degassing rate was also calculated from the change in solubility of CO_2 in freshwater at reservoir conditions to surface conditions. These calculations assume water emergence temperatures of 15°C, and thus CO_2 solubility in freshwater is approximately 0.042 molal at 10 m depth.

The results could then be informed by any permeability measurements from rocks that comprise the overburden, and area of CO_2 seepage at the surface. This area can be considered on two scales: the area of a single seep or the total area of a seep cluster (if relevant). It is important to note that the area of leakage in the subsurface could be much larger or smaller. The area of the CO_2 'cap' in the reservoir, and of high-permeability pathways offered by fault-related deformation, was also estimated to inform our results.

Results

Figure 1 shows the location of the 13 studied CO_2-bearing boreholes, neighbouring dry wells and all documented CO_2 seeps. Four additional boreholes in Figure 1 are known to contain CO_2 (Castelpagano 001, Vallauria, San Donato and Perugia 2), but their well logs are not publicly available and so are not studied here. Where two or more wells intercept what may be the same CO_2 reservoir, we refer to a CO_2 field. We classify the case studies into those with overburdens that successfully seal CO_2 (BS1, SAT1, Ben1/2 and Mu1), that leak CO_2 to surface (MF1 and PPS1), or are inconclusive – that is, it is not clear whether or not CO_2 is securely retained in the subsurface (Tr1 and MT1).

This section presents an overview of the broad structure and rock formations in the boreholes, and their relationships before detailing each case study (including observations from the boreholes, subsurface structure, any nearby geological structures and seeps, and whether the case study is interpreted to be leaking or sealed). Figure 2 shows the lines of cross-sections that describe the subsurface structure in Figures 3–6, and Table 2 summarizes the reservoir and overburden geology, pressure conditions, and proximity to the nearest surface CO_2 seeps for each CO_2-bearing borehole studied.

All but one of the CO_2-bearing boreholes are located in in the Central Apennines, and record CO_2 in anticline or horst structures in the Apulian Carbonate Platform units (Fig. 1c). Many of the well logs note that the Apulian Carbonate Platform units were associated with significant mud losses or loss of circulation, which can indicate a lower than anticipated pressure and/or increased rock permeability (e.g. the presence of a fracture system). A series of thrust-sheet deposits cap the CO_2 reservoirs. These nappes can be Middle Triassic–Miocene basinal flysch units of the Allochthonous Complex (including the basinal carbonates of the Lagonegro Formation or pelagic deposits of the Sannio Formation), or Miocene– Pleistocene-age sediments (turbidites, muds, sandstones and conglomerates). The thrusted contact between the reservoir and overburden is marked by a tectonic breccia or by a Messinian evaporite unit in some boreholes (indicated on the well logs in Figs 3–6). The thrust

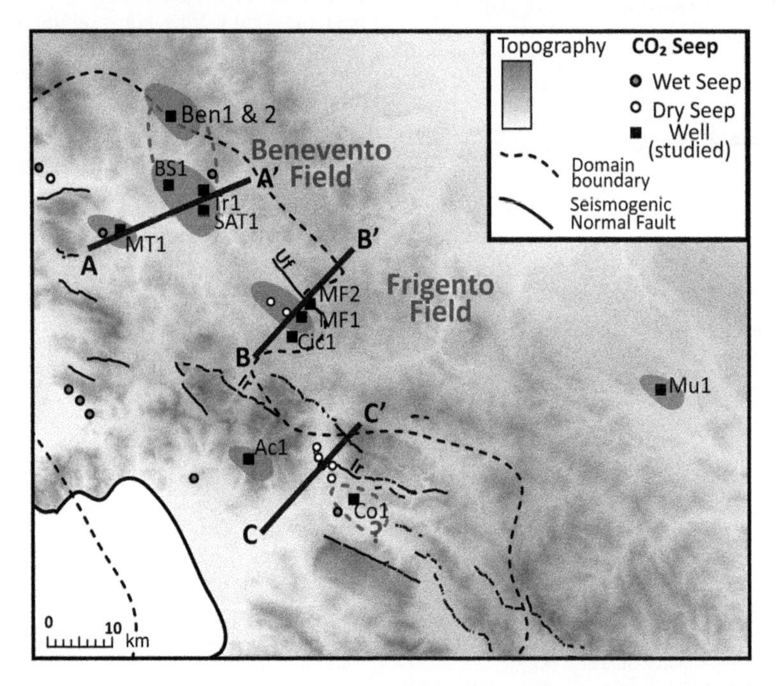

Fig. 2. Map of the Southern Apennine region, showing the location of studied CO_2-bearing wells, CO_2 seeps (shaded according to whether dry (vent, diffuse seeps) or wet (springs, bubbling water), mapped seismogenic normal faults (Roberts 2008) and lines of cross-sections in Figures 3–6.

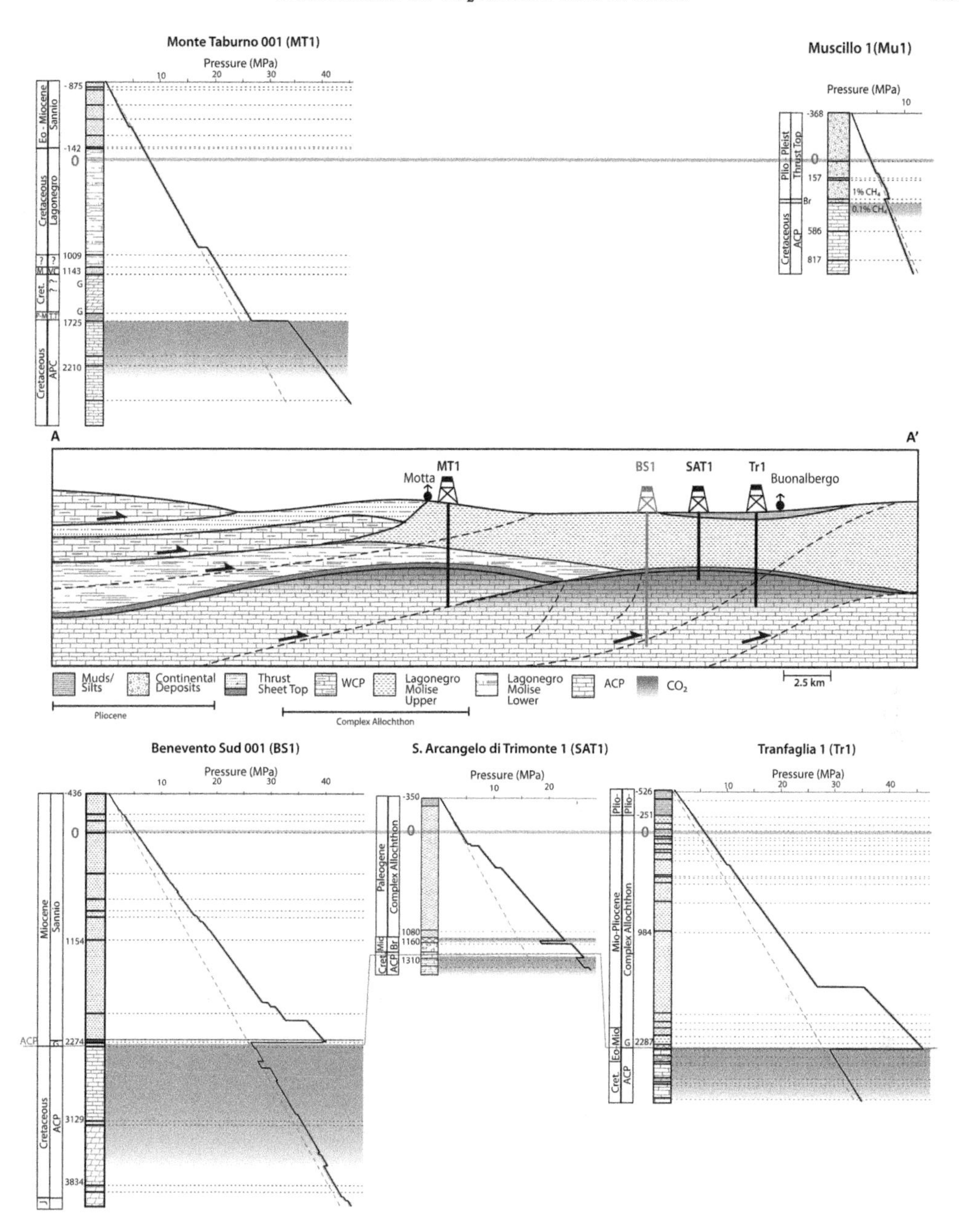

Fig. 3. Benevento and Monte Taburno structures (cross-section A–A′ in Fig. 2) and pressure–depth profiles for the Tranfaglia (Tr1), S. Arcangelo di Trimonte (SAT1), Benevento Sud (BS1), Monte Taburno (MT1) and Muscillo wells. CO_2-bearing formations are shaded in the depth profile of the well logs. It is unclear if the Motta and Buonalbergo CO_2 springs are related to the subsurface CO_2 fields, although they do exhibit significant deep CO_2 contributions. BS1, SAT1 and Tr1 show significant overpressure in the overburden. Data are from borehole logs, and from Improta *et al.* (2003*b*), Di Bucci *et al.* (2006), Nicolai & Gambini (2007) and Chiodini *et al.* (2010).

pile is now dissected by extensional structures relating to back-arc extension, including NW–SE- and east–west-trending high-angle faults, which tend to control seismogenesis in central Italy (Patacca *et al.* 2008). In Pieve Santo Stefano 1 (PSS1), CO_2 is hosted in the Burano Formation, a thick sequence

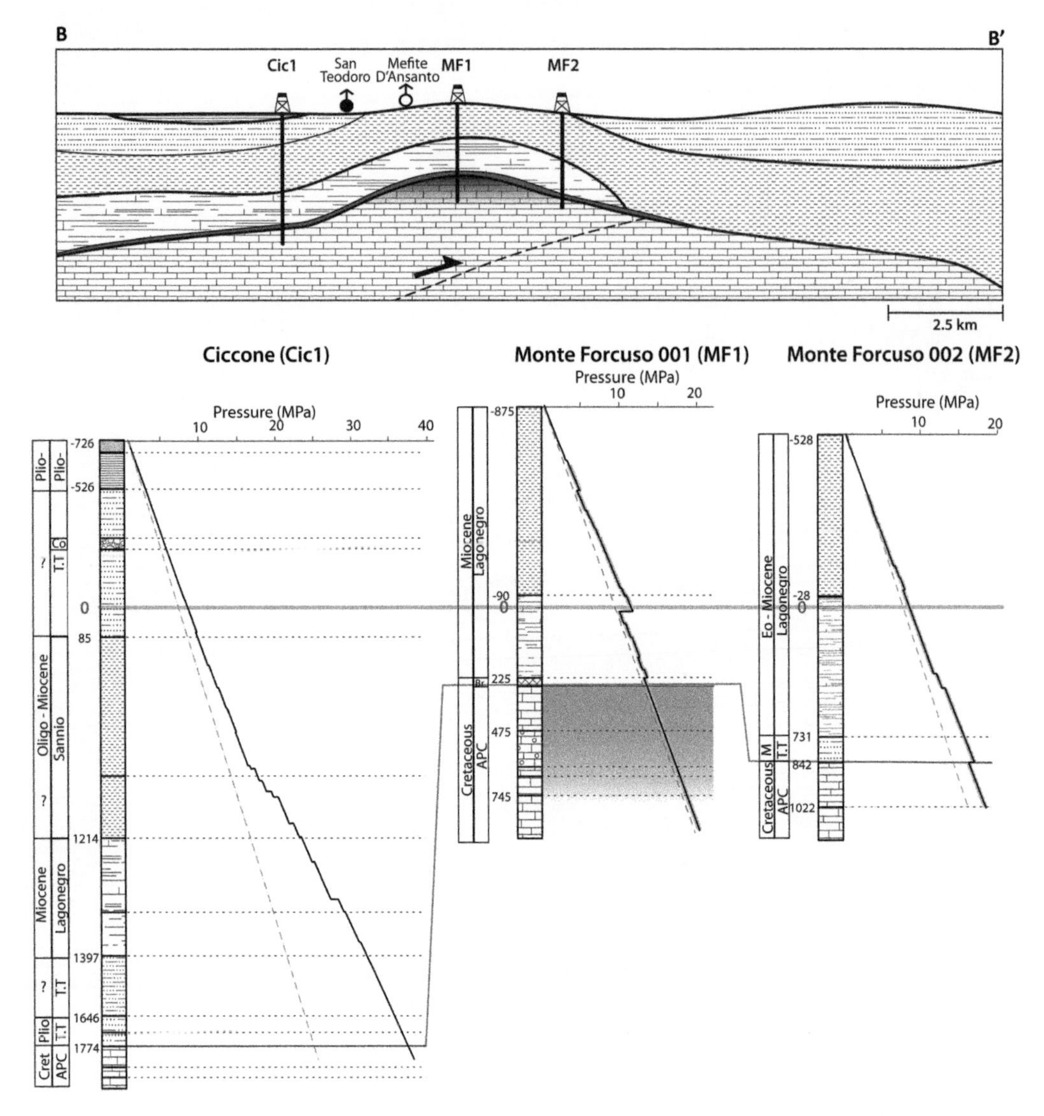

Fig. 4. Subsurface structure (cross-section B–B′ in Fig. 2) and boreholes that penetrate the Frigento Formation (Cic1, Ciccone; MF1, Monte Forcuso 1; MF2, Monte Forcuso 2). MF1 drills a CO_2 accumulation in the ACP at hydrostatic pressure (see the shaded horizon on the depth profile), but MF2 and Cic1 drill into the water leg. Mefite D'Ansanto (and Mefitiniella polla not shown here) are high-flux CO_2 vents. San Teodoro, located on the SW flank of the antiform, is a sulphurous thermal spring that does not degas CO_2. Data are from borehole logs, and from Improta *et al.* (2003*b*) and Di Bucci *et al.* (2006).

of Upper Triassic evaporitic carbonates which the Apulian Carbonate Platform overlies.

Benevento CO_2 field

Three boreholes, Benevento Sud 1 (BS1), San Arcangelo di Trimonte (SAT1) and Tranfaglia (Tr1), penetrate the large Benevento CO_2 field in the Campania region of Italy, as shown in Figure 3.

In BS1, CO_2 is continuously recorded in Apulian Carbonate Platform units from approximately 2707 to 4139 m (1432 m gross CO_2 column), and measurements show up to 98.5% CO_2 by volume, with small quantities of CH_4 (maximum 5.1%). Directly overlying the reservoir are 9 m of Cretaceous anhydrite and gypsum, overlain by 25 m of Messinian mudstone, and then 1 km of Miocene muds and marls. In SAT1, well tests record

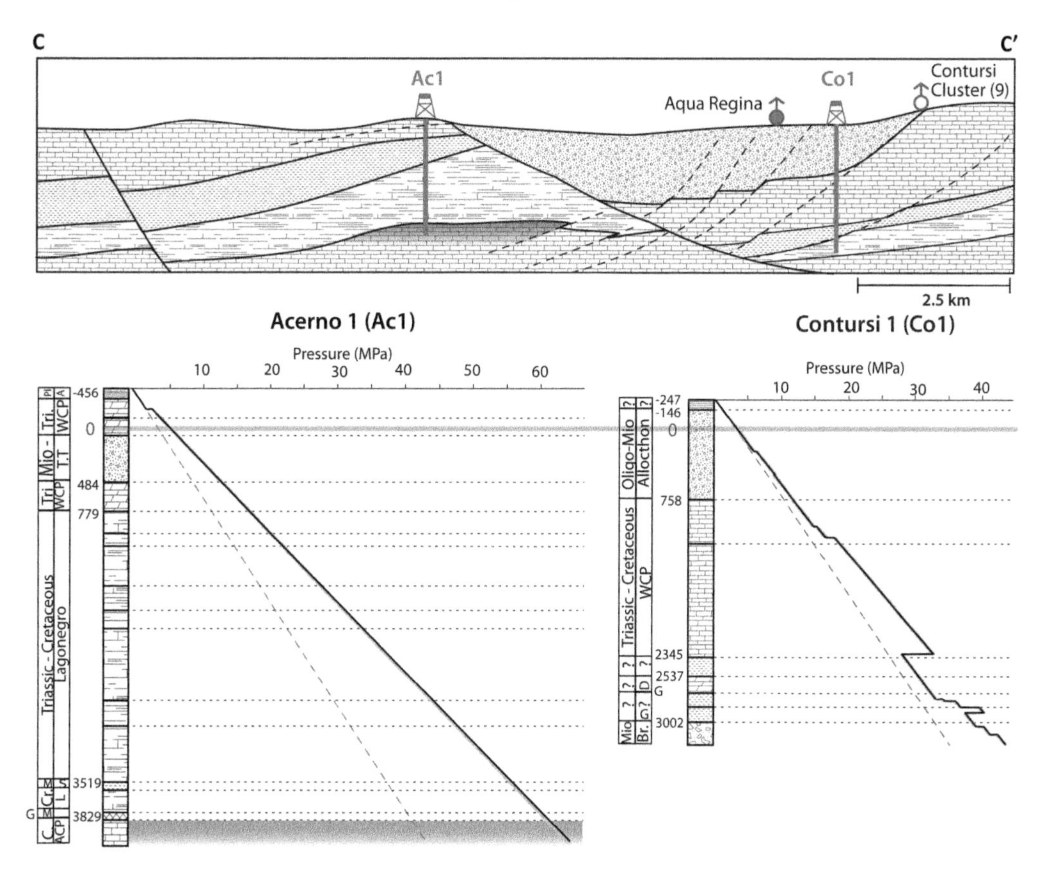

Fig. 5. Subsurface structure of the Acerno and Contursi reservoirs (cross-section C–C′ in Fig. 2) showing the location of the Acerno (Ac1) and Contursi (Co1) wells. Acerno (Ac1) contains CO_2 at 4263 m below surface (see the grey shaded horizon on the well log depth profile). The Contursi borehole (Co1) did not intercept any CO_2 accumulation at depth and so is not considered in our analysis. There are no known seeps above the Acerno horst structure. Data are from borehole logs, and from Improta *et al.* (2003*b*) and Scrocca *et al.* (2005).

CO_2 in the Apulian Carbonate Platform unit (at *c.* 1660 m depth) and also in a shallower, approximately 50 m-thick carbonate breccia, separated from the platform carbonate by approximately 80 m of muddy limestone breccia. It is not clear whether this represents continuous CO_2 from approximately 1520 m depth, or two distinct CO_2 shows.

The TR1 borehole intercepts a reservoir in the Apulian Carbonate Platform units at 2773 m depth containing 98% CO_2. Directly above lies 17 m of massive anhydrite, but there are shows of 2–17% CO_2 above the anhydrite for about a further 200 m and wet gas shows in the overburden all the way to surface. Without detailed geochemical data we cannot determine if the CO_2 documented above the reservoir in Tr1 represents CO_2 migration from the Apulian Carbonate Platform units through the anhydrite overburden, or *in situ* generation associated with the wet gas.

There are no formations in the TR1 borehole that qualify as a good seal; most of the overburden is comprised of siltstone–calcareous units of the Allochthonous Complex. This is in contrast to the SAT1 and BS1 boreholes, where the Allochthonous Complex overlying the CO_2 reservoir is comprised primarily of mudstones. The overburden is overpressured, exceeding 10 MPa in all three boreholes, however, which suggests that the units are low permeability.

The wells that intersect the Benevento field record different fluid pressure in the reservoir; BS1 and Tr1 show hydrostatic reservoir pressure, whereas in SAT1 the CO_2 is overpressured. As a result, the modelled CO_2 at reservoir conditions are different, finding that CO_2 is retained in the dense (BS1) and the light (Tr1) phase, and or close to the phase transition (SAT1). These differences in pressure and CO_2 properties may indicate compartmentalization of the reservoir.

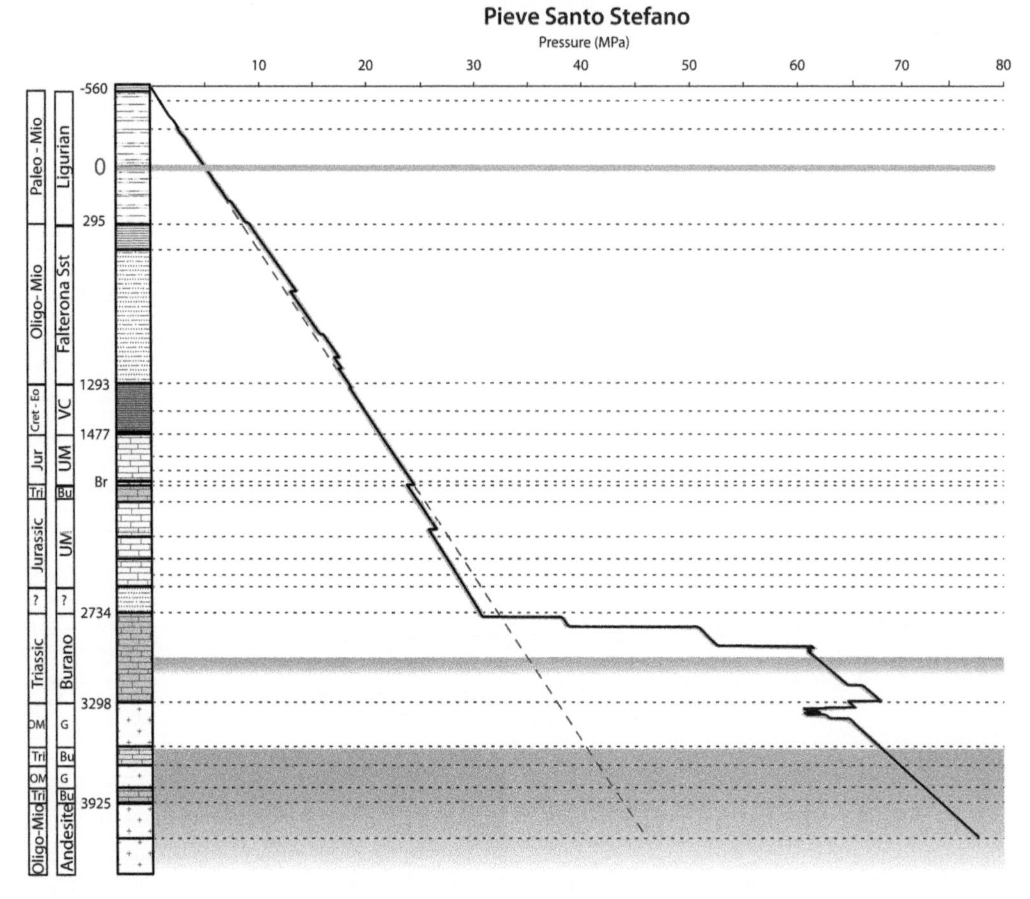

Fig. 6. Depth and pressure profile of the Pieve Santo Stefano borehole. This borehole is located in Tuscany, in the Northern Apennines (see Fig. 1). The multilayered CO_2 reservoir (shaded in grey in the well log) is hosted in thin dolomite layers (sandwiched between anhydrite layers) of the Burano Formation. The CO_2 is significantly above hydrostatic pressure conditions. The Caprese Michelangeo and Fungaia CO_2 vents are located near to the Pieve Santo Stefano well and are considered to represent surface seepage of the reservoir. Data are from borehole logs, and from Heinicke *et al.* (2006) and Bonini (2009*a*).

The Buonalbergo CO_2 seep is located 3.5 km from TR1, 5.3 km from SAT1 and 10.3 km from BS1. No further information, such as quantities of degassed CO_2, is available about this seep other that it is a CO_2 spring (Chiodini & Valenza 2008). Its visual appearance is unremarkable, suggesting that gas flux is not particularly high. The CO_2 dissolved in the spring could source from the Benevento CO_2 field or CO_2 in shallower formations (like those in Tr1 which show small quantities of CO_2 and hydrocarbons that could break down to CO_2). Otherwise it could source from carbonate rock dissolution (karstification), which is common where carbonate rocks form the shallow subsurface. Given the absence of a convincing seal in the Tr1 overburden and the presence of the Buonalbergo spring nearby, it is inconclusive whether the CO_2 documented in the Tr1 borehole is leaking to surface. The CO_2 reservoirs intercepted by BS1 and SAT1 are considered sealed.

To the north of these wells, the Benevento boreholes (Ben1 and Ben2) also encountered CO_2 at approximately 3 km depth. These boreholes drilled a broad structural high (see Fig. 1c), and were also found to contain some short-chain hydrocarbons (maximum of 6.2% CH_4 by volume) at approximately 3300 m (Ben2). These wells show significant overpressure in the overburden, which is comprised of several muddy units, but the reservoirs are hydrostatically pressured. There are no CO_2 seeps in the vicinity of these boreholes, so the reservoir is considered to be sealing.

Table 2. *A summary of the reservoir and overburden characteristics for each CO_2-bearing borehole, including the properties of the CO_2, the presence of nearby seeps or faults and whether the reservoir is interpreted to be sealed or leaking CO_2 to surface*

Field/ Reservoir	Borehole name; abbreviation	Reservoir conditions				CO_2 properties		Overburden conditions			Distance from borehole to:		Interpretation
		Depth below surface (m)	CO_2 (%v.v)	Over-pressure	Porefluid factor	Density (kg m^{-3}) (phase)	Buoyancy (MPa)	Evaporite (thickness)	Seal thickness (m; min – max)	Overpressure (ratio compared to hydrostatic)	Seep name (no. in Fig. 1); distance (km)	Fault distance (km) (name, sense)	
Benevento Sud	Benevento Sud 1; BS1	2710	98.5	N	0.4	624 ± 15 (dense – SC)	5	Y (9 m)	315–2710	Y (1.56)	None	5 (S. Matese, N); 15.7 (Telese Fault, N)	Sealing
	S. Arcangelo di Trimonte; SAT1	1520	–	Y	0.6	503 ± 11 (dense – SC)	–	N	1520–1520	Y (1.62)	None	4.8 (S. Matese, N); 23.2 (Telese Fault, N)	Sealing
	Tranfaglia; Tr1	2773	98	N	0.4	279 ± 6 (light – SC)	–	Y (17 m)	None	Y (1.67)	Buonalbergo (1); 3.5	7 (S. Matese, N); 20 (Ufita Fault, N)	Inconclusive
	Benevento 002; Ben2	3300	94	N	0.6	797 ± 9 (dense – SC)	1.8	N	30–600	Y (1.68)	None	12.7 (Boiano, N); 22 (Telese Fault, N)	Sealing
Monte Taburno	Monte Taburno 1; MT1	2093	>90	Y	0.5	569 ± 10 (dense – SC)	1.1	Y (2 m)	543–890	N (1.08)	Motta (2); 1.6	8 (Montesarchio, N)	Inconclusive
Muscillo	Muscillo; Mu1	694	97 (low sat)	N	0.4	139 ± 19 (light – gas)	2.2	N	305–305	N (1.09)	None	>40 km	Sealing
Frigento	Monte Forcuso 1; MF1	1128	99.7	N	0.4	200 ± 5 (light – SC)	3.5	N	168–1128	N (1.21)	Mefiteniella polla (4); 5.4 Mefite D'Ansanto (5); 1.8	4.3 (Ufita, N)	Leaking
Acerno	Acerno 1; Ac1	4263	97	Y	0.6	919 ± 8 (dense – SC)	–	Y (80 m)	70–228	Y (1.48)	San Benedetto (6); 11 Contursi cluster (7); 15.5	2.6 (Sabato Valley; N); 11.5 (Volturata, N)	Sealing
Caprese	Pieve Santo Stefano 1; PSS1	3600	92.2	Y	0.7	830 ± 8 (dense – SC)	1.1	Y (70 m)	150–550	N (1.08)	C. Michelangelo (9); 2.5 Fungaia (10); 3.6	7.9 (Upper Tiber Valley, N)	Leaking

If reservoir or overburden formation fluid pressures are 3 MPa above hydrostatic, then it is considered to be overpressured. For CO_2 density, 'dense' refers to CO_2 with densities greater than the critical density ($\rho_c = 464$ kg m^{-1}) and 'light' refers to CO_2 with densities below the critical density, and SC refers to the supercritical phase. Where column heights are not known, the buoyancy pressure, B, cannot be calculated. Further detail about the seeps (number, type, flux, temperature) are tabulated in Figure 1b. Modelled conditions in SAT1 are unreliable (see the main text for details).

There are two recently active SW-dipping normal fault systems near to these boreholes (BS1, SAT1, TR1 and Ben1/2). The Southern Matese Fault (less than 10 km from the boreholes) is a NW–SE-trending complex of faults that has been historically seismogenic (Di Bucci *et al.* 2006; ISPRA 2007). The Telese fault scarp, a north-dipping topographical break in slope, is between 16 and 22 km away from the boreholes (Roberts 2008). Previous earthquake sequences to the east show that extension is largely NW–SE (Milano *et al.* 2006).

Monte Taburno CO_2 reservoir

To the east of the Benevento CO_2 field, the Monte Taburno 1 (MT1) borehole cuts a separate structure containing >90% CO_2 at 2 km depth (Fig. 4). CO_2 is mostly found in the Apulian Carbonate Platform but also in the overlying thin layer of muddy Mio-Pliocene thrust-top deposits. This is capped by 511 m of dolomitic limestone, the bottom-most 2 m of which (i.e. capping the reservoir) is anhydrite-bearing, and then by over 1 km of low-permeability muds and sands of the Lagonegro and Sannio formations of the Allochthonous Complex. Overburden pressures are hydrostatic, but the CO_2 reservoir shows approximately 9 MPa over-pressure and under these conditions the CO_2 will be contained in its supercritical state. MT1 is 10 km from the Telese Fault, and 7 km from the SW-dipping Montesarchio and Ioannis seismogenic normal faults (Roberts 2008). The Motta thermal spring is 1.6 km away and has a small CO_2 emission (Chiodini & Valenza 2008), but no further information is available. Similar to the Buonalbergo spring near the Benevento field, there are several possible sources of CO_2 in a small spring that do not necessitate a subsurface CO_2 reservoir. Given the proximity of MT1 to the Motta thermal CO_2 spring, it is inconclusive whether the Monte Taburno reservoir is leaking to surface.

Muscillo CO_2–CH_4 reservoir

The Muscillo 1 (Mu1) borehole, located in the Basilicata region, penetrated a shallow accumulation of CO_2–CH_4 in the Apulian Platform Carbonate (Fig. 3). A CH_4 leg overlies a gas-phase CO_2 leg at 694 m below the surface, and both are low saturation. The reservoir top is marked by a thin breccia. The overburden is comprised of claystone and siltstone terrigenous deposits. In the outer thrust domain where the borehole is located, these sediments tend to represent rapid filling of structural depressions related to the development of extensional tectonics. Down-well pressures are hydrostatic and there are no CO_2 seeps in the vicinity of this well. No other subsurface information is available and so its structure is poorly constrained. The nearest faults are over 40 km from the well. We classify the Muscillo reservoir as sealing, although we note that low CO_2 saturation could indicate residual trapping of leaked CO_2.

Frigento CO_2 reservoir

Three boreholes penetrate the Frigento Antiform, located in a region of Campania: Monte Forcuso 001 (MF1), Monte Forcuso 002 (MF2) and Ciccone (Cic1) (Fig. 4). This structure in the Apulian Carbonate Platform correlates with a gravity and thermal anomaly (Improta *et al.* 2003*a*) (Fig. 1c). The geothermal gradient here reaches over 90°C km^{-1} at the crest of the anticline (Chiodini *et al.* 2010). The MF1 borehole is the only one to encounter CO_2. It intercepts an approximately 472 m gross CO_2 column in the Apulian Carbonate Platform at just over 1 km depth, above a freshwater leg. The absence of CO_2 in neighbouring boreholes constrains the extent of the CO_2 cap (<2 km radius). The overburden is mostly comprised of muds and marls of the Allochthonous Complex's Lagonegro Formation, and also brecciated and cemented sandstone. The overburden and the reservoir are at, or close to, hydrostatic pressure. The MF2 and Cic1 boreholes, located on the flanks of the anticline, do not contain free-phase CO_2, only freshwater and saline water, respectively. The differences in formation salinity and pressure in these boreholes may indicate compartmentalization of the reservoir.

The regional stress field (Barba *et al.* 2009) shows NW–SE extensional faulting which is currently active; the 1980 Irpinia earthquake (M 6.9) nucleated on the NE-dipping Irpinia Fault 32–35 km to the south of the reservoir. The SW-dipping Ufito normal fault scarp is located less than 1 km to the NE of the MF2 borehole, and 4.3 km from MF1 (Fig. 2) (Improta *et al.* 2003*b*; Roberts 2008). This is thought to be a splay from the Irpinia Fault (Brozzetti 2011), and thus the Frigento Antiform is located in the hanging wall of both faults.

Mefite D'Ansanto and Mefitiniella Polla CO_2 vents (seep nos 4 and 5 in Fig. 1c) are located above the structural high point of the NW–SE-trending Frigento Antiform (see Fig. 4). Mefite D'Ansanto emits more CO_2 than any other seep in Italy, releasing approximately 2000 t CO_2/day by venting and diffuse degassing over an area of 4000 m^2 (Chiodini *et al.* 2010). Mefitiniella Polla is a smaller CO_2 vent located 3.6 km NW from Mefite D'Ansanto and, although the seep rate has not been measured, field observations find that it vents CO_2 vigorously. On the flanks of the Frigento Antiform, less than 3 km ENE from Mefite, are the San Teodoro thermal springs. These springs do not release CO_2, but their emergence temperatures (*c.* 15–27°C, similar to the Mefite seeps) and geochemistry (Minissale 2004) indicate rapid fluid ascent of waters that have circulated in deep carbonate rocks, probably via fault-related flow paths (Duchi *et al.* 1995). Travertine deposits have been mapped within 5 km of the seeps (Roberts 2013) but these are no longer active, and there have been no geochemical investigations regarding their age or source.

Although there is currently insufficient geochemical information to irrefutably link the subsurface reservoir with the Mefite CO_2 seeps, or nearby thermal springs, it is reasonable to consider that the CO_2 released at the Mefite seeps could originate from the CO_2 reservoir located in the underlying anticline (Chiodini *et al.* 2010; Pischiutta *et al.* 2013). We therefore classify the Frigento CO_2 reservoir as leaking.

Acerno CO_2 reservoir

The Acerno 1 borehole (Ac1) penetrates the deepest studied CO_2 reservoir, located in a horst structure in the Apulian Carbonate Platform, 4363 m beneath Mount Picentini (Fig. 5), Campania region. The reservoir is overlain by a 305 m-thick evaporite and mud seal, then interlayered nappes of the Allochthonous Complex, basement carbonates and muds of the thrust-top deposits. Both the overburden and the reservoir are overpressured, with a pore fluid factor of 0.6 in the reservoir. Multiple mud losses were experienced when the well penetrated the Apulian Carbonate Platform, which suggests that the mud densities were too high for the reservoir properties (e.g. pressure or presence of pervasive fracture system); however, the mud densities were not adjusted. The borehole was plugged after drilling 300 m into the Apulian Carbonate Platform. The single drill stem test in this borehole yielded over 90% CO_2, which we model to be in the dense phase at reservoir conditions. The borehole is in the footwall of the east-dipping Sabato normal fault, which is considered to be seismogenic (ITHACA), and 11.5 km from the Volturara fault scarp (Roberts 2008). There are no CO_2 seeps located above the Acerno structure, but 11 km to the ENE is the San Benedetto CO_2 spring, which releases 10–100 t CO_2/day (Chiodini & Valenza 2008), and 15 km to the NW is the Contursi seep cluster (Fig. 1c, No. 6). The Acerno reservoir is classified as inconclusive.

Caprese CO_2 reservoir

The Pieve Santo Stefano 1 (PSS1) borehole, located in Tuscany, commercially exploits a multilayered CO_2 reservoir at approximately 3.6 km depth in the Caprese Antiform (Bicocchi *et al.* 2013) (Fig. 6). The main CO_2 reservoir is hosted within dolomites and evaporites of the Triassic Burano Group (Bonini 2009*b*) where thin reservoirs of fractured dolostone (porosity 2–6%) with high pore fluid pressures are sandwiched between sealing anhydrite layers (Trippetta *et al.* 2013). The CO_2 cap in the Caprese Antiform is likely to be elliptical in shape, with a maximum radius as great as 5 km (Bicocchi *et al.* 2013).

The reservoir brines are highly saline due to the interaction of meteoric waters with the evaporites (Bicocchi *et al.* 2013). Logging notes record significant mud losses while drilling through the overburden, which is multilayered and approximately hydrostatically pressured. Beside the anhydrites of the Burano Group, there are few low-permeability units in the cap rock that would offer a convincing very-low-permeability seal, although, in general, the Ligurian units that comprise the overburden are considered to be low permeability (Bicocchi *et al.* 2013).

The region around the Caprese Antiform is associated with CO_2 reservoirs and seeps. For example, approximately 40 km to the SE of PSS1, the San Donato and Perugia 2 boreholes penetrate the Monte Malbe structure (an anticline bounded by two active normal faults) and find pressurized CO_2 fluids in the Burano Group (Trippetta *et al.* 2013). Indeed, NE-trending, steep-dipping faults in the region form part of a regional transverse lineament known as the Arbia–Val Marecchia Line (AVML) which has been associated with CO_2 seepage (Bicocchi *et al.* 2013). More locally, the seismogenic Alto-Tiberina Fault is approximately 8 km SE of the PSS1 borehole and bounds the west side of the Quaternary Upper Tiber Basin (Collettini & Barchi 2004; Heinicke *et al.* 2006; ISPRA 2007).

The Caprese Michelangelo seeps and the Fungaia seeps are within 4 km of the PSS1 well. Caprese Michelangelo is a cluster of at least four seeps in an area of 400 m^2. The style of seeping is varied; there are CO_2 vents, bubbling water and diffuse degassing (seep No. 1 in Fig. 1c). The gas emission rate of two seeps has been measured: one seep in the Caprese cluster classifies as medium (1–10 t/day) and a seep in the Fungaia cluster classifies as high (10–100 t/day) (Chiodini & Valenza 2008). Here, we refer to the Caprese Michelangelo and Fungaia seeps collectively as the Caprese seeps. The rate and characteristics of these seeps (such as water content and area) are observed to vary with rainfall and following seismic events on the Alto-Tiberina Fault (Heinicke *et al.* 2006; Bonini 2009*b*).

CO_2 fluids from the PSS1 wellhead, the Caprese seeps and fluid inclusions from the PSS1 cores have a common origin (Bonini 2009*b*; Bicocchi *et al.*

2013; Trippetta *et al.* 2013). The seeps are aligned along NE–SW-trending faults that may connect to the deep CO_2 reservoir (Bonini 2009*b*; Bicocchi *et al.* 2013). On the basis of this information, we interpret that the Caprese CO_2 seeps source from the deep reservoir in the Caprese Antiform, and so this is classified as leaking.

Analysis: comparing the characteristics of leaking and sealing reservoirs

Four of the studied CO_2-bearing boreholes (PSS1, MF1 and Tr1, MT1) are located within 3 km laterally of documented surface CO_2 seeps. We interpret that two reservoirs are leaking: the Caprese (intercepted by the PSS1 borehole) and the Frigento (intercepted by MF1 borehole). Both reservoirs are hosted in antiform structures, and a number of CO_2 gas seeps with high rates of degassing are located within 3.5 km of the boreholes. In contrast, for Tr1 and MT1, very little is known about the small CO_2 springs located within 3.5 km of the boreholes, and so it is inconclusive whether the CO_2 in these reservoirs is leaking to surface. There are no seeps located within at least 10 km of the remaining boreholes (BS1, SAT1, Ben1/2 and Mu1) and so these are sealed.

Properties of the CO_2

Pressure, CO_2 density and, where possible, calculated CO_2 buoyancy pressure at the reservoir tops is shown in Figure 7. Most of the studied reservoirs contain CO_2 in the dense phase; MF1 and TR1 contain light-phase CO_2, and Mu1 is the only well to contain gaseous CO_2. No reservoirs contain liquid-phase CO_2. The physical properties of the CO_2 (phase or buoyancy) do not appear to be a first-order control on whether a CO_2 reservoir is leaking or sealed.

The sealed Benevento Sud reservoir has a higher estimated CO_2 buoyancy pressure at the reservoir–cap rock interface (5.0 MPa) than the seeping Monte Forcuso reservoir (3.5 MPa). The CO_2 column heights for SAT1 and Tr1 are unknown. If we assume the same CO_2–water contact in all three wells, the CO_2 buoyancy in SAT1 and Tr1 will be even higher than in BS1 because CO_2 is less dense. Despite this, unlike the Frigento Formation, the Benevento reservoirs are not obviously leaking. The Muscillo reservoir is the opposite; the net buoyancy pressure on the seal is effectively zero at the present day because gas saturation is so low. In this reservoir, CO_2 will also have extremely low relative permeability which will restrict its mobility.

CO_2 solubility in freshwater at reservoir conditions is typically between 1 and 1.5 molar

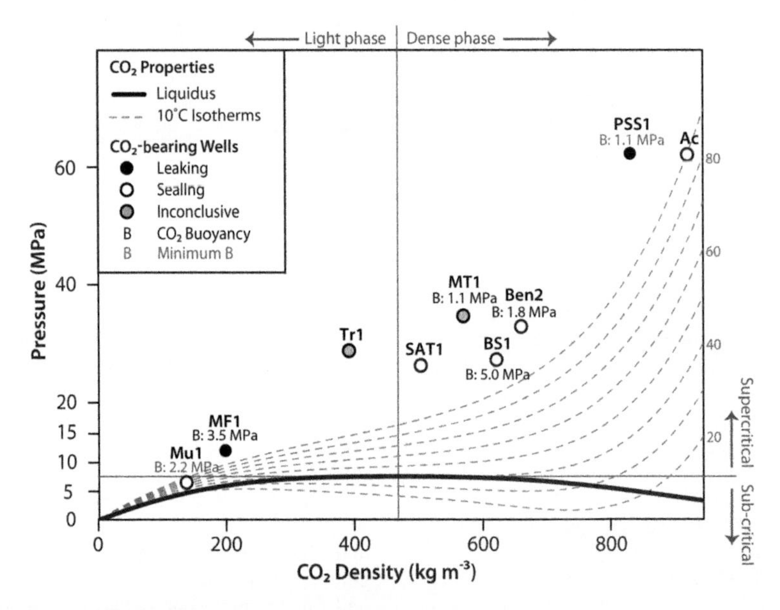

Fig. 7. CO_2 pressure–density phase diagram at the reservoir–seal boundary of CO_2-bearing reservoirs. Calculated CO_2 buoyancy is shown next to the data points where information is available. Critical density and critical pressure are shown as thin grey lines (and annotated). Only one reservoir, Mu1, contains CO_2 in the gas phase, all other case studies contain supercritical CO_2 in both the dense and light phase. Neither CO_2 density nor buoyancy determines whether a reservoir is sealing in these case studies.

(*c.* $40 - 60\,kg_{CO_2}\,m^{-3}{}_{(H_2O)}$) for all case studies. The formation waters in these reservoirs therefore have potential to dissolve significant quantities of CO_2, and have a greater solubility capacity than surface waters.

In most of the case studies, a CO_2 leg overlies a water leg in the reservoir, and CO_2 saturation in the cap is high. The exceptions are PSS1, where the reservoir is complex and CO_2 (in high saturation) is trapped within more permeable layers between evaporite layers and Mu1, where CO_2 saturation in the reservoir is low. Further, a unit overlying the primary CO_2 reservoir in Tr1 also has low CO_2 saturation. Low CO_2 saturation could result from several mechanisms. If the reservoir's seal has been breached, low CO_2 saturation confirms that the reservoir is leaking or has leaked in the past. If the cap rock is acting as a good seal but CO_2 saturation is low, then this could indicate that there was insufficient CO_2 charge to fill the reservoir. *In situ* generation of CO_2 may result in low saturation if the quantities of CO_2 generated are small. Similarly, CO_2 coming out of solution from formation waters as they depressurize during ascent may result in low CO_2 saturation. In the absence of further geochemical information on the CO_2 and formation waters, it is not possible to distinguish these scenarios.

Other gases which may affect the properties of the CO_2 mixture are present in small quantities in many of the CO_2 reservoirs, including short-chain hydrocarbons, such as CH_4, and H_2S. Small proportions of H_2S decrease the interfacial tension of CO_2 (Bennion & Bachu 2008; Savary *et al.* 2012), whereas CH_4 increases interfacial tension and decreases the fluid density (Naylor *et al.* 2011). Since only trace amounts (0.1% C v/v) of H_2S are recorded in some boreholes, its effects on CO_2 properties are likely to be negligible. In contrast, sealing reservoirs Ben2, BS1 and Mu1 contain over 5% CH_4 (% C v/v), and so the buoyancy of the CO_2–CH_4 mixture in these reservoirs will be greater than for pure CO_2. However, the effect of CH_4 on the interfacial tension will be more significant than the effect on the buoyancy (Naylor *et al.* 2011). As a result, relatively small quantities of CH_4 may be enhancing reservoir sealing at the Benevento reservoirs.

Properties of the CO_2 reservoir

The geological structures of all the reservoirs are broadly similar: CO_2 has accumulated in platform carbonate units, and the overburden is comprised of thick, heterogenous nappes. This is similar to hydrocarbon discoveries in central-southern Italy, many of which are hosted in fractured Apulian Carbonate Platform (Casero 2005). Whether the reservoir is hosted in an anticline or horst does not affect whether it leaks or seals.

The leaking Caprese and Frigento reservoirs are both hosted in thrust-related anticlines located in Quaternary graben structures. However, the depth of the reservoirs is very different (see Table 2), and so confining pressure is not a primary control on the seal quality. The Caprese reservoir is deep and pressured beyond hydrostatic; in this reservoir, CO_2 is in its dense phase. The Frigento reservoir is much shallower, hydrostatically pressured, and so CO_2 is in its light phase. The two reservoirs have similar temperatures, since the shallower Frigento formation is located in a region with an anomalously high geothermal gradient.

In three boreholes (Ben1/2, BS1 and Tr1) the reservoir carbonate units are close to hydrostatically pressured, in contrast with the significantly overpressured overburden. These reservoirs must be hydrologically connected to the surface; either by permeable faults or through surface outcrop. Examples of hydrocarbon reservoirs at hydrostatic fluid pressures overlain with high-pressure cap rock are common in overpressured basins (O'Connor *et al.* 2008). Isolated reservoir units will be in pressure equilibrium with the encasing low-permeability units (such as shales). However, if reservoirs are connected to surface via lateral outcrop or fracture/fault networks, fluids can escape and drain the overpressure in the reservoir, bypassing any buoyant fluids trapped in the overlying formation. The overburden can remain overpressured even though fluids may slowly bleed into adjacent lower-pressure reservoirs. In contrast, the Caprese and Monte Taburno structures contain overpressured reservoir fluids with a close to hydrostatically pressured overburden. This is often indicative of reservoir compartmentalization, which Trippetta *et al.* (2013) interpreted for the complex and multilayered Caprese structure.

The CO_2 contained in PSS1, MT1, Ac1 and SAT1 is overpressured. High fluid pressures can enhance or retard seal integrity, depending on the mechanism of seal failure. CO_2 density increases with reservoir pressure, which in turn decreases CO_2 buoyancy. CO_2 overpressure therefore reduces the likelihood of capillary seal failure. Indeed, reservoir overpressure in the leaking Caprese structure decreases CO_2 buoyancy by approximately 0.3 MPa compared to hydrostatic conditions. However, significant fluid overpressure can lead to seal failure by fluid-driven fracture propagation. For example, in the case of PSS1, Ac1 and SAT1, the reservoir pore fluid pressures are over 60% of lithostatic. These fluid pressures could jeopardize the integrity of the seal, particularly if the seal contains pre-existing fractures that are critically stressed. However, since only the Caprese reservoir leaks

CO_2, reservoir fluid pressure alone cannot control reservoir leakage. Regardless of the degree of overpressure in the reservoir, overpressure in the seal and in the overburden above the seal can act as a significant barrier as it increases the pressure required to drive CO_2 upwards and through the seal and overburden.

Properties of the overburden

Although the geological structure of all the cases studied is broadly similar, the overburden is variable in both rock type and thickness. Figure 8 shows the seal thickness, defined as the total thickness of units documented from drill cuttings that would be likely to be impermeable to CO_2.

There is no correlation between reservoir depth (overburden thickness) or seal quality/thickness and the presence of surface CO_2 seeps. Some well logs record thick low-permeability sequences in the overburden: for example, in SAT1, there are 1520 m of muds overlying the reservoir all the way to surface, and overlying the BS1 reservoir there are muds that, although becoming a little siltier towards the surface, remain low permeability. In contrast, TR1 records 17 m of massive anhydrite directly overlying the reservoir but no definable seal above this; the overlying (calcareous) siltstone records low-saturation CO_2 (for *c.* 200 m above the anhydrite) and wet natural gas all the way to surface. Similarly, PSS1 records 70 m of gypsum above the CO_2 reservoir overlain by sandy-marls (*c.* 160 m), but no other low-permeability formations above this.

The thrusted contact between reservoir and overburden is marked by a tectonic breccia in three boreholes (SAT1, MF1, Mu1), whereas Messinian anhydrite-bearing units (massive, or associated with muds) directly overlie the CO_2 reservoir in other boreholes (BS1, Tr1, Ac and MT1; see Table 2). Such low-permeability units may contribute to the sealing capability of the overburden at the sites. However, the Burano Triassic Evaporite Formation forms the reservoir–seal complex of the leaking Caprese reservoir. Thus, while evaporites often make a very effective seal, their presence or absence is not the only factor in determining overburden integrity.

The relationship between CO_2 seepage and overburden overpressure is summarized in Figure 9. CO_2 reservoirs that lack strong overpressure in overburden units (maximum pressure/hydrostatic pressure <1.3) are associated with surface seeps (boreholes MF1 and PSS1). In contrast, where the overburden shows significant overpressure (maximum pressure/hydrostatic pressure >1.3) there are no surface seeps within 10 km of the borehole (Ben2, BS1 and SAT1). The remaining boreholes are inconclusive (Ac1, Mu, Tr1 and MT1). The pressure conditions in the overburden seem to be a primary control on successful CO_2 retention.

Figure 10 shows the relationship between the fluid pressures in the overburden and the lateral distance from the wellbore to active normal faults

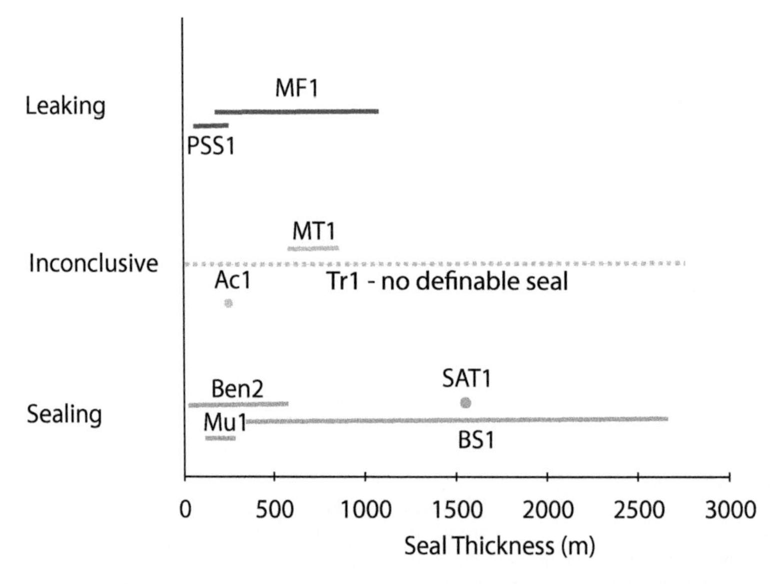

Fig. 8. Thickness of impermeable rock formations in the overburden, as interpreted from the well logs of CO_2 reservoirs, and the leaking–sealing classification of the reservoir. The thickness of low-permeability formations does not control whether or not the reservoir leaks CO_2 to surface.

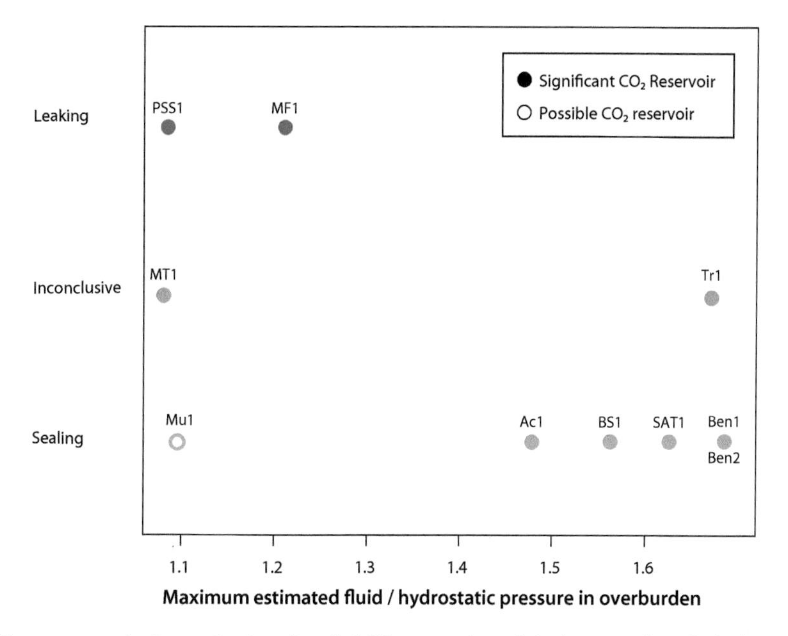

Fig. 9. Fluid overpressure in the overburden of studied CO_2 reservoirs and the interpretation of whether or not the reservoir is leaking CO_2 to surface. The degree of overpressure is indicated by the ratio of fluid pressure (P_f) to hydrostatic pressure as interpreted from the density of the drilling mud in the well log. The data points are coloured according to whether the reservoir is interpreted to be leaking (red), sealing (green) or is inconclusive (orange). CO_2 reservoirs with fluid overpressure (>3 MPa) in the overburden do not have CO_2 seeps located within at least 10 km of the borehole and so are considered to be sealing reservoirs (green). In contrast, deep CO_2 reservoirs that have little or no overpressure in the overburden are located close to CO_2 seeps (red). The shallow Mu1 reservoir shows low CO_2 saturation; hence, this is why it is not considered to be a significant CO_2 reservoir.

(faults with exposed scarps that are considered to pose a seismic hazard). The boreholes that penetrate the two leaking CO_2 reservoirs are located within 5–7 km of the surface trace of seismogenic normal faults. These boreholes record no overpressure in the overburden. Reservoirs located further from these faults show overpressure in the overburden rock units.

The exception to this trend is borehole Mu1, which is over 40 km from any known recent faults, and yet shows only minor deviations from hydrostatic pressure. However, this reservoir is at a relatively shallow burial depth compared to the other study sites (*c.* 700 m), and the overburden consists of sands, silts, clays and conglomerates which may be permeable even if not breached by faulting.

Analysis: characteristics of leakage and implications for risk management

As described in the sections above, the Frigento and Caprese antiforms are considered to be leaking. Both structures have a cluster of CO_2 seeps at the surface above the reservoir, and both have hydrostatically pressured overburden. However, in many respects, they are end-member case studies; the Frigento Antiform is one of the shallowest CO_2 reservoirs and has an anomalously high geothermal gradient, whereas the Caprese formation is the deepest CO_2-bearing structure in a region with relatively low geothermal gradient. The downhole conditions in the wells that penetrate these structures are therefore very different (as can be seen in Table 2). This has implications for the area-permeability criteria to leak a given mass of free-phase CO_2 from the reservoir. For example, for the same mass of CO_2 to leak from the reservoir, the volume of free-phase CO_2 that must leave the Caprese reservoir ($\rho_{CO_2} = 830$ kg m^{-3}) is a quarter of the volume that must leave from the Frigento reservoir ($\rho_{CO_2} = 200$ kg m^{-3}). As such, small volumes of free-phase CO_2 escaping from the Caprese structure would mean relatively high rates of CO_2 leakage. For both structures, much greater volumes of CO_2-saturated water must leak from the reservoir than free-phase CO_2; six times the volume of free-phase CO_2 for the Frigento Formation and up to 10 times for the Caprese Formation. Thus, larger permeabilities are needed for CO_2-saturated waters to transport the same rate of CO_2, unless the relative permeability to water is at least an order of magnitude higher for free-phase CO_2.

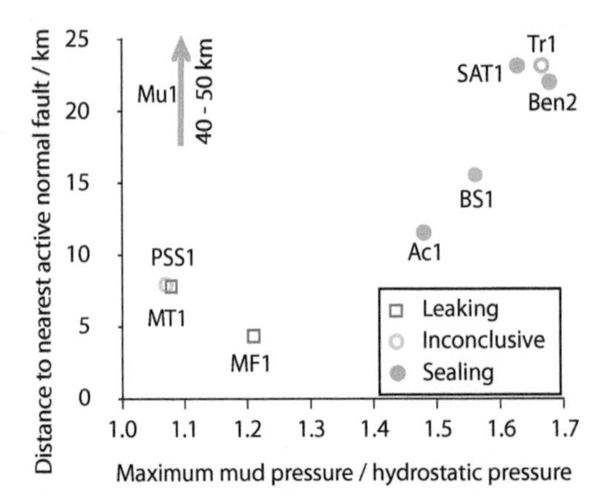

Fig. 10. Fluid overpressure in the overburden of studied CO_2 reservoirs and the lateral distance of the well to the nearest modern extensional fault structure. CO_2 structures that are located within 8 km of a fault leak CO_2 to surface. The degree of overpressure is indicated by the ratio of fluid pressure (P_f) to hydrostatic pressure as interpreted from the density of the drilling mud in the well log. Wells are coloured according to whether the reservoir is leaking CO_2 to surface (red) or not (green) or indeterminately so (orange). Overburden overpressure correlates with distance to the modern extensional faults mapped by Roberts (2008). CO_2 reservoirs that are classified as leaking (i.e. are located within 5 km of CO_2 seeps) show hydrostatic pressures in overburden formations, and are located within 8 km of a modern extensional fault. Mu1 does not fit this trend, possibly because it is so shallow; it is located over 40 km away from any mapped structures and so cannot fit on these axes.

Pathways of CO_2 leakage

CO_2 leakage from a reservoir could, in theory, occur over a large area (distributed flow through a large rock volume) or over a smaller area (focused by enhanced-permeability pathways such as those offered by faults, whether the fracture network is localized or more distributed). We use the Darcy flow (equation 1) to approximate flow through the rock volume and fracture networks in order to examine the area and permeability requirements to permit CO_2 leakage from the Caprese and Frigento reservoirs into the overburden (not through the overburden to surface). Figure 11 shows the combinations of overburden effective permeability (K_E) and area (A) for leaking free-phase or dissolved CO_2 at 100 and 2000 t CO_2/day from the Caprese and Frigento antiforms, respectively. These leak rates correspond to the maximum estimated CO_2 release rate at the Fungaia and Mefite D'Ansanto seeps, since there are no published estimates for CO_2 release from all the seeps in the Caprese and Mefite seep clusters. The permeability of formations measured from the PSS1 and MF1 well logs, and elsewhere, guides the possible cap-rock permeability. Further, reasonable possible leakage areas are indicated in Figure 11 for discrete and clustered seepage, the possible extent of a free-phase CO_2 caps in the antiforms, and the geometry of rock deformation related to faulting. Similar calculations are not performed for the case studies that are inconclusive (Tr1 and MT1) because we do not have information about seep rates, or seep area.

Figure 11 and its table inset shows that high leak rates of free-phase CO_2 can occur over smaller areas and lower permeability than for CO_2-saturated water. Enhanced-permeability pathways (i.e. faults) may not be necessary for free-phase CO_2 fluids to leak from the Caprese reservoir at 100 t/day. CO_2 could leak at this rate over an area smaller than that of the Caprese Michelangelo seep cluster if the permeabilities of the overlying rock formations are similar to measurements of the overburden recorded in the PSS1 well log. For the same CO_2 leak rate and permeability, CO_2 dissolved in water would need areas similar to the Caprese seep cluster, or faults. In contrast, to leak 2000 t/day from the Frigento reservoir, Darcy flow of free-phase CO_2 through mudstones (maximum permeability *c.* 0.8 mD) would require leakage over an area much larger than that estimated for the CO_2 reservoir top. For CO_2 leakage over smaller areas, such as those of faults, overburden permeabilities approaching 10^2 mD are necessary. At approximately 1.1 km depth, such permeability could only be provided by a network of open fractures, which could localized or distributed, and could be related to faulting. Fluid flow rates of 480 l s^{-1} of CO_2-saturated waters would transport 2000 t/day of CO_2 from the Frigento reservoir. Such flow rates are not impossible, since spring flow rates in Italy can exceed 800 l $^{-1}$ (Minissale 2004). However, rock permeabilities greater than 10^7 mD would be needed to enable these flow rates over a discrete area, which is difficult to achieve unless the rocks are karstified. Although karst environments are common in central and southern Italy (Santo *et al.* 2011), it is unlikely that karst in the overburden is responsible for CO_2 leakage from the reservoir, since karst environments are typically found in the region of the water table (current or historical). However, it is possible that karst could aid the rapid seepage of CO_2 from the near-surface.

Driving mechanism for CO_2 leakage

The results discussed above consider possible rock properties and geometries required to permit a given rate of fluid flow into the cap rock, not the mechanism driving the fluid flow. For free-phase or dissolved CO_2 to migrate from the reservoir and

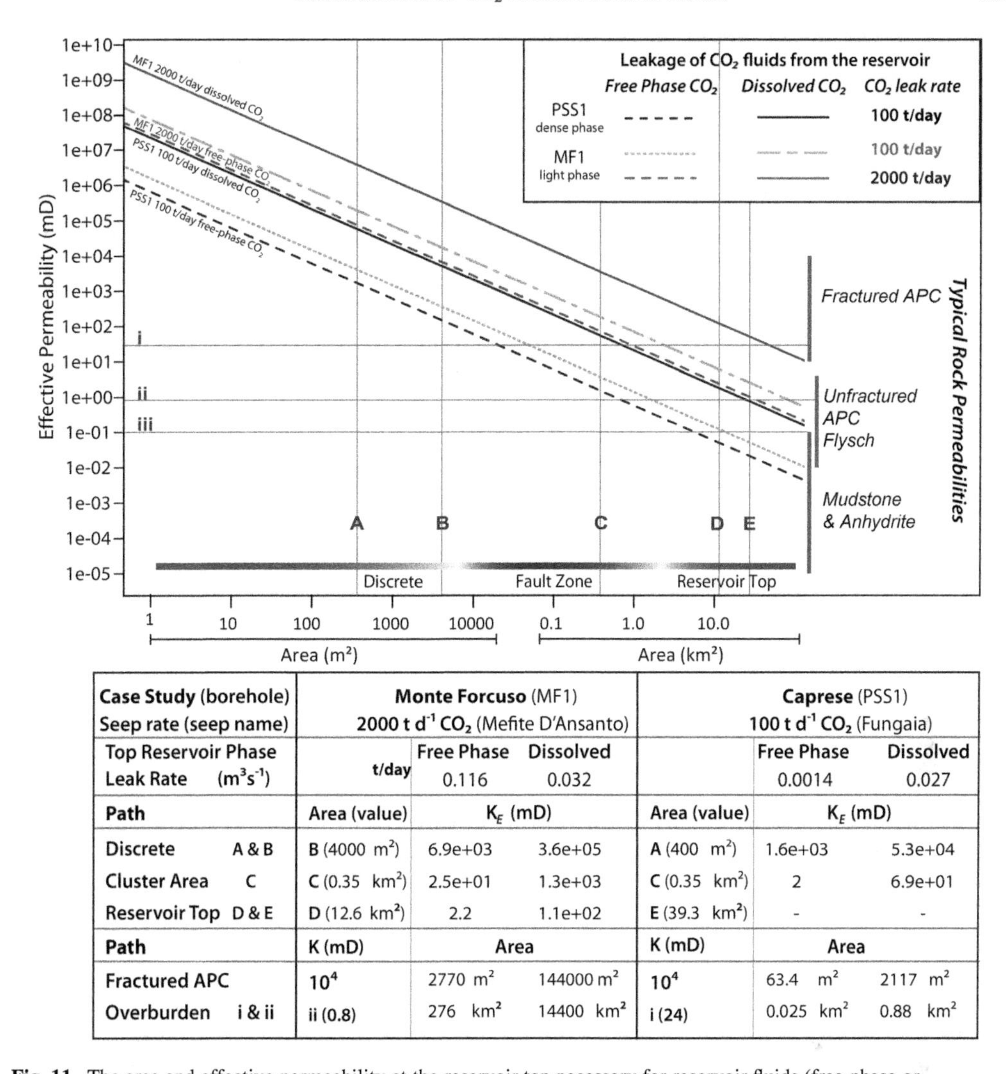

Case Study (borehole) Seep rate (seep name)	Monte Forcuso (MF1) 2000 t d^{-1} CO_2 (Mefite D'Ansanto)			Caprese (PSS1) 100 t d^{-1} CO_2 (Fungaia)		
Top Reservoir Phase Leak Rate (m^{3}s^{-1})	t/day	Free Phase 0.116	Dissolved 0.032		Free Phase 0.0014	Dissolved 0.027
Path	**Area (value)**	**K_E (mD)**		**Area (value)**	**K_E (mD)**	
Discrete A & B	B (4000 m^2)	6.9e+03	3.6e+05	A (400 m^2)	1.6e+03	5.3e+04
Cluster Area C	C (0.35 km^2)	2.5e+01	1.3e+03	C (0.35 km^2)	2	6.9e+01
Reservoir Top D & E	D (12.6 km^2)	2.2	1.1e+02	E (39.3 km^2)	-	-
Path	**K (mD)**	**Area**		**K (mD)**	**Area**	
Fractured APC	10^4	2770 m^2	144000 m^2	10^4	63.4 m^2	2117 m^2
Overburden i & ii	ii (0.8)	276 km^2	14400 km^2	i (24)	0.025 km^2	0.88 km^2

Fig. 11. The area and effective permeability at the reservoir top necessary for reservoir fluids (free-phase or dissolved CO_2) to seep at 100 t of CO_2 per day from PSS1 reservoir conditions, and 100 and 2000 t CO_2/day from MF1 conditions. For light- or dissolved-phase CO_2 to leak from the reservoir at these rates, high-permeability pathways in the overburden such as those offered by open fractures or faults are needed. In contrast, it is possible for dense-phase CO_2 to leak from PSS1 into low-permeability overburden formations at 100 t CO_2/day without the need for fracture permeability. Typical rock permeabilities and seepage area are annotated to the right of the plot. Permeabilities from well logs are annotated: i, Jurassic Umbria–Marche overburden in PSS1; ii, Allochthonous Complex overburden in MF1 well; and iii, Apulian Carbonate Platform units in MF2. Vertical lines A–D show estimates of minimum area of seepage at Caprese and Frigento case studies: (**a**) main area of degassing at the Caprese Michelangelo; (**b**) area of degassing at Mefite; (**c**) cluster area at the Caprese Michelangelo (0.2 × 1.52 km) and the Mefite and Mefitiniellapolla vents (3.5 × 0.1 km); and minimum area of seepage from (**d**) the Frigento CO_2 reservoir top (2 km radius circle) and (**e**) the Caprese CO_2 reservoir top (5 × 10 km ellipse). The table inset show calculated seep areas using relevant permeabilities, and permeability calculations using areas A–E. These illustrate that for dense-phase CO_2, high seep rates require only very small volumes of CO_2 to leak from the reservoir compared to light-phase CO_2. Similarly, for the same leak rates, much larger volumes of CO_2 must leak from the reservoir compared to free-phase CO_2.

into the overburden, there must be a driving force. This could be buoyancy pressure of free-phase CO_2, which is less dense than formation waters. Modelling of CO_2 properties at downhole conditions finds that CO_2 in the Frigento Formation is much more buoyant than CO_2 in the Caprese

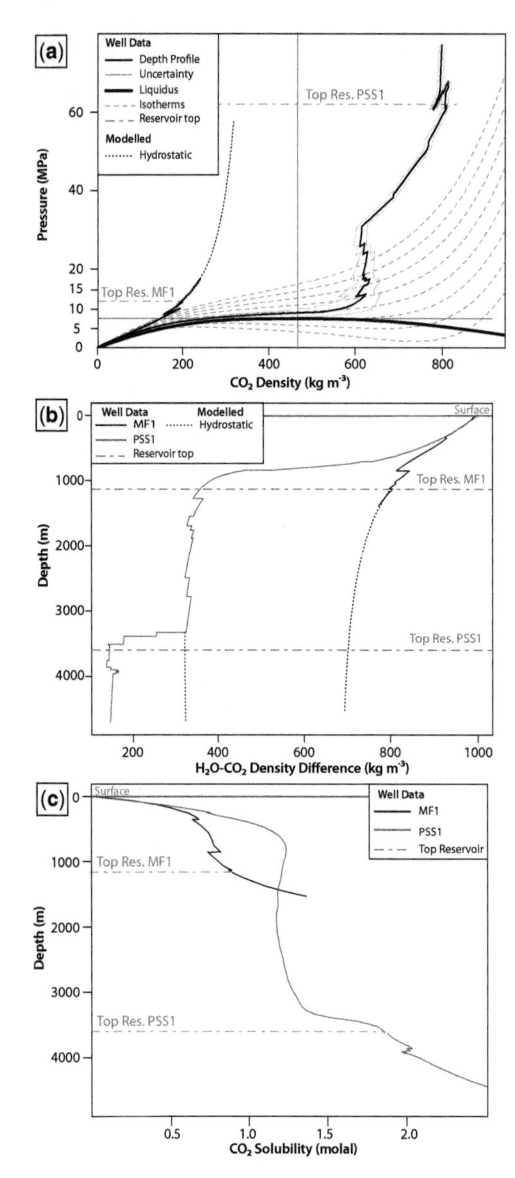

Fig. 12. Down-well pressure–ρ_{CO_2} profiles for MF1 and PSS1 plotted on a vapour pressure curve (**a**) and down-well depth and density difference ($\rho_{H_2O} - \rho_{CO_2}$) profiles (**b**), and change in solubility in freshwater (**c**). The depth of the reservoir top is shown on each graph. These graphs illustrate how changes in CO_2 properties, which will affect how CO_2 flows in geological formations, differ depending on the reservoir properties, and how changes are greatest towards the CO_2 phase transition. Increased reservoir pressure decreases the density difference between CO_2 and CO_2-saturated waters, thereby decreasing the buoyancy drive of the fluids. This effect is particularly enhanced in the MF1 reservoir. The vapour pressure curve (a) shows how the cooler PSS1 passes closer to the liquidus, which leads to more rapid changes in density

Fig. 12. (*Continued*) (b) as the fluids ascend to approach 1 km depth, and a pronounced change in solubility in freshwater (c). Note the rapid change in CO_2 solubility in both cases towards the phase transition. For leakage of dissolved CO_2, two-phase flow will become established towards the phase transition, decreasing the relative permeability of both the water and the CO_2 phase.

reservoir. Indeed, in PSS1, it finds that CO_2 will be in its dense phase (with low buoyancy) for several kilometres above the reservoir. Instead, fluid pressure in the Caprese reservoir could be driving CO_2 leakage, since the reservoir pore fluid pressure is much greater than hydrostatic.

Whether fluid pressure or buoyancy is driving fluid leakage, these forces will change during fluid ascent. For example, as shown in Figure 12a, CO_2 leaking from the Caprese reservoir will remain in its dense phase for a few kilometres and pass very close to the liquidus, where buoyancy will be lowest, during its ascent from 1 km depth, if the fluids are in thermal equilibrium with the geotherm. This means that CO_2 experiences a rapid increase in buoyancy as its density decreases approaching 800 m depth, and CO_2 solubility will concurrently decrease rapidly. These are depicted in Figure 12b, c, which also shows that, although CO_2 buoyancy is high in the Frigento reservoir, the buoyancy increases gradually and to a lesser degree during ascent to surface. For example, during the 500 m ascent between 1250 and 750 m, ρ_{CO_2} decreases by approximately 325 kg m^{-3} in PSS1 and approximately 75 kg m^{-3} in MF1. This could have a pronounced effect on the way that CO_2 leaks to surface. For PSS1, the area permeability of flow paths would need to rapidly increase to sustain the mass flux of leaking CO_2 since there will be a corresponding volume increase of the leaking fluids over this depth interval.

Effective permeability

Our calculations do not account for the effective permeability of CO_2 compared to water. The relative permeability of CO_2 can be very low when flow first establishes in water-wet rocks. However, the Caprese and Mefite seeps are long-established degassing sites. Due to drying-out effects, single-phase flow could now be established along the leak paths, and so effective permeability may approximate to rock permeability.

For CO_2-saturated waters migrating through water-wet rocks, the waters will initially behave as a single phase. However, two-phase flow may initiate towards the phase-transition depth, where solubility rapidly decreases (see Fig. 12c) causing CO_2 to exsolve. The resulting decrease in the effective permeability will impede flow of both phases, although the buoyancy of the water may increase as a result of 'gas lift' (the buoyant CO_2 bubbles). The exsolved CO_2 will redissolve if it comes into contact with unsaturated water, and so will only remain as a separate phase if its flow path is isolated from the ascending fluids (e.g. channelized flow in faults) or if the rocks through which it is flowing are not water-saturated. This is more common at shallower depths (vadose zone). If CO_2 remains as a separate phase, then its buoyancy and high interfacial tension could allow free-phase CO_2 to follow a different flow path to its parent waters.

It is also important to note that fracture flow is not accurately represented by Darcy's law. However, in the absence of further information about the fracture properties of the overburden, the simplified approach allows us to explore the constraints on the geological conditions that could enable leakage at the observed rates.

CO_2 mass transport

Our calculations assume conservative mass transport of leaked CO_2 (i.e. that there is minimal CO_2 loss during ascent), and so CO_2 leaks from the reservoir at the same rate that it reaches the Earth's surface regardless of its subsurface interactions. When CO_2 leaks first establish, or if leakage occurs through a large rock volume rather than a focused flow path, it is more reasonable to assume that CO_2 will disperse and attenuate as CO_2 becomes residually trapped or accumulates in secondary formations. Similarly, for many geological situations, the migrating CO_2 will encounter multiple barriers and cap rocks that will inhibit escape to surface. However, for long-established degassing sites, such as those studied here, the rocks and fluids that the CO_2 comes into contact with during ascent are probably saturated with CO_2. The quantity of CO_2 loss during ascent from the Caprese and Frigento reservoirs may therefore be limited. However, it is unlikely that the mass transport is truly conservative, and, in fact, geochemical studies at the Caprese reservoir and seeps find evidence of CO_2 mixing with shallow waters during ascent (Bicocchi *et al.* 2013).

Synthesis and discussion

Our study of CO_2 reservoirs in Italy identifies that reservoirs that are successfully sealed have low-permeability units and overpressured units in the overburden, and are located over 10 km from seismogenic normal faults.

The thrusted sediments that comprise the overburden of the studied reservoirs have experienced compressional tectonics, which is one mechanism of elevating pore fluid pressures beyond hydrostatic (Osborne & Swarbrick 1997). Overpressure is only preserved in low-permeability rocks, since the pressure will dissipate where there is sufficient permeability (whether due to the presence of slightly more permeable rock types in the overburden or a connected fracture and/or fault network, whether it is localized or distributed). While we find that there is no simple relationship between overpressure and the type of rock comprising the overburden, we do note that for many of the sealing reservoirs an evaporite-bearing formation caps the CO_2 reservoir. The presence of evaporites will contribute to the sealing capability of the overburden due to their low inherent permeability and the possibility that when mobilized they can cement pores or fractures (Trippetta *et al.* 2013). This may be the case for the Caprese reservoir where the CO_2-bearing horizons are overpressured and are confined by evaporites (Bicocchi *et al.* 2013), but there are no other evaporite layers in the overburden, and the Caprese reservoir is leaking. However, observing the borehole pressure profiles for leaking and sealing reservoirs finds that the most overpressured formations in a cap rock are rarely those that are evaporite-bearing (see Figs 3–6), and the boreholes that show greatest overpressure are not necessarily those that contain evaporite. Thus, the presence of evaporites does not systematically affect the overburden integrity or overpressure.

Several factors affect fracture connectivity in rocks, including confining pressure (corresponding to depth) and the regional stress regime. We find that confining pressure does not affect the maximum overpressure, but that proximity to active normal faults (as defined by Roberts 2008) does. Away from these faults, overpressure from the contractional tectonic regime could be preserved in the heterogeneous and compartmentalized thrust-top deposits. The primary control on overburden overpressure may, hence, be the hydraulic conductivity of localized or distributed fractures within the overburden; high connectivity resulting from either the presence of recent 'open' extensional faults or from high overpressures resulting in a reduction of overburden stress. For example, CO_2 leakage from the Frigento reservoir may be facilitated by the low confining pressures (from being relatively shallow) opening fractures in the overburden, and by permeability offered by extension and fault damage zones related to the nearby Ufito normal fault. In contrast, the leaking Caprese reservoir is overpressured, although its overburden is not. Faults in the region could have relieved any overpressure that once existed in the overburden units; however, the reservoir horizons are not in pressure communication with their overburden because they are interlayered with the low-permeability evaporites of the

Burano Formation. This reservoir is deeply buried and the resultant high confining pressure will have closed mesoscale fractures in the reservoir and much of the overburden, unless the high fluid pressures in the reservoir opens them locally or faults are critically stressed. Both scenarios are feasible. PSS1 is located <8 km from a seismically active fault, and also the pore fluid pressure in the Caprese reservoir could be sufficient to open fractures in the cap rock, locally enhancing rock fracture permeability and enabling CO_2 escape from the reservoir. Indeed, pressure pulses associated with seismicity have increased CO_2 degassing at the Caprese Michelangelo seeps (Heinicke *et al.* 2006; Bonini 2009*a*). These observations stress the need to understand the crustal stresses around potential storage sites.

Although we do not consider this here, the burial history might have affected the geomechanical properties and, as such, the fluid flow properties of the overburden, and therefore whether a reservoir leaks or seals. Further work could aim to resolve how the geomechanical context influences reservoir leakage.

The recorded overpressure in low-permeability units could be an artefact of deriving formation pressure from drilling mud weights. When drilling through low-permeability rocks, the borehole will not be in pressure communication with the rock and so high mud weights will be tolerated without affecting the well integrity. However, we assume that this is not so for two reasons: first, significant health and safety risk is associated with drilling with the incorrect mud weight, and it is considered poor practice to drill using mud weights that are not carefully calibrated to the subsurface conditions. Secondly, for many of the well logs, it is clear that the mud weights have been adjusted many times during drilling to reflect the complexity of the overburden formations.

Implications for storage site selection

Pressure seals are commonly observed in the overburden of hydrocarbon provinces. They are a highly effective seal for two reasons: first, they indicate the presence of very-low-permeability formations, like those proposed for cap rocks in sequestration operations. Secondly, where the overburden fluid pressure exceeds that of the reservoir, the net fluid pressure gradient over the interval between the reservoir and overpressured formation is directed downwards. Fluids would therefore flow into the reservoir rather than up from the reservoir into the overburden. Despite this, to date, little attention has been paid to the role of pressure seals in ensuring secure CO_2 storage. For the case studies in Italy that are presented here, it is not possible to determine which of these two retention mechanisms offered by the pressure seal is important for CO_2 security – if any.

Current industrial screening practices and the regulatory framework for site selection typically focus on possible mechanisms of CO_2 leakage from the reservoir into the overlying cap rock (capillary breakthrough, tensile fracturing of the cap rock or fault slip, and brine displacement) or necessary reservoir conditions, rather than the barriers to fluid flow offered in the overburden overlying the reservoir (Hannon & Esposito 2015). Multilayered reservoir–cap rock systems are identified as an effective barrier for leakage for storage site selection criteria (IEA-GHG 2009), but the only site selection guidance document to mention cap-rock fluid pressure gradients are those prepared by the World Resources Institute (2008), which note that the presence of a pressure differential between the reservoir and cap rock is one characteristic that may demonstrate the ability of the cap rock to prevent vertical migration of injected CO_2.

Table 3 summarizes the published criteria for storage site selection that will minimize the risks associated with the geological storage of CO_2, and how our case studies would perform against these criteria. All the reservoirs studied here, whether leaking or sealing, would not be deemed suitable for CO_2 storage. This suggests that site selection criteria are robust, and perhaps err on the side of caution. Table 3 shows how many of the reservoirs fulfill the most prescriptive criteria such as cap-rock thickness, and reservoir pressure and temperature conditions. Only one reservoir, Muscillo, would be deemed too shallow for storage, since it is less than 800 m deep. Most of the other case studies would be deemed too deep according to Chadwick *et al.* (2008) and Smith *et al.* (2011), but not according to IEA-GHG (2009), who provided no depth cut-off. Avoiding deep reservoirs does not minimize the risks of leakage, but, rather, the cost and ease of injection and monitoring, which at depths below 2500 m may become too difficult or expensive. In any case, the Aquistore CCS project in Canada is injecting at 3400 m (Rostron *et al.* 2014) and so clearly only the minimum depth criterion is prescriptive.

There is some uncertainty regarding the selection criteria for reservoir structures and cap-rock continuity. The leaking Frigento reservoir would fail several selection criteria (it is shallow, CO_2 in the reservoir is in is the light phase, see Table 3); however, the only criterion that the Caprese reservoir might fail regards proximity to faults. Site selection guidelines for CCS recommend that reservoirs selected for CO_2 storage should have no faults, or should at least have only small or a low density of faults. However, these are descriptive criteria; the constraints that define 'low fault frequency' or

Table 3. *A summary of how the natural CO_2 reservoirs in Italy studied in this paper would perform against published criteria for CO_2 storage site selection, for (A) Chadwick* et al. *(2008) (B) IEA-GHG (2009) and (C) Smith et al. (2011)*

	Property	CO_2 properties	Reservoir properties						Cap-rock property	
	Feature	CO_2 State	Structure		Depth (m)		Temp	Pressure	Thickness	Continuity:
Guidelines	Criteria	Dense phase	(i) Small or no faults; (ii) Low fault frequency	Multilayered system	800–2500	>800	>35°C	>7.5 MPa	10–100 m	(i) Uniform; (ii) Extensive
	Source	A	A, B, C	B	A, C	B	B	B	A, B, C	(i) A (ii) B
Case studies										
Leaks	MF1	N	N	N	Y	Y	Y	Y	Y	?
	PSS1	Y	N	Y	N	Y	Y	Y	Y	?
Inconclusive	TR1	N	N	N	N	Y	Y	Y	N	?
	MT1	Y	N	N	N	Y	Y	Y	Y	?
Seals	BS1	Y	N	N	N	Y	Y	Y	Y	?
	SAT1	N	N	Y	Y	Y	Y	Y	Y	?
	BEN 2	Y	?	N	N	Y	Y	Y	?	?
	Ac	Y	N	N	N	Y	Y	Y	?	?
	Mu	N	Y	N	N	N	Y	N	Y	?

All the case studies, whether leaking or sealing, would not be deemed suitable for CO_2 storage. Two of the features, reservoir structure and cap-rock continuity, are descriptive and therefore it is difficult to determine whether the case studies would fulfil these criteria or not.

'small faults', and whether this refers to fault length or fault throw, or only open faults, are not clear. Nor is it clear how their potential for storage integrity should be characterized; there are many examples from the hydrocarbon sector of sealing normal faults, and so the regional crustal stresses should also be considered. Further, the criteria refer mostly refer to faults in the reservoir (which our results indicate are not necessary for rapid CO_2 leakage from PSS1), rather than buried or surface faults in the overburden or nearby. Our results suggest that for dense-phase CO_2 to leak from a reservoir at a considerable rate (>100 t/day), faults do not need to connect from reservoir to surface; however, to seep CO_2 to surface, permeable faults are needed to provide flow paths for less dense CO_2. We would therefore argue that any faults in the overburden, as well as those that intersect the reservoir, should be characterized during site screening. Although the site selection criteria in Table 1 do not make it explicit, it is unlikely that sites located close to seismogenic faults would be considered for CO_2 storage.

The cap rocks of most case studies are suitably thick; however, it is difficult to determine if they would be considered 'uniform' or 'extensive' as required by Chadwick *et al.* (2008) and IEA-GHG (2009). This is because the well logs provide the only information about the case study overburden. Since most are comprised of thin interlayered nappes, the cap rocks may not be considered uniform on that basis. It is clear, though, that several case studies have interlayered cap rock–reservoir units comprising the overburden. This structure could be desirable above prospective CO_2 stores because interlayered reservoir units could, in the case of leakage, act as secondary or tertiary reservoirs and inhibit surface seepage. Our study suggests that CO_2 is securely retained in reservoirs with cap rocks that would be deemed unsuitable for storage according to current criteria. It might be reassuring to policy-makers and the public to learn that imperfect geosystems are capable of trapping large quantities of CO_2 in the reservoirs.

This work has identified two key controls on CO_2 retention: fluid pressure in the overburden and lateral distance of the reservoir from an active fault. The criteria for desirable properties of the cap rock and overburden above prospect CO_2 stores should therefore be improved. The regional stress regime and the overburden should be characterized during site assessment in order to identify the geological structure, pressure conditions, and possible fracture and fault properties (orientation, connectivity, stress state) in the overburden units. We recommend that the pressure seal becomes one of the first-order screening criteria for storage site selection. Furthermore, we support previous work proposing the artificial pressurization of overburden units as an effective remediation option should leakage from an engineered CO_2 storage reservoir occur, since this would decrease or reverse the normal fluid pressure gradient (Benson *et al.* 2003; Reveillere & Rohmer 2011).

The ascent of leaked CO_2

The Caprese Michelangelo and Mefite seeps are low-temperature CO_2 emissions, mostly characterized by CO_2 venting, where CO_2 is released above ambient pressure (Chiodini & Valenza 2008; Roberts *et al.* 2011). CO_2 is denser than air at surface temperature and pressure, and therefore subsurface pressure must be driving the escape of these fluids rather than buoyancy alone, otherwise gas would spread below surface in permeable soils. Pressurized CO_2 escape implies that flow is restricted below the surface. Previous work by Roberts *et al.* (2014) found that CO_2 vents in Italy tend to occur along faults in low-permeability rocks, and suggest that these rocks could be restricting CO_2 release from a more permeable (and CO_2-saturated) lithology beneath. Thus, CO_2 release through low-permeability rocks is limited to permeable pathways offered by open faults, and with minimal lateral CO_2 spread. As such, CO_2 flow could be restricted in the shallow subsurface.

Changes to fluid and rock properties encountered during ascent may also restrict CO_2 flow at depth. Our calculations find that as CO_2 density decreases during ascent, the seepage area or rock permeability must increase for mass transport to be conserved, unless fluids are not in pressure equilibrium with the rocks that they flow through. Baffles to flow are intrinsic to matrix and fracture complexities in geological units, and may encourage the channelling of ascending fluids. Fracture connectivity and rock permeability will not be continuous during fluid ascent from the reservoir. For example, there are several rock units in the Caprese overburden that have much lower permeability than that of the carbonate units directly overlying the reservoir, and so fracture permeabilities would be necessary for CO_2 transport through these units. What this amounts to is that, while free-phase CO_2 may not initially need fault-related rock permeability to leak from the Caprese reservoir, such pathways will become necessary for CO_2 transport to the surface. The location of CO_2 seeps in Italy is largely fault controlled (Ascione *et al.* 2014; Roberts *et al.* 2014) and, indeed, the Caprese Michelangelo seeps emerge along fault traces (Bonini 2009*b*). As such, natural CO_2 seeps illustrate the importance of considering the implications of fracture permeability for carbon capture and storage integrity (Bond *et al.* 2017).

Similarly, if CO_2 is migrating in its dissolved form, baffles to flow will arise from changes in the effective permeability when two-phase flow establishes towards the phase-transition depth, where CO_2 will start to exsolve from saturated waters. Flow rates will be inhibited as the effective permeability decreases, although gas lift may oppose this effect and, as discussed in our analysis, as the CO_2 and water phases have different properties they may follow different flow paths. If both phases subsequently reach the surface, several seep types will emerge in the seep cluster. Otherwise, if the hydraulic head driving the ascending waters is not great enough to enable the fluids to reach the surface, only dry CO_2 seeps will manifest. In this way, CO_2 can be transported from the reservoir in its dissolved phase and seep as a free phase at the surface. Conversely, CO_2 can leak from the reservoir as a free phase and dissolve into overlying aquifer units during ascent, and seep as a dissolved constituent in springs. Detailed geochemical studies could elucidate possible transport paths.

This is important for site selection. The likely style of CO_2 seep that might establish at the surface near a leaking store has implications for the design of subsurface and surface monitoring systems for both verification and for early warning systems. Additionally, if a leak or seep is detected, then the remediation strategies adopted would be dependent on the style of seep (Hepple & Benson 2003). Our work suggests that the characteristics of the overburden would allow some degree of forecasting of the risk and the potential risk-mitigation strategies.

Conclusions

We have studied nine boreholes in Italy that penetrate CO_2 reservoirs. Two reservoirs have high-flux surface CO_2 gas seeps within 2.5 km of the wellbore and are inferred to be leaking, whereas five have no surface seep expression and are inferred to be effectively sealed. The remaining two have small CO_2 springs located within 5 km of the borehole. These reservoirs are deemed to be inconclusively sealing, since the springs could originate from water circulation through carbonate rocks rather than from reservoir leakage.

The CO_2 reservoirs exhibit a range of subsurface structures and conditions. Reservoirs successfully retain CO_2 in the light or dense phase, and in some cases this CO_2 can be close to the critical point or exert high buoyancy pressures on the cap rock. The presence of surface CO_2 seeps is also unaffected by the structure or burial depth of the CO_2 reservoir, although the presence of evaporites may enhance its sealing capabilities. There are no seeps above reservoirs with fluid overpressure in the overburden; high fluid pressures may indicate the presence of an effective seal. The pressure seal could indicate the presence of a very-low-permeability formation, or where the net fluid pressure gradient between the reservoir and overpressured formation is directed downwards. Where there is a pressure seal, CO_2 buoyancy must be extraordinarily high to penetrate – or hydrofracture – the overpressured formation. CO_2 seeps are located at the surface above reservoirs with hydrostatically pressured overburden. These case studies are located near seismogenic extensional faults, which may be responsible for subsurface pressure connectivity at these sites, which, together with the higher permeability potentially offered by fault-related damage zones, may enable CO_2 to leak to surface.

We assess the geological conditions that could enable CO_2 leakage from the reservoir at the rates observed at the surface seeps. This finds that CO_2 is most likely to leak from the reservoir in a free phase. While formation waters have the potential to dissolve large quantities of CO_2, high leak rates of free-phase CO_2 can occur over smaller areas and lower permeability than those needed for the transport of CO_2-saturated water at the same rate. Significant (>100 t/day) leakage of dense-phase CO_2 from the reservoir can occur by flow through the overburden without the need for faults or enhanced-permeability pathways. In contrast, for the same mass flux of CO_2 leaking in its light phase, fault permeabilities are necessary since seepage through the overburden would otherwise have to occur over areas too large to be geological feasible. Changes in CO_2 properties during ascent from the leaking reservoir may therefore lead to the fluid channelling along high-permeability pathways such as faults. This leads to CO_2 venting and seep clustering observed at these sites in Italy.

This work informs the site selection of potential CO_2 stores, and the monitoring and leakage remediation strategies at selected sites. We find that all cases studied, leaking or sealing, would fail current storage site selection criteria. Although cap-rock thickness and reservoir conditions would be deemed suitable for most case studies, the proximity to faults would probably be considered detrimental to storage security. However, there is little guidance on the acceptable properties (density, scale, aperture) of fractures or faults, which is significant because our work suggests that, where the primary seal is breached, permeable fractures could permit significant leak rates from reservoirs containing dense-phase CO_2. We recommend that the overburden should be well characterized to inform the site selection process and monitoring design, and that more work is needed to detail the selection criteria for suitable overburden properties. The presence of a pressure seal could be used as a first-order screening criteria for potential stores, where this information is available. Monitoring should focus on high-permeability pathways, such faults. It must be borne in mind that faults do not need to connect the reservoir to the surface; even if they do not connect to the reservoir at depth, they could provide efficient fluid pathways to surface; or they could provide pathways through a cap rock into the overburden. Artificial pressurization of overburden units overlying a breached engineered CO_2 store could be an effective remediation option.

We thank Roberto Bencini (Independent Resources, Italy) for invaluable discussions, and, in addition, Alfredo Pugliese, Michele Impala, Francesco Bertello and Jonathon Craig at ENI for their hospitality and support. This research was funded by the University of Edinburgh Knowledge Transfer Partnership and the Scottish Carbon Capture and Storage (SCCS) consortium. SCCS is supported by the Scottish Funding Council, EPSRC, NERC and an industrial consortium of energy companies. RSH was funded by the Scottish Funding Council, NERC NE/H013474/1 and EPSRC EP/K000446/1. While completing this work, J.J. Roberts was funded by ClimateXChange, the Scottish Government-supported Centre for Expertise on Climate Change.

Correction notice: The original version was incorrect. This was due to an error in the Acknowledgements and Funding section, which omitted to list the funding bodies of RSH.

References

ALLIS, R., CHIDSEY, T., GWYNN, W., MORGAN, C., WHITE, S., ADAMS, M. & MOORE, J. 2001. Natural CO_2 reservoirs on the Colorado Plateau and Southern Rocky Mountains: Candidates for CO_2 sequestration. Paper presented at the DOE/NETL 1st National Conference of Carbon Sequestration, 14–17 May 2001, Washington, DC, USA.

ANNUNZIATELLIS, A., BEAUBIEN, S.E., BIGI, S., CIOTOLI, G., COLTELLA, M. & LOMBARDI, S. 2008. Gas migration along fault systems and through the vadose zone in the Latera caldera (central Italy): implications for CO_2 geological storage. *International Journal of Greenhouse Gas Control*, **2**, 353–372.

ASCIONE, A., BIGI, S. *ET AL.* 2014. The southern Matese active fault system: New geochemical and geomorphological evidence. Paper presented at the 33° Convegno Nazionale GNGTS, 25–27 Novembre 2014, Bologna, Italy.

BACHU, S. 2008. CO_2 storage in geological media: role, means, status and barriers to deployment. *Progress in Energy and Combustion Science*, **34**, 254–273.

BACHU, S. & BENNION, B. 2008. Effects of in-situ conditions on relative permeability characteristics of CO_2–brine systems. *Environmental Geology*, **54**, 1707–1722.

BARBA, S., CARAFA, M.M.C., MARIUCCI, M.T., MONTONE, P. & PIERDOMINICI, S. 2009. Present-day stress-field modelling of southern Italy constrained by stress and GPS data. *Tectonophysics*, **482**, 193–204, https://doi.org/10.1016/j.tecto.2009.10.017

BENNION, D. & BACHU, S. 2008. Drainage and imbibition relative permeability relationships for supercritical CO_2/brine and H_2S/brine systems in intergranular sandstone, carbonate, shale, and anhydrite rocks. *SPE Reservoir Evaluation & Engineering*, **11**, 487–496.

BENSON, S.M., APPS, J., HEPPLE, R., LIPPMANN, M., TSANG, C.F. & LEWIS, C. 2003. Health, safety and environmental risk assessment for geologic storage of carbon dixide: Lessons learned from industrial and natural analogues. *In*: GALE, J. & KAYA, Y. (eds) *Greenhouse Gas Control Technologies – 6th International Conference*. Pergamon, Oxford, 243–248.

BICOCCHI, G., TASSI, F. *ET AL.* 2013. The high pCO_2 Caprese Reservoir (Northern Apennines, Italy): relationships between present- and paleo-fluid geochemistry and structural setting. *Chemical Geology*, **351**, 40–56.

BOND, C.E., KREMER, Y. *ET AL.* 2017. The physical characteristics of a CO_2 seeping fault: the implications of fracture permeability for carbon capture and storage integrity. *International Journal of Greenhouse Gas Control*, **61**, 49–60, https://doi.org/10.1016/j.ijggc.2017.01.015

BONINI, M. 2009*a*. Mud volcano eruptions and earthquakes in the Northern Apennines and Sicily, Italy. *Tectonophysics*, **474**, 723–735.

BONINI, M. 2009*b*. Structural controls on a carbon dioxide-driven mud volcano field in the Northern Apennines (Pieve Santo Stefano, Italy): relations with pre-existing steep discontinuities and seismicity. *Journal of Structural Geology*, **31**, 44–54.

BROZZETTI, F. 2011. The Campania–Lucania Extensional Fault System, southern Italy: a suggestion for a uniform model of active extension in the Italian Apennines. *Tectonics*, **30**, TC5009, https://doi.org/10.1029/2010TC002794

BURNSIDE, N.M., NAYLOR, M., KIRK, K. & WHITTAKER, F. 2013. QICS Work Package 1: migration and trapping of CO_2 from a reservoir to the seabed or land surface. *Energy Procedia*, **37**, 4673–4681.

CASERO, P. 2004. Structural setting of petroleum exploration plays in Italy. *In*: CRESCENTI, U. (ed.) *Geology of Italy*. Società Geologica Italiana, Rome.

CASERO, P. 2005. Southern Apennines geologic framework and related petroleum systems. *Atti Ticinensi di Scienze della Terra*, **10**, 37–43.

CHADWICK, R.A., ARTS, R., BERNSTONE, C., MAY, F., THIBEAU, S. & ZWEIGEL, P. 2008. *Best Practice for the Storage of CO_2 in Saline Aquifers – Observations and Guidelines from the SACS and CO2STORE projects*. British Geological Survey, Occasional Publications, **14**.

CHIODINI, G. & VALENZA, M. 2008. *Googas – The Catalogue of Italian Gas Emissions [Online]*. Istituto Nazionale di Geofisica e Vulcanologia, Pisa, http://googas.ov.ingv.it [last accessed March 2009].

CHIODINI, G., CARDELLINI, C., AMATO, A., BOSCHI, E., CALIRO, S., FRONDINI, F. & VENTURA, G. 2004. Carbon dioxide Earth degassing and seismogenesis in central and southern Italy. *Geophysical Research Letters*, **31**, L07615, https://doi.org/10.1029/2004GL019480

Chiodini, G., Valenza, M., Cardellini, C. & Frigeri, A. 2008. A New Web-Based Catalog of Earth Degassing Sites in Italy. *Eos, Transactions of the American Geophysical Union*, **37**, 341–342.

Chiodini, G., Granieri, D., Avino, R., Caliro, S., Costa, A., Minopoli, C. & Vilardo, G. 2010. Non-volcanic CO_2 Earth degassing: Case of Mefite d'Ansanto (southern Apennines), Italy. *Geophysical Research Letters*, **37**, 11303.

Collettini, C. & Barchi, M.R. 2002. A low-angle normal fault in the Umbria region (Central Italy): a mechanical model for the related microseismicity. *Tectonophysics*, **359**, 97–115.

Collettini, C. & Barchi, M.R. 2004. A comparison of structural data and seismic images for low-angle normal faults in the Northern Apennines (Central Italy): constraints on activity. *In*: Alsop, G.I., Holdsworth, R.E., McCaffrey, K.J.W. & Hand, M. (eds) *Flow Processes in Faults and Shear Zones*. Geological Society, London, Special Publications, **224**, 95–112, https://doi.org/10.1144/GSL.SP.2004.224.01.07

Collettini, C., Cardellini, C., Chiodini, G., De Paola, N., Holdsworth, R.E. & Smith, S.A.F. 2008. Fault weakening due to CO_2 degassing in the Northern Apennines: short- and long-term processes. *In*: Wibberley, C.A.J., Kurz, W., Imber, J., Holdsworth, R.E. & Collettini, C. (eds) *The Internal Structure of Fault Zones: Implications for Mechanical and Fluid-Flow Properties*. Geological Society, London, Special Publications, **299**, 175–194, https://doi.org/10.1144/SP299.11

Di Bucci, D., Massa, B., Tornaghi, M. & Zuppetta, A. 2006. Structural setting of the Southern Apennine fold-and-thrust belt (Italy) at hypocentral depth: the Calore Valley case history. *Journal of Geodynamics*, **42**, 175–193.

Dixon, T., Mccoy, S.T. & Havercroft, I. 2015. Legal and regulatory developments on CCS. *International Journal of Greenhouse Gas Control*, **40**, 431–448.

Dockrill, B. & Shipton, Z.K. 2010. Structural controls on leakage from a natural CO_2 geologic storage site: central Utah, U.S.A. *Journal of Structural Geology*, **32**, 1768–1782.

Duchi, V., Minissale, A., Vaselli, O. & Ancillotti, M. 1995. Hydrogeochemistry of the Campania region in southern Italy. *Journal of Volcanology and Geothermal Research*, **67**, 313–328.

Elío, J., Ortega, M.F., Nisi, B., Mazadiego, L.F., Vaselli, O., Caballero, J. & Grandia, F. 2015. CO_2 and Rn degassing from the natural analog of Campo de Calatrava (Spain): implications for monitoring of CO_2 storage sites. *International Journal of Greenhouse Gas Control*, **32**, 1–14.

EU 2009. Directive 2009/31/EC on the geological storage of carbon dioxide. *Official Journal of the European Union*, L 140/114–L 140/135, http://eur-lex.europa.eu/legal-content/EN/ALL/?uri=CELEX%3A32009L0031

Frezzotti, M., Peccerillo, L., Panza, A. & May, G. 2009. Carbonate metasomatism and CO_2 lithosphere–asthenosphere degassing beneath the western mediterranean: an integrated model arising from petrological and geophysical data. *Chemical Geology*, **262**, 108–120.

Geothopica. 2010. *Geothermal Resources National Inventory [Online]*. CNR Geosciences and Earth Resources Institute, Pisa, http://unmig.sviluppoeconomico.gov.it/unmig/geotermia/inventario/inventario.asp [last accessed January 2010].

Ghisetti, F. & Vezzani, L. 2002. Normal faulting, transcrustal permeability and seismogenesis in the Apennines (Italy). *Tectonophysics*, **348**, 155–168.

Gilfillan, S., Ballentine, C., Lollar, B.S., Stevens, S., Schoell, M. & Cassidy, M. 2008. Quantifying the precipitation and dissolution of CO_2 within geological carbon storage analogues. *Geochimica et Cosmochimica Acta*, **72**, A309–A309.

Gilfillan, S.M.V., Lollar, B.S. *et al.* 2009. Solubility trapping in formation water as dominant CO_2 sink in natural gas fields. *Nature*, **458**, 614–618.

Gudmundsson, A., Berg, S.S., Lyslo, K.B. & Skurtveit, E. 2001. Fracture networks and fluid transport in active fault zones. *Journal of Structural Geology*, **23**, 343–353.

Hannon, M.J., Jr & Esposito, R.A. 2015. Screening considerations for caprock properties in regards to commercial-scale carbon-sequestration operations. *International Journal of Greenhouse Gas Control*, **32**, 213–223.

Heinicke, J., Braun, T., Burgassi, P., Italiano, F. & Martinelli, G. 2006. Gas flow anomalies in seismogenic zones in the Upper Tiber Valley, Central Italy. *Geophysical Journal International*, **167**, 794–806.

Hepple, R.P. & Benson, S.M. 2003. Implications of surface seepage on the effectiveness of geologic storage of carbon dioxide as a climate change mitigation strategy. eScholarship, Lawrence Berkeley National Laboratory, University of California, https://escholarship.org/uc/item/86c1t97r

Heptonstall, P., Markusson, N. & Chalmers, H. 2012. Pathways and branching points for CCS to 2030. *In*: *Carbon Capture and Storage: Realising the Potential?* Report UKERC/WP/ESY/2012/001. UK Energy Research Centre (UKERC), London, http://ukerc.rl.ac.uk/UCAT/PUBLICATIONS/Carbon_Capture_and_Storage_Realising_the_Potential_Pathways_and_branching_points_for_CCS_to_2030.pdf

Huang, F., Li, M., Le, L. & Starling, K. 1985. An accurate equation of state for carbon dioxide. *Journal of Chemical Engineering of Japan*, **18**, 490–496.

Hubbert, M.K. & Rubey, W.W. 1959. Role of fluid pressure in mechanics of overthrust faulting: i. Mechanics of fluid-filled porous solids and its application to overthrust faulting. *Geological Society of America Bulletin*, **70**, 115–166.

IEA-GHG. 2009. *CCS Site Characterisation Criteria*. IEA Greenhouse Gas R&D Programme, Global CCS Institute, Melbourne, https://www.globalccsinstitute.com/publications/ccs-site-characterisation-criteria

Improta, L., Iannaccone, G., Capuano, P., Zollo, A. & Scandone, P. 2000. Inferences on the upper crustal structure of Southern Apennines (Italy) from seismic refraction investigations and subsurface data. *Tectonophysics*, **317**, 273–297.

Improta, L., Bonagura, M., Capuano, P. & Iannaccone, G. 2003*a*. An integrated geophysical investigation of the upper crust in the epicentral area of the 1980, Ms = 6.9, Irpinia earthquake (Southern Italy). *Tectonophysics*, **361**, 139–169.

Improta, L., Zollo, A., Bruno, P.P., Herrero, A. & Villani, F. 2003*b*. High-resolution seismic tomography across the 1980 (Ms 6.9) Southern Italy earthquake fault scarp. *Geophysical Research Letters*, **30**, https://doi.org/10.1029/2003GL017077

Islam, A.W. & Carlson, E.S. 2012. Viscosity models and effects of dissolved CO_2. *Energy Fuels*, **26**, 5330–5336, https://doi.org/10.1021/ef3006228

ISPRA 2007. *ITaly HAzard from CApable faults (ITHACA) [Online]*. Istituto Superiore per la Protezione e la Ricerca Ambientale (ISPRA), Rome, http://www.isprambiente.gov.it/en/projects/soil-and-territory/italy-hazards-from-capable-faulting

Italiano, F., Martinelli, G. & Plescia, P. 2008. CO_2 degassing over seismic areas: the role of mechanochemical production at the study case of Central Apennines. *Pure and Applied Geophysics*, **165**, 75–94.

Jarvis, A., H.I.Reuter,, A.Nelson, & Guevara, E. 2008. *Hole-filled SRTM for the globe Version 4, available from the CGIAR-CSI SRTM 90 m database*. Consortium for Spatial Information (CGIAR-CSI), http://srtm.csi.cgiar.org/

Jones, D.G., Beaubien, S.E. *et al.* 2015. Developments since 2005 in understanding potential environmental impacts of CO2 leakage from geological storage. *International Journal of Greenhouse Gas Control*, **40**, 350–377.

Kampman, N., Burnside, N.M., Bickle, M., Shipton, Z.K., Ellam, R.M. & Chapman, H. 2010. Coupled CO_2-leakage and in situ fluid–mineral reactions in a natural CO_2 reservoir, Green River, Utah. *Geochimica et Cosmochimica Acta*, **74**, A492–A492.

Kirk, K. 2011. *Natural CO_2 Flux Literature Review for the QICS Project*. Energy Programme. British Geological Survey Commissioned Report CR/11/005.

Krevor, S., Blunt, M.J., Benson, S.M., Pentland, C.H., Reynolds, C., Al-Menhali, A. & Niu, B. 2015. Capillary trapping for geologic carbon dioxide storage – From pore scale physics to field scale implications. *International Journal of Greenhouse Gas Control*, **40**, 221–237.

Kuhlman, K.L., Malama, B. & Heath, J.E. 2015. Multiporosity flow in fractured low-permeability rocks. *Water Resources Research*, **51**, 848–860.

Lewicki, J.L., Birkholzer, J. & Tsang, C.F. 2007. Natural and industrial analogues for leakage of CO2 from storage reservoirs: identification of features, events, and processes and lessons learned. *Environmental Geology*, **52**, 457–467.

Likhachev, E. 2003. Dependence of water viscosity on temperature and pressure. *Technical Physics*, **48**, 514–515.

Lu, J.M., Wilkinson, M., Haszeldine, R.S. & Fallick, A.E. 2009. Long-term performance of a mudrock seal in natural CO_2 storage. *Geology*, **37**, 35–38.

Malagnini, L., Lucente, F.P., De Gori, P., Akinci, A. & Munafo, I. 2012. Control of pore fluid pressure diffusion on fault failure mode: insights from the 2009 L'Aquila seismic sequence. *Journal of Geophysical Research: Solid Earth*, **117**, B05302.

McGinnis, D.F., Schmidt, M., Delsontro, T., Themann, S., Rovelli, L., Reitz, A. & Linke, P. 2011. Discovery of a natural CO_2 seep in the German North Sea: implications for shallow dissolved gas and seep detection. *Journal of Geophysical Research: Oceans*, **116**, C03013.

Milano, G., Di Giovambattista, R. & Ventura, G. 2006. Seismicity and stress field in the Sannio-Matese area. *Annals of Geophysics*, **49,** (suppl.), 347–356.

Miller, S.A., Collettini, C., Chiaraluce, L., Cocco, M., Barchi, M. & Kaus, B.J.P. 2004. Aftershocks driven by a high-pressure CO_2 source at depth. *Nature*, **427**, 724–727.

Minissale, A. 2004. Origin, transport and discharge of CO_2 in central Italy. *Earth-Science Reviews*, **66**, 89–141.

Miocic, J.M., Gilfillan, S.M.V., Roberts, J.J., Edlmann, K., McDermott, C.I. & Haszeldine, R.S. 2016. Controls on CO_2 storage security in natural reservoirs and implications for CO_2 storage site selection. *International Journal of Greenhouse Gas Control*, **51**, 118–125.

Nara, Y., Meredith, P.G., Yoneda, T. & Kaneko, K. 2011. Influence of macro-fractures and micro-fractures on permeability and elastic wave velocities in basalt at elevated pressure. *Tectonophysics*, **503**, 52–59.

Naylor, M., Wilkinson, M. & Haszeldine, R.S. 2011. Calculation of CO_2 column heights in depleted gas fields from known pre-production gas column heights. *Marine and Petroleum Geology*, **28**, 1083–1093.

Nickschick, T., Kämpf, H., Flechsig, C., Mrlina, J. & Heinicke, J. 2015. CO_2 degassing in the Hartoušov mofette area, western Eger Rift, imaged by CO_2 mapping and geoelectrical and gravity surveys. *International Journal of Earth Sciences*, **104**, 2107–2129.

Nicolai, C. & Gambini, R. 2007. Structural architecture of the Adria-platform-and-basin system. *In*: Mazzotti, A., Patacca, E. & Scandone, P. (eds) *Results of the CROP Project, Sub-Project CROP-04, Southern Apennines (Italy). Bollettino Della Societa Geologica Italiana*, **7**, 21–37.

O'Connor, S.A., Swarbrick, R.E. & Jones, D. 2008. Where has all the pressure gone? Evidence from pressure reversals and hydrodynamic flow. *First Break*, **26**, 55–60.

Osborne, M.J. & Swarbrick, R.E. 1997. Mechanisms for generating overpressure in sedimentary basins; a reevaluation. *American Association of Petroleum Geologists Bulletin*, **81**, 1023–1041.

Patacca, E., Scandone, P., Di Luzio, E., Cavinato, G.P. & Parotto, M. 2008. Structural architecture of the central Apennines: interpretation of the CROP 11 seismic profile from the Adriatic coast to the orographic divide. *Tectonics*, **27**, TC2006, https://doi.org/10.1029/2005TC001917

Pischiutta, M., Anselmi, M., Cianfarra, P., Rovelli, A. & Salvini, F. 2013. Directional site effects in a non-volcanic gas emission area (Mefite d'Ansanto, southern Italy): evidence of a local transfer fault transversal to large NW–SE extensional faults? *Physics and Chemistry of the Earth, Parts A/B/C*, **63**, 116–123.

Pruess, K. 2008*a*. Leakage of CO_2 from geologic storage: role of secondary accumulation at shallow depth.

International Journal of Greenhouse Gas Control, **2**, 37–46.

PRUESS, K. 2008*b*. On CO_2 fluid flow and heat transfer behavior in the subsurface, following leakage from a geologic storage reservoir. *Environmental Geology*, **54**, 1677–1686.

REVEILLERE, A. & ROHMER, J. 2011. Managing the risk of CO_2 leakage from deep saline aquifer reservoirs through the creation of a hydraulic barrier. *Energy Procedia*, **4**, 3187–3194.

ROBERTS, G.P. 2008. Visualisation of active normal fault scarps in the Apennines, Italy: a key to assessment of tectonic strain release and earthquake rupture. *In*: DE PAOR, D. (ed.) *Google Earth Science. Journal of the Virtual Explorer*, **29**, paper 4, https://doi.org/10.3809/jvirtex.2008.00197

ROBERTS, J.J. 2013. *Natural CO_2 Fluids in Italy: Implications for the Leakage of Geologically Stored CO_2*. PhD thesis, University of Edinburgh.

ROBERTS, J.J., WOOD, R.A. & HASZELDINE, R.S. 2011. Assessing the health risks of natural CO_2 seeps in Italy. *Proceedings of the National Academy of Sciences of the United States of America*, **108**, 16545–16548.

ROBERTS, J.J., WOOD, R.A., WILKINSON, M. & HASZELDINE, S. 2014. Surface controls on the characteristics of natural CO_2 seeps: implications for engineered CO_2 stores. *Geofluids*, **15**, 453–463.

ROSTRON, B., WHITE, D., HAWKES, C. & CHALATURNYK, R. 2014. Characterization of the Aquistore CO_2 project storage site, Saskatchewan, Canada. *In*: DIXON, T., HERZOG, H. & TWINNING, S. (eds) *12th International Conference on Greenhouse Gas Control Technologies, GHGT-12. Energy Procedia*, **63**, 2977–2984.

SANTO, A., ASCIONE, A., PRETE, S.D., CRESCENZO, G.D. & SANTANGELO, N. 2011. Collapse sinkholes distribution in the carbonate massifs of central and southern Apennines. *Acta Carsologica*, **40**, 95–112.

SAVARY, V., BERGER, F., DUBOIS, M., LACHARPAGNE, J.-C., PAGES, A., THIBEAU, S. & LESCANNE, M. 2012. The solubility of $CO_2 + H_2S$ mixtures in water and 2 M NaCl at 120 deg C and pressures up to 35 MPa. *International Journal of Greenhouse Gas Control*, **10**, 123–133.

SCROCCA, D., CARMINATI, E. & DOGLIONI, C. 2005. Deep structure of the southern Apennines, Italy: thin-skinned or thick-skinned? *Tectonics*, **24**, TC2005, https://doi.org/10.1029/2004TC001634

SHAFFER, G. 2010. Long-term effectiveness and consequences of carbon dioxide sequestration. *Nature Geoscience*, **3**, 464–467.

SMITH, M., CAMPBELL, D., MACKAY, E. & POLSON, D. (eds). 2011. *CO_2 Aquifer Storage Site Evalutation and Monitoring: Understanding the Challenges of CO_2 storage: Results of the CASSEM Project*. Scottish Carbon Capture and Storage, Edinburgh.

SPAN, R. & WAGNER, W. 1996. A new equation of state for carbon dioxide covering the fluid region from the triple point temperature to 1100 K at pressures up to 800 MPa. *Journal of Physical and Chemical Reference Data*, **25**, 1509–1596.

SPYCHER, N., PRUESS, K. & ENNIS-KING, J. 2003. CO_2–H_2O mixtures in the geological sequestration of CO_2. I. Assessment and calculation of mutual solubilities from 12 to 100°C and up to 600 bar. *Geochimica et Cosmochimica Acta*, **67**, 3015–3031.

STREIT, J.E. & COX, S.F. 2001. Fluid pressures at hypocenters of moderate to large earthquakes. *Journal of Geophysical Research*, **106**, 2235–2243.

TALUKDER, A.R. 2012. Review of submarine cold seep plumbing systems: leakage to seepage and venting. *Terra Nova*, **24**, 255–272.

TRIPPETTA, F., COLLETTINI, C., BARCHI, M.R., LUPATTELLI, A. & MIRABELLA, F. 2013. A multidisciplinary study of a natural example of a CO_2 geological reservoir in central Italy. *International Journal of Greenhouse Gas Control*, **12**, 72–83.

WILKINSON, M., GILFILLAN, S.M.V., HASZELDINE, R.S. & BALLENTINE, C.J. 2009. Plumbing the depths – testing natural tracers of subsurface CO2 origin and migration, Utah, USA. *In*: GROBE, M., PASHIN, J.C. & DODGE, R.L. (eds) *Carbon Dioxide Sequestration in Geological Media – State of the Science*. American Association of Petroleum Geologists, Studies in Geology, **59**, 619–636.

WILKINSON, M., HASZELDINE, R.S., MACKAY, E., SMITH, K. & SARGEANT, S. 2013. A new stratigraphic trap for CO_2 in the UK North Sea: appraisal using legacy information. *International Journal of Greenhouse Gas Control*, **12**, 310–322.

WRI. 2008. *CCS Guidelines: Guidelines for Carbon Dioxide Capture, Transport, and Storage*. World Resources Institute (WRI), Washington, DC.

YANG, C.-Y. & MANGA, M. 2009. *Earthquakes and Water*. Springer, Berlin.

ZWAAN, B. & GERLAGH, R. 2009. Economics of geological CO_2 storage and leakage. *Climatic Change*, **93**, 285–309.

ZWEIGEL, P., LINDEBERG, E., MOEN, A. & WESSEL-BERG, D. 2004. Towards a methodology for top seal efficacy assessment for underground CO_2 storage. *In*: RUBIN, E.S., KEITH, D.W., GILBOY, C.F., WILSON, M., MORRIS, T., GALE, J. & THAMBIMUTHU, K. (eds) *Greenhouse Gas Control Technologies 7. Proceedings of the 7th International Conference on Greenhouse Gas Control Technologies*, 5 September 2004, Vancouver, Canada. Elsevier, Amsterdam, 1323–1328, https://doi.org/10.1016/B978-008044704-9/50145-2

An improved procedure for pre-drill calculation of fracture pressure

RICHARD W. LAHANN[1]* & RICHARD E. SWARBRICK[2]

[1]*Lahann Geoservices, 980 Timbercrest Road, Nashville, IN 47448, USA*

[2]*Swarbrick Geopressure Consultancy, 43 Ancroft Garth, Durham DH1 2UD, UK*

**Correspondence: rlahann@indiana.edu*

Abstract: Pre-drill modelling of fracture pressure (FP) is an essential part of well planning, reserve estimation and evaluation of the potential for inducing seismicity as the result of fluid injection. Estimation of stress ratio or Poisson's ratio values or compaction state with depth is required in frequently used FP models. A new method to estimate FP is proposed which is based on Leak Off (LOT) and pore fluid pressure (Pp) data from offset wells and vertical stress (S_v)–depth relationships. LOT/S_v ratios observed in intervals of offset wells that are normally pressured (hydrostatic) are used to define an expected FP/S_v ratio for hydrostatic Pp conditions for all depths. Typical FP/S_v ratios for hydrostatic conditions derived using LOT data range from 0.81 to 0.89. Observed LOT values associated with Pp greater than hydrostatic (overpressured) in offset wells are used to quantify the rate of increase in FP with increasing overpressure (OP). The expected FP for hydrostatic conditions is compared with observed LOT values from depths where the pore fluid is overpressured and a relationship of increased FP, relative to the expected FP for hydrostatic conditions (residual FP (FPr)) with increasing OP is defined. The FPr:OP ratio typically ranges from 0.24 to 0.43. Fracture pressure models developed by this procedure may be used to predict FP for wells in different water depths and with Pp conditions different from those in the offset wells. The use of the model is demonstrated in three case studies taken from different geological settings: the Scotian shelf (offshore Nova Scotia), offshore Central Gulf of Mexico and the chalk interval from the Central North Sea.

The creation of an anticipated fracture pressure (FP)–depth profile is an integral part of well planning. During drilling and other well-site operations, the pressure exerted by the drilling fluid on the formation exposed in the wellbore is maintained at a pressure above that of the fluids in the pore space of the rocks (Pp) and below the FP. The FP–depth profile of the formations exposed in the wellbore places limits on the density of drilling fluid that can be circulated during drilling.

Drilling fluid circulation systems are designed to prevent large fluxes of fluid either into or out of the formation while maintaining the ability to cool the drill string and remove the cuttings. To accomplish these goals, circulation systems combine mud weight and pump pressures to be at the same pressure as the Pp ('balanced'), slightly below Pp ('underbalanced') or above Pp ('overbalanced'). If the pressure of the drilling fluid is below the Pp in the formation, an influx of fluid into the wellbore will occur, and where the formation has sufficient permeability, the influx can affect drilling operations. Where the pressure of the drilling fluid is greater than the FP of any of the formations exposed in the wellbore, then fracturing of the weak formation can occur. The formation fracturing may be relatively minor and result in fluid flow-back to the wellbore when the borehole pressure is reduced by drilling operations (borehole breathing). If the fracturing is more severe, substantial fluid losses to the formation can occur, which can lead to cross-reservoir communication, propagation of fractures to the surface (e.g. Champion Field, Brunei, Tingay *et al.* 2005; LUSI mud volcano, Davies *et al.* 2007) or blowout of the well.

An accurate pre-drill FP model is therefore essential to the creation of an appropriate casing programme which will allow for safe and cost-effective drilling. The challenge of maintaining static and drilling wellbore fluid pressure between Pp and FP (drilling window) is substantially greater when the Pp is much greater than hydrostatic Pp (see Fig. 1). Pp greater than hydrostatic pressure is termed 'overpressure' (OP) and is a frequent characteristic of formations found below deep water or buried to great depths.

LOTs are often performed after casing has been set in a wellbore. The LOT identifies the pressure, at the depth of the casing shoe, at which increasing fluid pressure in the wellbore causes the formation to begin to accept fluid at an increasing rate. Ideally, an LOT pressure indicates the initiation of fractures in the formation; a LOT should be a larger pressure than a formation integrity test (FIT) or a minimum horizontal stress (S_h) or a fracture propagation pressure. Gaarenstroom *et al.* (1993) provides a more

From: TURNER, J. P., HEALY, D., HILLIS, R. R. & WELCH, M. J. (eds) 2017. *Geomechanics and Geology*. Geological Society, London, Special Publications, **458**, 213–225.
First published online May 30, 2017, https://doi.org/10.1144/SP458.13

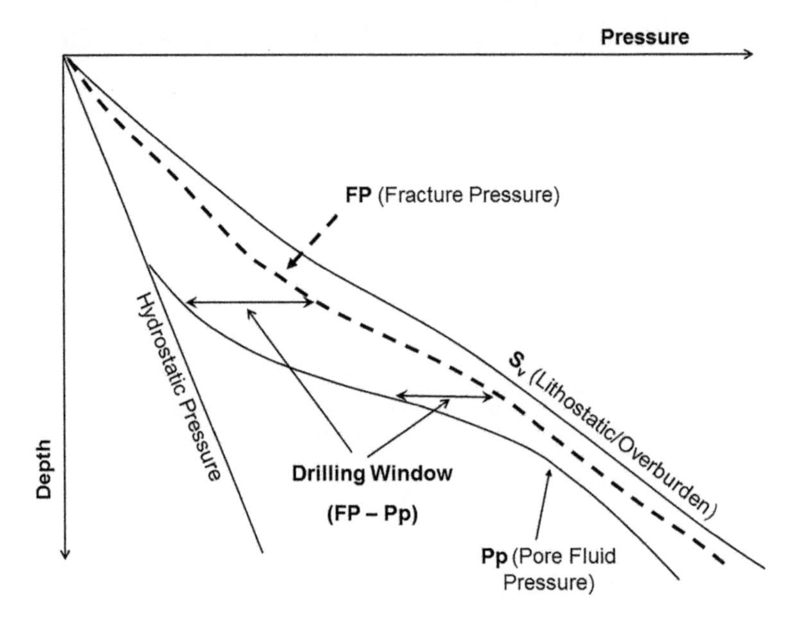

Fig. 1. Relationship of pore fluid pressure (Pp), vertical stress (S_v), fracture pressure (FP) and drilling window.

detailed discussion of these pressures. LOTs are a common basis for estimating fracture strength of the formation at the casing seat. Compilation of multiple LOTs from multiple wells, referenced to a common water depth or seabed surface, can form the basis for an empirical FP–depth model. Care should be taken to ensure that all the compiled data represent LOTs and that other pressure tests, particularly FITs, are not included in the dataset. Inclusion of FITs in a FP study can lead to underestimation of the FP.

FP models (based on LOTs) are commonly used as part of a seal-breach risk evaluation. In this application, the difference between FP and Pp, at the same depth, is an estimate of the capacity of the reservoir to trap hydrocarbons. Decreased difference between Pp and FP is associated with increased risk of seal breach (Gaarenstroom *et al.* 1993; Seldon & Flemings 2005; Swarbrick *et al.* 2010). In this way an accurate FP model becomes important as part of an exploration risk evaluation strategy.

In a similar way, an accurate FP model can be a useful supplement or independent check on basin modelling/hydrocarbon migration models. If these models generate Pp in either reservoirs or source rocks that exceed modelled or observed FPs, then the basin model results may be viewed as suspect.

A number of FP models are described in the petroleum geology/engineering literature and are used by the energy and underground waste repository industries to create FP–depth models prior to and during drilling. This paper introduces a new FP model that explicitly links FP to a fraction of S_v and to OP and offers advantages over traditional methods. Key features of a successful FP model are considered to be utility, accuracy and ease of calibration. The data required to calibrate a local model using the new method is available from a pressure–depth plot (relative to seabed) which contains an S_v profile, offset well LOT data, a hydrostatic Pp profile and Pp data for the offset wells.

Existing fracture pressure models

General relationship to depth and overpressure

The data in Figure 2 demonstrate that in formations that are hydrostatically pressured FP can often be represented by a pseudolinear increase with depth or a constant relationship relative to the vertical stress, S_v. A linear increase in FP with depth neglects the increased density of shallow sediments as compaction with increasing effective stress occurs. The increased density increases the S_v gradient. Grauls (1999) proposed representing the FP in the hydrostatic section as a fraction of S_v. Estimating FP as a fraction of S_v is only accurate in the hydrostatically pressured portion of a well profile and will substantially underestimate the FP in intervals with large amounts of OP. In intervals with significant OP, the Pp may exceed an FP predicted

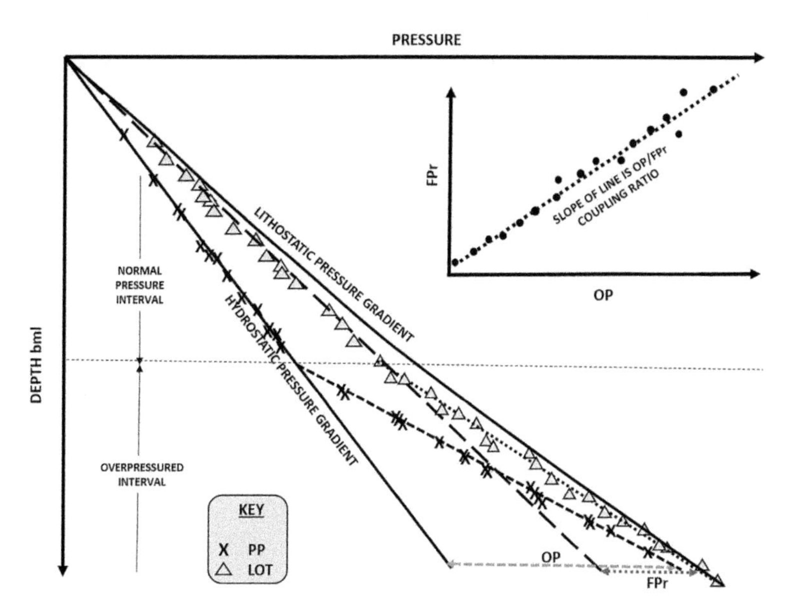

Fig. 2. Example from Nile Delta of change in FP gradient relative to changes in Pp, plotted relative to S_v. The shallow interval is hydrostatically pressured and the deeper interval is overpressured. Figure modified after Nashaat (1998) and reprinted with permission of the AAPG, whose permission is required for further use.

by a trend that was established in the hydrostatically pressured interval (Fig. 2).

The three FP models discussed in this paper and the new model require identical datasets for calibration and application. Offset well data are required for generation of hydrostat and S_v pressure–depth models. A Pp interpretation for the calibration wells is required as well as LOT data. These datasets are used to determine k_i for the Matthews & Kelly (1967) model and Poisson's ratio for the Eaton (1969) model, unless some other source is available. These same datasets are used to calibrate a hydrostatic fracture model and an OP influence for the Breckles & van Eekelen (1982) model and the model described in this paper. Despite the extreme similarities in the required data, the various models treat the inputs differently and thus provide different FP interpretations.

Matthews & Kelly (1967) and Eaton (1969)

These models are considered together because they share the same equation:

$$FP = Pp + k*(S_v - Pp)$$

The k variable is called the stress ratio (k_i) by Matthews & Kelly (1967) and is Poisson's ratio in the Eaton (1969) model (Fig. 3). Both models address the issue of underestimating the FP in overpressured intervals or sections by ensuring that the minimum value of the FP is always greater than Pp. Matthews & Kelly (1967) indicate that the k_i value employed in an overpressured section should be the k_i value appropriate for the compaction condition of the overpressured shale and should be determined by reference to shales in the hydrostatically pressured section. Matthews & Kelly (1967) provide two depth plots of k_i for regions of the Gulf of Mexico. The plots are for hydrostatic conditions since the k_i values increase monotonically with depth. The utility of these plots is limited by the Gulf of Mexico calibration and using a S_v profile of 1.0 psi/ft. The k_i values published by Matthews & Kelly for the shallow section are artificially low, relative to deeper depths, in order to compensate for the overestimation of S_v in the shallow section (Swarbrick & Lahann 2016). Calibration of k_i for a well with OP requires a model of k_i as a function of effective stress (S_v–Pp) so an appropriate value can be entered into the equation.

The Eaton (1969) method has constraints similar to those noted for Matthews & Kelly (1967) method. Poisson's ratio must be available for the entire depth range of the well. Eaton (1969) and Eaton & Eaton (1977) have provided Poisson's ratio/depth curves but these are influenced by the local depositional history, porosity preservation and OP development and should ideally be related to effective stress. Efforts have been made (Dutta 2002) to extract Poisson's ratio from sonic or seismic velocity data for use in the Eaton (1969) model. The Dutta (2002)

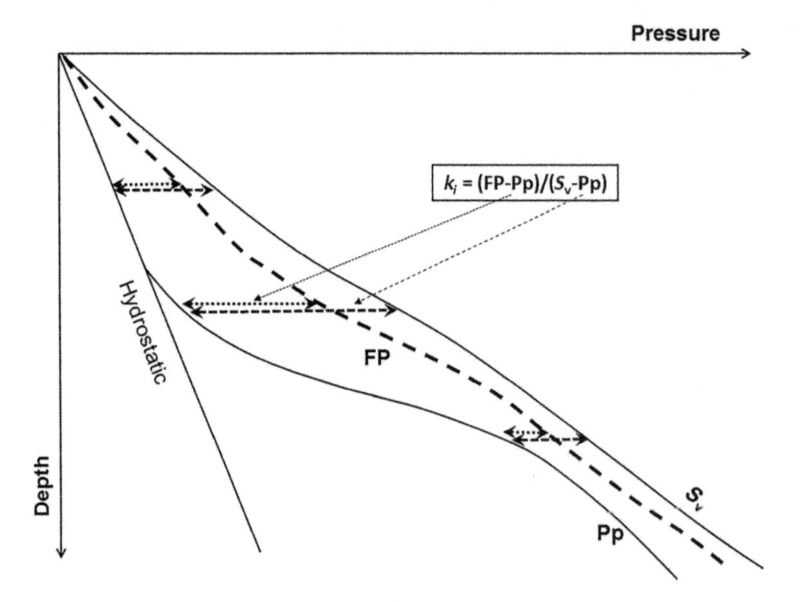

Fig. 3. Graphical display of the *k* term in the Matthews & Kelley and the Eaton methods.

approach requires shear velocity for the Poisson's ration calculation; shear velocity data may not be available in many situations requiring FP calculation.

Both of these methods place great emphasis on having the proper Pp as an entry point into the calculation. A miscalibration of k_i can create significant error at low OP regimes; miscalibration of k_i is less serious in highly overpressured settings because the (S_v−Pp) term is small in these pressure regimes (Fig. 3). Errors in Pp estimation can result in quite large errors in FP estimates because the Pp term appears in the FP equation.

Breckles & van Eekelen (1982)

The FP modelling method of Breckles & van Eekelen is quite different from those of Matthews & Kelly (1967) and Eaton (1969) in that the variations in FP with OP are not dealt with by the $k*(S_v-\mathrm{Pp})$ term. Breckles & van Eekelen (1982) developed FP equations for the US Gulf Coast, Venezuela and Brunei with the form:

$$\mathrm{FP} = X*D^y + Z*(\mathrm{overpressure}),$$

where D is depth, and X, Y and Z are locally derived constants. The coupling term of FP to OP (Z) ranged from 0.46 to 0.56. For deeper intervals Breckles & van Eekelen used a hydrostatic FP term which increased linearly as a function of depth. The key factors in this approach are the power functions, which provide a depth-based estimate of FP for a hydrostatic fluid pressure, and the inclusion of a term that relates the deviation from the hydrostatic FP model to the amount of OP developed. The power function creates an FP profile which is non-linear with depth and is progressively offset to slightly higher FP with depth, relative to a linear model. The use of a single power function for a region neglects local effects on the S_v such as low-density salt layers or high-density carbonate intervals.

New model

We propose a new model for modelling FP using data from offset wells. The key equation to the new model is:

$$\mathrm{FP} = A*S_v + B*\mathrm{Overpressure},$$

where A and B are empirical constants derived from available well data. The form of the equation is similar to that of Breckles & van Eekelen (1982), differing in that the A term is multiplied by S_v, rather than depth raised to a power. The approach is indicated schematically in Figure 4. LOT test data associated with hydrostatic formations are used to define a relationship of FP with S_v in hydrostatic conditions. Increasing FP gradient with depth for hydrostatic samples is automatically incorporated into the model by the use of S_v, rather than depth, to calculate the hydrostatic FP. Typical values of A are from 0.81 to 0.89 (Swarbrick & Lahann 2016). For many basins, especially in deep water, the top of

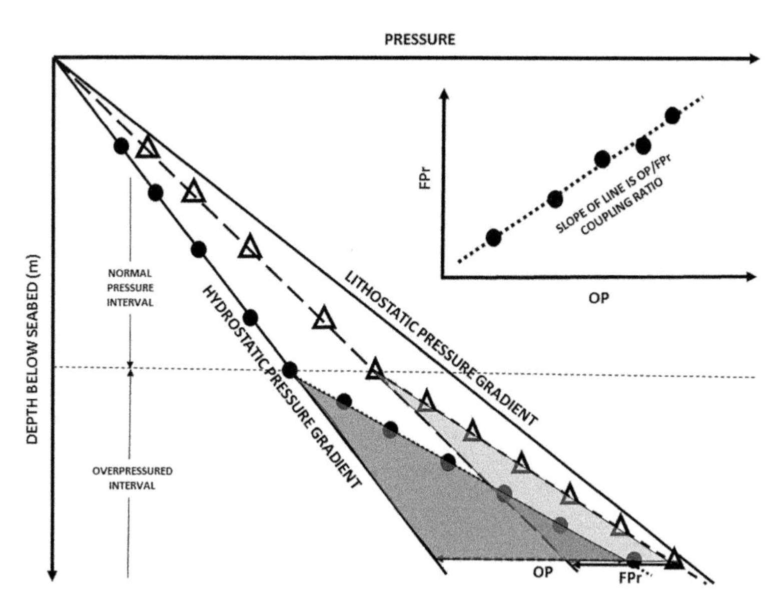

Fig. 4. Schematic representation of the procedure for determination of A and B in the Lahann & Swarbrick fracture pressure model, modified from Swarbrick & Lahann (2016) and reproduced here with permission from *Marine and Petroleum Geology*. The term FPr refers to the fracture pressure residual which is determined by subtraction of the fracture pressure model for hydrostatic conditions from the observed Leak Off (LOT) value.

OP occurs within 2000 m of the seabed. Hence the data available to calibrate A frequently occur only in the shallowest portion of the control wells. The linkage of FP to $A*S_v$ allows local variations in S_v profiles to influence the calculated FP. Although the general form of the new model equation is similar to that of Breckles & van Eekelen (1982), the new model contains only two constants to be calibrated, rather than the three constants in the Breckles & van Eekelen equation.

The Matthews & Kelly (1967) and Eaton (1969) models may be calibrated to pressure data expressed as either pressure (e.g. psi or MPa) or by pressure gradients (pound/gallon or MPa/km) or density (specific gravity). The model presented here should only be calibrated by pressure data expressed as pressure, rather than as a gradient. As shown in Figure 5a on a dataset (from seabed to 2000 m below seabed) from the Scotian shelf (Bell 1990), a reliable relationship between FP and $A*S_v$ can often easily be obtained. However, because of the shallow depth, relatively small differences in the LOT pressures result in large differences in equivalent mud weight (Fig. 5b). For this example no reliable estimate of A is possible from the shallow LOT when expressed as a gradient. An inaccurate determination of A will result in an inaccurate determination of B and, subsequently, an inaccurate FP model. An inappropriate fracture gradient interpretation (from Fig. 5b), generated from the shallow/hydrostatic interval, will generate substantial errors (in psi) when extrapolated to greater depths.

The data in Figure 5a are from a shallow-cool interval which is hydrostatically pressured. Despite these factors, which would be expected to create a consistent LOT trend with S_v, the average difference between observed and model LOT, for the data in Figure 5a, is 1.5 MPa (225 psi). This result demonstrates a challenge to successful (accurate) pressure modelling using LOTs. Significant deviation of observation from a model is a characteristic of LOT modelling. Some of this variation may be associated with lithological variations within the sedimentary section. Additionally, variations in borehole size, orientation or drilling practices (e.g. pump rate) used in the collection of LOT data are also likely contributors to the variability. A further influence which would increase data scatter is inadvertent inclusion of FIT data in the analysis. Inaccurate Pp is unlikely to be a significant contributor to the model error in this case because the Pp is near hydrostatic for all the LOT depths. The average variation in observed LOT compared with model data in the dataset in Figure 5b, for which the LOTs are expressed as gradients, is 1.2 pounds/gallon, an unacceptably large error for LOT modelling or for well planning.

The value of B (Fig. 4), the coupling term linking OP and FP, is determined by employing the A value

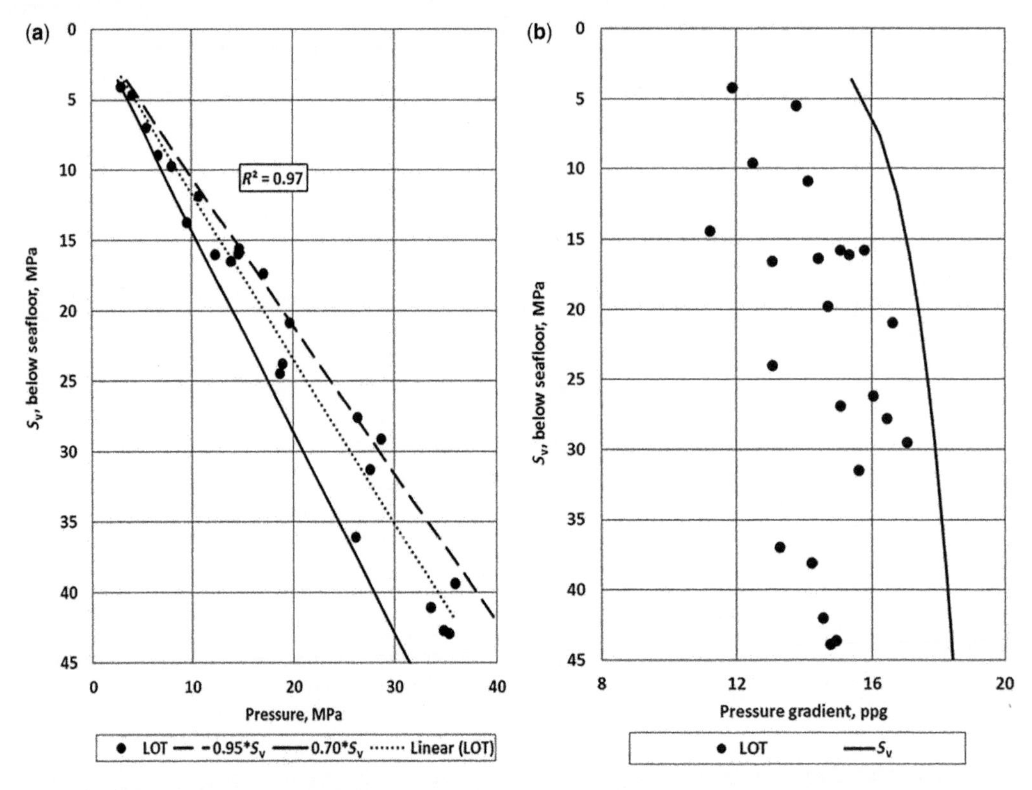

Fig. 5. (**a** and **b**) Comparison of reliability of *A* value determined by depth plot v. pressure (a) and by depth plot v. pressure gradient (b). Data from Bell (1990).

and S_v to calculate the expected value of FP at depths with known LOT and S_v values. The predicted value of FP from the hydrostatic model ($A*S_v$) is subtracted from the observed FP. This value, termed the fracture pressure residual (FPr), is the excess fracture pressure in the observation relative to the hydrostatic model. The FPr is plotted against the observed OP, as in the insert on Figure 4. The regressed slope of the residual (the FPr:OP ratio) is the value of *B*, which has been shown to vary from 0.24 to 0.43 based on 11 global case studies (Swarbrick & Lahann 2016).

The range of *B* values reported by Swarbrick & Lahann (2016) is lower than the range 0.46–0.56 derived by Breckles & van Eekelen (1982) using depth rather than S_v. However, we note that Breckles & van Eekelen (1982) derived their constants using LOT and OP data which are depth-normalized (pressure gradients, not pressures). Their regression to determine a coupling factor also included depleted reservoirs in which they were plotting underpressure rather than OP. Pp:FG coupling over the lifetime of an oil or gas field may be at a different rate than found in geological timeframes (Swarbrick & Lahann 2016 and references therein), and hence combining underpressure and OP values may introduce artefacts not captured in the method described in this paper. Other possible causes for the variation in coupling coefficient between this study and Breckles & Van Eekelen are the same as were discussed previously as causes for variation in LOT values from the hydrostatic model, specifically variations in lithology, borehole size and orientation, drilling practices and inclusion of FIT data.

We also note that the typical *B* values from the FPr/OP plot (Swarbrick & Lahann 2016) are substantially different from the values determined by Engelder & Fischer (1994) and Hillis (2003) in gradient/gradient plots of Pp and FP (i.e. 0.60–0.75). Engelder & Fischer refer to the relationship between increased OP and FP (at constant depth) as a poro-elastic response of the formation to increasing Pp in the pore system. The differences in *B* values are discussed by Swarbrick & Lahann (2016), who conclude that the differences in *B* values are mainly due to inappropriate treatment of FP data associated with hydrostatic Pp data in the FP/Pp gradient plots created by Engelder & Fischer (1994) and Hillis (2003).

Swarbrick & Lahann (2016) present representative A and B values from 11 global studies. The applications reported to date have all been for offshore locations, although a range of ages, lithologies and tectonic settings have been investigated. The authors see no reason why the methodology could not be applied to onshore locations. These authors also discuss application of the new FP model to a compressive basin setting in Southeast Asia but despite the substantial differences in tectonic setting, the new FP model was used successfully to model FP in the province.

Case study applications

Scotian Shelf

The Scotian Shelf data (Bell 1990) are from 22 wells with 64 LOTs distributed over a water depth ranging from 24 to 1477 m. The development of OP is largely confined to depths greater than 4000 m (Fig. 6a) where the LOTs in Mesozoic clastic sediments depart from the shallow trend, reflecting FP coupling with OP. The application of the model to these data is shown in Figure 6a and b. The derived A from the LOT and S_V associated with hydrostatic pore fluid pressures was 0.82 (derived from data in Fig. 5a). The calculated B value was 0.39 with an r^2 value of 0.82 (Fig. 6b). Comparison between FP predicted by the new method and actual LOTs (Fig. 7) reveals close agreement at all depths. At depth 4805 m on Figure 7, the S_V calculated relative to seabed is smaller than the reported LOT and the LOT model. One of the unique features of the current model is that the LOT is not constrained to remain between Pp and S_V. This feature allows modelling, via the coupling with OP, of LOTs that are greater than S_V, presumably owing to tensile strength (lithification).

The Bell (1990) dataset demonstrates some of the challenges associated with obtaining accurate FP models. The data reported for Pp were based

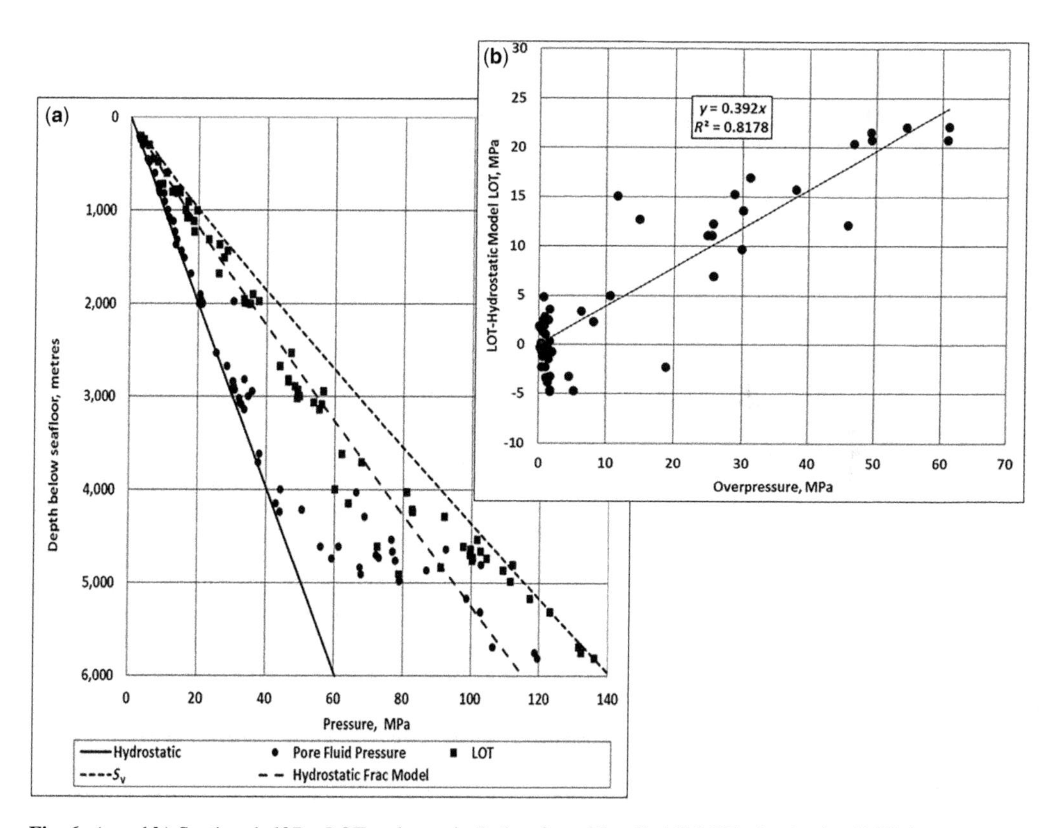

Fig. 6. (**a** and **b**) Scotian shelf Pp, LOT and generic S_V data from 25 wells (65 LOT). An A value (0.82) is determined from the LOT associated with near hydrostatic Pp. Data in (a) replotted from Bell (1990). The insert (b) demonstrates the relationship between measured LOT and the hydrostatic model, plotted against OP. The derived FPr:OP value for B (coupling coefficient) is 0.39. (b) modified from Swarbrick & Lahann (2016) with permission from *Marine and Petroleum Geology*.

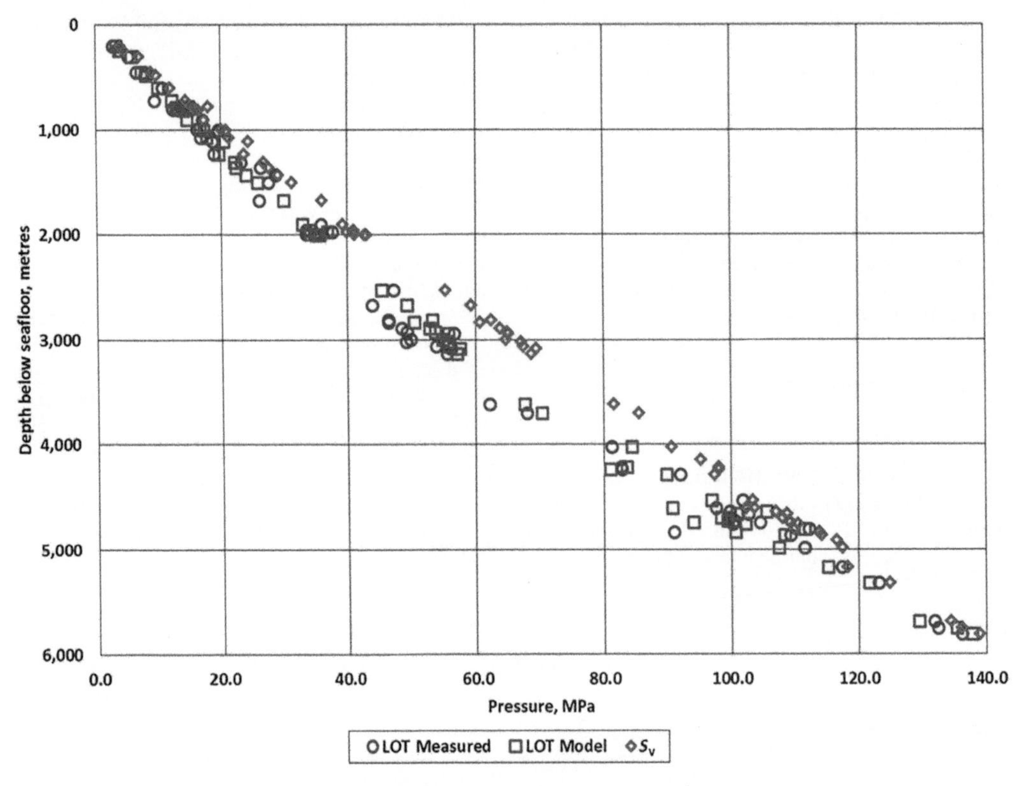

Fig. 7. Plot of S_v, observed LOT and modelled FP values for the Scotian Shelf dataset (Bell 1990), relative to depth below seabed.

on drill-stem tests, repeat formation tests and often mud weights without drill-stem or repeat formation test confirmation. Some of the LOT data were reported as 'less than' values, indicating that the tests were probably FITs not LOTs. Uncertainty in the input data, especially the Pp data, probably contribute to the scatter observed in Figure 6a and b.

The calculation of A and B is subject to multiple technical challenges. The input value for overpressure, which may have considerable uncertainty, clearly influences the FPr:OP coupling term. The exact value of A also has a direct influence on B since $A*S_v$ is the hydrostatic model that is subtracted from the actual S_v. S_v values are commonly reported relative to sea-level and must be converted, accurately, to a seabed reference depth. Errors in S_v estimation will influence both the A and B values because of their effect on FPr. A too high value of A will yield a too low B value because of the reduction in FPr (a too low value for A will increase B). The inadvertent inclusion of an FIT test in the A determination will lower A and in turn increase B. The value for B calculated for the Scotian dataset reported in Swarbrick & Lahann (2016) was 0.36. The B value for the Scotian dataset was redetermined in this study to be 0.39. The authors consider this variance to be due to minor changes in the calculation and regression process.

Both Matthews & Kelly calculations and the current method calculations of FP were performed as a comparison of methods. The wells used to define A for the current method were used to define k_i for Matthews & Kelly calculation. The k_i calibration showed no relationship of k_i with depth or effective stress. An average k_i of 0.72 was therefore used for the Matthews & Kelly interpretation. The average error values for the new method and the Matthews & Kelly interpretation were calculated for 1000 m depth intervals for both methods. Very little difference exists in the two models over the shallow, near-hydrostatic depth range (Table 1). From 2000 m (bsf) and deeper, the current method provided significantly better FP estimates, ranging from 1.0 to 1.7 MPa (145–247 psi) average improvement over a 1000 m interval. Individual wells and LOT tests would have greater or lesser values of improvement. The improved FP values with the new method occur in a depth range over which the Matthews & Kelly method is relatively

Table 1. *Compilation of average error and depth ranges from interpretation of fracture pressure (FP) by Matthews & Kelly and Lahann & Swarbrick FP models applied to data from the Scotian Shelf and Central Gulf of Mexico*

Application	Depth, m bsf	Average error, MPa	Average error, MPa
		Matthews & Kelly	Lahann & Swarbrick
Scotian Shelf	0–999	1.1	1.1
	1000–1999	1.9	1.9
	2000–2999	4.3	3.3
	3000–3999	4.3	2.6
	4000–4999	3.1	2.8
	5000–5814	3.8	2.5
	0–5814	2.8	2.3
Central Gulf of Mexico	0–999	0.8	0.8
	1000–1999	1.2	1.2
	2000–2999	2.1	1.9
	3000–3999	3.1	2.4
	4000–4999	2.1	1.9
	5000–5999	3.0	2.9
	6000–6072	2.7	2.6
	0–7072	1.9	1.8

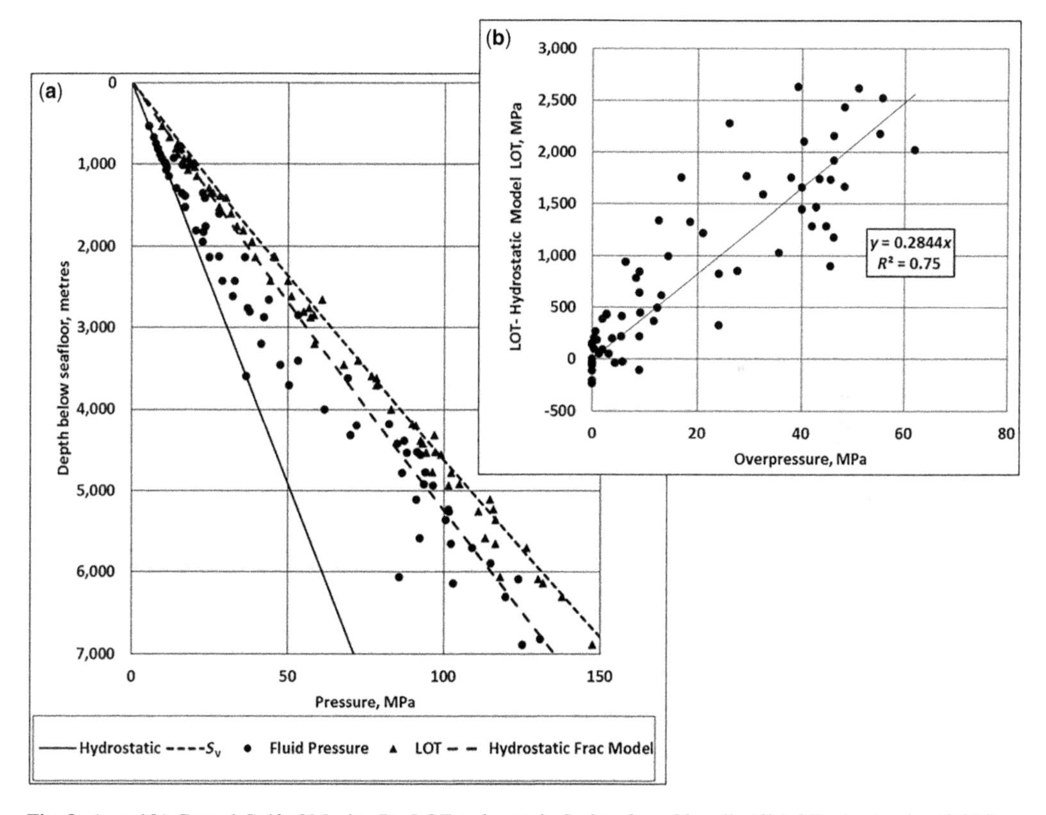

Fig. 8. (**a** and **b**) Central Gulf of Mexico Pp, LOT and generic S_v data from 20 wells (69 LOT). An A value (0.875) is determined from the LOT associated with near hydrostatic Pp. The insert (b) demonstrates the relationship between measured LOT and the hydrostatic model, plotted against OP. The derived FPr:OP value for B (coupling coefficient) is 0.28.

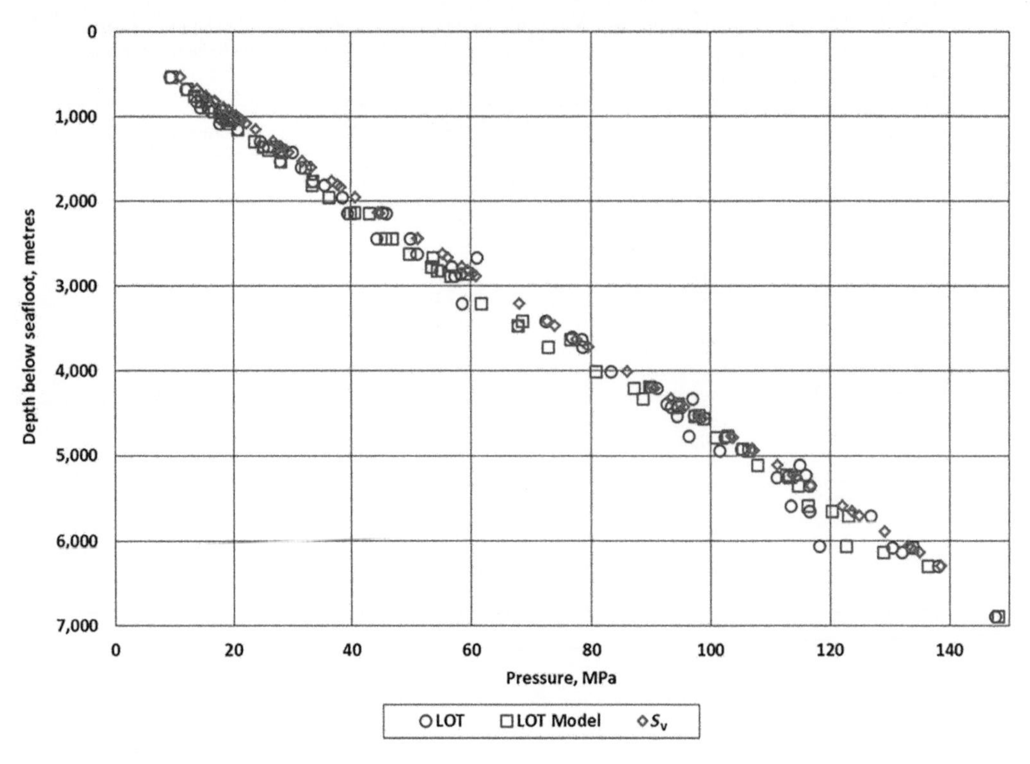

Fig. 9. Plot of observed LOT and modelled FP values for the Central Gulf of Mexico dataset, relative to depth below seabed.

insensitive to k_i value because of the relatively small difference between Pp and S_v.

Central Gulf of Mexico

The Central Gulf of Mexico data are from 20 deep-water wells and with 69 LOTs distributed over four offshore deep-water Gulf of Mexico blocks (Mississippi Canyon, Ewing Bank, Green Canyon and Walker Ridge). The area is characterized by Upper Tertiary, clastic sediments dominated by claystones in which OP starts typically about 1300 m below the seabed (Fig. 8a). The derived A from the LOT and S_v associated with hydrostatic Pp is 0.871 (illustrated by dotted line in Fig. 8a). The calculated B value is 0.28 (regression through data shown in Fig. 8b) with an associated r^2 of 0.75. Pp values, sourced from Pp log interpretation calibrated to direct Pp measurements and mud weights, are a probable source of scatter in Figure 8b. Figure 9 shows a comparison plot of S_v, observed LOTs vs modelled FP values.

Both Matthews & Kelly calculations and the current method calculations of FP were performed for the Gulf of Mexico dataset. The wells used to define A for the current method were used to define k_i for the Matthews & Kelly calculation. The k_i calibration showed no relationship of k_i with depth or effective stress. An average k_i of 0.74 was used for the Matthews & Kelly interpretation. As was noted for the previous application, at depth less than 2000 m bsf, the two models yield very similar answers (Table 1). At depths greater than 2000 m bsf, the new model provides a better match to observed data by 0.1–0.7 MPa (15–102 psi) with reduced error over a 1000 m interval. Individual LOTs or wells would have more or less improvement than the average value calculated. The improved FP interpretation, relative to the Matthews & Kelly model, occurs in a depth range over which the Matthews & Kelly method is relatively insensitive to k_i value because of the small difference between Pp and S_v.

North Sea chalk

Carbonate lithologies provide substantial challenges to FP modelling, in part because of the wide range of sedimentary and diagenetic textures. The Central North Sea offers an opportunity to test the new model on fine-grained carbonate chalk with permeability low enough to form part of the pressure seal

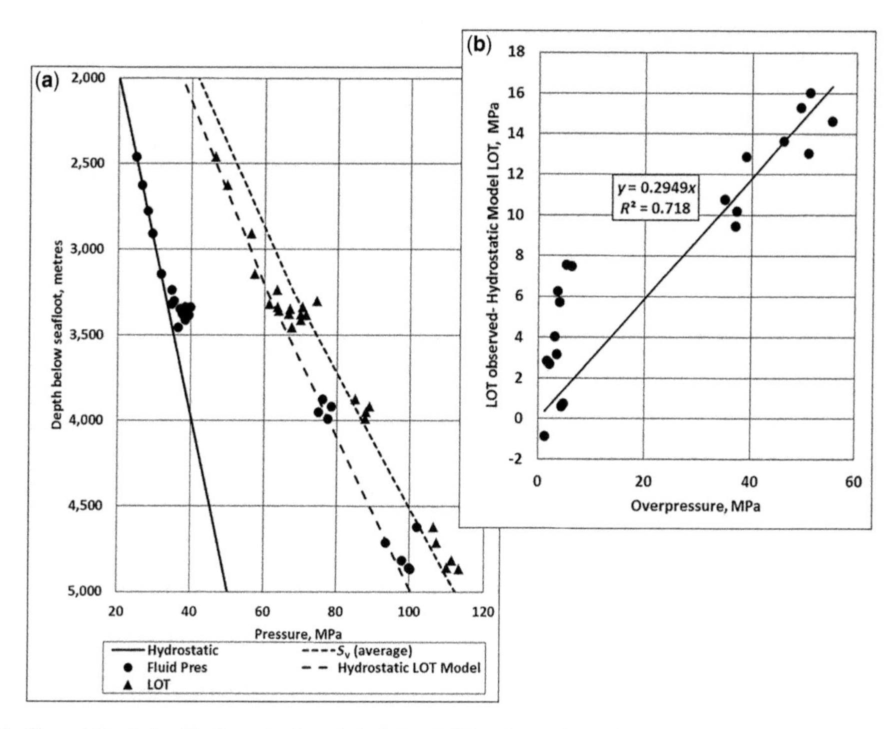

Fig. 10. Central North Sea Tertiary clastic and chalk Pp, LOT and generic S_v data from 20 wells (23 LOT). An *A* value (0.89) determined from the LOT associated with near hydrostatic pore pressure. The insert demonstrates the relationship between measured LOT and the hydrostatic model, plotted against overpressure. The derived FPr:OP value for *B* (coupling coefficient) is 0.29.

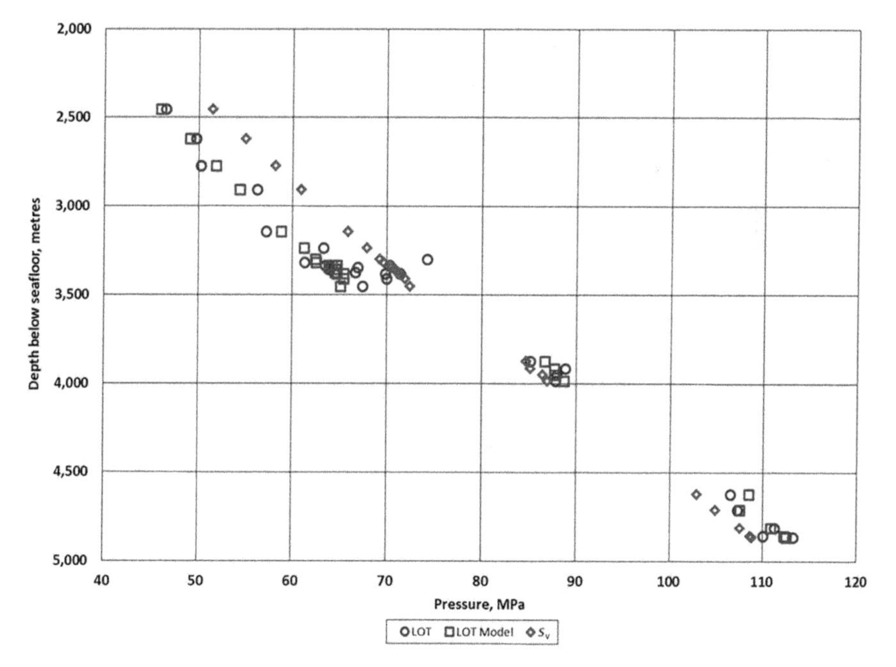

Fig. 11. Plot of S_v, observed LOT and modelled FP values for the Central North Sea Tertiary clastic/chalk dataset, relative to depth below seabed.

(Swarbrick *et al.* 2010), through which OP increases from near-normal to high values from shallow to deep. The data examined in this application are from 18 chalk LOTs and five Tertiary clastic LOTs collected from 20 wells; the wells are distributed over three offshore UK Quadrangles in the Central North Sea. Figure 10a and b display the data used to calculate an A value of 0.89 based on hydrostatic LOTs and a B value of 0.29 with an r^2 value of 0.72. The presence of the chalk results in a rapid increase in Pp between 3200 and 3900 m (Fig. 10). A linear increase in OP within the chalk is assumed for each well (Swarbrick *et al.* 2010). The FP model (Fig. 11) described in this paper provides an adequate model when compared with measured LOT in North Sea chalk, employing the same procedures as described above for the clastic rocks of the Scotian shelf and the Central Gulf of Mexico.

In the Central North Sea, LOTs can exceed the best estimates of S_v (based on integrated density data from wells with almost continuous log coverage), as shown in Figures 10a and 11. This relationship could be explained most simply as caused by LOTs which incorporate the tensile strength (which is estimated to be on the order of 7 MPa for North Sea caprocks from unpublished research results known to authors). However LOT above S_v may also be a function of: (1) bias to low density in determination of the S_v; and (2) elevated LOT associated with generation of a vertical fracture within a vertical borehole when S_h is similar to or greater than S_v. The results of the FP model are to predict fracture strength, as calibrated to LOT, and the relationship between LOT and S_v is therefore secondary.

Conclusions

A new procedure for FP calculation has been shown to provide a good match of model to observed data in offset wells in three substantially different geological settings: the Scotian Shelf, the Central Gulf of Mexico and the chalk interval of the Central North Sea. The new procedure provides improved estimation of FP, relative to a Matthews & Kelly method, at depths greater than 2000 m. The new procedure utilizes the same inputs (hydrostatic pressure, S_v, Pp and LOT data for calibration) as required for application of the most commonly used algorithms (e.g. Matthews & Kelly 1967; Eaton 1969). All necessary data are shown to be easily extracted from a pressure–depth plot of Pp, S_v and LOT from offset wells (adjusted to seabed reference depth).

The new FP model utilizes an FP trend associated with hydrostatic Pp conditions and supplements the hydrostatic FP value with an increment in FP associated with the extent of OP present at the depth of interests. The method does not limit FP values to occur between Pp and S_v, as is characteristic of most other FP models.

Application of the model to a proposed or exploratory well requires predicted depth profiles of S_v and Pp. The expected FG:S_v ratios in the hydrostatic portion of the well are in the region of 0.81–0.89 and the FPr:OP ratios are in the region of 0.24–0.43, based on current experience of application of this approach in both clastic and carbonate (chalk) environments (Swarbrick & Lahann 2016). The provision of probable ranges for FG:S_v ratio and FPr:OP ratios provides guidelines for calculating ranges for fracture pressure in the absence of reliable offset data.

The authors thank Nexen Petroleum USA Inc. for permission to use fracture and fluid pressure information used in the Central Gulf of Mexico discussion in this paper. The authors also thank the Indiana Geological Survey for support in creating the figures in this paper, and two anonymous referees for their valuable comments.

References

Bell, J.S. 1990. The stress regime of the Scotian Shelf offshore Eastern Canada to 6 kilometers depth and implications for rock mechanics and hydrocarbon migration. *In*: Maury, V. & Fourmaintraux, E. (eds) *Rock at Great Depth*. Balkema, Rotterdam, **3**, 1243–1265.

Breckles, I.M. & van Eekelen, H.A.M. 1982. Relationship between horizontal stress and depth in sedimentary basins. *Journal of Petroleum Technology*, **34**, 2191–2199.

Davies, R.W., Swarbrick, R.E., Evans, R.J. & Huuse, M. 2007. Birth of a mud volcano, East Java: 29 May 2006. *GSA Today*, **17**, 4–9.

Dutta, N.C. 2002. Deepwater geohazard prediction using prestack inversion of large offset P-wave data and rock model. *The Leading Edge*, February, 193–198.

Eaton, B.A. 1969. Fracture gradient prediction and its application in oil field operations. *Journal of Petroleum Technology*, **21**, 1353–1360.

Eaton, B.A. & Eaton, T.L. 1977. Fracture gradient prediction for the new generation. *World Oil*, October, 93–100.

Engelder, T. & Fischer, M.P. 1994. Influence of poroelastic behavior on the magnitude of minimum horizontal stress, Sh, in overpressured parts of sedimentary basins. *Geology*, **22**, 949–952.

Gaarenstroom, L., Tromp, R.A.J., De Jong, M.C. & Brandenburg, A.M. 1993. Overpressures in the Central North Sea: implications for trap integrity and drilling safety. *In*: Parker, J.R. (ed.) *Petroleum Geology of NW Europe: Proceedings of the 4th Conference*. Geological Society, London, 1305–1313, https://doi.org/10.1144/0041305

Grauls, D. 1999. Overpressures: causal mechanisms, conventional and hydromechanical approaches. *Oil & Gas Science and Technology*, **54**, 667–678.

Hillis, R.R. 2003. Pore pressure/stress coupling and its implications for rock failure. *In*: Van Rensbergen,

P., HILLIS, R.R., MALTMAN, A.J. & MORLEY, C.K. (eds) *Subsurface Sediment Mobilization*. Geological Society, London, Special Publications, **216**, 359–368, https://doi.org/10.1144/GSL.SP.2003.216.01.23

MATTHEWS, W.R. & KELLY, J. 1967. How to predict formation pressure and fracture gradient. *The Oil and Gas Journal*, February 20, 92–106.

NASHAAT, M. 1998. Abnormally high formation pressure and seal impacts on hydrocarbon accumulations in the Nile Delta and North Sinai Basins, Egypt. *In*: LAW, B.E., ULMISHEK, G.F. & SLAVIN, V.I. (eds) *Abnormal Pressures in Hydrocarbon Environments*. AAPG, Memoirs, **70**, Chapter 10.

SELDON, B.J. & FLEMINGS, P.B. 2005. Reservoir pressure and seafloor venting; predicting trap integrity in a Gulf of Mexico deepwater turbidite minibasin. *AAPG Bulletin*, **89**, 193–209.

SWARBRICK, R.E. & LAHANN, R.W. 2016. Estimating pore fluid pressure–stress coupling. *Marine and Petroleum Geology*, **78**, 562–574.

SWARBRICK, R.E., LAHANN, R.W., O'CONNOR, S.A. & MALLON, A.J. 2010. Role of the chalk in development of deep overpressure in the Central North Sea. *In*: VINING, B.A. & PICKERING, S.C. (eds) *Petroleum Geology: From Mature Basins to New Frontiers – Proceedings of the 7th Petroleum Geology Conference*. Geological Society, London, 493–507, https://doi.org/10.1144/0070493

TINGAY, M.R.P., HILLIS, R.R., MORLEY, C.K., SWARBRICK, R.E. & DRAKE, S.J. 2005. Present-day stress orientation in Brunei: a snapshot of 'prograding tectonics' in a Tertiary delta. *Journal of the Geological Society, London*, **162**, 39–49, https://doi.org/10.1144/0016-764904-017

Relationships between geomechanical properties and lithotypes in NW European chalks

FANNY DESCAMPS[1]*, OPHÉLIE FAŸ-GOMORD[2], SARA VANDYCKE[1], CHRISTIAN SCHROEDER[3], RUDY SWENNEN[2] & JEAN-PIERRE TSHIBANGU[1]

[1]*UMONS, University of Mons, 20 place du Parc, 7000 Mons, Belgium*

[2]*Department of Earth and Environmental Sciences, KU Leuven, Katholieke Universiteit Leuven, Geology, Celestijnenlaan 200E, 3001 Heverlee, Belgium*

[3]*ULB, Université Libre de Bruxelles, 50 avenue Roosevelt, 1050 Brussels, Belgium*

**Correspondence: fanny.descamps@umons.ac.be*

Abstract: As a result of increasing interest in unconventional reservoirs, a wide range of sedimentary systems are now being investigated with regard to petroleum applications, including various tight chalk formations. We examined a wide variety of chalk samples from NW Europe (micritic, grainy, argillaceous, marl seam, cemented and silicified chalks) and investigated the relationships between their petrophysical properties, mechanical properties and associated microtextures and how diagenesis can affect these properties. A diagenesis index based on an evaluation of textural and diagenetic parameters was used to quantify the effect of global porosity-reducing diagenesis on the microtexture of chalks. We used petrographic and petrophysical measurements to determine the petrography, density, porosity, permeability and sonic velocity of the chalk samples and uniaxial compression experiments to assess their mechanical behaviour. Our dataset of >30 samples covers a wide range of values for these properties. We determined a linear porosity–permeability relationship controlled by the diagenesis index. Porosity influences the unconfined compressive strength and Young's modulus, but our analyses suggest that the diagenesis of the studied lithologies provides us with a further understanding of the mechanical behaviour of chalks. Micritic and grainy chalks are associated with the lowest diagenesis index and exhibit the lowest strength, whereas the higher diagenesis indices observed for other microtextures correspond to higher compressive strengths.

High-porosity pure white chalks have been extensively studied in relation to oil and gas production for >40 years (e.g. Schroeder 1995, 2002; Delage *et al.* 1996; Papamichos *et al.* 1997; Homand *et al.* 1998; Risnes & Flaageng 1999; Homand & Shao 2000*a*, *b*, *c*; Gommesen & Fabricius 2001; Risnes 2001; Collin *et al.* 2002; DeGennaro *et al.* 2003, 2005; Risnes *et al.* 2003; Nguyen *et al.* 2008). Recently, however, there has been increasing interest in unconventional reservoirs, such as tight chalk formations, leading to a need to investigate a wider range of sedimentary and diagenetic systems.

Several studies have focused on the characterization of microtextures and pore networks within microporous carbonate reservoirs (Cantrell & Hagerty 1999; Richard *et al.* 2005; Vincent *et al.* 2011; Brigaud *et al.* 2014; Regnet *et al.* 2014; Kaczmarek *et al.* 2015). Classifications based on the morphology of the micritic matrix have been developed for microcrystalline calcite (Lambert *et al.* 2006; Deville de Periere *et al.* 2011; Kaczmarek *et al.* 2015). However, the proposed classifications are not applicable to chalk because they do not include the nanobioclast component. Mortimore & Fielding (1990) attempted to classify chalk microtextures based on scanning electron microscopy (SEM) observations, with a classification applicable to pure chalks only.

Fritsen *et al.* (1996) proposed a classification of chalks based on macroscopic observations from North Sea cores. Mallon & Swarbrick (2002, 2008) focused on the petrographic and petrophysical properties of non-reservoir low-permeability chalk lithologies. These deposits were defined by the Joint Chalk Research (JCR) group (Bailey *et al.* 1999) as tight chalks and include all chalks with a matrix permeability <0.2 mD. They are of interest to the petroleum industry (Fabricius 2001; Røgen & Fabricius 2002; Strand *et al.* 2007; Lindgreen & Jakobsen 2012) because they might be underexplored reservoirs or may have a crucial role in hydrocarbon migration, acting as seals or fluid conduits depending on their fracture pattern (Gennaro *et al.* 2013).

Depositional and diagenetic processes are known to control chalk microtextures (Anderskouv

From: TURNER, J. P., HEALY, D., HILLIS, R. R. & WELCH, M. J. (eds) 2017. *Geomechanics and Geology*. Geological Society, London, Special Publications, **458**, 227–244.
First published online May 25, 2017, https://doi.org/10.1144/SP458.9

& Surlyk 2011). The size and connectivity of the pore network may be enhanced by dissolution or reduced by cementation and compaction. Hence microtexture appears to be the link between the current behaviour of chalk formations and their geological history. Rashid *et al.* (2015) examined the factors affecting the distribution of porosity, permeability and reservoir quality in the Kometan Formation (northern Iraq). Faÿ-Gomord *et al.* (2016*a*) proposed an in-depth understanding of the microtexture of tight chalks and highlighted the controlling role of the non-carbonate content and the degree of diagenesis on the petrophysical properties.

Few studies are currently available on the mechanical properties of tight chalks. This is, however, essential for designing hydraulic fracturing in tight formations, such as the Niobrara plays in the USA, where diagenetic changes have been proved to increase the brittleness of chalk (Pollastro 2010; Maldonado *et al.* 2011). Bell *et al.* (1999) reviewed the engineering properties of English chalk, including some tight chalks. Using indentation experiments, Faÿ-Gomord *et al.* (2016*b*) highlighted the role of microtextures in the mechanical behaviour of chalk and underlined the distinct behaviour of tight chalk.

The study reported here investigated how diagenesis can affect the petrophysical and mechanical properties of chalk by studying the relationships between these properties and the associated microtextures. This will help our understanding of the behaviour of tight chalk formations in terms of storage capacity, transport mechanisms and mechanical behaviour. It is also of interest when searching for good analogues from outcrops because diagenesis has proved to be a key issue for the characterization of reservoir chalk from chalk outcrops (Hjuler & Fabricius 2009).

Sample areas

NW European chalks were investigated from several outcrops in Belgium, France and the UK (Fig. 1). The Belgian samples were from the Harmignies quarry, where pure white chalks from the Mons Basin are exploited. The samples came from the following Campanian formations (Marlière 1949):

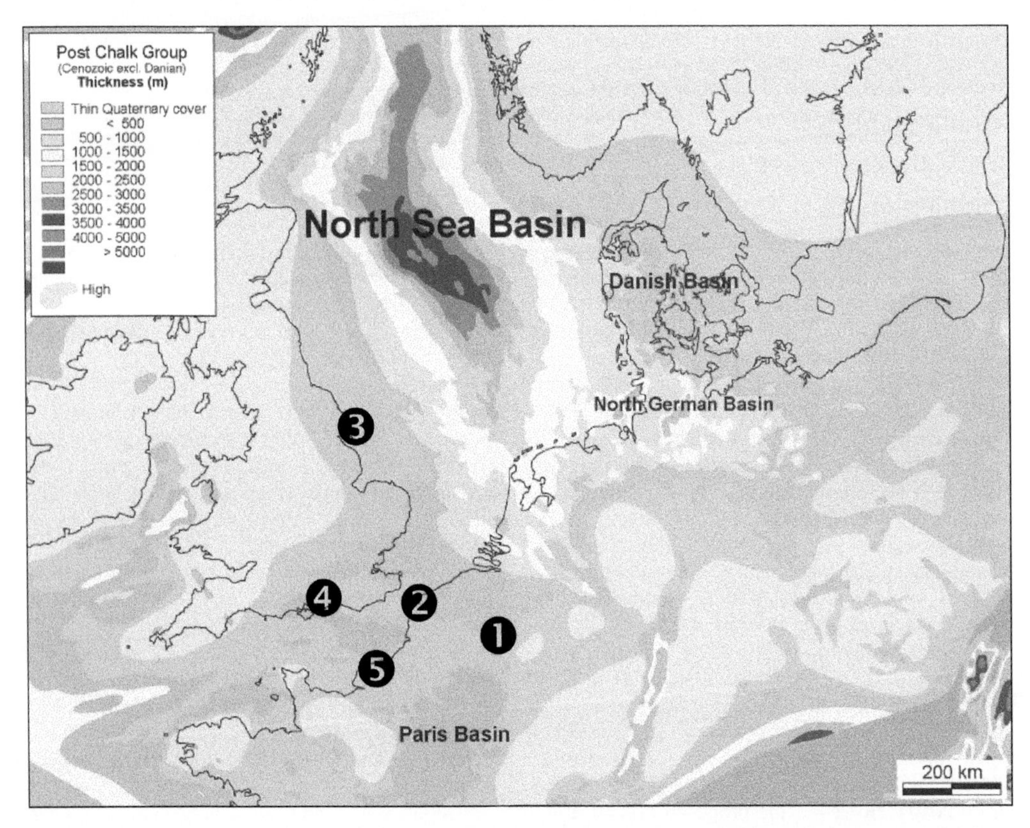

Fig. 1. Location of outcrop samples (modified from Hjuler & Fabricius 2009). 1, Harmignies Quarry (Belgium); 2, Boulonnais (France); 3, Yorkshire (UK); 4, Sussex (UK); and 5, Upper Normandy (France).

Obourg (CH01), Nouvelles (CH02) and Spiennes (CH03). In Belgium, Campanian pure white chalk is exploited in different quarries in the Mons Basin and in the East Chalk district near Lixhe village (Robaszynski *et al.* 2001). These pure white chalks contain a majority of intact coccoliths and are considered as analogous to chalk reservoirs; they have been extensively studied (Schroeder 2002; Papamichos *et al.* 2012; Megawati *et al.* 2015). The Belgian white chalk is affected by normal and strike-slip faulting with a large number of major joints (Vandycke *et al.* 1991; Vandycke 2002).

In France, outcrops were studied in the Boulonnais region (Cap Blanc Nez site) and in Upper Normandy at different locations along the coast in Cenomanian to Turonian chalks (Table 1). The Cenomanian deposits of Cap Blanc Nez have been described as evolving from silicate-rich chalk in the Lower Cenomanian towards white chalk at the top (Robaszynski & Amédro 1986; Amédro & Robaszynski 2001). The argillaceous chalk is related to a strong detrital input, most probably from the Brabant Massif (Deconinck *et al.* 1991); the input decreased as the depositional environment deepened during the Late Cretaceous. The Normandy Basin is well known and has been well studied (Juignet 1974; Kennedy & Juignet 1975; Quine & Bosence 1991; Mortimore & Pomerol 1997; Robaszynski *et al.* 1998; Lasseur *et al.* 2009).

Samples were taken from two sites in the British chalk district. The Flamborough Head samples are clean Santonian chalk (Whitham 1993), which has been deeply buried (Menpes & Hillis 1996) and has thus undergone strong burial diagenesis (Faÿ-Gomord *et al.* 2016*a*). The samples from Sussex (southern England) were only buried by up to 700 m (Law 1998). Several formations display very different lithotypes, from clean chalk with flint bands in the Birling Gap Turonian New Pit Formation, to argillaceous Cenomanian chalk from Eastbourne.

This first overview of sampling sites (Table 1; Fig. 1) shows the wide variety of chalk materials considered in this study. Their selection was governed by our search for a diversity of lithotypes and geological burial histories. Samples were also specifically selected in zones that were not influenced by faults (Gaviglio *et al.* 2009). As indicated on Figure 1, the chalk formations represent different burial depths, ranging from 200 to 250 m for the Harmignies chalk (Dupuis & Vandycke 1989) to >1200 m for the Flamborough Head chalk (Menpes & Hillis 1996). This study focused on chalks already referenced in terms of lithostratigraphy (Table 1) for all the different sites. Some of the mechanical properties are currently known, but no clear relationship has been established between the diagenetic features and mechanical properties.

Methodology

Petrographic analysis: development of a diagenesis index

Classification systems for chalk microtextures and the degree of diagenesis are often subjective. To develop a quantitative approach allowing comparison with other properties, the microtexture and diagenesis need to be assessed using a numerical value. This will establish a key link between the geology and the petrophysical and geomechanical properties.

We obtained microphotographs from careful SEM observations of the samples to document the microtextures at different magnifications. The diagenesis index, developed by Faÿ-Gomord *et al.* (2016*a*), was assessed for each sample. This index is based on seven diagenetic criteria, which are each graded from 0 (low diagenesis) to 10 (intense diagenesis); the average value, calculated from the seven grades for each studied sample, determines the diagenesis index (Fig. 2). The seven diagenetic criteria are: (1) the micritic microtexture; (2) grain contacts; (3) coccolith disintegration; (4) cemented zones; (5) authigenic calcite crystals; (6) coccolith grain overgrowth; and (7) intraparticle cement. Figure 3 shows typical micrographs corresponding to extreme cases encountered for each of the criteria assessed in the diagenesis index. Each criterion is described below.

Micritic microtexture. The micritic microtexture corresponds to the general arrangement of particles in the matrix. The micritic fraction is defined as particles <10 μm. The microtexture can be loose (with a grading 0–3), tight (4–7) or anhedral compact (8–10). A similar classification was used by Lambert *et al.* (2006) and Deville de Periere *et al.* (2011). The micritic microtexture is often closely related to compaction because the arrangement of grains depends on both mechanical and chemical compaction during burial diagenesis. However, the arrangement of grains that affects the overall microtexture can also be affected by eogenesis during early lithification, as is seen with hardgrounds.

Grain contacts. The types of contact between micritic particles range from punctic contacts (0–2), serrate contacts (3–4), meshed contacts (5–6), coalescent contacts (7–8) to fused contacts (9–10). A punctic contact means that the contacts between grains are punctual and the grains seem to only lie on each other. A serrate contact refers to adjoined grains, connected to each other by a surface. A meshed contact occurs when the grains show indentation of adjacent grains; they are partly nested together. A coalescent contact refers to grains that

Table 1. *Geographical and stratigraphic location of the samples*

Sample no.	Sampling site	Formation	Stratigraphy	Lithotype	References*
CH01	Harmignies Quarry, Belgium	Obourg	Middle Campanian	Micritic	3, 10
CH02	Harmignies Quarry, Belgium	Nouvelles	Middle Campanian	Micritic	3, 10
CH04	Harmignies Quarry, Belgium	Spiennes	Upper Campanian	Micritic	3, 10
CO01	Coquelles Quarry, Boulonnais, France	Caffier	Santonian	Micritic	1, 14
NH03	Newhaven. Sussex, UK	Newhaven	Campanian	Micritic	4, 8, 12, 13
RA01	Ramsgate. Kent, UK	Seaford	Santonian	Micritic	4, 8, 12, 13
BG01	Birling Gap. Sussex, UK	New Pit	Turonian	Grainy	4, 8, 12, 13
CM02	Mimoyecques Quarry. Boulonnais, France	Guet	Turonian	Grainy	1
CM06	Mimoyecques Quarry. Boulonnais. France	Guet	Turonian	Grainy	1
ETR33	Etretat, Upper Normandy, Francc	Saint Pierre en Port	Coniacian	Grainy	7, 9, 12, 15
SS01	Seven Sisters, Sussex, UK	Chalk mudstone	Coniacian	Grainy	4, 8, 12, 13
CB14	Cap Blanc Nez, Boulonnais, France	Escalles	Upper Cenomanian	Cemented	2, 6, 14
FA15	Flamborough Head, Yorkshire, UK	Flamborough	Santonian	Cemented	10, 16
FA39B	Flamborough Head, Yorkshire, UK	Flamborough	Santonian	Cemented	10, 16
FH11	Flamborough Head, Yorkshire, UK	Flamborough	Santonian	Cemented	10, 16
SC01	Saint Martin en Campagne, Upper Normandy, France	Tilleul	Turonian	Cemented	7, 9, 12, 15
CB13	Cap Blanc Nez, Boulonnais, France	Escalles	Upper Cenomanian	Marl seams	2, 6, 14
CB16	Cap Blanc Nez, Boulonnais, France	Grand Nez	Base Turonian	Marl seams	2, 6, 14
EA02	Eastbourne. Sussex, UK	Hollywell Nodular	Base Turonian	Marl seams	4, 5, 8, 12, 13
ETR21	Senneville, Upper Normandy, France	Senncville	Middle Turonian	Marl seams	7, 9, 12, 15
ETR47	Senneville, Upper Normandy, France	Senneville	Middle Turonian	Marl seams	7, 9, 12, 15
SC02	Saint Martin en Campagne, Upper Normandy, France	Tilleul	Turonian	Marl seams	7, 9, 12, 15
SC03	Saint Martin en Campagne, Upper Normandy, France	Tilleul	Turonian	Marl seams	7, 9, 12, 15
CB02	Cap Blanc Nez, Boulonnais, France	Strouanne	Lower Cenomanian	Argillaceous	2, 6, 14
CB04	Cap Blanc Nez, Boulonnais, France	Strouanne	Lower Cenomanian	Argillaceous	2, 6, 14
CB06	Cap Blanc Nez, Boulonnais, France	Strouanne	Lower Cenomanian	Argillaceous	2, 6, 14
CB07	Cap Blanc Nez, Boulonnais, France	Petit Blanc-Nez	Lower Cenomanian	Argillaceous	2, 6, 14
CB09	Cap Blanc Nez, Boulonnais, France	Petit Blanc-Nez	Mid-Cenomanian	Argillaceous	2, 6, 14
CB10	Cap Blanc Nez, Boulonnais, France	Petit Blanc-Nez	Mid-Cenomanian	Argillaceous	2, 6, 14

(*Continued*)

Table 1. (*Continued*)

Sample no.	Sampling site	Formation	Stratigraphy	Lithotype	References*
CB11	Cap Blanc Nez, Boulonnais, France	Cran	Mid-Cenomanian	Argillaceous	2, 6, 14
CB23	Cap Blanc Nez, Boulonnais, France	Petit Blanc-Nez	Lower Cenomanian	Argillaceous	2, 6, 14
CB24	Cap Blanc Nez, Boulonnais, France	Petit Blanc-Nez	Lower Cenomanian	Argillaceous	2, 6, 14
CB25	Cap Blanc Nez, Boulonnais. France	Petit Blanc-Nez	Lower Cenomanian	Argillaceous	2, 6, 14
EA01	Eastbourne. Sussex, UK	Zig-Zag	Cenomanian	Argillaceous	4, 5, 8, 12, 13
BR01	Bruneval, Upper Normandy, France	Glauconieuse	Cenomanian	Silicified	7, 9, 12, 15

*Studies performing geological logs on the studied outcrops: (1) Amédro & Robaszynski (2000); (2) Amédro & Robaszynski (2001); (3) Boulvain & Pingot (2012); (4) Bristow *et al.* (1997); (5) Gale *et al.* (2005); (6) Gräfe (1999); (7) Juignet (1974); (8) Kennedy (1969); (9) Lasseur *et al.* (2009); (10) Marlière (1949); (11) Mitchell (1994); (12) Mortimore & Pomerol (1997); (13) Mortimore (2011); (14) Robaszynski & Amédro (1986); (15) Robaszynski *et al.* (1998); (16) Whitham (1993). From Faÿ-Gomord *et al.* (2016*a*, *b*).

are difficult to identify because several grains are nested together. A fused contact corresponds to a contact where it is impossible to define clearly the boundaries of the grains. The contact between grains depends on both mechanical compaction and pressure dissolution, particularly grain-to-grain contact dissolution processes, which are often enhanced in the presence of clays.

Coccolith disintegration. When sediments are buried, gravitational forces induce mechanical compaction. The grains are brought closer together and the mechanical breakage of delicate coccolith tests increases as the overburden pressure increases. The crushing of microfossils requires stress levels corresponding to significant depths, even if some species break more easily than others, depending on the thickness of their tests. This evidence of burial diagenesis can be graded from 0, when all the coccoliths are well preserved, to 10, when the coccolith tests are broken apart into very small – sometimes hardly recognizable – tests.

Cemented zones. The cemented zones are defined as homogeneous surfaces of calcite cement for which the longest surface trace observed on the micrograph is >10 μm. These zones result from calcite cementation either during eogenesis (hardgrounds) or burial diagenesis. This criterion is ranked from 0, when cemented zones are absent, to 10 when they represent at least 50% of the chalk sample.

Authigenic calcite crystals. Authigenic calcite crystals generally occur as euhedral crystals of apparently non-biogenic origin, usually measuring 1–5 μm. These authigenic crystals have previously been reported as associated with both early diagenetic processes (Faÿ-Gomord *et al.* 2016*a*) and burial diagenesis (Fabricius 2003). This criterion is rated from 0 (no authigenic crystals) to 10 (a high density of authigenic crystals).

Coccolith grain overgrowth. Coccolith fragments often exhibit calcite cement overgrowths, but the degree of occurrence is significantly variable. This diagenetic criterion ranges from 0 for none to very few overgrowth cements, to 10 when almost all the nanofossils show overgrowths. Cement overgrowth results from the dissolution of calcite from less stable surfaces and reprecipitation on stable

Criteria	From	1 … 10	To	Mark
Micritic matrix texture	Microrhombic		Anhedral	7
Grain contact	Punctic		Coalescent	3
Coccolith disintegration	Low		High	6
Cemented zone	Absent		Common	7
Automorphous cement	Absent		Common	5
Coccolith grain overgrowth	Absent		Common	5
Intraparticle cementation	Absent		Common	3
			TOTAL	5

Fig. 2. Example of diagenesis index assessment from the evaluation of seven criteria.

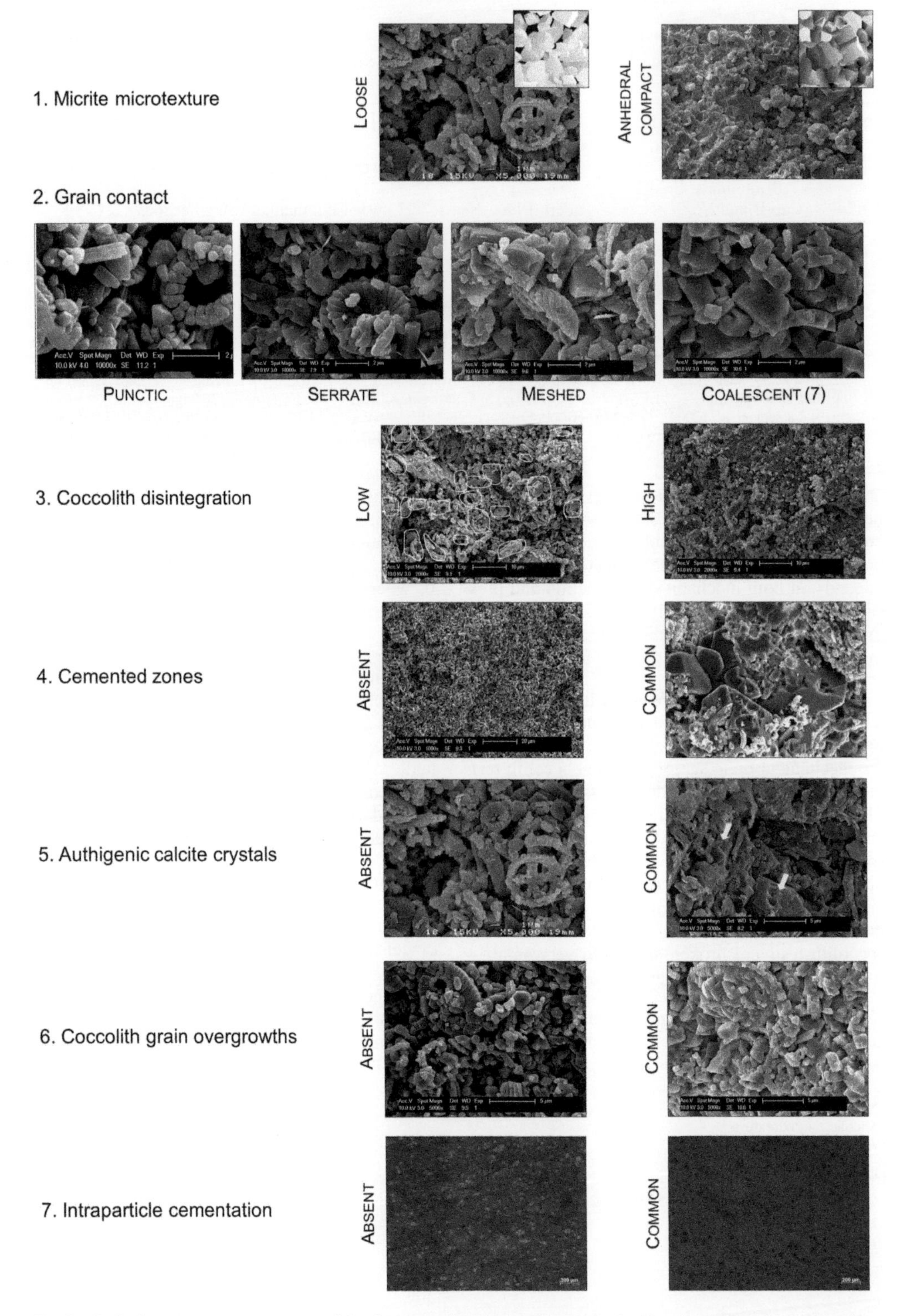

Fig. 3. Typical extreme cases encountered for the seven diagenesis-related criteria. Sketches in Part 1 from Lambert *et al.* (2006).

surfaces (Hjuler & Fabricius 2009). Overgrowths derive from diagenetic processes either by recrystallization or cementation and tend to be more pronounced as the burial depth increases.

Intraparticle calcite cement. The intraparticle cement is characterized from thin section observations under fluorescent light. Intraparticle cementation in chalk essentially refers to the cementation inside forams and calcisphere tests. The cementation of intrafossil porosity indicates active pressure dissolution and thus significant burial depth (Hjuler & Fabricius 2009). This parameter ranges from 0, when there is no intraparticle cementation, to 10, when the whole intraparticle porosity is filled by cement. In this case the cement is often found in the form of large sparitic calcite crystals.

Petrophysical and mechanical properties

The experimental procedure aimed to determine as many properties as possible from the cored plugs available. Therefore non-destructive testing techniques were used to determine the porosity, permeability, dry and saturated P-wave velocities before performing uniaxial compression tests.

Porosity and permeability. Porosity was determined both by water saturation (φ_{water}) and helium expansion (φ_{He}) techniques. In the first method, the void volume was directly measured by a water saturation stage followed by a drying stage; the bulk volume was deduced from the core dimensions. Saturation and drying operations were performed according to American Petroleum Institute (API) standard methods (API 1998). In the second method, a Boyle's law EPS porosimeter was used to determine the grain volume (API 1998). Those porosity measurements correspond to connected pore space.

The gas permeability of the samples was determined with a Vinci nitrogen permeameter. The plugs were mounted in a Hassler-type core-holder at a confining pressure of 28 bar and a steady state gas flow was established through the samples. The permeabilities were corrected for gas slippage using the Klinkenberg empirical correlation (API 1998).

P-wave velocity. Ultrasonic measurement is a non-destructive method used to determine the velocity of ultrasonic waves in materials. Velocity is influenced by the rock type, density, porosity, water content and defects and is therefore closely related to the rock properties (Kahraman 2007). In this study, a PUNDIT Plus system with 54 kHz P-wave transducers was used to determine both the dry and saturated velocities. The velocities were computed from the measured transit time (resolution 0.1 μs) and length (resolution 0.01 mm) of the samples in a pulse transmission arrangement (Rummel & van Heerden 1978). Vaseline was used as the coupling fluid.

Unconfined compressive strength tests. The unconfined compressive strength (UCS) test is a widespread measurement used to mechanically characterize rock materials. The International Society for Rock Mechanics (Fairhurst & Hudson 1999) recommends cylindrical samples with a height to diameter ratio between 2.0 and 3.0. However, smaller height to diameter ratios can also give acceptable results (Dzulinski 1969; Thuro *et al.* 2001). As a result of chalk's very small grain size, smaller samples are still representative. For instance, Duperret *et al.* (2005) used length to diameter ratios ranging from only 1.1 to 1.4. In this study, the tested samples had a length to diameter ratio between 1.2 and 1.6.

Tests were performed on a stiff frame with a servo-controlled loading rate (Fig. 4). Pressure transducers were used to measure the axial stress (range 0–35 MPa or 0–493 MPa depending on the transducer used). Inductive displacement transducers (range ± 1.5 mm) were used to compute the axial strains, leading to a full record of the stress–strain curves. The UCS and Young's modulus (the average modulus of the linear portion of the axial stress–strain curve) could then be computed (Fairhurst & Hudson 1999).

Experimental results

The experimental workflow was applied to 35 samples. Detailed results for the petrographic description are given in Table 1 and the petrophysical and mechanical properties are given in Table 2.

Petrographic results

The dataset acquired covers a wide range of chalks, especially tight chalks, characterized by a spectrum of sedimentary textures, non-carbonate content, various degrees of cementation and compaction. Six lithotypes could be defined based on sedimentary features, colour and apparent strength: micritic, grainy, marl seam, argillaceous, cemented and silicified chalk. However, rather than the lithotype, it is the microtexture (i.e. the description of the arrangement of the matrix micrograins observed under SEM) that is expected to constrain the petrophysical properties of chalk. For each lithotype, a brief description of the associated microtexture is given in the following text. For more details, see Faÿ-Gomord *et al.* (2016*a*), where the impact of microtexture on petrophysical properties is discussed and typical SEM images are shown.

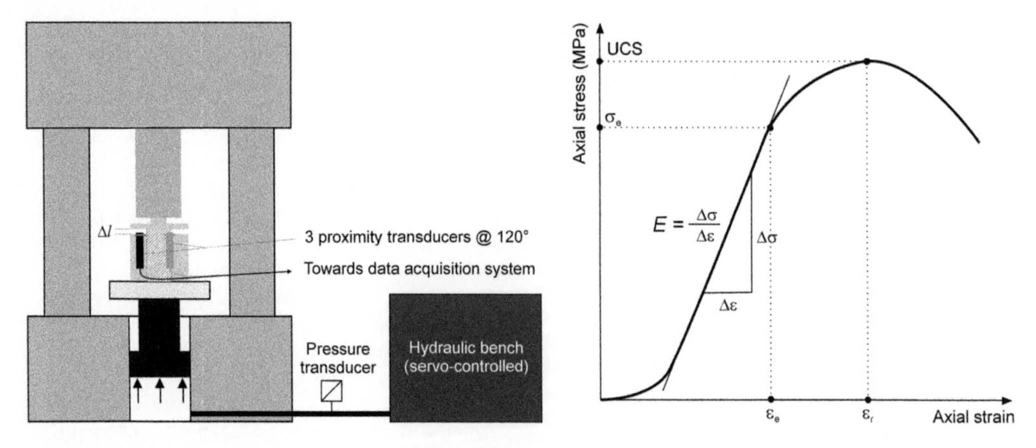

Fig. 4. Sketch of the uniaxial compression test equipment and parameters determined from a typical stress–strain curve.

Micritic chalks have a mud-dominated texture (often mudstone, occasionally wackestone) according to the Dunham (1962) classification. Micritic chalks are mainly composed of coccolith fragments, leading to a micro-rhombic matrix texture, with rare authigenic calcite crystals. Contacts between coccolith fragments are mostly punctic and the interparticle porosity is well preserved.

Grainy chalks are grain-dominated and display a packstone texture. Macroscopically, they cannot always be distinguished from micritic chalks, as they also correspond to pure chalks. However, they show a higher foram and bioclastic content in thin section, giving rise to their packstone texture (Dunham 1962), with 32–43% of grains occurring in a micritic matrix. Under SEM, the micritic matrix of grainy chalks is similar to that of micritic chalk.

Marl seams in chalks are mostly interpreted as the result of pressure solution resulting from burial diagenesis (Lind 1993). Marl seam chalks have a mudstone–wackestone texture, with the percentage of grains ranging from 6 to 17%. Fluorescence microscopy highlights the presence of intraparticle porosity inside calcispheres, forams and various bioclasts. Unlike other chalk lithotypes, there is no one typical texture associated with marl seam chalks. In clay-rich seams, the microtexture is dominated by clay flakes, which may show a preferential orientation. A few millimetres away from the clay seams, the texture is not dominated by the clay content and microtextural features include authigenic cement crystals and grain overgrowths, which do not develop in clay-rich seams. Marl seams seem to develop preferentially in nodular chalk and are affected by early diagenesis and the clays concentrated between the nodules.

Argillaceous chalks are clearly identifiable macroscopically by their light to dark grey colour. They initially formed during the Cenomanian stage where specific sedimentological conditions occurred (with a great detrital input, as described by Deconinck & Chamley 1995; Faÿ-Gomord *et al.* 2016*a*), leading to petrographic characteristics somewhere between those of marls and chalk. Thin section observations show that the clays are dispersed in a brownish matrix. The texture of argillaceous chalk varies from mudstone to wackestone, where the grains mostly include forams, bioclasts and calcispheres. Under SEM, argillaceous chalks are easy to identify, with clay flakes dispersed in the matrix; mechanical compaction tends to align the flakes on a parallel plane to the bedding.

Cemented chalks exhibit a mudstone texture, with 5–8% grain content, and the grains are either cemented with sparite crystals or are micritic. All cemented chalks show similar microtextures with coalescent grain contacts, many authigenic calcite crystals and grain overgrowths. Deeply buried chalks show a higher disintegration of coccoliths and seemingly more grain overgrowths than early cemented chalks.

Outcrops of silicified chalk are scarce, but these chalks have been described in the Ekofisk and South Arne oilfields in the North Sea (Jakobsen *et al.* 2000; Lindgreen *et al.* 2010; Gennaro *et al.* 2013). Only one sample of silicified chalk was investigated in this study, from a 30 cm thick bed from Brunneval (Normandy, France). The rock still had a chalk texture with distinct bioclasts, but its strength was increased compared with the surrounding chalk. Patches of amorphous silica were present in higher porosity zones, whereas lower porosity zones appeared completely silicified, with chalcedony within what used to be the chalk matrix.

Table 2. *Petrophysical and mechanical properties*

Lithotype	Sample no.	Diagenetic index	φ_{water} (%)	φ_{He} (%)	k (mD)	Grain density ($g \cdot cm^{-3}$)	$V_{P\ dry}$ ($m\ s^{-1}$)	$V_{P\ sat}$ ($m\ s^{-1}$)	UCS (MPa)	E (MPa)
Micritic	CH01	1	44.14	41.90	3.18	2.70	2742	3298	7	9469
	CH02	0.5	43.01	44.60	4.48	2.71	2434	2921	4	1224
	CH04	0.5	41.38	42.40	2.85	2.71	2563	2919	5	–
	CO01	1	42.31	42.90	5.83	2.70	2873	2924	5	2585
	NH03	2.5	26.69	28.50	1.84	2.71	4931	4256	–	–
	RA01	1.5	–	45.90	5.95	2.70	2594	2541	4	1187
Grainy	BG01	1	–	37.80	3.09	2.68	2704	2961	5	5455
	CM02	1	39.18	39.70	3.67	2.70	3525	3550	8	9607
	CM06	1	42.41	43.60	4.32	2.71	2602	2728	4	358
	ETR33	2	30.48	33.80	13.30	2.70	2813	2966	3	1068
	SS01	1.5	36.65	40.20	3.10	2.68	2930	2723	4	1345
Argillaceous	CB02	3	26.4	21.60	0.10	2.69	3544	2154	–	–
	CB04	4	–	18.80	0.06	2.69	3544	3111	–	–
	CB06	3.5	–	23.40	0.11	2.68	3095	2476	20	40 542
	CB07	5.5	20.96	21.40	0.13	2.70	4752	4265	21	53 062
	CB09	4	24.09	22.10	0.08	2.68	4310	3012	29	22 442
	CB10	6	23.83	18.90	0.10	2.70	3553	2481	24	35 799
	CB11	3.5	–	23.80	0.10	2.69	4797	2863	–	–
	CB23	5	23.29	21.00	0.06	2.68	3976	2806	27	–
	CB24	3.5	19.03	18.20	0.08	2.67	4536	3724	24	26 401
	CB25	4	22.34	20.10	0.06	2.69	3795	2753	–	–
	EA01	5	15.39	14.20	0.06	2.72	4762	3788	–	–
Marl seam	CB13	3	32.85	31.40	0.30	2.68	3696	2343	15	2674
	CB16	4.5	23.68	20.70	0.19	2.71	4107	3739	–	–
	EA02	5	16.86	14.50	0.25	2.70	5618	4274	24	38 958
	ETR21	1	36.37	36.40	1.32	2.70	3494	3269	12	7855
	ETR47	1.5	38.94	40.40	2.68	2.68	2964	2592	10	–
	SC02	3.5	33.03	33.00	0.48	2.68	3229	2447	–	–
	SC03	3	24.62	25.40	0.40	2.69	4630	3788	19	14 090
Cemented	CB14	6.5	26.89	27.10	0.43	2.70	4090	3514	20	–
	FA15	6.5	15.81	17.60	0.10	2.71	7232	6653	20	50 789
	FA39B	7	15.20	16.20	0.14	2.71	6944	6716	30	23 444
	FH11	7.5	15.98	19.30	0.16	2.71	6526	6200	31	41 935
	SC01	6	23.69	23.80	0.44	2.70	5208	4673	21	23 021
Silicified	BR01	8	26.21	26.40	0.04	2.45	4550	4960	51	25 076

E, Young's modulus; k, empirical Klinkenberg permeability; UCS, unconfined compressive strength; $V_{P\ dry}$, dry P-wave velocity; $V_{P\ sat}$, saturated P-wave velocity; φ_{water}, water saturation porosity; φ_{He}, helium porosity.

Petrophysical and mechanical results

As shown in Figure 5, the porosity determined either by water saturation or helium porosimetry gave very similar values, ranging between 14 and 46%. The porosity in marl seam chalk ranged between 14 and 40% due to variations in the intensity of compaction in those rocks.

The measured permeabilities ranged between 0.04 and 13 mD. Following the definition of the JCR group, we defined tight chalks as having a matrix permeability <0.2 mD (Bailey *et al.* 1999) and therefore almost half of the samples tested in this study (16/35) can be considered as tight chalks based on this criterion. All the argillaceous chalk samples were tight, as well as the deeply buried cemented chalk samples from Flamborough Head (UK samples FA15, FA39 and FA11), the nodular marl seam chalk sample from the base of the Turonian at Cap Blanc Nez (CB16) and the silicified chalk sample (BR01). As the permeabilities are low, fracture porosity is absent and the porosity is associated with interparticle voids. Therefore the measured connected porosity can be considered as the total porosity.

Mercury injection capillary pressure measurements were performed on the same samples. They showed pore throat diameters ranging from 25 to 1100 nm (Faÿ-Gomord *et al.* 2016*a*). The largest pores were found in micritic (510 nm) and

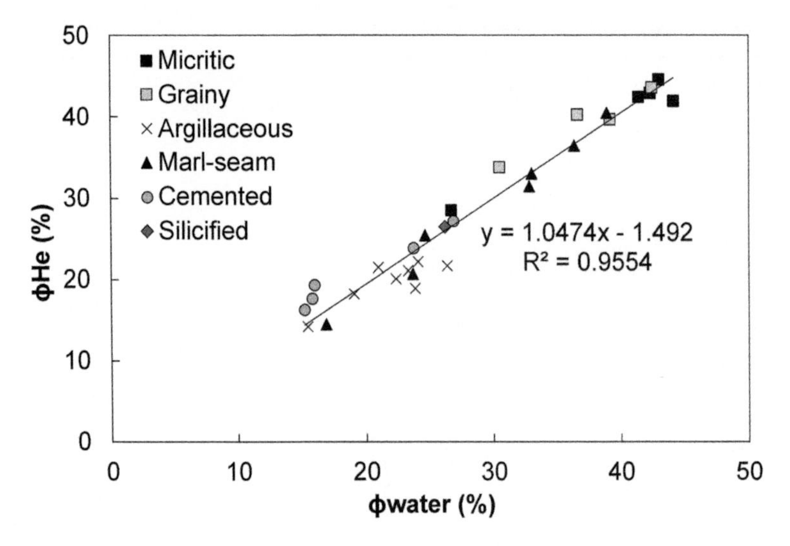

Fig. 5. Correlation between water and helium porosities.

grainy chalk (760 nm), whereas the smallest pores occurred in argillaceous (100 nm) and silicified chalk (25 nm). The pore size distribution was generally unimodal.

UCS tests were performed on dry samples. Despite the variety of sedimentary and diagenetic systems investigated in this study, the mechanical behaviour of chalk under atmospheric conditions was generally characterized by brittle failure. A wide range of values for the UCS was observed, from a few MPa for micritic and grainy chalks to several tens of MPa for argillaceous, cemented and silicified chalks (Fig. 6). The computed Young's moduli varied between 350 and 53 000 MPa. The upper limit may seem high, but the type of chalk investigated in this work can be very different from traditional pure chalk. In some cases, plastification was observed before failure, mainly in the grainy and argillaceous chalks.

Discussion

This study investigated the impact of diagenesis on the petrophysical and mechanical properties of chalk. This will contribute to our understanding of the effect of diagenesis on the storage capacity, transport mechanisms and mechanical behaviour of tight chalk formations. It may also be helpful for identifying outcrop analogues. The following discussion focuses first on the typical diagenetic indices associated with the six lithotypes. The relationships between the petrophysical and mechanical properties are then investigated and related to diagenetic considerations and lithotypes.

Overview of lithotypes and associated diagenesis index

Six lithotypes were identified within the dataset: micritic, grainy, marl seam, argillaceous, cemented and silicified chalks. They were associated with typical ranges of the diagenesis index. The lowest diagenesis indices (<2.5) correspond to micritic and grainy chalks. The loose matrix texture, punctic contacts between grains, rare authigenic calcite crystals, a lack of coccolith grain overgrowth and intraparticle cementation encountered in these microtextures showed the limited cementation of the chalks.

The marl seam chalks exhibited a wider range of diagenesis indices (1–5) because of their heterogeneous nature (clay seams and surrounding chalk). SEM observations indicated a wide range of microtextures relating to the intensity of the burial pressure solution or the initial sedimentary clay content.

The highest diagenesis indices were found in argillaceous, cemented and silicified chalks. In the argillaceous chalks the contacts between coccolith fragments were reduced and very limited grain-to-grain contact dissolution or grain overgrowths developed as a result of the dispersed clay flakes in the matrix. The matrix appeared to be more compact in the argillaceous chalks than in the micritic chalks as a result of compaction by clay-rich chalks, resulting in tighter particles with a higher degree of coccolith disintegration. In the cemented chalks, the high diagenesis index can be explained by the coalescence of grain contacts, the development of many authigenic calcite crystals and grain overgrowths. For the silicified chalk

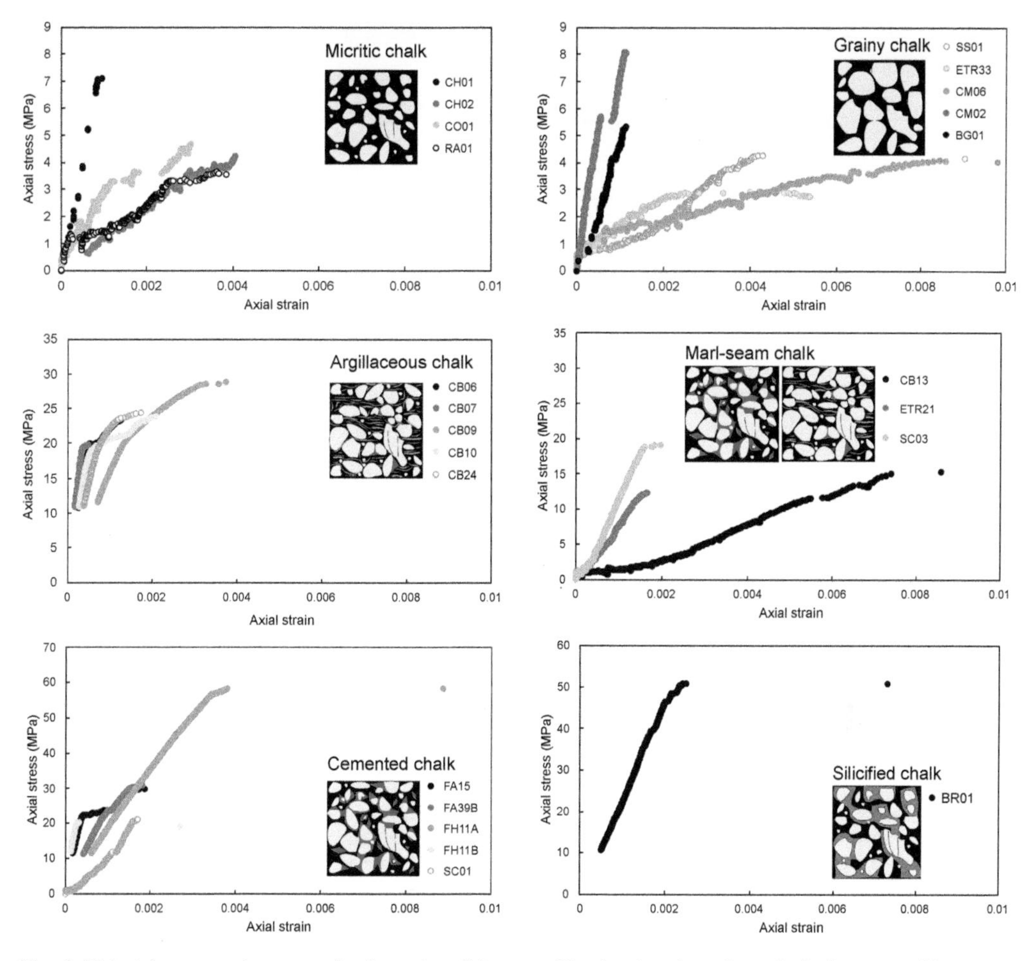

Fig. 6. Uniaxial compression curves for the various lithotypes. The sketches show the typical microtexture(s) associated with each lithotype.

sample, the insoluble residue increased to 76%, with silica dispersed in the chalk matrix and developed as a cement in forams and calcisphere chambers.

Relationships between petrophysical and mechanical properties

Porosity–permeability relationships are often considered in petrophysical studies, particularly to establish correlations within one given formation (Tiab & Donaldson 2004). The relationship is not straightforward because high-porosity rocks may show very low permeabilities and highly permeable rocks may have a low porosity. Among the studied samples, covering a wide range of chalk microtextures, the porosity and the logarithm of the permeability followed a linear relationship (Fig. 7a). The defined microtextures corresponded to typical areas within the cross-plot: micritic and grainy chalks showed the highest porosities and permeabilities, whereas argillaceous, cemented and silicified chalks were less porous (<30%) and less permeable (<1 mD). The porosity–permeability values are related to the intensity of the diagenetic processes because the highest porosities and permeabilities are associated with the lowest diagenesis indices. Figure 7a also compares the data with porosity–permeability values in the JCR database. These data are from North Sea chalks (the Tor, Hod and Ekofisk formations) and their analogues. They are generally in good agreement with our values, although there is more scatter in the JCR data.

The P-wave velocities of both the dry and saturated samples ranged between 2 and 7 km s^{-1} (Fig. 7b). The saturated P-wave velocities were generally higher than the corresponding dry values in the micritic, grainy and silicified chalks; water filled

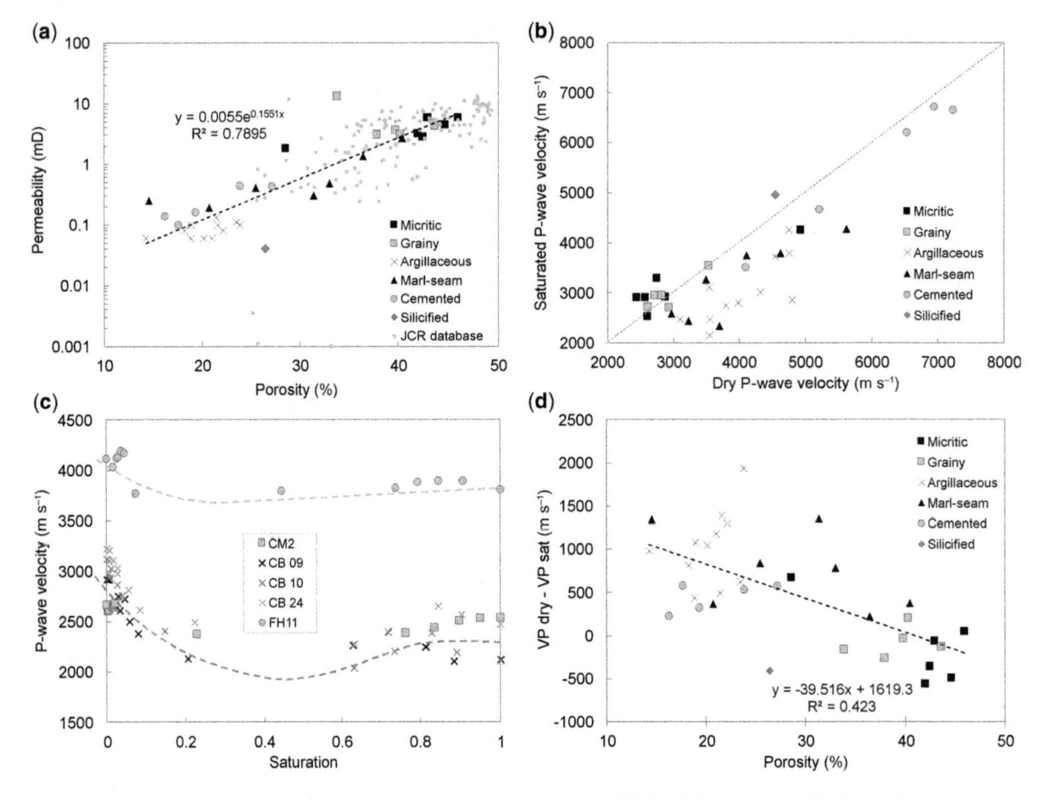

Fig. 7. (**a**) Porosity–permeability plot of data from this study classified by lithotype and comparison with porosity–permeability data from the JCR database. (**b**) P-wave velocities measured on saturated and dry samples. (**c**) P-wave velocity v. saturation plot. CM2 is a grainy chalk; CB09, CB10 and CB24 are argillaceous chalks. FH11 is a cemented chalk. (**d**) Relationship between dry and saturated P-wave velocities and porosity.

voids are more difficult to compress than air-filled voids and tend to increase the P-wave velocities (Bourbié *et al.* 1986).

However, for almost half of the samples, the dry P-wave velocities were higher than the velocities in the saturated samples; this mainly occurred in the argillaceous and marl seam chalks, but also in cemented chalks. To better understand the physical mechanism for this, the P-wave velocity was measured at various saturation values for several samples. Figure 7c shows that the velocity was at a maximum for dry rocks and rapidly decreased for a small water saturation. At high saturation states, the P-wave velocity stabilized or slightly increased, but did not reach the value of the dry material. The introduction of water to a dry sample first increases its density, resulting in a decrease in the velocity (Gassmann 1976; Bourbié *et al.* 1986). With increasing saturation, the apparent rigidity of the material decreases as well as the velocity. Beyond a limit of saturation, water compressibility is important and tends to harden the material, as predicted by Gassmann (1976).

The evolution of P-wave velocity with fluid saturation has already been observed for pure chalk (Schroeder 2002), emphasizing a minimum velocity at partial saturation. In that case, the dry velocities were lower than the fully saturated velocities. Murphy (1982) observed higher dry velocities for the Massillon sandstone (23% porosity). The presence of clay minerals may explain the difference between dry and saturated P-wave velocities. Longitudinal waves show lower velocities in dry clays than in saturated clays. Ghorbani *et al.* (2009) have shown a desiccation-driven hardening when measuring the elastic wave velocities of clay rocks.

As clay minerals were not found in all the samples, other parameters were also investigated. A point to consider is the relationship between P-wave velocities and porosity. Gregory (1976) proposed three characteristic behaviours for the P-wave velocity–saturation relationship of consolidated sediments depending on their porosity. Low porosity sediments (<10%) exhibit dry P-wave velocities that are much smaller than the saturated velocities, with a sigmoidal evolution. For medium

(10–25%) to high-porosity sediments (>25%), the P-wave velocity decreases when a small amount of water is introduced; the decrease is steeper for more porous sediments. An increase in velocity at higher saturation levels is also observed and linked to the compressibility of the fluid. Gregory (1976) observed that the difference between dry and saturated P-wave velocities depends on the porosity, but did not propose a mathematical relationship. For the chalk samples tested in this study, Figure 7d also suggests a relationship between porosity and the difference between dry and saturated P-wave velocities. A linear correlation was attempted, but the complexity of the chalks produced scatter in the data ($R^2 = 0.423$).

In terms of mechanical behaviour, a linear relationship is generally considered to show the influence of porosity (or intact dry density) on UCS (Mortimore & Fielding 1990; Duperret *et al.* 2005). This study confirmed a correlation between porosity and UCS (Fig. 8a). However, the data, as well as previously reported data, showed some scatter from the straight correlation line, meaning that porosity alone cannot be considered as the intrinsic parameter governing the mechanical strength of chalk. The mineralogical composition (e.g. silica and clay minerals) also has a non-negligible effect on the mechanical behaviour of chalk (Monjoie *et al.* 1985; Schroeder 2002).

The features observed and rated through the diagenesis index are the result of both eogenetic and mesogenetic processes affecting chalk; they quantify the diagenetic alteration of chalk. During eogenesis, a low sedimentation rate may result in the early formation of indurated surfaces on the seafloor and thus the early cementation of chalky sediments. Later, mesogenetic processes occur during burial diagenesis and lead to the formation of cements, either by grain overgrowth or authigenic calcite crystals in the matrix. Hence the diagenesis index is a means of quantifying cementation, which strengthens grain contacts and increases the UCS (Fig. 8b). This cross-plot indicates a higher correlation coefficient between the UCS and the diagenesis index than between the UCS and porosity. Several factors can reduce porosity, such as the presence of clay minerals, but these factors do not necessarily reinforce the microstructure, whereas cementation

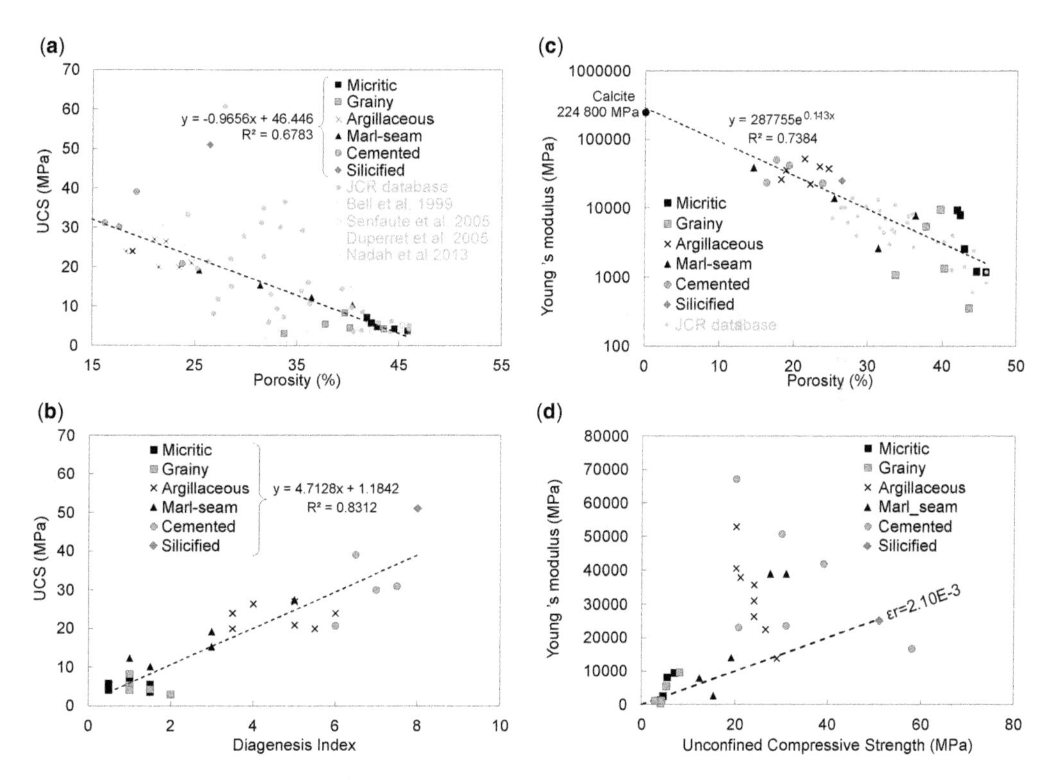

Fig. 8. (**a**) UCS data from this study as a function of porosity, classified by lithotype, and comparison with previously reported UCS–porosity data. (**b**) UCS as a function of diagenesis index. (**c**) Relationship between Young's modulus and porosity, according to lithotype, and comparison with data from the JCR database. (**d**) Relationship between Young's modulus and UCS.

strengthens the mechanical properties of chalk. This result is also important because it emphasizes how features observed at the SEM scale influence mechanical behaviour, whereas the depositional facies, as observed from thin section, only poorly constrain the petrophysical and geomechanical properties (Faÿ-Gomord *et al.* 2016*a*). The microtexture, essentially defined by the non-carbonate content and the degree of diagenesis, controls the mechanical properties of the sample.

The microtextural families were associated with specific mechanical behaviour (Fig. 6). The micritic and grainy chalks underwent less cementation during diagenesis and were the weakest and more deformable materials in the dataset. It has previously been shown that the presence of clays in chalk from deposition enhances chemical compaction (Mallon & Swarbrick 2002, 2008; Fabricius *et al.* 2008). This is why argillaceous chalk has a higher diagenesis index than pure chalk (micritic or grainy) with a similar burial history. Argillaceous chalks are therefore stronger and stiffer than micritic or grainy chalks. In cemented chalk, the coalescence of grain contacts and the development of many authigenic calcite crystals and grain overgrowths explain the high diagenesis index and subsequent high compressive strength. Cementation is controlled by a range of factors; texture, burial history and fluids (Schroeder 2002; Hjuler & Fabricius 2009) can all affect the mechanical properties.

The Young's modulus was found to be an exponential function of porosity (Fig. 8c). The data in this study are in good agreement with values in the JCR database, but cover a wider spectrum of chalk types and properties because the tested material was not limited to reservoir chalks. Engstrøm (1992) used a similar correlation for Danish chalk and proposed an extrapolation towards the Young's modulus of pure calcite (224 800 MPa) for zero porosity:

$$E = 224\,800\, e^{-0.112\varphi} \tag{1}$$

where E is the Young's modulus (MPa) and φ is the porosity. This principle was also applied in this work and gave good results for a wider dataset. Effective media models have been proposed to explain the relations between elastic properties and porosity. The modified upper Hashin–Shtrikman model (Nur *et al.* 1998; Walls *et al.* 1998; Anderson 1999) considers a mixture of hollow spherical shells of one component filling the space, whereas the other component fills the spheres. A first end-member corresponds to zero porosity and the elastic properties are those of the solid phase (mainly calcite); a second end-member depends on a critical porosity estimated to be 50 (Bhakta & Landro 2013) or 70% (Fabricius 2007) for chalk. Fabricius (2003) proposed the iso-frame model, which considers mixtures of suspended solids in the spherical pores of a solid. This modification estimates that not all the grain materials take part in building the frame in the Hashin–Shtrikman model and some will stay in suspension in the pore space. In other words, part of the solid is in suspension in the pore fluid and this suspension is embedded in the supporting frame of calcite and silicates (Fabricius *et al.* 2005).

The relationship between Young's modulus and the UCS was also examined (Fig. 8d). Our results were less clear than previously reported results (Schroeder 2002), which were limited to pure chalk. The idea of a constant brittle failure strain in the shear mode ($\varepsilon_r = 0.002$) seems to be valid for micritic and grainy chalks and some reservoir chalk (JCR database), but argillaceous and cemented chalk are far from this trend. If diagenesis plays a part in the deformability of these chalks, other factors will also affect the deformability, such as the occurrence of clay minerals, pore size and shape (Faÿ-Gomord *et al.* 2016*a*). It can explain the scatter in the cross-plot for argillaceous, marl seam and cemented chalk samples.

Conclusions

Chalk is usually defined as a pure, highly porous and low-permeability carbonate rock. This definition only gives a limited insight into the wide variety of existing chalk materials. Considering the increasing interest in unconventional reservoirs, including tight chalk formations, this study broadens the sedimentary and diagenetic systems investigated and showed how diagenesis can affect the petrophysical and the mechanical properties of chalk by studying the relationships between these properties and the associated microtextures.

Several outcrops in NW Europe were sampled and characterized. The petrographic description included the assessment of a diagenesis index that quantified diagenesis using seven criteria. Each criterion was observed on SEM micrographs of the samples at various scales. Petrophysical and mechanical tests were also conducted, including the determination of porosity, permeability, P-wave velocity and UCS tests.

Six lithotypes were defined: micritic, grainy, marl seam, argillaceous, cemented and silicified chalk. They were characterized by a typical range of values for the diagenesis index. The determined petrophysical and mechanical properties cover a wide range of values, with half of the samples being considered as tight chalks. A linear relationship between porosity and the logarithm of permeability was obtained. Porosity and permeability are linked with diagenetic processes, with the highest values associated with the lowest diagenetic indices where compaction and cementation are less developed.

P-wave velocity was found to be dependent on the saturation state of the samples with, in some cases, higher values for the dry velocity than for the saturated velocity. This can be explained by a combination of two mechanisms: density increases when water is introduced into a dry sample, but the material hardens when the saturation exceeds a threshold value. The predominant mechanism seems to be influenced by the presence of clay and the porosity of the rock.

From a mechanical point of view, a linear correlation between UCS and porosity was confirmed, even when a wide variety of chalks was considered. However, a much better relationship was found between the UCS and the diagenetic index, which quantified the degree of cementation. Argillaceous chalk, in particular, exhibited a higher diagenetic index than pure chalk (micritic or grainy) with a similar burial history. In other words, as a result of their higher degree of cementation, some clay-rich chalks may be stronger than pure chalks subject to a similar burial history. Young's modulus was exponentially linked with porosity; this can be explained by effective medium models. The stiffer chalk samples were also associated with the highest diagenetic indices. Hence typical behaviours in terms of deformability and strength were observed for the six lithotypes. We therefore showed that diagenetic processes, quantified by means of a newly developed diagenesis index, govern the microtextural features of chalk observed at the SEM scale and, in turn, affect the petrophysical and mechanical properties of these rocks.

The authors acknowledge TOTAL for its financial support as part of a Fractured Tight Chalk project, including the PhD of Ophélie Faÿ-Gomord. Our thanks go to Yves Leroy and Bruno Caline, TOTAL, for support granted throughout this study.

References

Amédro, F. & Robaszynski, F. 2000. Les craies à silex du Turonien supérieur au Santonien du Boulonnais (France) au regard de la stratigraphie événementielle. Comparaison avec le Kent (U.K.). *Géologie de la France*, **4**, 39–56.

Amédro, F. & Robaszynski, F. 2001. Les craies cénomaniennes du Cap Blanc-Nez (France) au regard de la stratigraphie évènementielle. Extension géographique de niveaux repères du bassin anglo-parisien (Boulonnais, Kent, Normandie) à l'Allemagne du Nord. *Bulletin Trimestrial la société géologique Normandie Amis Muséum du Havre*, **87**, 9–29.

Anderskouv, K. & Surlyk, F. 2011. Upper Cretaceous chalk facies and depositional history recorded in the Mona-1 core, Mona Ridge, Danish North Sea. *Geological Survey of Denmark and Greenland Bulletin*, **25**, 3–60.

Anderson, J.K. 1999. The capabilities and challenges of the seismic method in chalk exploration. *In*: Fleet, A.J. & Boldy, S.A.R. (eds) *Petroleum Geology of Northwest Europe. Proceedings of the 5th Conference.* Geological Society, London, 939–947, https://doi.org/10.1144/0050939

API 1998. *Recommended Practices for Core Analysis*. 2nd edn. American Petroleum Institute, Washington DC.

Bailey, H., Gallagher, L. et al. 1999. *Joint Chalk Research Phase V: A Joint Chalk Stratigraphic Framework.* Norwegian Petroleum Directorate, Stavanger.

Bell, F.G., Culshaw, M.G. & Cripps, J.C. 1999. A review of selected engineering geological characteristics of English Chalk. *Engineering Geology*, **54**, 237–269, https://doi.org/10.1016/S0013-7952(99)00043-5

Bhakta, T. & Landro, M. 2013. Comparison of different rock physics models for chalk reservoirs. *Proceedings of 10th Biennial International Conference & Exposition*, 23–25 November 2013, Kochi, India, Society of Petroleum Geophysicists (SPG), https://www.spgindia.org/10_biennial_form/P295.pdf

Boulvain, F. & Pingot, J-L. 2012. *Genèse du sous-sol de la Wallonie*. Classe des Sciences, Académie royale de Belgique, Brussels.

Bourbié, T., Coussy, O. & Zinszner, B. 1986. *Acoustique des milieux poreux*. Editions Technip, Paris.

Brigaud, B., Vincent, B. et al. 2014. Characterization and origin of permeability-porosity heterogeneity in shallow-marine carbonates: from core scale to 3D reservoir dimension (Middle Jurassic, Paris Basin, France). *Marine and Petroleum Geology*, **57**, 631–651, https://doi.org/10.1016/j.marpetgeo.2014.07.004

Bristow, R., Mortimore, R.N. & Wood, C.J. 1997. Lithostratigraphy for mapping the Chalk of southern England. *Proceedings of the Geologists' Association*, **108**, 293–315.

Cantrell, D.L. & Hagerty, R.M. 1999. Microporosity in Arab Formation carbonates, Saudi Arabia. *GeoArabia*, **4**, 129–154.

Collin, F., Cui, Y.J., Schroeder, C. & Charlier, R. 2002. Mechanical behavior of Lixhe chalk partly saturated by oil and water: experiment and modelling. *International Journal for Numerical and Analytical Methods in Geomechanics*, **26**, 897–924, https://doi.org/10.1002/nag.229

Deconinck, J.F. & Chamley, H. 1995. Diversity of smectite origins in Late Cretaceous sediments; example of chalks from northern France. *Clay Minerals*, **30**, 365–379.

Deconinck, J.F., Amedro, F., Fiolet-Piette, A., Juignet, P., Renard, M. & Robaszynski, F. 1991. Contrôle paléogéographique de la sédimentation argileuse dans le Cénomanien du Boulonnais et du Pays de Caux. *Annales de la Société géologique du Nord*, **1**, 57–66.

DeGennaro, V., Delage, P., Cui, Y.-J., Schroeder, C. & Collin, F. 2003. Time-dependent behavior of oil reservoir chalk: a multiphase approach. *Soils and Foundations*, **43**, 131–147, https://doi.org/10.3208/sandf.43.4_131

DeGennaro, V., Sorgi, C. & Delage, P. 2005. Air–water interaction and time dependent compressibility of a subterranean quarry chalk. Paper presented at the Symposium Post mining, November 2005, Nancy, France, https://hal.archives-ouvertes.fr/ineris-00972513/document

Delage, P., Schroeder, Ch. & Cui, Y.-J. 1996. Subsidence and capillary effects in chalk. *In*: Barla, G. (ed.) *Proceedings of the Eurock 1996 Conference, Prediction and Performance in Rock Mechanics and Rock Engineering*, 2–5 September 1996, Turin, Italy, Balkema, Rotterdam, 1291–1298.

Deville de Periere, M., Durlet, C., Vennin, E., Lambert, L., Bourillot, R., Caline, B. & Poli, E. 2011. Morphometry of micrite particles in Cretaceous microporous limestones of the Middle East: influence on reservoir properties. *Marine and Petroleum Geology*, **28**, 1727–1750, https://doi.org/10.1016/j.marpetgeo.2011.05.002

Dunham, R.J. 1962. Classification of carbonate rocks according to depositional texture. *In*: Ham, W.E. (ed.) *Classification of Carbonate Rocks--A Symposium*. American Association of Petroleum Geologists Memoirs, **1**, 108–121.

Duperret, A., Taibi, S., Mortimore, R.N. & Daigneault, M. 2005. Effect of groundwater and sea weathering cycles on the strength of chalk rock from unstable coastal cliffs of NW France. *Engineering Geology*, **78**, 321–343, https://doi.org/10.1016/j.enggeo.2005.01.004

Dupuis, C. & Vandycke, S. 1989. Tectonique et karstification profonde: un modèle de subsidence original pour le Bassin de Mons. *Annales de la Société géologique Belgique*, **112**, 479–487.

Dzulinski, M. 1969. L'essai de compression simple des matériaux pierreux. *Memoires CERES (nouvelle Série)*, **28**, 60–78.

Engstrøm, F. 1992. Rock mechanical properties of Danish North Sea Chalk. *In*: Joint Chalk Research Program (ed.) *Proceedings of the 4th North Sea Chalk Symposium*, 21–23 September 1992, Deauville, France.

Fabricius, I.L. 2001. Compaction of microfossil and clay-rich sediments. *Physics and Chemistry of the Earth, Part A: Solid Earth and Geodesy*, **26**, 59–62, https://doi.org/10.1016/S1464-1895(01)00023-0

Fabricius, I.L. 2003. How burial diagenesis of chalk sediments controls sonic velocity and porosity. *American Association of Petroleum Geologists Bulletin*, **87**, 1755–1778.

Fabricius, I.L. 2007. Chalk: composition, diagenesis and physical properties. *Bulletin of the Geological Society of Denmark*, **55**, 97–128.

Fabricius, I.L., Prasad, M. & Olsen, C. 2005. Iso-frame modeling of marly chalk and calcareous shale. Search and Discovery Article **#40151**.

Fabricius, I.L., Gommesen, L., Krogsbøll, A. & Olsen, D. 2008. Chalk porosity and sonic velocity v. burial depth: influence of fluid pressure, hydrocarbons, and mineralogy. *American Association of Petroleum Geologists Bulletin*, **92**, 201–223, https://doi.org/10.1306/10170707077

Fairhurst, C.E. & Hudson, J.A. 1999. Draft ISRM suggested method for the complete stress-strain curve for intact rock in uniaxial compression. *International Journal of Rock Mechanics and Mining Science & Geomechanics Abstracts*, **36**, 279–289.

Faÿ-Gomord, O., Soete, J. *et al.* 2016*a*. New insight into the microtexture of chalks from NMR analysis. *Marine and Petroleum Geology*, **75**, 252–271, https://doi.org/10.1016/j.marpetgeo.2016.04.019

Faÿ-Gomord, O., Descamps, F., Tshibangu, J.-P., Vandycke, S. & Swennen, R. 2016*b*. Unraveling chalk microtextural properties from indentation tests. *Engineering Geology*, **209**, 30–43.

Fritsen, A., Crabtree, B. *et al.* 1996. *Description and Classification of Chalks: North Sea Central Graben, Joint Chalk Research Phase IV*. Norwegian Petroleum Directorate, Stavanger.

Gale, A.S., Kennedy, W.J., *et al.* 2005. Stratigraphy of the Upper Cenomanian–Lower Turonian Chalk succession at Eastbourne, Sussex, UK: ammonites, inoceramid bivalves and stable carbon isotopes. *Cretaceous Research*, **26**, 460–487.

Gassmann, F. 1976. Uber die Elastizität Poröser Medien. *Vierteljahrsschrift der Naturforschenden Gesellschaft in Zürich*, **96**, 1–23.

Gaviglio, P., Bekri, S. *et al.* 2009. Faulting and deformation in Chalk. *Journal of Structural Geology*, **31**, 194–207.

Gennaro, M., Wonham, J.P., Sælen, G., Walgenwitz, F., Caline, B. & Faÿ-Gomord, O. 2013. Characterization of dense zones within the Danian chalks of the Ekofisk Field, Norwegian North Sea. *Petroleum Geoscience*, **19**, 39–64, https://doi.org/10.1144/petgeo2012-013

Ghorbani, A., Zamora, M. & Cosenza, P. 2009. Effects of desiccation on the elastic wave velocities of clay-rocks. *International Journal of Rock Mechanics and Mining Science & Geomechanics Abstracts*, **46**, 1267–1272.

Gommesen, L. & Fabricius, I.L. 2001. Dynamic and static elastic moduli of North Sea and deep sea chalk. *Physics and Chemistry of the Earth, Part A: Solid Earth and Geodesy*, **26**, 63–68, https://doi.org/10.1016/S1464-1895(01)00024-2

Gräfe, K.-U. 1999. Foraminiferal evidence for Cenomanian sequence stratigraphy and palaeoceanography of the Boulonnais (Paris Basin, N-France). *Palaeogeography, Palaeoclimatology, Palaeoecology*, **153**, 41–70.

Gregory, A.R. 1976. Fluid saturation effects on dynamic elastic properties of sedimentary rocks. *Geophysics*, **41**, 895–921.

Hjuler, M.L. & Fabricius, I.L. 2009. Engineering properties of chalk related to diagenetic variations of Upper Cretaceous onshore and offshore chalk in the North Sea area. *Journal of Petroleum Science and Engneering*, **68**, 151–170, https://doi.org/10.1016/j.petrol.2009.06.005

Homand, S. & Shao, J.F. 2000*a*. Mechanical behavior of a porous chalk and effect of saturating fluid. *Mechanics of Cohesive-frictional Materials*, **5**, 583–606, https://doi.org/10.1002/1099-1484(200010)5:7<583::AID-CFM110>3.0.CO;2-J

Homand, S. & Shao, J.F. 2000*b*. Comportement mécanique d'une craie poreuse et effets de l'interaction eau/craie.

Première partie: Résultats expérimentaux. *Oil & Gas Science and Technology – Revue d'IFP*, **55**, 591–598.

HOMAND, S. & SHAO, J.F. 2000*c*. Comportement mécanique d'une craie poreuse et effets de l'interaction eau/craie Deuxième partie : Modélisation numérique. *Oil & Gas Science and Technology – Revue d'IFP*, **55**, 599–609.

HOMAND, S., SHAO, J.F. & SCHROEDER, C. 1998. Plastic modelling of compressible porous chalk and effect of water injection. Paper SPE-47585-MS, presented at SPE/ISRM Rock Mechanics in Petroleum Engineering Conference, 8–10 July 1998, Trondheim, Norway, https://doi.org/10.2118/47585-MS

JAKOBSEN, F., LINDGREEN, H. & SPRINGER, N. 2000. Precipitation and flocculation of spherical nano-silica in North Sea chalk. *Clay Minerals*, **35**, 175–175, https://doi.org/10.1180/000985500546567

JUIGNET, P. 1974. *La transgression crétacée sur la bordure orientale du Massif armoricain. Aptien, Albien, Cénomanien de Normandie et du Maine. Le stratotype du Cénomanien.* Thesis, Université de Caen.

KACZMAREK, S.E., FULLMER, S.M. & HASIUK, F.J. 2015. A universal classification scheme for the microcrystals that host limestone microporosity. *Journal of Sedimentary Research*, **85**, 1197–1212, https://doi.org/10.2110/jsr.2015.79

KAHRAMAN, S. 2007. The correlations between the saturated and dry P-wave velocity of rocks. *Ultrasonics*, **46**, 341–348.

KENNEDY, W.J. 1969. The correlation of the Lower Chalk of south-east England. *Proceedings of the Geologists' Association*, **80**, 459–560, https://doi.org/10.1016/S0016-7878(69)80033-7

KENNEDY, W.J. & JUIGNET, P. 1975. Carbonate banks and slump beds in the Upper Cretaceous (Upper Turonian – Santonian) of Haute Normandie, France. *Sedimentology*, **21**, 1–42.

LAMBERT, L., DURLET, C., LOREAU, J.P. & MARNIER, G. 2006. Burial dissolution of micrite in Middle East carbonate reservoirs (Jurassic–Cretaceous): keys for recognition and timing. *Marine and Petroleum Geology*, **23**, 79–92, https://doi.org/10.1016/j.marpetgeo.2005.04.003

LASSEUR, E., GUILLOCHEAU, F., ROBIN, C., HANOT, F., VASLET, D., COUEFFE, R. & NERAUDEAU, D. 2009. A relative water-depth model for the Normandy Chalk (Cenomanian–Middle Coniacian, Paris Basin, France) based on facies patterns of metre-scale cycles. *Sedimentary Geology*, **213**, 1–26, https://doi.org/10.1016/j.sedgeo.2008.10.007

LAW, A. 1998. Regional uplift in the English Channel: quantification using sonic velocity. *In*: UNDERHILL, J.R. (ed.) *Development and Evolution of the Petroleum Geology of the Wessex Basin.* Geological Society, London, Special Publications, **133**, 187–197, https://doi.org/10.1144/GSL.SP.1998.133.01.08

LIND, I.L. 1993. Stylolites in Chalk from Leg 130 Ontong Java Plateau. *In*: BERGER, W.H., KROENKE, J.W. & MAYER, L.A. (eds) *Proceedings of the ODP, Scientific Results*, **130**. Ocean Drilling Program, College Station, TX, 445–451.

LINDGREEN, H. & JAKOBSEN, F. 2012. Marine sedimentation of nano-quartz forming flint in North Sea Danian chalk. *Marine and Petroleum Geology*, **38**, 73–82, https://doi.org/10.1016/j.marpetgeo.2012.08.007

LINDGREEN, H., JAKOBSEN, F. & SPRINGER, N. 2010. Nano-size quartz accumulation in reservoir chalk, Ekofisk Formation, South Arne Field, North Sea. *Clay Minerals*, **45**, 171–182, https://doi.org/10.1180/claymin.2010.045.2.171

MALDONADO, A., BATZLE, M. & SONNENBERG, S. 2011. Mechanical properties of the Niobrara Formation. Paper presented at the AAPG-RMS 2011 Annual Meeting 25–29 June 2011, Cheyenne, Wyoming, USA, Search and Discovery Article #50465.

MALLON, A.J. & SWARBRICK, R.E. 2002. A compaction trend for non-reservoir North Sea Chalk. *Marine and Petroleum Geology*, **19**, 527–539, https://doi.org/10.1016/S0264-8172(02)00027-2

MALLON, A.J. & SWARBRICK, R.E. 2008. Diagenetic characteristics of low permeability, non-reservoir chalks from the Central North Sea. *Marine and Petroleum Geology*, **25**, 1097–1108, https://doi.org/10.1016/j.marpetgeo.2007.12.001

MARLIÈRE, R. 1949. Le site géologique du Captage d'Hainin-Hautrage (Hainaut). *Annales de la Société géologique de Belgique*, **T73**, B55–B90.

MEGAWATI, M., MADLAND, M.V. & HIORTH, A. 2015. Mechanical and physical behavior of high-porosity chalks exposed to chemical perturbation. *Journal of Petroleum Science and Engineering*, **133**, 313–327, https://doi.org/10.1016/j.petrol.2015.06.026

MENPES, R.J. & HILLIS, R.R. 1996. Determining apparent exhumation from Chalk outcrop samples, Cleveland Basin/East Midlands Shelf. *Geological Magazine*, **133**, 751, https://doi.org/10.1017/S0016756800024596

MITCHELL, S.F. 1994. New data on the biostratigraphy of the Flamborough Chalk Formation (Santonian, Upper Cretaceous) between South Landing and Danes Dyke, North Yorkshire. *Proceedings of the Yorkshire Geological Society*, **50**, 113–118, https://doi.org/10.1144/pygs.50.2.113

MONJOIE, A., SCHROEDER, Ch., HALLEUX, L., DA SILVA, F., DEBANDE, G., DETIÈGE, Cl & POOT, B. 1985. Mechanical behaviour of chalks. *In*: Amoco Norway Oil Company (ed.) *Proceedings of the 2nd North Sea Chalk Symposium*, May 1985, Stavanger, Norway.

MORTIMORE, R. & POMEROL, B. 1997. Upper Cretaceous tectonic phases and end Cretaceous inversion in the Chalk of the Anglo–Paris Basin. *Proceedings of the Geologists' Association*, **108**, 231–255.

MORTIMORE, R.N. 2011. A Chalk Revolution: what have we done to the Chalk of England? *Proceedings of the Geologists' Association*, **122**, 232–297.

MORTIMORE, R.N. & FIELDING, P.M. 1990. The relationship between texture, density and strength of chalk. *In*: BURLAND, J.B. (ed.) *Chalk, Proceedings of the International Chalk Symposium, Brighton Polytechnic*, 4–7 September 1989. Thomas Telford, London.

MURPHY, W.F., III. 1982. Effects of partial water saturation on attenuation in sandstones. *Journal of the Acoustical Society of America*, **71**, 1458–1468.

NADAH, J., BIGNONNET, F., DAVY, C.A., SKOCZYLAS, F., TROADEC, D. & BAKOWSKI, S. 2013. Microstructure and poro-mechanical performance of Haubourdin chalk. *International Journal of Rock Mechanics and*

Mining Science & Geomechanics Abstracts, **58**, 149–165, https://doi.org/10.1016/j.ijrmms.2012.11.001

Nguyen, H.D., DeGennaro, V., Sorgi, C. & Delage, P. 2008. Experimental and modelling investigation on the behaviour of a partially saturated mine chalk. Paper presented at the Symposium Post-Mining 2008, February 2008, Nancy, France. ASGA, Vandoeuvre-les-Nancy.

Nur, A., Mavko, G., Dvorkin, J. & Galmudi, D. 1998. Critical porosity: a key to relating physical properties to porosity in rocks. *The Leading Edge*, **17**, 357–362.

Papamichos, E., Brignoli, M. & Santarelli, F.J. 1997. An experimental and theoretical study of a partially saturated collapsible rock. *Mechanics of Cohesive-frictional Materials*, **2**, 251–278.

Papamichos, E., Berntsen, A.N., Cerasi, P., Vandycke, S., Baele, J-M. & Fuh, G.F. 2012. Solids production in chalk. *In*: Bobet, A. (ed.) *46th US Rocks Mechanics/Geomechanics Symposium*, 24–27 June 2012, Chicago, IL, USA. American Rock Mechanics Association, ARMA Conference Paper 2012–479.

Pollastro, R.M. 2010. Natural fractures, composition, cyclicity, and diagenesis of the Upper Cretaceous Niobrara Formation, Berthoud Field, Colorado. *The Mountain Geologist*, **47**, 135–149.

Quine, M. & Bosence, D. 1991. Stratal geometries, facies, and sea-floor erosion in Upper Cretaceous Chalk, Normandy, France. *Sedimentology*, **38**, 1113–1152.

Rashid, F., Glover, P.W.J., Lorinczi, P., Collier, R. & Lawrence, J. 2015. Porosity and permeability of tight carbonate reservoir rocks in the north of Iraq. *Journal of Petroleum Science and Engineering*, **133**, 147–151, https://doi.org/10.1016/j.petrol.2015.05.009

Regnet, J.B., Robion, P., David, C., Fortin, J., Brigaud, B. & Yven, B. 2014. Acoustic and reservoir properties of microporous carbonate rocks: implication of micrite particle size and morphology. *Journal of Geophysical Research: Solid Earth*, **120**, 790–811, https://doi.org/10.1002/2014JB011313

Richard, J., Sizun, J.P. & Machhour, L. 2005. Environmental and diagenetic records from a new reference section for the Boreal realm: the Campanian chalk of the Mons basin (Belgium). *Sedimentary Geology*, **178**, 99–111, https://doi.org/10.1016/j.sedgeo.2005.04.001

Risnes, R. 2001. Deformation and yield in high porosity outcrop chalk. *Physics and Chemistry of the Earth, Part A: Solid Earth and Geodesy*, **26**, 53–57.

Risnes, R. & Flaageng, O. 1999. Mechanical properties of chalk with emphasis on chalk-fluid interactions and micromechanical aspects. *Oil & Gas Science and Technology*, **54**, 751–758, https://doi.org/10.2516/ogst:1999063

Risnes, R., Haghighi, H., Korsnes, R.I. & Natvik, O. 2003. Chalk–fluid interactions with glycol and brines. *Tectonophysics*, **370**, 213–226, https://doi.org/10.1016/S0040-1951(03)00187-2

Robaszynski, F. & Amédro, F. 1986. The Cretaceous of the Boulonnais (France) and a comparison with the Cretaceous of Kent (United Kingdom). *Proceedings of the Geologists' Association*, **97**, 171–208.

Robaszynski, F., Gale, A.S., Juignet, P., Amédro, F. & Hardenbol, J. 1998. Sequence stratigraphy in the Upper Cretaceous series of the Anglo-Paris Basin: exemplified by the Cenomanian stage. *In*: Graciansky de, P.-C., Hardenbol, J., Jacquin, T. & Vail, P.R. (eds) *Mesozoic and Cenozoic Sequence Stratigraphy of European Basins*. Society of Economic Paleontologists and Mineralogists, Special Publications, **60**, 363–386.

Robaszynski, F., Dhondt, A. & Jagt, J.W.A. 2001. Cretaceous lithostratigraphic units (Belgium). *Geologica Belgica*, **4**, 121–134.

Røgen, B. & Fabricius, I.L. 2002. Influence of clay and silica on permeability and capillary entry pressure of chalk reservoirs in the North Sea. *Petroleum Geoscience*, **8**, 287–293, https://doi.org/10.1144/petgeo.8.3.287

Rummel, F. & van Heerden, W.L. 1978. Suggested Methods for Determining Sound Velocity. *International Journal of Rock Mechanics and Mining Science & Geomechanics Abstracts*, **15**, 53–58.

Schroeder, C. 1995. Le pore collapse: aspect particulier de l'interaction fluide-squelette dans les craies? *In*: *Proceedings of the Colloque international du Groupement Belge de Mécanique des Roches*. Brussels, 1.1.53–1.1.60.

Schroeder, C. 2002. *Du coccolithe au réservoir pétrolier*. PhD thesis, University of Liège.

Senfaute, G., Amitrano, D., Lenhard, F. & Morel, J. 2005. Etude en laboratoire par méthodes acoustiques de l'endommagement des roches de craie et corrélation avec des résultats in situ. *Revue Française de Géotechnique*, **110**, 9–18.

Strand, S., Hjuler, M.L., Torsvik, R., Pedersen, J.I., Madland, M.V. & Austad, T. 2007. Wettability of chalk: impact of silica, clay content and mechanical properties. *Petroleum Geoscience*, **13**, 69–80, https://doi.org/10.1144/1354-079305-696

Tiab, D. & Donaldson, E.C. 2004. *Petrophysics*. 2nd edn. Gulf Professional Publishing, Houston, TX.

Thuro, K., Plinninger, R.J., Zäh, S. & Schütz, S. 2001. Scale effects in rock strength properties. Part 1: unconfined compressive test and Brazilian test. *In*: Särkkä, P. & Eloranta, P. (eds) *Rock Mechanics – a Challenge for Society*. Taylor & Francis.

Vandycke, S. 2002. Paleostress records in Cretaceous formations in NW Europe: extensional and strike-slip events in relationships with Cretaceous-Tertiary inversion tectonics. *Tectonophysics*, **357**, 119–136.

Vandycke, S., Bergerat, F. & Dupuis, C. 1991. Meso-Cenozoic faulting and inferred paleostresses of the Mons basin (Belgium). *Tectonophysics*, **192**, 261–271.

Vincent, B., Fleury, M., Santerre, Y. & Brigaud, B. 2011. NMR relaxation of neritic carbonates: an integrated petrophysical and petrographical approach. *Journal of Applied Geophysics*, **74**, 38–58, https://doi.org/10.1016/j.jappgeo.2011.03.002

Walls, J., Divorcing, J. & Smith, B.A. 1998. Modeling seismic velocity in Ekofisk chalk. Expanded Abstract presented at the 68th SEG Meeting, Society of Exploration Geophysicists.

Whitham, F. 1993. The stratigraphy of the Upper Cretaceous Flamborough Chalk Formation north of the Humber, north-east England. *Proceedings of the Yorkshire Geological Society*, **49**, 235–258, https://doi.org/10.1144/pygs.49.3.235

Mechanical constraints on kink band and thrust development in the Appalachian Plateau, USA

PAUL GILLESPIE* & GÜNTHER KAMPFER

Statoil ASA, Forusbeen 50, 4035, Stavanger, Norway

**Correspondence: pgil@statoil.com*

Abstract: The internal deformation of the Appalachian Plateau décollement sheet has a distinctive style involving kink bands and thrusts. In areas where the décollement sheet is underlain by thin salt, the dominant structures are thrusts developed at shallow levels, underlain by a series of steep kink bands that terminate downwards at the Silurian salt décollement. Where the salt is thick, large asymmetrical anticlines developed with hinterland-verging kinks on their back-limbs that deformed the entire supra-salt sequence. In order to understand the constraints on deformation, we have used analytical mechanical modelling based on the maximum strength theorem. The simplified model consists of three layers: two are fluids and the third, intervening layer is a stratified competent material. The model is compressed horizontally and the predictions made are based on the kinematic approach of classical limit analysis. Two modes of deformation are investigated: the thrust and the kink band. The modelling shows that kink bands dominate deformation at large burial depth. At shallower depth and small regional bedding dip, the dominant mode is thrusting. In areas of open folding it is predicted that through-going hinterland-verging kink bands will form at a critical limb dip angle of about 10°.

Supplementary material: Technical details of the mechanical theory behind this article are available at https://doi.org/10.6084/m9.figshare.c.3799492

The Appalachian Basin, in the eastern United States, is well known for hosting organic-rich shale units with prodigious hydrocarbon potential (Fig. 1). The recent surge in production activity in the region has led to the acquisition of large quantities of subsurface data, including both 3-D seismic and well data. Interpretation of the newly available data from Pennsylvania has shown that the contractional structural style is quite distinctive and includes large-scale folds, thrusts and kink bands underlain by an evaporite décollement (Mount 2014; Gillespie *et al.* 2015). The structural deformation poses difficulties for steering of horizontal wells in the organic-rich shale and so it is important to be able to predict the occurrence of the different kinds of structures. We have used analytical mechanical modelling based on the maximum strength theorem in order to understand the constraints on deformation, and to develop a predictive understanding of the variations in structural style.

Stratigraphy and tectonics

A giant saline basin developed in the eastern United States during the late Silurian, with the development of evaporites including basinal halite and marginal gypsum (the Salina Gp evaporites). During subsequent deformation the evaporite sequence respresented a mobile layer, or décollement. In the mid Devonian, loading by the Acadian mountains caused the development of a foreland flexure that defined the Appalachian Basin. The basin was filled with a laterally continuous shale-rich sequence punctuated by ramp carbonate development (Fig. 2). The sequence includes marine organic-rich shales such as the Marcellus shale and the Geneseo shale, which form prominent detachment horizons, i.e. bedding parallel faults (Evans 1994; Aydin & Engelder 2014; Gillespie *et al.* 2015). In the Upper Devonian, progradation and filling of the basin led to the development of the Catskill delta and the sequence became more sand prone (Ettensohn 2008).

The succeeding Alleghanian Orogeny was of Late Carboniferous to Permian in age (Lash & Engelder 2007; Ettensohn 2008), but constraint of the cessation of the orogeny is poor because the syn-orogenic sediments have been largely eroded by later uplift. Alleghanian shortening is expressed in the well-developed folds and thrusts of the Valley and Ridge province. In the Appalachian Plateau, towards the foreland, the macroscopic contractional deformation is weaker, with the development of large open folds, kink bands and thrusts, and the deformed rocks are underlain by the Salina Gp evaporites that form an effective décollement. The sequence above the décollement is therefore referred to as the Appalachian Plateau décollement sheet.

From: Turner, J. P., Healy, D., Hillis, R. R. & Welch, M. J. (eds) 2017. *Geomechanics and Geology*. Geological Society, London, Special Publications, **458**, 245–256.
First published online June 12, 2017, https://doi.org/10.1144/SP458.12

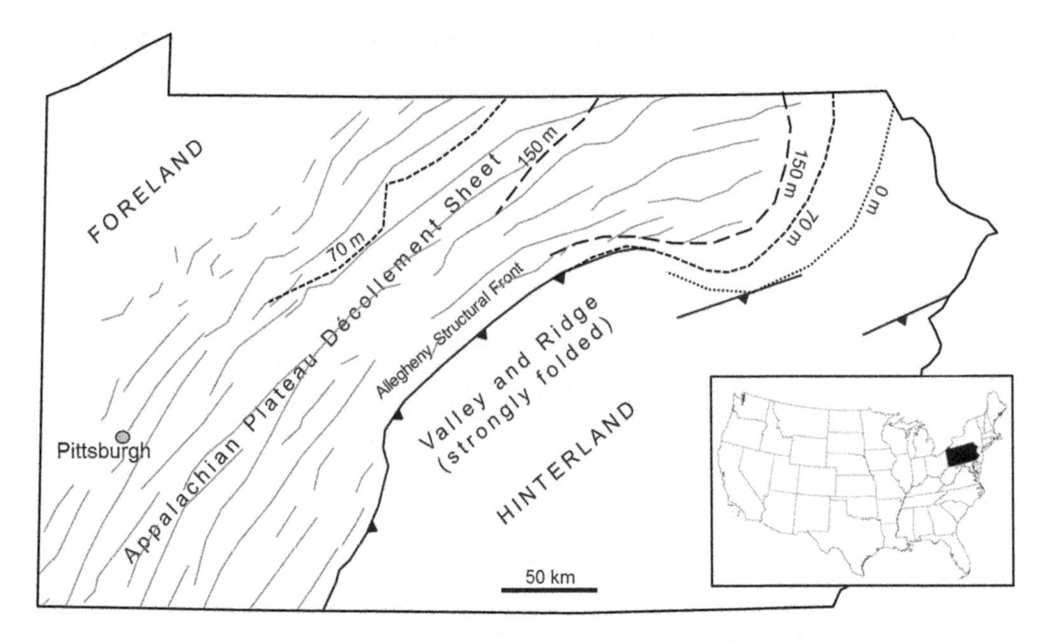

Fig. 1. Structural map of Pennsylvania showing the anticlinal axes within the Appalachian Plateau dćcollement sheet (grey lines, from Faill 2011). Contours of Salina Group halite shown from Rickard (1969). Zone of thickest halite is the zone of highest amplitude folding.

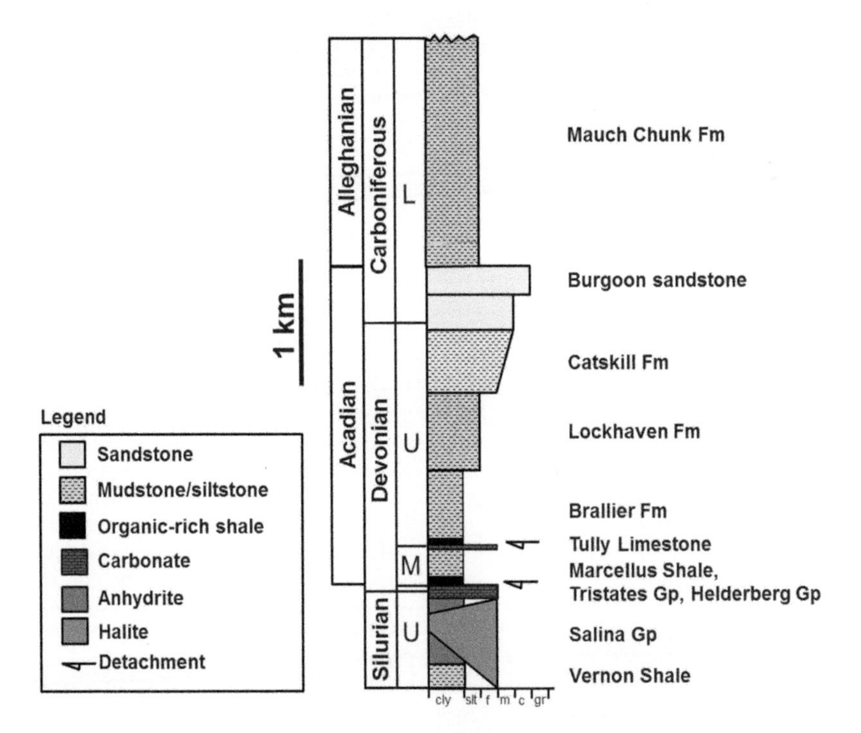

Fig. 2. Simplified stratigraphic section from the Appalachian Plateau, Pennsylvania, showing some of the main units. The halite of the Salina Gp has very variable thickness owing both to primary lateral variation and to salt tectonics. Compiled from the Shoemaker Well (Cathcart & Myers 1934) and Mount (2014).

Analysis of fluid inclusion data indicates that the maximum burial depth of the Marcellus shale in the central Appalachian Plateau was about 5.2 km (Evans 1995) and occurred during the Alleghenian Orogeny. It is estimated that 3.1 km of the overburden has been eroded.

Borehole data from Pennsylvannia indicate that the Devonian sediment of the Appalachian Basin has regional dip of 0.5° to the SE. This regional dip had its origin in the Acadian and Alleghanian foreland flexures. The regional dip of the sediment at the present day surface is approximately zero. The overall taper of the wedge is therefore very low, an observation that has been attributed to the very weak décollement formed by the Syracuse Salt (Chapple 1978; Davis & Engelder 1985).

Structural style

The subsurface structural style within the Appalachian Plateau décollement sheet has been determined from 3D seismic data and well data including core, borehole image logs and geosteering data (Gillespie *et al.* 2015). The structural style varies significantly according to the thickness and facies of the underlying evaporite (Mount 2014).

Areas of thin salt

In areas of the décollement sheet underlain by thin salt, the dominant structures are thrusts developed at shallow levels, underlain by a series of steep kink bands that terminate downwards at the Silurian salt décollement (Gillespie *et al.* 2015). The kink bands have a mean dip of 61°. The kink bands verge towards both the foreland and hinterland and typically are reflected downwards (Cobbold 1976) at the level of organic-rich shale to form box synclines that protrude into the underlying salt (Fig. 3). Above the kinked units, contractional deformation is taken up by a series of planar thrust ramps with mean dip of 21° and which verge towards the foreland. The lower termination of the thrust planes frequently coincides with the position of convergence of paired kink bands (Fig. 3). However, the details of the interaction between kink bands and thrusts cannot be readily resolved.

On a regional scale, the transition between thrusts and kink bands appears to occur at fixed depth and cuts across stratigraphy. Thus in the foreland, where the sequence is shallower, the transition occurs typically at the Marcellus shale. Further towards the hinterland, the sequence is deeper and the transition occurs higher in the stratigraphy, typically at the Geneso shale, immediately above the Top Tully horizon (Fig. 3). The transition depth is at about 2.1 km below the present day surface.

Areas of thick salt

Where the salt is thicker than about 150 m, in the more central part of the evaporite basin (Fig. 1), structural style is very different. Here, large open anticlines are developed that fold the entire preserved sequence of the Appalachian Plateau décollement sheet (Frey 1973; Wiltschko & Chapple 1977; Scanlin & Engelder 2003). The folds are salt cored and have wavelength of *c.* 11 km and amplitude of up to 500 m, although amplitude diminishes towards the foreland. Between the anticlines are flat areas of low structural elevation that represent areas underlain by salt welds, i.e. areas in which the salt has been partially or totally evacuated (Wagner & Jackson 2011). Within the areas of thick salt there is little sign of smaller-scale kink bands that are observed in areas of thin salt.

Vergence of anticlines

The anticlines in areas of thick salt are asymmetrical and typically verge towards the hinterland

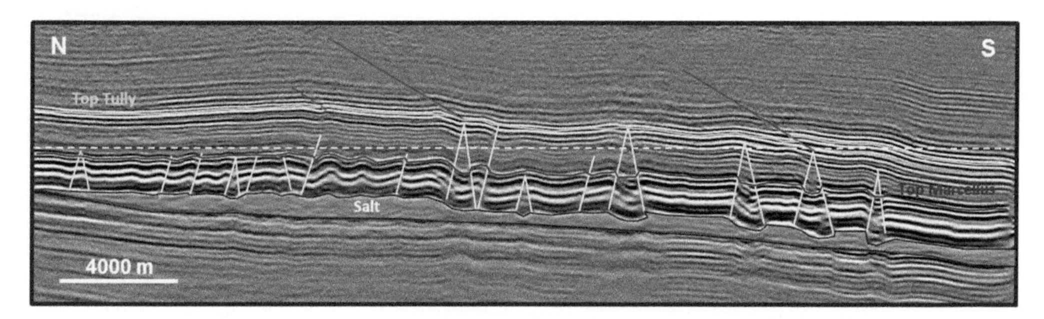

Fig. 3. Seismic cross section from area of relatively thin salt (pink), showing thrusts (red lines), kink bands (yellow) and the approximate depth of the transition between kink bands and thrusts (dashed white line). Evidence that the steep structures are kink bands comes both from seismic data and from well data. Seismic section in depth domain, vertical exaggeration ×3. Seismic data courtesy of Geophysical Pursuit and Geokinetics.

(Sherrill 1934, 1941; Gwinn 1964). Detailed examination shows that the steep limbs are formed by giant kink bands that extend from the salt up to the present day surface (Fig. 4). The vertical offset (or throw) on the kink bands is up to 1 km. The origin of the hinterlandward vergence in the Appalachian Plateau décollement sheet has been long debated. Willis (1894) and Chamberlin (1931) suggested that the asymmetry arose from underthrusting. Bain (1931) and Cathcart & Myers (1934) argued that the vergence was the result of gravity gliding towards the hinterland. However, Sherrill (1934) dismissed this possibility on the grounds that there is little or no sign of extension in the foreland, as would be expected in a gravitational system. More recently, Mount (2014) suggested that the asymmetrical folds developed as the result of buttressing above local stratigraphic pinchouts of the salt. Our own interpretation of the seismic data is that, although there are disturbances in the reflectors underneath the asymmetrical folds, these are the result of seismic velocity effects rather than being due to local stratigraphic pinchouts (Fig. 4).

Observation of small-scale structures

Small-scale structures are described below that give insight into the mechanisms that control structural development within the Appalachian décollement sheet.

Bedding-parallel slip

Examination of outcrops and core from the Devonian shale has shown that there is abundant evidence of bedding-parallel slip in the form of slickensided surfaces and small-scale duplexes (Bosworth 1984; Nickelsen 1986; Evans 1994: Aydin & Engelder 2014; Gillespie *et al.* 2015). Bedding-parallel slip surfaces occur through the shale, but are most abundant in the organic-rich facies. The high intensity of bedding-parallel slip in the organic-rich shale may be due to either high pore fluid pressure induced by organic matter cracking (Evans 1994; Aydin & Engelder 2014) or the low frictional strength of organic-rich shale (Crawford *et al.* 2008; Rutter *et al.* 2013; Kohli & Zoback 2013).

Bedding-parallel slip is a common feature of all core; however, it becomes more intense in strongly deformed regions, closer to the Allegheny structural front. In an example core from the Marcellus shale (Fig. 5), the bedding dip changes abruptly from 27° to 63° across a very sharp hinge; this is interpreted to be the upper fold axial plane of a large kink band. The paired axial planes of kink bands are termed the kink planes (Faill 1969). The shale is divided by numerous bedding-parallel slip surfaces marked by polished slickensides and shear fibre veins. Within the kink band the average spacing of the bedding-parallel slip surfaces measured perpendicular to bedding is 2.5 cm.

Accommodation of deformation in the kink planes

A small example of a kink band in a carbonate-rich facies of the Marcellus shale shows that the kink band is internally made from a series of clockwise rotated blocks (or lithons) that are separated by bedding-parallel slip planes (Fig. 6). The deformation within the kink band is similar to a deck of cards that is tilted, causing slip between the cards. Within the kink planes, individual lithons pivot

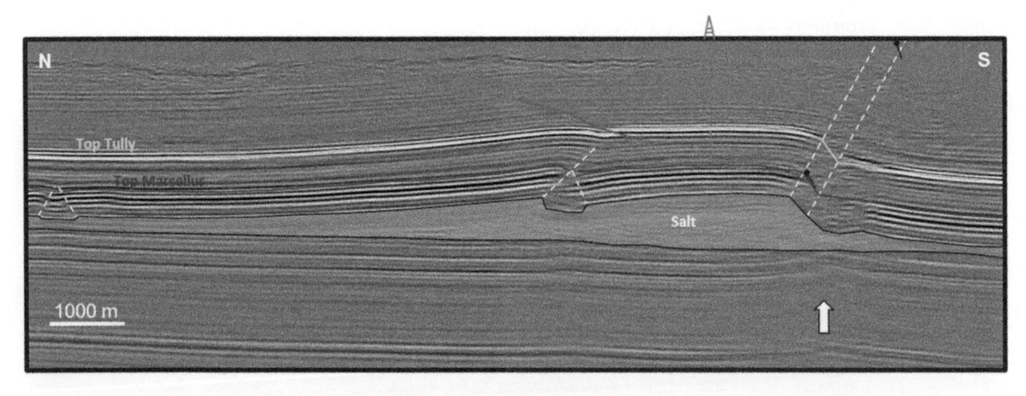

Fig. 4. Seismic section in region of relatively thick salt showing asymmetrical anticline with a kink band (yellow dashed lines) on its southern, hinterland, limb. Black tadpoles give structural dip at outcrop and from borehole image log. Data in time domain, but scaled so that vertical and horizontal scales are approximately equal. High beneath the kink band (arrowed) is a velocity pull-up. Seismic data courtesy of Geophysical Pursuit and Geokinetics.

Fig. 5. Core from a vertical well in the Marcellus shale showing a kink plane (dashed line) and multiple bedding-parallel slip surfaces. Image used with permission of Core Lab Marcellus Joint Industry Project.

about points such that triangular veins occur on one side of the pivot point, and there is compaction of the lithon and the wall rock on the other side of the pivot point. The compaction is thought to be accommodated by pressure solution in this case. Thus the kink plane is accommodated by a combination of dilation and volume loss.

Model development

To understand the mechanical controls on the deformation within the Appalachian Plateau, we have used an analytical mechanical method based on limit analysis. Limit analysis has its origin in civil and structural engineering. Its main principle is to construct lower and upper bounds of the load that a structure can sustain using either a static or a kinematic approach. The static approach is based on the construction of statically admissible stress fields providing a maximum load which a structure can sustain. Souloumiac *et al.* (2009) applied limit analysis to an accretionary wedge under lateral compression to study the onset of thrusting. Cubas *et al.* (2008) and Pons & Leroy (2012) investigated accretionary wedges applying the external approach, which uses a kinematically admissible velocity field to determine the minimum load which is required for a specific failure mode.

The chosen approach in this paper builds on the theoretical work by Maillot & Leroy (2006) and extended by Kampfer & Leroy (2009), which investigates the onset and the development of a kink fold with the help of the maximum strength theorem. The maximum strength theorem is the external approach of limit analysis as described by Chen & Liu (1990) and Salençon (2002). The main ingredients for this theory are mechanical equilibrium and the condition that the material can be described with a convex

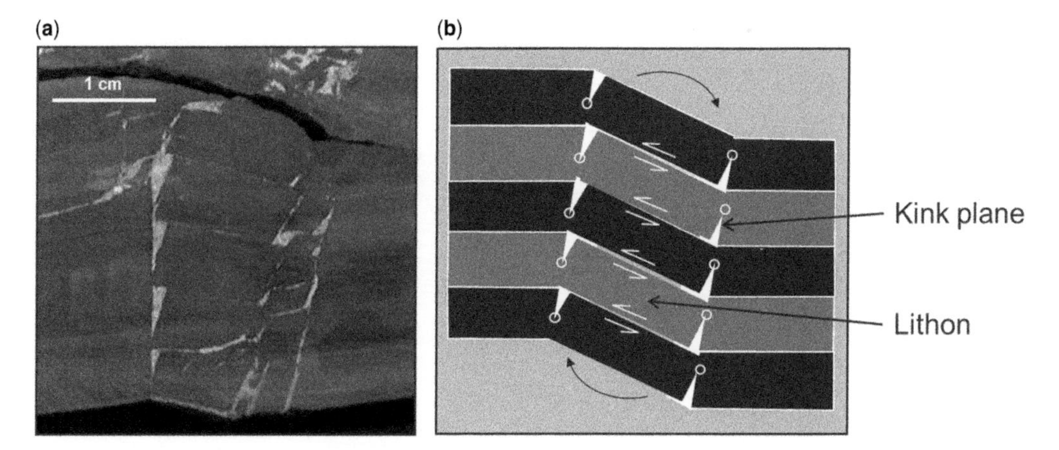

Fig. 6. (**a**) Small kink band in core from carbonate-rich part of the Marcellus shale. (**b**) Cartoon to show the rotation of the lithons within the kink bands and the opening of triangular veins within the kink planes. Circles represent the transition from dilation to compaction along the kink planes. Image in (**a**) used with permission of Core Lab Marcellus Joint Industry Project.

strength domain. More theoretical details of this application of limit analysis are given in the Supplementary material.

Application to the structures of the Appalachian Plateau

We have applied the analytical method to answer the following questions:

(1) What controls the depth at which thrusts and kink bands develop?
(2) Why are large folds characterized by a single large kink band on the back limb?

Applying the maximum strength theorem, two modes of failure – thrusts and kink bands – have been investigated for a multilayered structure exposed to regional compression as determined by the burial depth (H_o) and the sheet dip, treated as a geometrical imperfection (angle ν in Fig. 7a). A detailed explanation on the construction of admissible velocity fields, derivation of the equations and a discussion of the onset and development of thrusts and kink bands can be found in Kampfer & Leroy (2009) and Kampfer (2010).

The setup for the analytical approach consists of a stiff frame and a right sidewall, which is pushed from the right towards the left with the force Q in order to compress the multilayer in the frame (Fig. 7a). To keep the setup simple enough to be solved analytically, the multilayered system consists of three layers. Two are fluids and surround the third, which is a layered competent material. The lower fluid represents the evaporitic décollement and the layered competent material represents the cohesive rocks of the décollement sheet, while the upper fluid represents unconsolidated synorogenic sediment. The competent layers are tilted towards the right sidewall, which enables the activation of bed parallel slip. The tilt angle ν is further referred to as the sheet dip.

The individual competent layers of the rock are assumed to be separated by weak interfaces whose frictional strength is described by the Mohr–Coulomb criterion. In addition to cohesive and frictional strength, the competent material has a limited strength in compression owing to a compaction mechanism. The compaction pressure, P^*, is the strength under a hydrostatic load, and in porous rocks it is also known as the grain crushing pressure (Wong *et al.* 1997). The compaction pressure can be experimentally evaluated and closes the failure envelope for rocks exposed to compressive stresses. At the compaction pressure, deformation may be accommodated by mechanical collapse of void

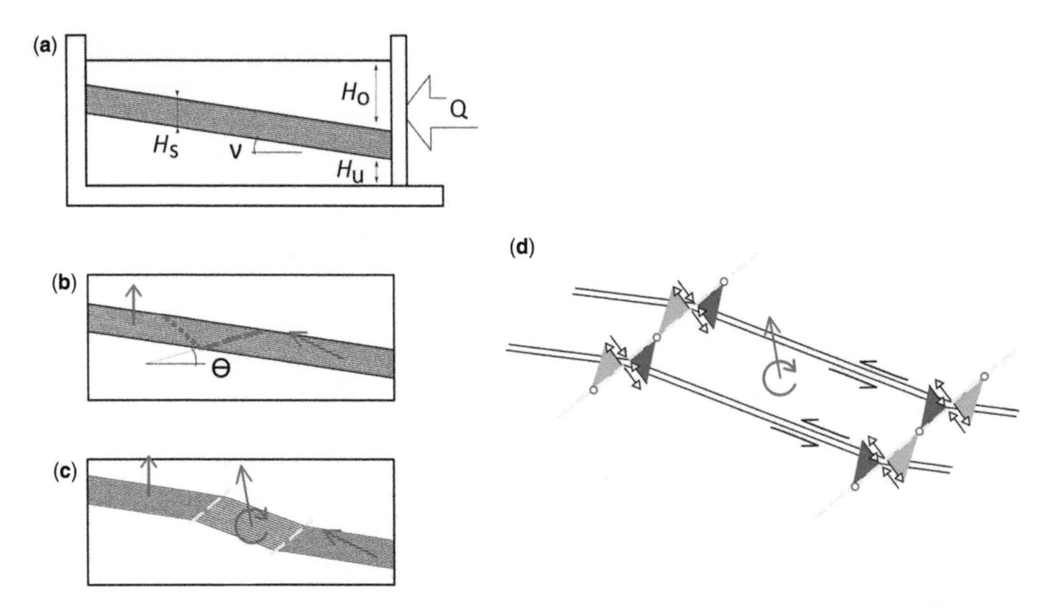

Fig. 7. (**a**) Set-up with tilted multilayer surrounded by an inviscid fluid above and below; (**b**) thrusting and (**c**) kink band mode of failure, with red arrows indicating virtual velocities. The red dotted line in (b) is the conjugate thrust orientation; yellow dashed lines in (c) are the kink planes. (**d**) Close-up of a single layer of the kink band. The green and blue regions represent regions of opening and compression/compaction along the kink planes. The red circles mark the transition from compaction to opening. The black arrows along the bedding planes indicate bedding-parallel slip, which is activated during the kink band development owing to the rotation of the lithons.

space in the matrix structure, grain crushing or pressure solution.

Sonic logs show that the carbonate units Tristates and Helderberg Groups are stiffer than the overlying formations. It can be assumed therefore that these carbonate units had a controlling influence on kink initiation. Steep pressure solution cleavage is a common feature of the limestones of the Appalachian Plateau (Engelder & Geiser 1979) and structures in the Helderberg Group are accommodated by intense stylolites (Marshak & Engelder 1985). Therefore pressure solution is another possible compaction mechanism. Pressure solution can be translated into a strain rate-dependent compaction pressure using Rutter & Elliott's (1976) deformation mechanism map for calcite.

In the following study, two modes of large-scale deformation under compression are investigated: the thrust and the kink band. The thrust (Fig. 7b) has a single failure plane, which divides the solid, layered structure into a left and a right part. The setup restricts any horizontal movement of the left side of the faulted layer. Therefore, the kinematically admissible velocity is only in the vertical direction, and may be upwards or downwards. The right-hand side of the fault can move upwards or downwards but also has a horizontal component, equal to the velocity of the sidewall. The velocity jump from the right- to the left-hand side of the fault is defined as the velocity jump vector, which is used for calculating the dissipated energy for creating the thrust plane.

The kink band has two hinges (the kink planes), shown in Figure 7c. For each layer along the kink planes there is contraction and/or dilation (Fig. 7d), owing to the rotation of the lithons. Therefore, the velocity jump vector is oriented perpendicular to the kink planes. In addition to the deformation in the kink planes, we account for bedding-parallel slip, indicated by the shear arrows along the bedding planes (Fig. 7d). Care needs to be taken of the translational movement of the three individual blocks as well as of the velocity jump vector owing to lithon rotation. The left-most part of the block can only move up- and downwards and the right-most part has an additional horizontal component. Kinematic concepts are then used to describe the complete set of velocities within the compressed system.

Once the virtual velocity fields have been described, they are used in the maximum strength theorem to obtain the least upper bound of the tectonic force for the onset of thrusting and for the onset of a kink band. The lower of these two forces is used to determine the dominant failure mode. This comparison in terms of forces can be made for any combination of regional dips, overburden, height of the structure, number of interfaces and even for determining the geometry, as the setup is shortened. In the following, we highlight the results of the onset of thrusting and kinking as a function of sheet dip and overburden. A more detailed parameter study of the analytical model is given in Kampfer & Leroy (2009).

Parameters

The parameters used in the analysis are given in Table 1. The stack thickness chosen ($H_s = 300$ m) corresponds to the stratigraphic interval between the top of the Salina Gp evaporites and the Tully Limestone (Fig. 2). This used is an estimate of the thickness of the mechanical unit undergoing kinking and does not represent the thickness of the entire sedimentary sequence. The spacing between adjacent bedding parallel slip surfaces observed in the core in Figure 5 was *c.* 2.5 cm. As the Marcellus shale typically has a higher frequency of bedding-parallel slip than other units, this is thought to be an underestimate and so the spacing was set to 5 cm.

The compaction strength of the stratified formation is quite uncertain and a value of 300 MPa was chosen in order to provide a match between the transition depth between kink bands and thrusts in simulation results and seismic observations.

Table 1. *Parameter values used in the setup of the limit analysis*

Symbol	Parameter	Value	Unit
H_s	Stack thickness	300	m
	Spacing between slip surfaces	0.05	m
φ_i	Friction angle across the interface	15	degree
φ_s	Angle of internal friction of the solid	30	degree
φ^*	Compaction angle	89	degree
C	Cohesion	0	MPa
P^*	Compaction strength	300	MPa
ρ	Density	2300	kg m^{-3}
e	Initial strain	2×10^{-7}	–

Results

The bounds for the thrust and the kink band are now compared in the space spanned by the sheet dip and the burial depth. The resulting deformation mechanism map (Fig. 8a) indicates that, at low initial dips, a critical depth exists at which there is a transition from a deeper zone of kink bands to a shallow zone of thrusting. The value of the critical depth is relatively insensitive to initial bedding dips. The deformation map also shows that, above critical dip of about 10°, the kink band is the favoured mode of deformation at all burial depths and thrusts are not developed.

The kink mechanism requires accommodation structures along the kink planes in order for layer-parallel slip to be activated. The accommodation structures are represented by gradients of deformation along the kink planes (Fig. 7d). At low sheet dips accommodation is not possible with opening mode failure alone, but some compaction mode failure is required. However, compaction in the kink planes expends a lot of internal power. At large depth, thrusts need to lift a lot of overburden and it becomes more economical to introduce the otherwise power-intensive compaction, rather than developing thrusting that would involve power-intensive lifting.

Kink bands are only possible at shallow depths if the critical sheet dip is exceeded. This is because, above this sheet dip, the angle between the orientation of the bedding planes and the tectonic force is sufficient to allow bedding-parallel slip to occur and opening along the kink planes is activated. The dip of the kink planes is predicted by limit analysis. For a burial depth of 5.5 km and a sheet dip of 0.5°, representing the regional dip of the décollement, the predicted kink plane dip is 64°. According to the analysis, the kink planes steepen gradually upwards, developing dips of up to 74° at the surface (Fig. 9).

Limit analysis determines the optimum orientations for the thrust plane with respect to the horizontal axis (angle θ, Fig. 7b), which is $\pi/4-\varphi_s/2$ (φ_s is the friction angle of the bulk material) when the layers are horizontal, corresponding to 30° in this case. This is the same solution one obtains from the Mohr circle construction. As a sheet dip is introduced, the angle of the failure plane is reduced by approximately $\nu/2$ (Kampfer & Leroy 2009).

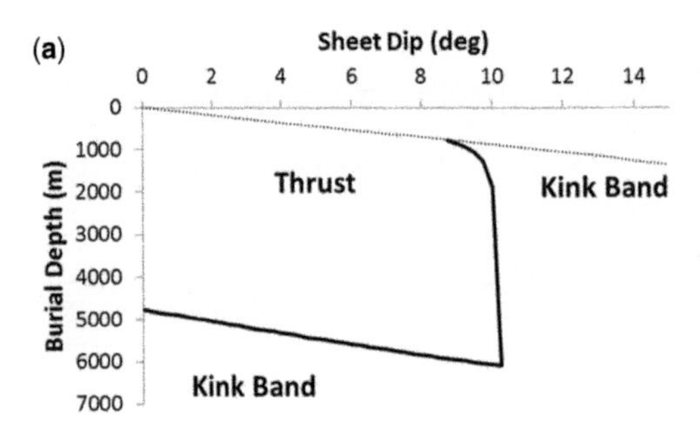

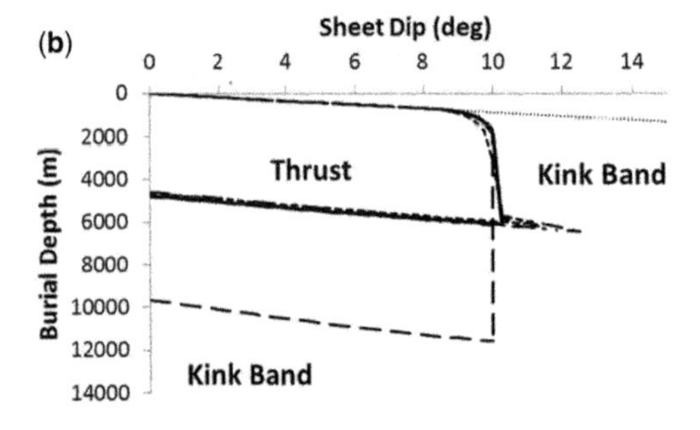

Fig. 8 (**a**) Deformation mechanism map, showing the dominance of thrusts compared with kink bands at low regional dip and shallow burial depth. At depth larger than 5000 m and sheet dip of <10°, kink bands dominate. The dotted line defines a minimum burial depth such that the shallow part of the structure does not crop out. (**b**) Variability of the results when changing H_s to $2H_s$ (short dashed line), doubling the individual layer thickness (dotted dashed line), doubling P^* (long dashed line), changing to 1×10^{-9} initial strain (double dotted dashed line). The influence of these parameter changes is small so that the curves superpose, with the exception of the curve for P^*.

Parameter sensitivity

Figure 8b presents the results of a simple parameter study. The stack height was changed to $2H_s$, the individual layer thickness was doubled and the initial strain was changed to 1×10^{-9}. The individual graphs for the results cannot easily be distinguished from each other. However, doubling P^* (long dashed line) causes a linear increase in the transition depth between thrusts and kink bands. Hence, the critical depth of the transition from thrusts to kink bands is very sensitive to the value of P^*.

The displacement of the right sidewall of the box in Figure 7a towards the left introduces compression to the layered structure. At small enough displacements of this wall, there is a limited possibility for a change in the geometry when the kink band is included and therefore the deformation mechanism maps look identical. Once the displacement of the sidewall reaches a normalized value of 10^{-6}, the locus of the kink band in the deformation mechanism map starts to extend, which means that it becomes increasingly easy to initiate kink bands compared with thrusts. In this study we use a very small value of displacement, which means that we are comparing the forces at the very early stage of layer parallel shortening. The onset of the kink

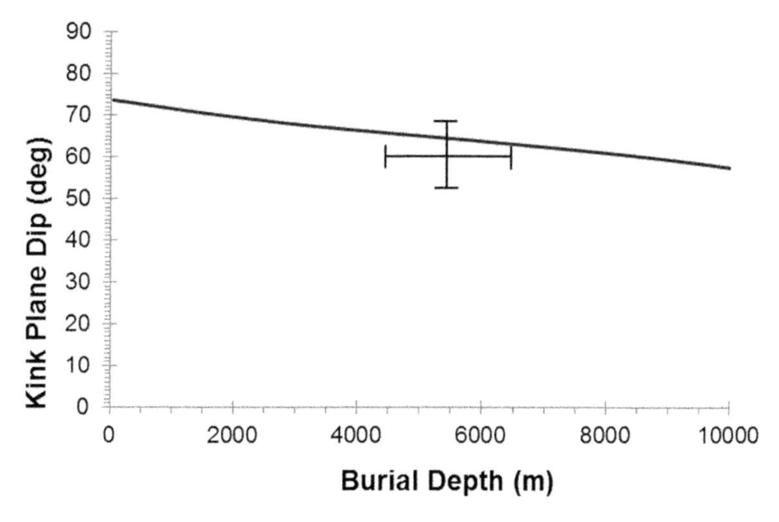

Fig. 9. Calculated relationship between kink plane dip and burial depth for a sheet dip of 0.5°. Error bars show the dip of kink bands from areas of thin salt measured from depth converted 3-D seismic data (±1 standard deviation, $n = 36$) and the estimated depth at time of deformation (±1000 m).

band as a function of the initial shortening and the further development of deformation is presented in Kampfer & Leroy (2009).

Discussion

The value of $P^* = 300$ MPa was chosen as a matching parameter resulting in a transition depth of about 5–6 km, which accords with our empirical estimate of the transition depth at the time of deformation. Some discussion is required of the physical meaning of this parameter. The strongest units of the sequence are the carbonates below the Marcellus shale, which probably controlled the initiation of the kink bands. Therefore, the compaction strength can be interpreted assuming that calcite was the dominant mineral phase. At temperatures of less than 400°C, the dominant deformation mechanism of calcite is pressure solution and the compaction strength of calcite is a function of the strain rate (Rutter & Elliott 1976). Using the deformation map of Rutter & Elliott (1976) it was found that the compaction strength of 300 MPa used in the model corresponds to a strain rate of $1 \times 10^{-13}\ \mathrm{s}^{-1}$.

Estimated regional strain rates from tectonic fold belts with evaporitic décollements and preserved synorogenic sediment are in the range of 6×10^{-17} to $5 \times 10^{-15}\ \mathrm{s}^{-1}$ (calculated for the south Pyrenees, Spain, Fars, Iran, Salt Range, Pakistan and Tien Shan, China from Vergés *et al.* 1992; Cotton & Koyi 2000; Scharer *et al.* 2004; Mouthereau *et al.* 2007) and therefore much lower than the value estimated from the model. However, as kink bands represent strain localization, the strain rates estimated from regional cross-sections are far lower than strain rates within individual structures. If we assume that the kink bands are the main component of shortening and that they represent only 1% of the section length, then the estimated strain rate within the kink bands is of the order of $10^{-13}\ \mathrm{s}^{-1}$.

In areas of thin salt, there was no large-scale buckling and initial dips were of the order of 0.5°. This regional dip was established during the flexing of the foreland in the Acadian and Alleghanian events as a result of the orogenic loads. The low dip constrains the deformation to the left-hand side of the deformation mechanism map, where thrusts occur above kink bands, separated by a critical depth. The mechanical modelling is therefore consistent with the observed change in structural style with depth in areas of thin salt (Fig. 3). In detail, the exact position of the transition also depends on the presence of available detachments in the form of organic-rich shale units. The predicted angle of dip of the kink planes is in good agreement with the measured kink plane dips, assuming a depth at time of deformation of 5.5 km (Fig. 9).

The presence of a thick mobile layer is a prerequisite for the development of décollement folds (Stewart 1996), and so décollement folds are best developed where the salt is thickest (Wiltschko & Chapple 1977). Yet why did the structural style change from open sinusoidal folding to kinking? As the folds developed, their limbs reached a dip of about 10°, at which point the kink band mechanism was favoured at all depths. A kink band

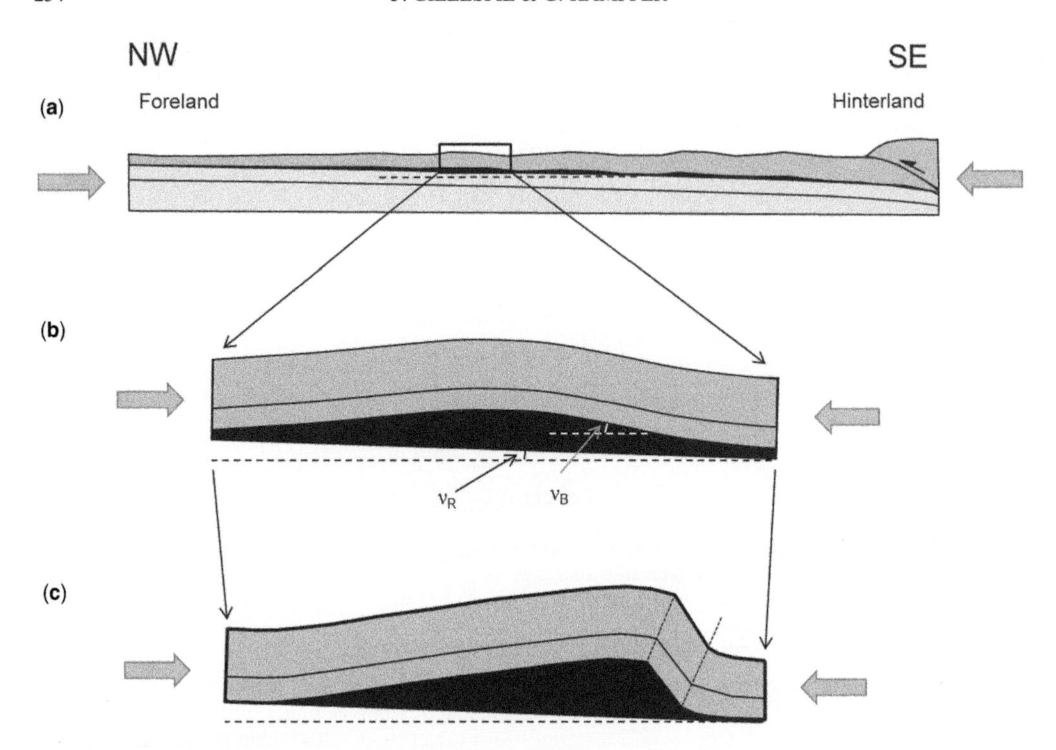

Fig. 10. (**a**) Regional sketch section through the Appalachian Plateau showing the development of weak contractional deformation above the salt (black unit). (**b**) Initial development of open buckle folds in area of thick salt; regional dip ν_R, back-limb dip ν_B. (**c**) Development of a through-going kink band once the back limb develops a critical dip.

therefore developed that propagated throughout the entire décollement sheet. Owing to the regional foreland flexure, the hinterland-ward limb reached the critical angle before the foreland-ward limb. Therefore the kink initiated on the back limb of the fold and a hinterland-ward vergence was established (Fig. 10). Observations of the maximum (un-kinked) limb dips of folds of Pennsylvannia indicate that they are in the range of 1–10° (Fig. 11). The lack of limb dips of greater than 10°

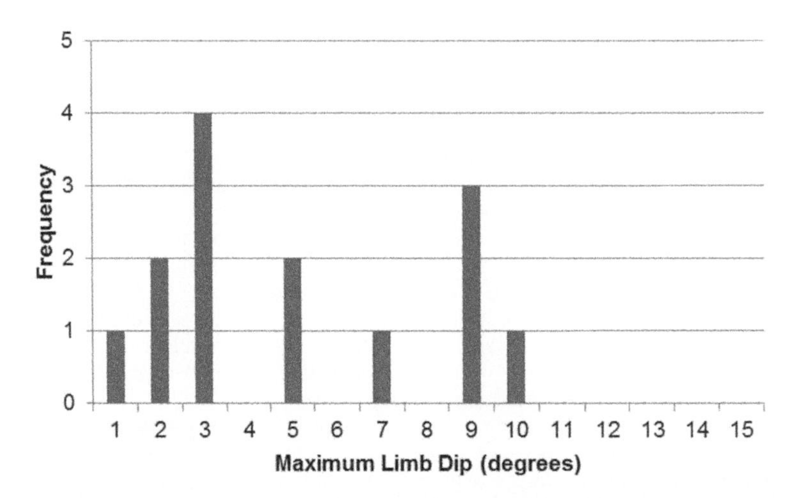

Fig. 11. Histogram of maximum limb dip on folds from the Appalachian Plateau in Pennsylvannia, based on Kindle (1904) and Wiltschko & Chapple (1977), $n = 14$. Local steep dips (>45°) related to kink bands were excluded.

is entirely consistent with the results of limit analysis.

A limitation of the analysis presented here is the assumption the surface topography was horizontal at the time of deformation. In a mature orogeny the surface topography will tend to dip towards the foreland owing to over-filling of the basin. The sedimentary geometry may alter the symmetry of the deformation (Duerto & McClay 2009). A more complete analysis would combine both the kinking layered medium presented here and the wedge geometry described in Cubas *et al.* (2008). This advance would require further theoretical development.

Conclusions

Application of the analytical model has helped to explain the development of the structures of the Appalachian Plateau décollement sheet, improving the understanding of the observed spatial variation in structural style.

- The results of the mechanical modelling using the maximum strength theorem are in good agreement with observations, including the observed occurrence of the different structures at different depth and the dip of the kink planes.
- The mechanical modelling gives us insight into the variations of structural style seen in the Appalachian Plateau and improves our ability to predict the structural style in poorly known regions of the basin. Décollement folds were favoured in areas of thick primary salt, whereas small kink bands and thrusts were favoured where primary salt was thinner.
- In regions of thin salt and low initial sheet dip, the thrusting mechanism was favoured at shallow depths, but below a critical depth of a few kilometres, the kink band mechanism was predominant.
- In regions of thick primary salt, deformation began with the development of broad symmetrical décollement folds. However, once the limbs reached a critical dip of about 10°, kink bands developed in the back limbs that folded the entire succession and a hinterlandwards vergence was established.
- The results have implications for the correct placement and steering of wells through kink bands and thrusts.
- As thrusts are brittle structures and kink bands are essentially folds, the two kinds of structures are expected to have very different hydraulic characteristics. Modelling of the structures therefore has an impact on prediction of reservoir integrity and induced seismicity during fluid injection.

We would like to thank Scott Wessels for insightful discussions. Silvan Hoth and Alexander Rozhko are thanked for their helpful and detailed comments on the manuscript. Jaume Vergés is thanked for information on shortening rates in orogenic belts. Comments from Terry Engelder and an anonymous reviewer significantly improved the manuscript.

References

Aydin, M.G. & Engelder, T. 2014. Revisiting the Hubbert–Rubey pore pressure model for overthrust faulting: Inferences from bedding-parallel detachment surfaces within Middle Devonian gas shale, the Appalachian Basin, USA. *Journal of Structural Geology*, **69**, 519–537.

Bain, G.W. 1931. Flowage folding. *American Journal of Science*, **22**, 503–530.

Bosworth, W. 1984. Foreland deformation in the Appalachian Plateau, central New York: the role of small scale detachment structures in regional overthrusting. *Journal of Structural Geology*, **6**, 73–81.

Cathcart, S.H. & Myers, T.H. 1934. *Gas in Tioga County, Pennsylvania: PA*. Topographic and Geologic Survey, Bulletins, **107**.

Chamberlin, R.T. 1931. Isostasy from the geological point of view. *The Journal of Geology*, **39**, 1–23.

Chapple, W.M. 1978. Mechanics of thin-skinned fold-and-thrust belts. *Geological Society of America Bulletin*, **89**, 1189–1198.

Chen, W.F. & Liu, X.L. 1990. *Limit Analysis in Soil Mechanics*. Developments in Geotechnical Engineering, **52**. Elsevier.

Cobbold, P.R. 1976. Fold shapes as a function of progressive strain. *Philosophical Transactions of the Royal Society, London, A*, **283**, 129–138.

Cotton, J.T. & Koyi, H.A. 2000. Modeling of thrust fronts above ductile and frictional detachments: application to structures in the Salt Range and Potwar Plateau, Pakistan. *Geological Society of America Bulletin*, **112**, 351–363.

Crawford, B.R., Faulkner, D.R. & Rutter, E.H. 2008. Strength, porosity, and permeability development during hydrostatic and shear loading of synthetic quartz–clay fault gouge. *Journal of Geophysical Research: Solid Earth*, **113**, https://doi.org/10.1029/2006JB004634

Cubas, N., Leroy, Y.M. & Maillot, B. 2008. Prediction of thrusting sequences in accretionary wedges. *Journal of Geophysical Research*, **113**, B12412, https://doi.org/10.1029/2008JB005717

Davis, D.M. & Engelder, T. 1985. The role of salt in fold-and-thrust belts. *Tectonophysics*, **119**, 67–88.

Duerto, L. & McClay, K. 2009. The role of syntectonic sedimentation in the evolution of doubly vergent thrust wedges and foreland folds. *Marine and Petroleum Geology*, **26**, 1051–1069.

Engelder, T. & Geiser, P. 1979. The relationship between pencil cleavage and lateral shortening within the Devonian section of the Appalachian Plateau, New York. *Geology*, **7**, 460–464.

Ettensohn, F.R. 2008. The Appalachian foreland basin in eastern United States. *Sedimentary Basins of the*

World, **5**, 105–179, https://doi.org/10.1016/S1874-5997(08)00004-X

Evans, M.A. 1994. Joints and décollement zones in Middle Devonian shales; evidence for multiple deformation events in the central Appalachian Plateau. *Geological Society of America Bulletin*, **106**, 447–460.

Evans, M.A. 1995. Fluid inclusions in veins from the Middle Devonian shales: a record of deformation conditions and fluid evolution in the Appalachian Plateau. *Geological Society of America Bulletin*, **107**, 327–339.

Faill, R.T. 1969. Kink band structures in the Valley and Ridge province, central Pennsylvannia. *Geological Society of America Bulletin*, **80**, 2539–2550.

Faill, R.T. 2011. *Folds map of Pennsylvania*. Pennsylvania Geological Survey Open-File Report OFGG 11-01.0.

Frey, M.G. 1973. Influence of the Salina salt on structure in New York-Pennsylvania part of the Appalachian Plateau. *AAPG Bulletin*, **57**, 1027–1037.

Gillespie, P., van Hagen, J., Wessels, S. & Lynch, D. 2015. Hierarchical kink band development in the Appalachian Plateau décollement sheet. *AAPG Bulletin*, **99**, 51–76.

Gwinn, V.E. 1964. Thin-skinned tectonics in the Plateau and northwestern Valley and Ridge provinces of the central Appalachians. *Geological Society of America Bulletin*, **75**, 863–900.

Kampfer, G. 2010. *Plis et fractures d'extension dans les roches stratifiées (Folds and tension fractures in stratified rocks)*. PhD thesis, École Normale Supérieure, Paris.

Kampfer, G. & Leroy, Y.M. 2009. Imperfection and burial-depth sensitivity of the initiation and development of kink folds in laminated rocks. *Journal of Mechanics and Physics of Solids*, **57**, 1314–1339.

Kindle, E.M. 1904. A series of gentle folds on the border of the appalachian system. *Journal of Geology*, **12**, 281–289.

Kohli, A.H. & Zoback, M.D. 2013. Frictional properties of shale reservoir rocks. *Journal of Geophysical Research: Solid Earth*, **118**, 5109–5125.

Lash, G. & Engelder, T. 2007. Jointing within the outer arc of a forebulge at the onset of the Alleghanian Orogeny. *Journal of Structural Geology*, **29**, 774–786.

Maillot, B. & Leroy, Y.M. 2006. Kink-fold onset and development based on the maximum strength theorem. *Journal of the Mechanics and Physics of Solids*, **54**, 2030–2059.

Marshak, S. & Engelder, T. 1985. Development of cleavage in limestones of a fold-thrust belt in eastern New York. *Journal of Structural Geology*, **7**, 345–359.

Mount, V.S. 2014. Structural style of the Appalachian Plateau fold belt, north-central Pennsylvania. *Journal of Structural Geology*, **69**, 284–303.

Mouthereau, F., Tensi, J., Bellahsen, N., Lacombe, O., De Boisgrollier, T. & Kargar, S. 2007. Tertiary sequence of deformation in a thinskinned/thick skinned collision belt: the Zagros Folded Belt (Fars, Iran). *Tectonics*, **26**, TC5006, https://doi.org/10.1029/2007TC002098

Nickelsen, R.P. 1986. Cleavage duplexes in the Marcellus Shale of the Appalachian foreland. *Journal of Structural Geology*, **8**, 361–371.

Pons, A. & Leroy, Y.M. 2012. Stability of accretionary wedges based on the maximum strength theorem for fluid-saturated porous media. *Journal of the Mechanics and Physics of Solids*, **60**, 643–664, https://doi.org/10.1016/j.jmps.2011.12.011

Rickard, L.V. 1969. *Stratigraphy of the Upper Silurian Salina Group, New York, Pennsylvania, Ohio, Ontario*. New York State Museum and Science Service, Albany, NY, Map and Chart Series, **12**.

Rutter, E.H. & Elliott, D. 1976. The kinetics of rock deformation by pressure solution [and discussion]. *Philosophical Transactions of the Royal Society of London A. Mathematical, Physical and Engineering Sciences*, **283**, 203–219.

Rutter, E.H., Hackston, A.J., Yeatman, E., Brodie, K.H., Mecklenburgh, J. & May, S.E. 2013. Reduction of friction on geological faults by weak-phase smearing. *Journal of Structural Geology*, **51**, 52–60.

Salençon, J. 2002. *De l'élasto-plasticité au calcul à la rupture*. Editions École Polytechnique, Palaiseau and Ellipses, Paris.

Scanlin, M.A. & Engelder, T. 2003. The basement v. the no-basement hypotheses for folding within the Appalachian plateau detachment sheet. *American Journal of Science*, **303**, 519–563.

Scharer, K.M., Burbank, D.W., Chen, J., Weldon, R.J., Rubin, C., Zhao, R. & Shen, J. 2004. Detachment folding in the Southwestern Tian Shan–Tarim foreland, China: shortening estimates and rates. *Journal of Structural Geology*, **26**, 2119–2137.

Sherrill, R. 1934. Symmetry of Appalachian foreland folding. *Journal of Geology*, **42**, 225–247.

Sherrill, R.E. 1941. Some problems of Appalachian structure. *AAPG Bulletin*, **25**, 416–423.

Souloumiac, P., Leroy, Y.M., Maillot, B. & Krabbenhoft, K. 2009. Predicting stress distributions in fold-and-thrust belts and accretionary wedges by optimization. *Journal of Geophysical Research*, **114**, B09404, https://doi.org/10.1029/2008jb005986

Stewart, S.A. 1996. Influence of detachment layer thickness on style of thin-skinned shortening. *Journal of Structural Geology*, **18**, 1271–1274.

Vergés, J., Muñoz, J.A. & Martínez, A. 1992. South Pyrenean fold and thrust belt: role of evaporitic levels in thrust geometry. *In*: McClay, K.R. (ed.) *Thrust Tectonics*. Chapman & Hall, London, 255–264.

Wagner, B.H. & Jackson, M.P. 2011. Viscous flow during salt welding. *Tectonophysics*, **510**, 309–326.

Willis, B. 1894. *The Mechanics of Appalachian Structure*. US Government Printing Office, Washington.

Wiltschko, D.V. & Chapple, W.M. 1977. Flow of weak rocks in Appalachian Plateau folds. *AAPG Bulletin*, **61**, 653–670.

Wong, T.F., David, C. & Zhu, W. 1997. The transition from brittle faulting to cataclastic flow in porous sandstones: Mechanical deformation. *Journal of Geophysical Research: Solid Earth*, **102**, 3009–3025.

Opening-mode fracture systems: insights from recent fluid inclusion microthermometry studies of crack-seal fracture cements

JOSEPH M. ENGLISH[1]* & STEPHEN E. LAUBACH[2]

[1]*Stellar Geoscience Limited, Dublin, Ireland*

[2]*Bureau of Economic Geology, Jackson School of Geosciences, The University of Texas at Austin, Austin, TX 78758, USA*

Correspondence: je@stellargeoscience.com

Abstract: Overpressuring, tectonic stretching and thermoelastic contraction are all processes that can drive the formation of opening-mode fractures in the subsurface. Recent studies on crack-seal quartz deposits in opening-mode fractures have yielded fluid inclusion microthermometric data, which for the first time allow us to constrain the pressure–temperature conditions under which these fractures formed. Here, we utilize the results from studies in the Lower Cretaceous Travis Peak Formation in the East Texas Basin and the Upper Cretaceous Mesaverde Group in the Piceance Basin to construct stress history models based on mechanical properties, burial history and tectonic setting to evaluate the various driving mechanisms for opening-mode fracture formation. Our results show progress towards separating and independently evaluating these mechanisms. Although high fluid pressure and tectonic stretching can play a major part in the formation of opening-mode fractures, our results suggest that the persistence of fracture growth during uplift could have been strongly influenced by thermoelastic contraction associated with exhumation and cooling. For sandstone reservoirs, thermoelastic contraction will be more pronounced for stiffer, high Young's modulus rocks with higher quartz contents. These models can therefore be used to provide additional insights into the distribution of opening-mode fractures in exhumed basins.

Opening-mode fracture systems have an important bearing on regional fluid flow in the subsurface. Constraining their origin and distribution is of great interest in the fields of hydrogeology, petroleum geology and economic geology. In nearly flat-lying rocks, these natural fracture systems are generally characterized by regionally extensive sets of sub-parallel vertical fractures that form as a result of brittle deformation in the Earth's crust (e.g. Pollard & Aydin 1988). Natural fractures form and propagate when failure criteria are met and these failure criteria are defined in terms of states of stress. *In situ* stress in the subsurface is typically described using three principal orthogonal compressive stress components: vertical stress (S_v), minimum horizontal stress (S_{hmin}) and maximum horizontal stress (S_{Hmax}). Different stress regimes are defined on the basis of the relative magnitudes of these principal stresses (Anderson 1951): extensional faulting ($S_v > S_{Hmax} > S_{hmin}$), strike-slip faulting ($S_{Hmax} > S_v > S_{hmin}$) and contractional faulting ($S_{Hmax} > S_{hmin} > S_v$). Opening-mode fractures form in the subsurface when the effective least compressive principal stress (σ_3) across a plane becomes tensile (i.e. negative) and exceeds the tensile strength (T_0) of the rock (e.g. Jaeger *et al.* 2007; Fjær *et al.* 2008). Hence the opening-mode failure criterion is given by

$$\sigma_3 = S_3 - P_p = -T_0 \quad (1)$$

where S_3 is the magnitude of the least compressive principal stress and P_p is the pore fluid pressure. Rock is much weaker in tension than in compression and the tensile strength is typically only of the order of a few MPa (e.g. Lockner 1995; Fjær *et al.* 2008). The tensile strength of rock can also sometimes be assumed to be zero because of the presence of pre-existing flaws within an appreciable volume of rock mass (e.g. Zoback 2007). When considering vertical opening-mode fractures, it can be deduced that they must form when the least compressive principal stress is horizontal. Hence vertical opening-mode fractures in the subsurface form orthogonal to the S_{hmin} direction in an extensional faulting or strike-slip faulting regime. In summary, if the tensile strength of the rock is assumed to be negligible, the failure criterion for vertical opening-mode fractures in an extensional faulting or strike-slip faulting stress regime can be simplified to

$$S_{hmin} = P_p \quad (2)$$

From: TURNER, J. P., HEALY, D., HILLIS, R. R. & WELCH, M. J. (eds) 2017. *Geomechanics and Geology*. Geological Society, London, Special Publications, **458**, 257–272.
First published online May 24, 2017, https://doi.org/10.1144/SP458.1

When assessing the likelihood of opening-mode failure in the subsurface, it is important to also consider whether shear failure of the rock will occur before the opening-mode failure criterion is met (Hillis 2001, 2003; Olson *et al.* 2009). In general, shear failure will occur before opening-mode failure at greater depth; the transition from one failure mode to another occurs at greater depth for stronger rocks and under conditions of lower effective maximum compressive principal stress (i.e. higher pore pressure and/or lower magnitude of maximum principal stress). A more detailed discussion of this is provided in English (2012) and the discussion here will be restricted to opening-mode failure.

Using equation (2), and acknowledging that stable conditions exist when the minimum horizontal stress (S_{hmin}) is greater than the pore fluid pressure (P_p), it can be deduced that there are two different categories of mechanisms to drive opening-mode failure in the subsurface: (1) increase P_p and (2) decrease S_{hmin}. Increased pore fluid pressure is generally accepted to be the primary driver behind the formation of opening-mode fractures at depth (Hubbert & Rubey 1959; Handin *et al.* 1963; Secor 1965, 1969; Engelder & Lacazette 1990; Lacazette & Engelder 1992). Elevated pore fluid pressures in the subsurface can arise via a number of different processes, including disequilibrium compaction, kerogen transformation and gas generation, hydrocarbon buoyancy, smectite dehydration, smectite to illite transformation, lateral compressive stress, aquathermal expansion, chemical compaction, osmosis and hydraulic head (Bradley 1975; Osborne & Swarbrick 1997; Swarbrick & Osborne 1998; Swarbrick *et al.* 2002). In general, the most significant overpressure-generating mechanisms are associated with increasing burial depth and temperature. However, a number of recent studies have demonstrated that fracture formation can also occur during periods of rock uplift and exhumation (Perez & Boles 2005; Becker *et al.* 2010; Hooker *et al.* 2015; Lander & Laubach 2015; English *et al.* 2016; Laubach *et al.* 2016). Driving mechanisms that would act to decrease the minimum horizontal stress gradient during exhumation include thermoelastic contraction due to cooling (e.g. Voight & St Pierre 1974; Haxby & Turcotte 1976; Narr & Currie 1982; Warpinski 1989; English 2012) and extensional lateral horizontal strain (e.g. the 'parallel uplift' model of Price 1966; Haxby & Turcotte 1976). A more detailed review of each of these mechanisms is provided by English (2012).

Recent studies on the Travis Peak Formation in the East Texas Basin (Fig. 1a; Becker *et al.* 2010) and the Mesaverde Group in the Piceance Basin (Fig. 1b; Fall *et al.* 2012, 2015) have yielded sequences of fluid inclusion microthermometric data from crack-seal quartz cement in opening-mode

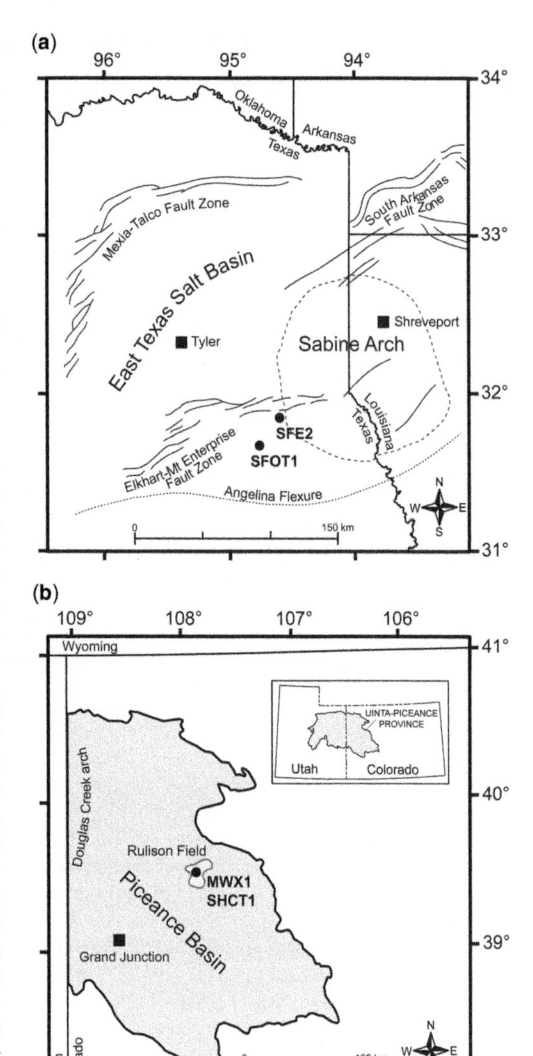

Fig. 1. Location of study areas. (**a**) SFE2 well located on the western flank of the Sabine Arch in the East Texas Basin (modified from Becker *et al.* 2010); (**b**) MWX and SHCT wells located in the Rulison Field in the southern Piceance Basin (modified from Fall *et al.* 2012).

fractures in the subsurface. These pioneering studies allow us, for the first time, to constrain the pressure–temperature conditions under which these fractures formed. Quartz bridges are isolated, pillar-shaped deposits within the opening-mode fractures that accumulated contemporaneous with fracture opening (Laubach 1988; Laubach *et al.* 2004*b*, *c*; Lander & Laubach 2015). Bridges contain multiple cement bands bounded by fracture surfaces and surrounding fluid inclusion assemblages parallel to the fracture walls. Cross-cutting relations visible in cement textures are consistent with incremental crack-seal opening. High-resolution scanning electron microscopy–cathodoluminescence images can be used to identify cross-cutting relationships and hence establish the sequence of incremental quartz cement growth within the fracture fill (see example in Fig. 2).

Homogenization temperatures from each stage of the fracture fill are gathered using fluid inclusion microthermometry analysis and hence the thermal history recorded within a single fracture fill can be established. These thermal records are then compared with independently derived burial and thermal histories for the formations in question to place the fracture fill record in a temporal context. Using Raman spectroscopy of fluid inclusions, the fluid pressure conditions during the fracture fill history can be broadly constrained based on gas concentration estimates and equation-of-state calculations. Taken together, these analyses allow for the reconstruction of the pressure–temperature–composition history of fluids present during opening-mode fracture formation.

Here we integrate geomechanical and *in situ* stress data with the timing constraints from previous fluid inclusion based studies of the Travis Peak Formation (Becker *et al.* 2010) and the Mesaverde Group (Fall *et al.* 2012, 2015) and discuss what each of these datasets tells us about the driving mechanisms for opening-mode fracture formation in the subsurface. In particular, we review the extent of overpressuring recorded during initial opening-mode fracture formation in each of the two study areas and investigate what role, if any, thermoelastic contraction had in the formation and growth of opening-mode fractures during exhumation and cooling.

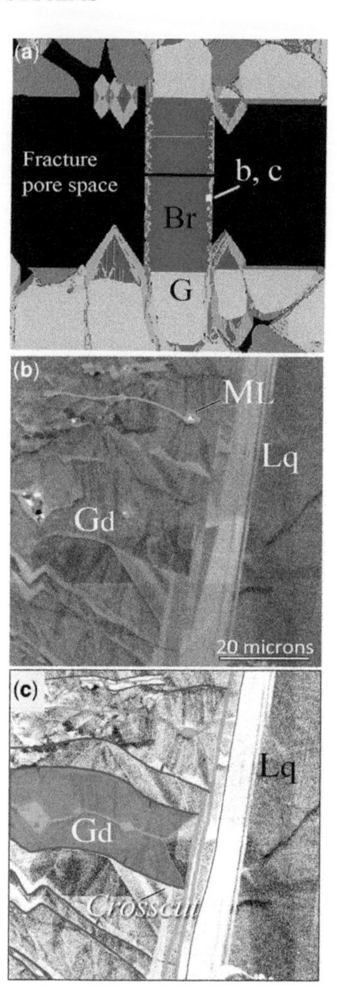

Fig. 2. Examples of how fluid inclusion assemblages are sequenced within cement bridges using the procedure described in Laubach *et al.* (2004*c*). (**a**) Quartz bridge (Br) simulation model from Lander & Laubach (2015) showing east–west oriented opening-mode fracture within a quartz grain (G); black is porosity; red is rapid quartz accumulation in gap deposits; blue is slow accumulation in lateral deposits; yellow square shows the location of the bridge texture examples shown in (b) and (c) along the contact between rapidly accumulated gap deposits and more slowly accumulated lateral cement (modified from Lander & Laubach 2015). (**b**) Cathodoluminescence (CL) image of gap deposits (Gd) and lateral quartz deposits (Lq) within a quartz bridge; ML, median line where fluid inclusions (visible as oblong, bright areas on CL image) are trapped (modified from Laubach *et al.* 2016). (**c**) Annotated CL interpretation where red lines with inwards-facing ticks represent the walls of the gap deposits (Gd) that define the crack-seal texture. Note cross-cutting relations within the gap deposits and the progressive overlapping (yellow, tan and blue lines) of the lateral quartz (Lq) deposits (modified from Laubach *et al.* 2016).

Travis Peak Formation, East Texas Basin

The Lower Cretaceous Travis Peak Formation was deposited during a thermal subsidence phase in the East Texas Basin (Fig. 1a), part of the northern Gulf of Mexico Basin, following Late Triassic to Middle Jurassic rifting (Buffler *et al.* 1980; Pindell 1985). The Travis Peak Formation is *c.* 600 m thick and the present day depth to the top of the formation ranges from 1800 to 2900 m (Laubach 1988). The Travis Peak sandstones are typically quartz-arenitic in composition, with quartz contents >95% (Table 1; Dutton & Diggs 1990; Plumb *et al.* 1992). Extensive quartz cementation reduced the matrix permeability of much of the Travis Peak Formation to <0.1 mD (Dutton 1987; Dutton & Diggs 1990; Jackson & Laubach 1991) and the presence of partially sealed opening-mode fractures is locally

Table 1. *Representative X-ray diffraction data from the Travis Peak and Cozzette sandstones*

Unit	Well	Depth (m)	Lithology	Quartz (wt%)	K-feldspar (wt%)	Plagioclase feldspar (wt%)	Calcite (wt%)	Dolomite (wt%)	Ankerite (wt%)	Chlorite (wt%)	Illite (wt%)	Total clay (wt%)	Source
Travis Peak	SFE-2	3030	Sandstone	97.5	0.0	0.5	0.0	0.0	0.0	1.0	1.0	2.0	Plumb *et al.* (1992)
Cozzette	MWX-2	2394	Sandstone	66.0	0.0	8.0	0.0	3.0	0.0	2.5	20.5	23.0	Pollastro (1984); Pitman *et al.* (1989)
Williams Fork	MWX-2	2171	Sandstone	54.0	0.0	5.0	0.0	7.0	9.0	0.0	25.0	25.0	Pollastro (1984); Pitman *et al.* (1989)

important for reservoir permeability (Laubach 1989, 2003).

The north–south-trending, basement-cored Sabine Uplift is the major structural feature within the basin (Laubach & Jackson 1990; Nunn 1990).

About 550 m of stratigraphic section were removed from this uplift at the beginning of the Late Cretaceous (Halbouty & Halbouty 1982; Jackson & Laubach 1988) and subsidence resumed at *c.* 90 Ma with the deposition of the Austin Chalk (Fig. 3a). The onset of dry gas generation in the underlying Bossier shale in East Texas is interpreted at *c.* 57 Ma (Dutton 1987). Maximum burial of the Travis Peak Formation occurred during the middle Eocene (*c.* 41 Ma) just prior to a period of erosion that removed *c.* 450 m of stratigraphic section

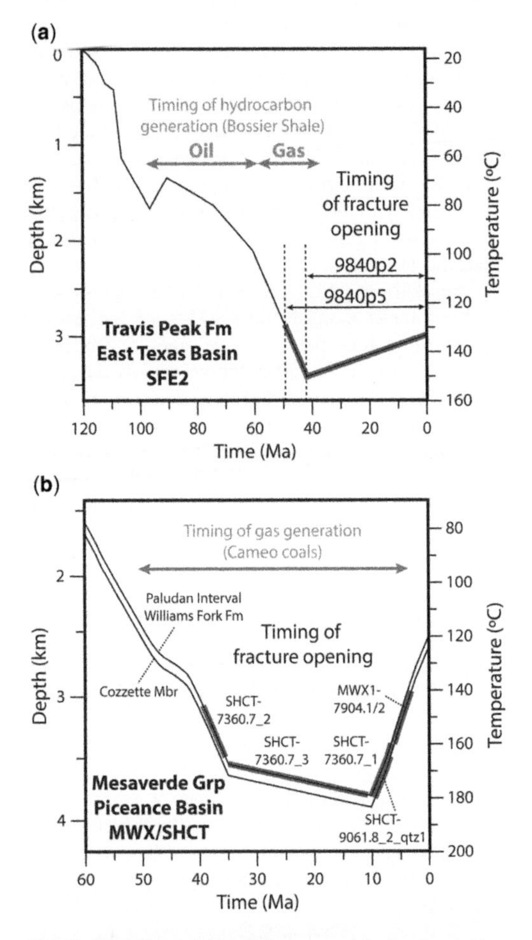

Fig. 3. Burial and thermal history models for the study areas. (**a**) Travis Peak fracture fills in the SFE2 core, adapted from the burial history for the nearby Ashland SFOT1 well (modified from Becker *et al.* 2010). (**b**) Mesaverde fracture fills in the MWX1 and SHCT cores (modified from Fall *et al.* 2012).

(Dutton 1987; Dutton & Land 1988; Jackson & Laubach 1988). The estimated magnitude of Cenozoic exhumation is based on the maturity of Eocene sub-bituminous coals of the Wilcox Formation that are presently at or near to the surface (Mukhopadhyay 1989). The exact rate and duration of the Cenozoic exhumation is unconstrained by stratigraphy, but the rates and durations used by Becker *et al.* (2010) are compatible with quartz accumulation models dependent on thermal history (Lander & Laubach 2015).

Natural fractures within the Travis Peak Formation are described as being vertical sub-parallel opening-mode fractures that are oriented normal to bedding (e.g. Laubach *et al.* 2009; Becker *et al.* 2010). These fractures generally trend ENE and are sub-parallel to the maximum horizontal stress direction (S_{Hmax}), to regional fault trends and to the continental margin (Laubach *et al.* 2009, 2004*a*). Fracture intensity is observed to vary as a function of lithology and the greatest intensity is observed in the quartz-rich channel sandstones of the lower Travis Peak Formation that are characterized by a higher Young's modulus and lower Poisson's ratio (Laubach *et al.* 2009). Published static and dynamic measurements (Thiercelin & Plumb 1994; Walls & Dvorkin 1994; Laubach *et al.* 2009) for the lower Travis Peak Formation in Staged Field Experiment well no. 2 (SFE2) from along the western flank of the Sabine Uplift (Fig. 1a) indicate that representative values for Young's modulus and Poisson's ratio can be taken as 62 GPa and 0.15 respectively. The Biot–Willis coefficient for these sandstones is estimated at 0.55 (Thiercelin & Plumb 1994) and the coefficient of internal friction (μ_i) is estimated at 1.4 based on limited core data (Thiercelin & Plumb 1994). The unconfined compressive strength is of the order of 100 MPa, although values for the lower Travis Peak sandstones are commonly >200 MPa (Thiercelin & Plumb 1994). In contrast with the opening-mode fractures commonly observed in the sandstone, sub-parallel fractures with evidence of shear movement and <45° dips are present in the mudstones (Laubach 1988). This indicates that shear failure preferentially occurred in the weaker mudstones and opening-mode failure preferentially occurred in the stronger sandstones (English 2012).

Becker *et al.* (2010) have described sub-millimetre-scale quartz cement deposits in two natural opening-mode fractures in lower Travis Peak Formation core from a depth of 3000 m in the SFE2 well. The present day temperature at 3000 m depth in SFE2 is *c.* 130°C, whereas the peak palaeotemperature is estimated at *c.* 154°C based on fluid inclusion data (Becker *et al.* 2010). The maximum burial depth (at 41 Ma) is estimated at 3450 m for this interval (Table 2). Assuming an average surface temperature of 19°C, the corresponding present day and peak palaeotemperature gradients are 37 and 39°C km^{-1}, respectively. Independently, the temperature at maximum burial is estimated to be *c.* 150°C based on the present day geothermal gradient (38°C km^{-1}) and the burial history models from a nearby well (Ashland SFOT No. 1 in Dutton 1987; Figs 1a & 3a).

The decrease in formation temperature from 154 to 130°C is interpreted to have resulted from exhumation from maximum burial depth. The interpretation that the fracture system remained in thermal equilibrium with the host formation throughout its crack-seal opening history is supported by (1) the lack of sporadic thermal pulses in the fluid inclusion record and (2) the internal consistency of the salinities within the fluid inclusions and their similarity with the present day formation water salinity (Becker *et al.* 2010). Pore fluid pressures in the vicinity of SFE2 are close to hydrostatic present day values (Bartberger *et al.* 2002); a reservoir pressure of *c.* 34 MPa has been reported from SFE2 at *c.* 3000 m depth (Thiercelin & Plumb 1994; Walls & Dvorkin 1994), giving a pore pressure gradient of 11.7 kPa m^{-1}.

Scanning electron microscopy–cathodoluminescence mapping and fluid inclusion analyses from the SFE2 fracture fill have identified a systematic change in temperature and methane concentration during the opening of individual fractures (Becker *et al.* 2010). The thermal history recorded in a single fracture fill ranges from initial temperatures of *c.* 130°C up to *c.* 150°C and back down to *c.* 134°C (Fig. 3a). The pore fluid pressure during maximum burial is estimated at *c.* 55 MPa based on methane concentrations in the fluid inclusions (Becker *et al.* 2010), giving a pore pressure gradient of 15.9 kPa m^{-1}. The methane concentrations are observed to systematically vary with the homogenization temperatures of the fluid inclusions, indicating that the pore pressure decreased continuously from *c.* 55 to 35 MPa as the formation temperature decreased from 154 to 130°C (Becker *et al.* 2010). Based on these observations, Becker *et al.* (2010) interpreted that opening-mode fracturing initially occurred during hydrocarbon charge and overpressuring at close to maximum burial, followed by incremental crack-seal fracture opening events during subsequent exhumation and cooling (Fig. 3a). An independent quartz accumulation model replicates the quartz bridge as well as quartz cement abundances and internal textures and morphologies in the sandstone host rock and fracture zone using the same kinetic parameters while honouring fluid inclusion and thermal history constraints (Lander & Laubach 2015).

As this area is characterized as an extensional faulting stress regime, the vertical stress (S_v) is the

Table 2. *Summary of geomechanical parameters for the Travis Peak and Cozzette sandstones*

Stratigraphic unit	Travis Peak	Cozzette	Data source
Basin	East Texas	Piceance	
Well	SFE-2	MWX and SHCT wells	
Lithology	Sandstone	Siltstone/sandstone	
Present depth (m TVD)	3000	2409	TP, Becker *et al.* (2010); Cz, Fall *et al.* (2012)
Magnitude of exhumation (m)	450	1433	TP, Dutton (1987), Dutton & Land (1988); Cz, Nuccio & Roberts (2003)
Maximum burial depth (m)	3450	3842	TP, 450 m overburden; Cz, 1433 m overburden
Temperature (°C)	130	127	TP, Becker *et al.* (2010); Cz, Warpinski & Lorenz (2008)
Peak temperature (°C)	154	180	TP, Becker *et al.* (2010); Cz, Nuccio & Roberts (2003), Fall *et al.* (2012)
Pore pressure P_p (MPa)	34.1	43.4	TP, Thiercelin & Plumb (1994) (SFE2 3014 m); Cz, Warpinski *et al.* (1985) (MWX2, 2406 m)
Pore pressure gradient (kPa m^{-1})	11.4	18.0	Calculated
Pore pressure at maximum burial depth (MPa)	55	85	TP, Becker *et al.* (2010); Cz, Fall *et al.* (2012) (mid-range of estimates from SHCT1-9061.8)
Pore pressure gradient at maximum burial depth (kPa m^{-1})	15.9	22.1	Calculated (Cz similar to Green River Basin, Spencer 1987)
Vertical stress (S_v) (MPa)	72.3	57.2	TP, Thiercelin & Plumb (1994) (SFE2: 3014 m); Cz, Warpinski *et al.* (1985) (MWX2: 2406 m)
Vertical stress gradient (kPa m^{-1})	24.1	23.7	Calculated
*S*v at maximum burial depth (MPa)	83.1	91.1	TP, 450 m overburden; Cz, 1433 m overburden
Minimum horizontal stress (S_{hmin}) (MPa)	41.9	47.1	TP, Thiercelin & Plumb (1994) (SFE2: 3014 m); Cz, Warpinski *et al.* (1985) (MWX2: 2406 m)
S_{hmin} gradient (kPa m^{-1})	14.0	19.6	Calculated
Young's modulus (GPa)*	62	30	TP, Thiercelin & Plumb (1994) (SFE2: 3012 m); Cz, MWX Project Team (1987) (MWX1: 2408 m)
Poisson's ratio*	0.15	0.23	TP, Thiercelin & Plumb (1994) (SFE2: 3012 m); Cz, MWX Project Team (1987) (MWX1: 2408 m)
Biot–Willis coefficient	0.55	0.5	TP, Thiercelin & Plumb (1994) (SFE2: 3012 m); Cz, assumption
Unconfined compressive strength (MPa)*	260	94	TP, Thiercelin & Plumb (1994) (SFE2: 3012 m); Cz, MWX Project Team (1987) (MWX1: 2408 m)
Coefficient of internal friction	1.4	1.2	TP, Thiercelin & Plumb (1994) (SFE2: 3012 m); Cz, assumption
Quartz content (%)	98	66	TP, Plumb *et al.* (1992) (SFE2: 3030 m); Cz, Pollastro (1984) (MWX2: 2394 m)

Cz, Cozzette; TP, Travis Peak.
*The mechanical properties given here are from a Cozzette siltstone in MWX1, which contains one of the fractures studied in Fall *et al.* (2012). The Cozzette sandstones typically have a Young's modulus of *c.* 35 GPa, a Poisson's ratio of *c.* 0.23 and an unconfined compressive strength of *c.* 150–200 MPa (MWX Project Team 1987).

maximum principal compressive stress. The S_v magnitude at a present day depth of 3000 m in SFE2 is *c.* 72 MPa (Thiercelin & Plumb 1994), giving an average gradient of 24 kPa m^{-1}. Assuming that the 450 m of eroded stratigraphy had the same average density, the S_v at maximum burial of 3450 m is estimated as 82.8 MPa. The subvertical orientation of the opening-mode fractures indicates that the minimum principal compressive stress was horizontal during fracture formation. The present day minimum horizontal stress (S_{hmin}) magnitude at 3000 m depth in SFE2 is *c.* 42 MPa based on the estimation of closure stress from small volume hydraulic fractures (Thiercelin & Plumb 1994). This present day S_{hmin} magnitude corresponds to a gradient of 14 kPa m^{-1} and a minimum horizontal to vertical stress ratio (S_{hmin}/S_v) of 0.58. These values are similar to the average values (14.1 kPa m^{-1} and 0.59) from a series of 13 different stress tests conducted in the SFE wells (Thiercelin & Plumb 1994). A summary of all the key geomechanical data for the lower Travis Peak Formation in the SFE2 well is presented in Table 2.

Mesaverde Group, Piceance Basin

The Piceance Basin is an intracratonic foreland basin in northwestern Colorado (Fig. 1b) formed during the Late Cretaceous to Eocene Laramide orogeny (Johnson & Nuccio 1986). Rapid subsidence resulted in a thick Upper Cretaceous to Palaeogene sedimentary sequence including the Upper Cretaceous Mesaverde Group, which is 1280 m thick at the Multiwell Experiment (MWX) site in the southern Piceance Basin (Law & Johnson 1989). The Mesaverde Group at this location consists of the Corcoran, Cozzette and Rollins Sandstone members of the Iles Formation and the overlying Williams Fork Formation. The sandstones of the Iles Formation are typically subarkosic in composition with quartz contents of *c.* 65% (Table 1), while the sandstones of the Williams Fork Formation tend to have lower quartz contents (46–62%) and are arkosic to litharenitic in composition (Pitman *et al.* 1989). The Mesaverde Group sandstones are characterized by low porosities and matrix permeabilities in the microdarcy range (Ozkan *et al.* 2011; Stroker *et al.* 2013) and the presence of regionally extensive natural fractures (Finley & Lorenz 1988; Lorenz & Finley 1991) is important for reservoir productivity (Lorenz & Finley 1989; Lorenz *et al.* 1989*b*). Subsidence and burial continued until *c.* 20 Ma, when the formations reached their maximum burial depth, and regional uplift occurred from 10 Ma to present (Fig. 3b) (Yurewicz *et al.* 2005). The amount of exhumation and erosion has been estimated at 1433 m at the MWX1 well (Nuccio & Roberts 2003). Thermal history modelling of the MWX1 well indicates that significant gas generation from the Cameo coals within the Mesaverde Group began at 50 Ma during burial and ceased at 6 Ma during uplift and exhumation (Fig. 3b) (Yurewicz *et al.* 2005).

The low-permeability Mesaverde Group sandstones frequently contain open subvertical natural fractures in the subsurface (Lorenz & Finley 1991; Lorenz & Hill 1994; Hooker *et al.* 2009; Fall *et al.* 2012, 2015). Although a number of different types of fractures were observed in cores from the MWX and SHCT wells (Fig. 1b), vertical opening-mode fractures are the most abundant and exert the dominant control on reservoir permeability (Finley & Lorenz 1989; Lorenz & Finley 1991). The natural fractures are oriented normal to bedding and the dominant strike is WNW–ESE (Lorenz & Finley 1991), which is sub-parallel to the orientation of the present day maximum horizontal stress (Warpinski & Teufel 1989). The fractures are predominantly found within the sandstone and siltstone units and often terminate at boundaries with adjacent mudstones (Lorenz & Finley 1991), most likely due to contrasts in the mechanical properties. The Young's modulus for the sandstones and mudstones generally ranges from about 20 to 40 and 10 to 30 GPa, respectively, and Poisson's ratio is typically in the range 0.15–0.30 (Lorenz *et al.* 1989*a*). The compressive strength (at 10 MPa confining pressure) of the siltstones and sandstones is in the range 100–300 MPa, with the mudstones having lower strengths of 50–150 MPa (Lorenz *et al.* 1989*a*).

Fall *et al.* (2012, 2015) investigated fluid inclusion assemblages within crack-seal quartz in opening-mode fractures in the Mesaverde Group across the Piceance Basin. Herein, we focus on the data from the Cozzette sandstone at a true vertical depth of *c.* 2409 m in the MWX and SHCT wells in the southern Piceance Basin (Fig. 1b) because there is a multitude of additional geomechanical data available for these cores. The present day temperature in the Cozzette sandstone in MWX is 127°C (Warpinski & Lorenz 2008) and the peak palaeotemperature at the base of the Mesaverde Group is estimated at *c.* 180°C based on fluid inclusion data (Fall *et al.* 2012). Maximum burial depth (at 20–10 Ma) is estimated at 3842 m for this interval (Table 2). Assuming an average surface temperature of 10°C, the corresponding present day and peak palaeotemperature gradients are 49 and 44°C km^{-1}, respectively, and these gradients bracket the estimate of 46°C km^{-1} used by Fall *et al.* (2012).

The peak palaeotemperature estimates from the fluid inclusion data of Fall *et al.* (2012) are broadly consistent with the MWX-1 thermal history model of Nuccio & Roberts (2003) and with fluid inclusion homogenization temperatures of 150–175°C reported from previous studies in the MWX area (Barker 1989; Lorenz & Finley 1991). The decrease in formation temperature from 180 to 127°C is interpreted to have resulted from exhumation from maximum burial depth. Present day pore fluid pressure gradients in the MWX area increase gradually from 9.9 kPa m^{-1} (hydrostatic) at the top of the Mesaverde Group to *c.* 18 kPa m^{-1} (overpressured) at the base of the Mesaverde Group (Spencer 1987). Fluid inclusion data from the Cozzette Member and the Paludan Interval of the Williams Fork Formation in the MWX/SHCT wells (Fall *et al.* 2012) record a palaeotemperature increase from *c.* 145 to 180°C and then a decrease to *c.* 140°C over time (Fig. 3b). The palaeo-pore fluid pressure is estimated to have ranged from 55 to 110 MPa based on methane concentrations in the fluid inclusions, indicating that fracture opening occurred under near-lithostatic conditions during maximum burial (Fall *et al.* 2012). The observed lack of systematic trends of pressure v. temperature or time has been interpreted to reflect dynamic conditions with episodic gas charge from deeper source rocks along fractures and faults (Fall *et al.* 2012, 2015). Based on their data from across the Piceance Basin, Fall *et al.* (2012, 2015) suggest that overpressuring and fracture formation were intimately associated with gas charge during maximum burial conditions (at palaeotemperatures >140°C).

The present day horizontal stresses in Mesaverde Group sandstones in the MWX area are anisotropic, with the ratio of minimum and maximum horizontal stresses to vertical stress being 0.82 and 0.96, respectively (Teufel 1986). This indicates that the area is currently characterized by a transitional extensional faulting to strike-slip faulting stress regime. The S_v magnitude at a present day depth of 2406 m in MWX-2 is *c.* 57.2 MPa (Warpinski *et al.* 1985), giving an average gradient of 23.8 kPa m^{-1}. Assuming that the eroded stratigraphy had the same average density, the S_v at maximum burial is estimated at 91.1 MPa. The subvertical orientation of the opening-mode fractures indicates that the minimum principal compressive stress was horizontal during fracture formation. The present day minimum horizontal stress (S_{hmin}) magnitude at 2406 m depth in MWX-2 is *c.* 47.1 MPa based on instantaneous shut-in pressures from small volume hydraulic fractures (Warpinski *et al.* 1985). This present day S_{hmin} magnitude corresponds to a gradient of 19.6 kPa m^{-1} and a minimum horizontal to vertical stress ratio (S_{hmin}/S_v) of 0.82. These relatively high S_{hmin} gradients (compared with the Travis Peak study) are probably driven by the high pore fluid pressure gradient at these depths (18.0 kPa m^{-1}) and the associated pore pressure–stress coupling. It is worth noting that the minimum horizontal stresses are strongly a function of lithology within the Mesaverde Group in the MWX area and that the S_{hmin} gradients approach lithostatic in the mudstone intervals (Warpinski *et al.* 1985; Teufel 1986). A summary of all the key geomechanical data for the Cozzette Sandstone Member of the Mesaverde Group in the MWX wells is presented in Table 2.

Role of overpressure in fracture formation

It is clear from the studies of natural fractures in the Travis Peak Formation (Becker *et al.* 2010) and the Mesaverde Group (Fall *et al.* 2012, 2015) that significant overpressure in these basins played a part in reducing the effective stresses during the initiation and growth of opening-mode fractures in the deep subsurface. In both cases, fluid inclusion data indicate that the formation water was saturated with respect to methane during fracture formation and, although disequilibrium compaction may have played a part in developing some initial overpressure, gas generation and charge from adjacent or underlying source rocks during deep burial is interpreted to have been the primary trigger for the onset of opening-mode fracture formation (Fig. 3; Becker *et al.* 2010; Fall *et al.* 2012, 2015). The dramatic increase in pore fluid volume associated with primary gas generation, or secondary cracking of oil to gas, has been acknowledged as a possible mechanism to generate regional overpressure (Barker 1990; Osborne & Swarbrick 1997; Swarbrick & Osborne 1998) and hydrocarbon generation has been proposed as the primary mechanism for generating the overpressure observed in the deeper parts of a number of Rocky Mountain basins, including the Piceance Basin (Spencer 1987).

Opening-mode fracture growth occurred at a palaeo-pressure gradient of *c.* 16 kPa m^{-1} during maximum burial of the Travis Peak Formation (Becker *et al.* 2010) in a relatively 'relaxed' extensional tectonic setting with an interpreted S_{hmin}/S_v ratio of 0.66 (Table 2). Fracture growth continued during exhumation and cooling, even as the palaeo-pressure gradient dropped down towards normal hydrostatic conditions (a present day pore pressure gradient of 11.4 kPa m^{-1}). This indicates that there must also be an additional driving mechanism for opening-mode failure in the Travis Peak sandstones.

The palaeo-pressure data from the Mesaverde Group in the Piceance Basin is more complex, but indicates that opening-mode fracture growth occurred at palaeo-pressure gradients in the range 16–24 kPa m^{-1} during maximum burial (Fall *et al.* 2012, 2015). In any event, the present day pore pressure gradient of 18 kPa m^{-1} in the Piceance Basin (Spencer 1987) is higher than any of the palaeo-pressure gradients interpreted in the Travis

Peak study. The amount of overpressure required to create vertical opening-mode fractures not only depends on rock properties (Poisson's ratio and the Biot–Willis coefficient), but also on the initial magnitude of the minimum horizontal stress (English 2012). All else being equal, tectonically relaxed basins require less overpressure than more compressive basins to generate opening-mode fractures at depth. Therefore one possible explanation for the higher palaeo-pressure gradients recorded in the Mesaverde Group fluid inclusions in the Piceance Basin, a foreland basin formed during the Laramide orogeny, is that this basin was subjected to higher horizontal tectonic stresses than the extensional East Texas Basin.

On a regional scale, horizontal *in situ* stress magnitudes in the Earth's crust are believed to be limited by critically stressed faults and the ratio of the maximum and minimum effective stresses are frequently observed to correspond to a crust in frictional failure equilibrium, with the coefficient of friction in the range of 0.6–1.0 (Zoback & Healy 1984; Townend & Zoback 2000; Zoback *et al.* 2002). Although frictional sliding on optimally oriented faults should always occur before opening-mode failure in the subsurface (English 2012), the formation of opening-mode fractures in the subsurface necessitates that *in situ* stresses can locally (spatially or stratigraphically) or temporarily violate the frictional sliding criterion. It is noteworthy, however, that the present day *in situ* stresses in the Travis Peak Formation in SFE2 correspond to frictional failure equilibrium along optimally oriented faults with a coefficient of friction of *c.* 0.88. Recent movement along the eastwards-trending Elkhart/Mt Enterprise Fault (Fig. 1a) (Collins *et al.* 1980) also supports the interpretation that faults oriented at low angles to the maximum horizontal stress direction are critically stressed in this region. These critically stressed fault systems may have played a part in facilitating the bleed-off of overpressure from the SFE2 area after the cessation of hydrocarbon generation (see discussion of how exhumation can lead to critical stress conditions in English *et al.* 2017).

Role of thermoelastic contraction in fracture formation

Opening-mode fracture formation in the Travis Peak Formation in SFE2 initiated during overpressured conditions at close to maximum burial depth. However, the fact that these fractures continued to grow as the pore pressure gradient reduced down towards hydrostatic conditions suggests that there must have been an additional driving mechanism active during exhumation. English (2012) attributed this to thermoelastic contraction, whereby volumetric shrinkage of rock during exhumation and cooling can act to reduce the magnitude of the horizontal stresses. By contrast, Fall *et al.* (2012, 2015) observed that opening-mode fracture formation in the Mesaverde Group sandstones in the Piceance Basin generally ceased after uplift from maximum burial conditions, and interpreted that opening-mode fracture growth stopped when hydrocarbon (and overpressure) generation in the basin ceased. Why might there be a difference between these two areas?

Following the approach of English (2012), stress history models can be used to assess the possible role of thermoelastic contraction in the formation of opening-mode fractures during exhumation and cooling from maximum burial conditions. These models are based on the assumption that the rock behaves as an isotropic and linearly elastic medium and that the calculated minimum horizontal stress path is a function of the elastic properties of the rock and any temporal changes in overburden stress, temperature, pore pressure and horizontal strain. For the purposes of the model, the mechanical properties of the rock are assumed to be locked in at maximum burial and do not change during exhumation. The starting point for the model is at the onset of exhumation from maximum burial depth and temperature. In the cases under consideration here, opening-mode fracture growth is demonstrated to have been occurring during maximum burial conditions and, hence, the initial stress state for the model lies on the opening-mode failure criterion (i.e. $S_{hmin} = P_p$). No far-field horizontal strain is modelled during the exhumation event (i.e. $\varepsilon_{hmin} = \varepsilon_{Hmax} = 0$). The tensile strength of the rock is assumed to be negligible because natural fractures already exist at the starting point of the model. Inherent uncertainties in this approach include: (1) the variation of elastic properties with differential stress and confining stress; (2) the variation of elastic properties due to further diagenesis; (3) uncertainty in burial, thermal and pore pressure history; (4) uncertainty in tectonic stress history and the possibility of far-field extensional or compressive horizontal strain; and (5) the extent of viscoelastic relaxation over geologically long time periods (English 2012).

According to linear elastic theory, the change in horizontal stress (ΔS_h) due to a change in temperature (ΔT) is given by:

$$\Delta S_h = \frac{E}{1 - \nu} \alpha_T \Delta T \qquad (3)$$

where ν is Poisson's ratio, E is Young's modulus and α_T is the coefficient of linear thermal expansion. Typical values for the coefficient of linear thermal expansion for sedimentary rocks are of the order of 1×10^{-5}°C^{-1} (Fjær *et al.* 2008). The value of α_T for silica (*c.* 10^{-5}°C^{-1}) is an order of magnitude higher than that of most other rock-forming minerals (Zoback 2007) and the bulk value of α_T appears to be a function of the silica content in the rock (Fig. 4a) (data from Griffith 1936). Available data from clastic sedimentary rocks also indicate that the bulk rock α_T is highly dependent on the quartz content, with values ranging from 1.2×10^{-5}°C^{-1} for quartz-rich sandstones and quartzites down to 0.7×10^{-5}°C^{-1} for more argillaceous lithologies (Fig. 4b) (data from Wong & Brace 1979; O'Connor *et al.* 2001; Hoffmann 2006; Stück *et al.* 2011; Siegesmund & Dürrast 2011).

The S_{hmin} exhumation stress path for the Travis Peak Formation has been modelled previously by English (2012) and the full details of the model are presented therein. A slightly modified version of the model is presented in Figure 5a using the properties outlined in Table 2. If the thermoelastic effect is excluded from the model, the conditions for opening-mode failure would cease immediately at the onset of exhumation. However, if thermoelastic contraction is incorporated into the model (the range of stress paths shown correspond to a range of α_T from 0.7×10^{-5} to 1.1×10^{-5}°C^{-1}), the predicted effective minimum horizontal stress becomes increasingly tensile during exhumation, despite the reduction in the pore pressure gradient. Therefore thermoelastic contraction provides a mechanism for driving the continuous crack-seal fracture opening in the Travis Peak sandstones documented by Becker *et al.* (2010). It is noteworthy that continuous opening-mode failure during exhumation is predicted for the entire range of α_T used in the model. This is in large part due to the high Young's moduli of these rocks. As the lower Travis Peak Formation sandstones are composed of *c.* 95% quartz grains (Table 1; Dutton & Diggs 1990), the appropriate coefficient of linear thermal expansion is probably closer to the 1.1×10^{-5}°C^{-1} end of the range (Fig. 4b).

As a result of the high strength of the Travis Peak sandstones, opening-mode failure will occur before the Mohr–Coulomb shear failure criterion is met (Fig. 5a). English (2012) also modelled stress paths for weaker mudstone units within the Travis Peak Formation and concluded that shear failure is likely to occur first in rocks with a lower strength; this prediction is consistent with observations from core samples (Laubach 1988). Any tensile effective stress generated during exhumation will be continuously relieved by opening-mode fracture growth and the S_{hmin} stress path is predicted to follow the pore pressure curve (i.e. the opening-mode failure criterion in Fig. 5a). As discussed ealier, the present day magnitude of S_{hmin} is significantly greater than the pore pressure and this may reflect the return of the *in situ* stresses to a state of frictional failure equilibrium after the end of the exhumation event. As a result of compressive effective stresses acting on natural fractures in the present day subsurface, cement deposited in the rock mass during fracture opening and bridges spanning between the fracture walls are both of paramount importance in keeping these Travis Peak fractures open and permeable (Laubach *et al.* 2004*a*).

An exhumation stress history model has also been generated for the Cozzette Sandstone Member of the Mesaverde Group (Fig. 5b) using the properties outlined in Table 2. In this case, the field of stress paths resulting from the assumed range of α_T used in the model straddles the opening-mode failure criterion. Although there are significant uncertainties in this model, particularly regarding the assumed evolution of pore pressure during exhumation, this model indicates that there are scenarios in which the thermoelastic effect is not sufficient to drive continuous opening-mode failure during exhumation. This is in part due to the lower Young's moduli of these rocks compared with the Travis Peak sandstones (Table 2). As the Cozzette sandstones have lower quartz contents that the Travis Peak sandstones (Table 1), the appropriate coefficient of linear thermal expansion is more likely to be in the 0.8×10^{-5} to 0.9×10^{-5}°C^{-1} part of the range (Fig. 4b), although the data to constrain this are not currently available. The prediction is that the more quartz-rich sandstones with a higher Young's modulus are more likely to have experienced continued opening-mode fracture growth during exhumation. This is consistent with the observations from the Cozzette sandstone samples from the MWX-1 and SHCT wells (Fig. 3b).

Sandstones of the overlying Williams Fork Formation of the Mesaverde Group tend to have even lower quartz contents (Pitman *et al.* 1989) and lower Young's moduli (Lorenz *et al.* 1989*a*), which suggests that continuous opening-mode fracture growth during exhumation is less likely in these stratigraphic units. Of particular note, sandstones of the Paludan Interval of the Williams Fork Formation have quartz contents in the range 29–55% (Table 1; Pitman *et al.* 1989) and Young's moduli in the range 20–25 GPa (Lorenz *et al.* 1989*a*). It appears that exhumation-related fracturing is less well developed or continuous in these rocks compared with the underlying Cozzette sandstone (Fig. 3b).

In both the Travis Peak and Cozzette examples, our results are compatible with previous results, implying that high fluid pressure probably played a key part in initiating and sustaining opening-mode fracture growth. However, our results suggest that, during uplift, the persistence of fracture growth in

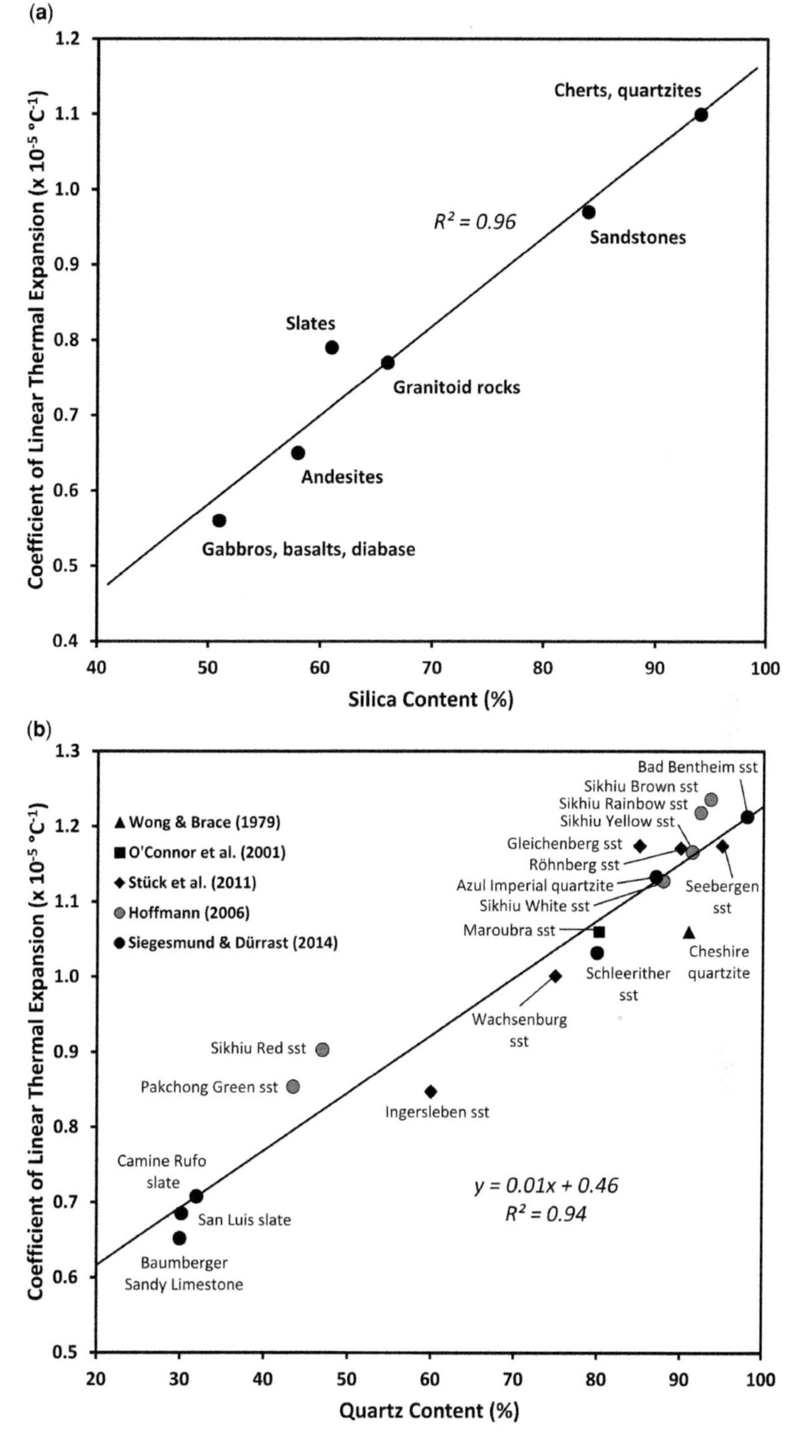

Fig. 4. Controls on thermal expansivity of rocks. (**a**) Coefficient of linear thermal expansion for a variety of rocks as a function of the percentage of silica (data from Griffith 1936; modified from Zoback 2007). (**b**) Coefficient of linear thermal expansion for a variety of clastic sedimentary rocks as a function of the percentage of quartz (data from Wong & Brace 1979; O'Connor *et al.* 2001; Hoffmann 2006; Stück *et al.* 2011; Siegesmund & Dürrast 2011).

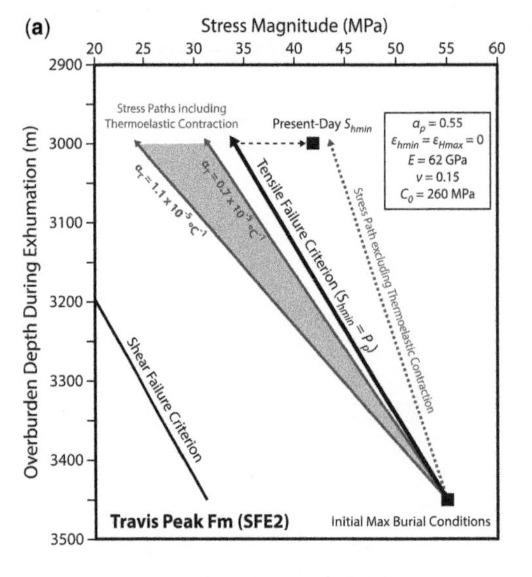

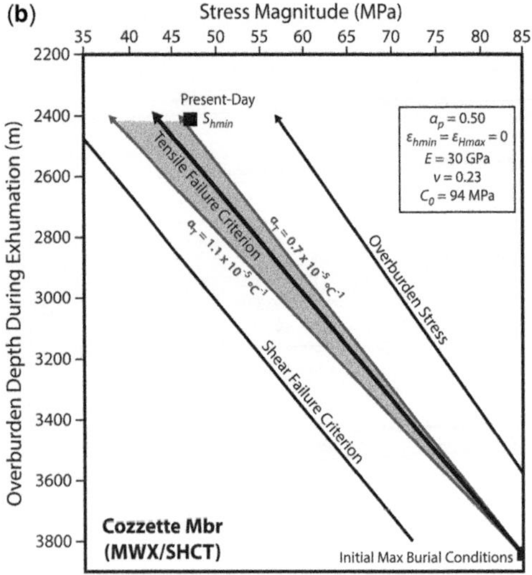

Fig. 5. Stress history models during exhumation from maximum burial conditions assuming linear elastic behaviour. (**a**) Sandstones of the Travis Peak Formation in the SFE2 well in the East Texas Basin. (**b**) Sandstones of the Cozzette Member of the Mesaverde Group in the MWX/SHCT wells in the southern Piceance Basin. The initial S_{hmin} magnitude for the sandstone model is taken to be on the tensile failure criterion (i.e. $P_{\mathrm{p}} = S_{\mathrm{hmin}}$) on the basis of the existence of opening-mode fractures at maximum burial in both studies. The grey shaded area represents the predicted stress paths using a range of coefficients of linear thermal expansion. A stress path to the left of the tensile failure criterion on the graph predicts continuous opening-mode failure during exhumation. The models are especially sensitive to the coefficient of linear thermal expansion and Young's modulus (see equation 3). No far-field horizontal strain is included in these models.

the Travis Peak Formation in East Texas and in the Cozzette Member in the Piceance Basin could have been strongly influenced by thermoelastic contraction during exhumation and cooling. In other basins, sequenced microthermometric data and isotopic information have been used to infer a dominant role for tectonic stretching and unloading (Hooker *et al.* 2015). Elsewhere, fracture cement deposits record episodic opening separated by long stretches of time when fractures were quiescent, with late-stage reactivation correlated with uplift and exhumation (Laubach *et al.* 2016). Geomechanical properties and *in situ* stress data integrated with the timing constraints from the geochemical evidence for timing allow the various driving mechanisms for opening-mode fracture formation to be evaluated. Overpressuring, tectonic stretching and thermoelastic contraction can all drive the formation of opening-mode fractures. Our results show progress towards separating and independently evaluating these mechanisms.

Conclusions

Characterization and the accurate prediction of natural fracture systems is challenging. Recent progress in unravelling the timing of fractures using cements deposited contemporaneously with fracture opening more closely constrains the pressure–temperature conditions under which these fractures formed. These data allow for a more meaningful application of stress history models to investigate the main driving mechanisms for fracture formation and the geomechanical controls on their distribution. In both examples reviewed here, in agreement with previous work, high fluid pressure probably played a key part in initiating and sustaining opening-mode fracture growth during burial, but our results suggest that the persistence of fracture growth during exhumation and cooling could have been strongly influenced by thermoelastic contraction. For sandstone reservoirs, thermoelastic contraction will be more pronounced for stiffer, high Young's modulus rocks with higher quartz contents. This concept can be tested further in other exhumed basins where fluid inclusions within crack-seal fracture cement deposits record the pressure–temperature history of the fracture fluid during fracture opening.

SEL's work is funded by grant DE-FG02-03ER15430 from the Chemical Sciences, Geosciences and Biosciences Division, Office of Basic Energy Sciences, Office of Science, US Department of Energy and by the Fracture Research and Application Consortium. The manuscript has benefited from input from Siegfried Siegesmund and from the constructive feedback of two anonymous reviewers.

References

Anderson, E.M. 1951. *The Dynamics of Faulting*. Oliver and Boyd, Edinburgh.

Barker, C. 1990. Calculated volume and pressure changes during the thermal cracking of oil to gas in reservoirs. *American Association of Petroleum Geologists Bulletin*, **74**, 1254–1261.

Barker, C.E. 1989. Fluid inclusion evidence for paleotemperatures within the Mesaverde Group, Multiwell Experiment site, Piceance Basin, Colorado. *In*: Law, B.E. & Spencer, C.W. (eds) *Geology of Tight Gas Reservoirs in the Pinedale Anticline Area, Wyoming, and at the Multiwell Experiment Site, Colorado*. US Geological Survey Bulletin, **1886**, M1–M7.

Bartberger, C.E., Dyman, T.S. & Condon, S.M. 2002. Potential for a basin-centred gas accumulation in Travis Peak (Hosston) Formation, Gulf Coast Basin, USA. *In*: Nuccio, V.F. & Dyman, T.S. (eds) *Geologic Studies of Basin-Centred Gas Systems*. US Geological Survey Bulletin, 2184-E, https://pubs.usgs.gov/bul/b2184-e/B2184E-508.pdf

Becker, S.P., Eichhubl, P., Laubach, S.E., Reed, R.M., Lander, R.H. & Bodnar, R.J. 2010. A 48 m.y. history of fracture opening, temperature, and fluid pressure: Cretaceous Travis Peak Formation, East Texas basin. *Geological Society of America Bulletin*, **122**, 1081–1093, https://doi.org/10.1130/B30067.1

Bradley, J.S. 1975. Abnormal formation pressure. *American Association of Petroleum Geologists Bulletin*, **59**, 957–973, https://doi.org/10.1306/83D91EFC-16C7-11D7-8645000102C1865D

Buffler, R.T., Watkins, J.S., Worzel, J.L. & Shaub, F.J. 1980. Structure and early geologic history of the deep central Gulf of Mexico. *In*: Pilger, R. (ed.) *Proceedings of a Symposium on the Origin of the Gulf of Mexico and the Early Opening of the Central North Atlantic*. Louisiana State University, Baton Rouge, LA, 3–16.

Collins, E.W., Hobday, D.K. & Kreitler, C.W. 1980. *Quaternary Faulting in East Texas*. Bureau of Economic Geology, Geological Circulars, 80-1.

Dutton, S.P. 1987. *Diagenesis and Burial History of the Lower Cretaceous Travis Peak Formation, East Texas*. Bureau of Economic Geology, Austin, Report of Investigations 164.

Dutton, S.P. & Diggs, T.N. 1990. History of quartz cementation in the Lower Cretaceous Travis Peak Formation, East Texas. *Journal of Sedimentary Research*, **60**, 191–202, https://doi.org/10.1306/212F914C-2B24-11D7-8648000102C1865D

Dutton, S.P. & Land, L.S. 1988. Cementation and burial history of a low-permeability quartzarenite, Lower Cretaceous Travis Peak Formation, East Texas. *Geological Society of America Bulletin*, **100**, 1271–1282, https://doi.org/10.1130/0016-7606(1988)100<1271:CABHOA>2.3.CO;2

Engelder, T. & Lacazette, A. 1990. Natural hydraulic fracturing. *In*: Barton, N. & Stephansson, O. (eds) *Rock Joints*. Balkema, Rotterdam, 35–43.

English, J.M. 2012. Thermomechanical origin of regional fracture systems. *American Association of Petroleum Geologists Bulletin*, **96**, 1597–1625, https://doi.org/10.1306/01021211018

English, J.M., Finkbeiner, T., English, K.L. & Yahia Cherif, R. 2017. State of stress in exhumed basins and implications for fluid flow – insights from the Illizi Basin, Algeria. *In*: Turner, J.P., Healy, D., Hillis, R.R. & Welch, M.J. (eds) *Geomechanics and Geology*. Geological Society, London, Special Publications, **458**. First published online May 30, 2017, https://doi.org/101144/SP458.6

English, K.L., Redfern, J., Corcoran, D.V., English, J.M. & Yahia Cherif, R. 2016. Constraining burial history and petroleum charge in exhumed basins: new insights from the Illizi Basin, Algeria. *American Association of Petroleum Geologists Bulletin*, **100**, 623–655, https://doi.org/10.1306/12171515067

Fall, A., Eichhubl, P., Cumella, S.P., Bodnar, R.J., Laubach, S.E. & Becker, S.P. 2012. Testing the basin-centered gas accumulation model using fluid inclusion observations: southern Piceance Basin, Colorado. *American Association of Petroleum Geologists Bulletin*, **96**, 2297–2318, https://doi.org/10.1306/05171211149

Fall, A., Eichhubl, P., Bodnar, R.J., Laubach, S.E. & Davis, J.S. 2015. Natural hydraulic fracturing of tight-gas sandstone reservoirs, Piceance Basin, Colorado. *Geological Society of America Bulletin*, **127**, 61–75, https://doi.org/10.1130/B31021.1

Finley, S.J. & Lorenz, J.C. 1988. *Characterization of Natural Fractures in Mesaverde Core from the Multiwell Experiment*. Sandia National Laboratories, Albuquerque, NM.

Finley, S.J. & Lorenz, J.C. 1989. Characterization and significance of natural fractures in Mesaverde Reservoirs at the Multiwell Experiment site. Paper SPE-19007-MS, presented at the Low Permeability Reservoirs Symposium, 6–8 March 1989, Denver, CO, USA, https://doi.org/10.2118/19007-MS

Fjær, E., Holt, R.M., Horsrud, P., Raaen, A.M. & Risnes, R. 2008. *Petroleum Related Rock Mechanics*. Elsevier, Amsterdam.

Griffith, J.H. 1936. *Thermal Expansion of Typical American Rocks*. Iowa State College of Agriculture and Mechanic Arts, Iowa Engineering Experiment Bulletin, **128**.

Halbouty, M.T. & Halbouty, J.J. 1982. Relationships between East Texas Field region and Sabine Uplift in Texas. *American Association of Petroleum Geologists Bulletin*, **66**, 1042–1054.

Handin, J., Hager, R.V., Jr, Friedman, M. & Feather, J.N. 1963. Experimental deformation of sedimentary rocks under confining pressure: pore pressure tests. *American Association of Petroleum Geologists Bulletin*, **47**, 717–755.

Haxby, W.F. & Turcotte, D.L. 1976. Stresses induced by the addition or removal of overburden and associated thermal effects. *Geology*, **4**, 181–184, https://doi.org/10.1130/0091-7613(1976)4<181:SIBTAO>2.0.CO;2

Hillis, R.R. 2001. Coupled changes in pore pressure and stress in oil fields and sedimentary basins. *Petroleum Geoscience*, **7**, 419–425, https://doi.org/10.1144/petgeo.7.4.419

Hillis, R.R. 2003. Pore pressure/stress coupling and its implications for rock failure. *In*: Van

Rensbergen, P., Hillis, R.R., Maltman, A.J. & Morley, C.K. (eds) *Subsurface Sediment Mobilization*. Geological Society, London, Special Publications, **216**, 359–368, https://doi.org/10.1144/GSL. SP.2003.216. 01.23

Hoffmann, A. 2006. *Naturwerksteine Thailands: Lagerstättenerkundung Und Bewertung*. Universität zu Göttingen, Göttingen, Germany.

Hooker, J.N., Gale, J.F.W., Gomez, L.A., Laubach, S.E., Marrett, R. & Reed, R.M. 2009. Aperture-size scaling variations in a low-strain opening-mode fracture set, Cozzette Sandstone, Colorado. *Journal of Structural Geology*, **31**, 707–718, https://doi.org/10. 1016/j.jsg.2009.04.001

Hooker, J.N., Larson, T.E., Eakin, A., Laubach, S.E., Eichhubl, P., Fall, A. & Marrett, R. 2015. Fracturing and fluid flow in a sub-décollement sandstone; or, a leak in the basement. *Journal of the Geological Society, London*, **172**, 428–442, https://doi.org/10.1144/ jgs2014-128

Hubbert, M.K. & Rubey, W.W. 1959. Role of fluid pressure in mechanics of overthrust faulting: I. Mechanics of fluid-filled porous solids and its application to overthrust faulting. *Geological Society of America Bulletin*, **70**, 115–166, https://doi.org/10.1130/0016-7606 (1959)70[115:ROFPIM]2.0.CO;2

Jackson, M.L.W. & Laubach, S.E. 1988. Cretaceous and Tertiary compressional tectonics as the cause of the Sabine Arch, East Texas and northwest Louisiana. *Gulf Coast Association of Geological Societies Transactions*, **38**, 245–256.

Jackson, M.L.W. & Laubach, S.E. 1991. *Structural History and Origin of the Sabine Arch, East Texas and Northeast Louisiana*. Bureau of Economic Geology, Austin, Geological Circulars, 91-3.

Jaeger, J.C., Cook, N.G.W. & Zimmerman, R. 2007. *Fundamentals of Rock Mechanics*. Wiley-Blackwell, New York.

Johnson, R.C. & Nuccio, V.F. 1986. Structural and thermal history of the Piceance Creek Basin, Western Colorado, in relation to hydrocarbon occurrence in the Mesaverde Group. *In*: Spencer, C.W. & Mast, R.F. (eds) *Geology of Tight Gas Reservoirs*. AAPG Studies in Geology, **24**, 165–205.

Lacazette, A. & Engelder, T. 1992. Fluid-driven cyclic propagation of a joint in the Ithaca Siltstone, Appalachian Basin, New York. *In*: Evans, B. & Wong, T.F. (eds) *Fault Mechanics and Transport Properties of Rocks*. Academic Press, San Diego, 297–323, https://doi.org/10.1016/S0074-6142(08)62827-2

Lander, R.H. & Laubach, S.E. 2015. Insights into rates of fracture growth and sealing from a model for quartz cementation in fractured sandstones. *Geological Society of America Bulletin*, **127**, 516–538, https://doi. org/10.1130/B31092.1

Laubach, S.E. 1988. Subsurface fractures and their relationship to stress history in East Texas basin sandstone. *Tectonophysics*, **156**, 37–49, https://doi.org/10. 1016/0040-1951(88)90281-8

Laubach, S.E. 1989. Paleostress directions from the preferred orientation of closed microfractures (fluid-inclusion planes) in sandstone, East Texas basin, U.S.A. *Journal of Structural Geology*, **11**, 603–611, https://doi.org/10.1016/0191-8141(89)90091-6

Laubach, S.E. 2003. Practical approaches to identifying sealed and open fractures. *American Association of Petroleum Geologists Bulletin*, **87**, 561–579.

Laubach, S.E. & Jackson, M.L.W. 1990. Origin of arches in the northwestern Gulf of Mexico basin. *Geology*, **18**, 595–598, https://doi.org/10.1130/0091-7613(1990) 018<0595:OOAITN>2.3.CO;2

Laubach, S.E., Olson, J.E. & Gale, J.F.W. 2004*a*. Are open fractures necessarily aligned with maximum horizontal stress? *Earth and Planetary Science Letters*, **222**, 191–195, https://doi.org/10.1016/j.epsl.2004. 02.019

Laubach, S.E., Reed, R.M., Olson, J.E., Lander, R.H. & Bonnell, L.M. 2004*b*. Coevolution of crack-seal texture and fracture porosity in sedimentary rocks: cathodoluminescence observations of regional fractures. *Journal of Structural Geology*, **26**, 967–982, https://doi.org/10.1016/j.jsg.2003.08.019

Laubach, S.E., Lander, R.H., Bonnell, L.M., Olson, J.E. & Reed, R.M. 2004*c*. Opening histories of fractures in sandstone. *In*: Cosgrove, J.W. & Engelder, T. (eds) *The Initiation, Propagation, and Arrest of Joints and Other Fractures*. Geological Society, London, Special Publications, **231**, 1–9, https://doi.org/ 10.1144/GSL.SP.2004.231.01.01

Laubach, S.E., Olson, J.E. & Gross, M.R. 2009. Mechanical and fracture stratigraphy. *American Association of Petroleum Geologists Bulletin*, **93**, 1413–1426, https://doi.org/10.1306/07270909094

Laubach, S.E., Fall, A., Copley, L.K., Marrett, R. & Wilkins, S.J. 2016. Fracture porosity creation and persistence in a basement-involved Laramide fold, Upper Cretaceous Frontier Formation, Green River Basin, USA. *Geological Magazine*, **153**, 897–910, https:// doi.org/10.1017/S0016756816000157

Law, B.E. & Johnson, R.C. 1989. Structural and stratigraphic framework of the Pinedale Anticline, Wyoming, and the Multiwell Experiment site, Colorado. *In*: Law, B.E. & Spencer, C.W. (eds) *Geology of Tight Gas Reservoirs in the Pinedale Anticline Area, Wyoming, and at the Multiwell Experiment Site, Colorado*. US Geological Survey Bulletin, **1886**, B1–B11.

Lockner, D.A. 1995. Rock failure. *In*: Ahrens, T.J. (ed.) *Rock Physics and Phase Relations – A Handbook of Physical Constants*. American Geophysical Union, Washington DC, 127–147.

Lorenz, J.C. & Finley, S.J. 1989. Differences in fracture characteristics and related production: Mesaverde Formation, northwestern Colorado. *SPE Formation Evaluation*, **4**, 11–16, https://doi.org/10.2118/16809-PA

Lorenz, J.C. & Finley, S.J. 1991. Regional fractures II: fracturing of Mesaverde reservoirs in the Piceance Basin, Colorado. *American Association of Petroleum Geologists Bulletin*, **75**, 1738–1757.

Lorenz, J.C. & Hill, R.E. 1994. Subsurface fracture spacing: comparison of inferences from slant/horizontal core and vertical core in Mesaverde Reservoirs. *SPE Formation Evaluation*, **9**, 66–72, https://doi. org/10.2118/21877-PA

Lorenz, J.C., Sattler, A.R. & Stein, C.L. 1989*a*. *Differences in Reservoir Characteristics of Marine and Nonmarine Sandstones of the Mesaverde Group, Northwestern Colorado*. Sandia National Laboratories, Albuquerque, NM.

LORENZ, J.C., WARPINSKI, N.R., BRANAGAN, P.T. & SATTLER, A.R. 1989*b*. Fracture characteristics and reservoir behavior of stress-sensitive fracture systems in flat-lying lenticular formations. *Journal of Petroleum Technology*, **41**, 615–622, https://doi.org/10.2118/15244-PA

MUKHOPADHYAY, P.K. 1989. *Organic Petrography and Organic Geochemistry of Tertiary Coals from Texas in Relation to Depositional Environments and Hydrocarbon Generation*. Bureau of Economic Geology, Austin, Report of Investigations 188.

MWX PROJECT TEAM 1987. Multiwell experiment final report: I. The marine interval of the Mesaverde Formation. Sandia National Laboratories Report **SAND87-0327**, https://www.osti.gov/scitech/servlets/purl/6823230

NARR, W. & CURRIE, J.B. 1982. Origin of fracture porosity – example from Altamont Field. *American Association of Petroleum Geologists Bulletin*, **66**, 1231–1247.

NUCCIO, V.F. & ROBERTS, L.N.R. 2003. Thermal maturity and oil and gas generation history of petroleum systems in the Uinta-Piceance province, Utah and Colorado. *In*: *Petroleum Systems and Geologic Assessment of Oil and Gas in the Uinta-Piceance Province*, Utah and Colorado. US Geological Survey, Digital Data Series, DDS-69-B, https://pubs.usgs.gov/dds/dds-069/dds-069-b/REPORTS/Chapter_4.pdf

NUNN, J.A. 1990. Relaxation of continental lithosphere: an explanation for Late Cretaceous reactivation of the Sabine Uplift of Louisiana-Texas. *Tectonics*, **9**, 341–359, https://doi.org/10.1029/TC009i002p00341

O'CONNOR, J., RAY, A., FRANKLIN, B. & STUART, B. 2001. Changes in the physical and chemical properties of weathered Maroubra Sandstone in Sydney. *AICCM Bulletin*, **26**, 20–25, https://doi.org/10.1179/bac.2001.26.1.004

OLSON, J.E., LAUBACH, S.E. & LANDER, R.H. 2009. Natural fracture characterization in tight gas sandstones: integrating mechanics and diagenesis. *American Association of Petroleum Geologists Bulletin*, **93**, 1535–1549, https://doi.org/10.1306/08110909100

OSBORNE, M.J. & SWARBRICK, R.E. 1997. Mechanisms for generating overpressure in sedimentary basins: a reevaluation. *American Association of Petroleum Geologists Bulletin*, **81**, 1023–1041.

OZKAN, A., CUMELLA, S.P., MILLIKEN, K.L. & LAUBACH, S.E. 2011. Prediction of lithofacies and reservoir quality using well logs, Late Cretaceous Williams Fork Formation, Mamm Creek field, Piceance Basin, Colorado. *American Association of Petroleum Geologists Bulletin*, **95**, 1699–1723, https://doi.org/10.1306/01191109143

PEREZ, R.J. & BOLES, J.R. 2005. Interpreting fracture development from diagenetic mineralogy and thermoelastic contraction modeling. *Tectonophysics*, **400**, 179–207, https://doi.org/10.1016/j.tecto.2005.03.002

PINDELL, J.L. 1985. Alleghenian reconstruction and subsequent evolution of the Gulf of Mexico, Bahamas, and Proto-Caribbean. *Tectonics*, **4**, 1–39, https://doi.org/10.1029/TC004i001p00001

PITMAN, J.K., SPENCER, C.W. & POLLASTRO, R.M. 1989. Petrography, mineralogy, and reservoir characteristics of the Upper Cretaceous Mesaverde Group in the East-Central Piceance Basin, Colorado. *In*: *Evolution of Sedimentary Basins – Uinta and Piceance Basins*. US Geological Survey Bulletin, **1787**, G1–G31.

PLUMB, R.A., HERRON, S.L. & OLSEN, M.P. 1992. Composition and texture on compressive strength variations in the Travis Peak Formation. Paper SPE-24758-MS, presented at the SPE Annual Technical Conference and Exhibition, 4–7 October 1992, Washington, DC, USA, 985–998, https://doi.org/10.2118/24758-MS

POLLARD, D.D. & AYDIN, A. 1988. Progress in understanding jointing over the past century. *Geological Society of America Bulletin*, **100**, 1181–1204, https://doi.org/10.1130/0016-7606(1988)100<1181:PIUJOT>2.3.CO;2

POLLASTRO, R.M. 1984. Mineralogy of selected sandstone/shale pairs and sandstone from the Multiwell Experiment; interpretations from X-ray diffraction and scanning electron microscopy analyses. *In*: SPENCER, C.W. & KEIGHIN, C.W. (eds) *Geologic Studies in Support of the U.S. Department of Energy Multiwell Experiment, Garfield County, Colorado*. US Geological Survey Open-File Report **84–757**, 67–74.

PRICE, N.J. 1966. *Fault and Joint Development in Brittle and Semibrittle Rock*. Pergamon Press, Oxford.

SECOR, D.T. 1965. Role of fluid pressure in jointing. *American Journal of Science*, **263**, 633–646, https://doi.org/10.2475/ajs.263.8.633

SECOR, D.T. 1969. Mechanics of natural extension fracturing at depth in the Earth's crust. *In*: BAER, A.J. & NORRIS, D.K. (eds) *Research in Tectonics*. Geological Survey of Canada, Papers, **68–52**, 3–47, http://ftp.maps.canada.ca/pub/nrcan_rncan/publications/ess_sst/102/102928/pa_68_52.pdf

SIEGESMUND, S. & DÜRRAST, H. 2011. Physical and mechanical properties of rocks. *In*: SIEGESMUND, S. & SNETHLAGE, R. (eds) *Stone in Architecture*. Springer, Berlin, 97–225, https://doi.org/10.1007/978-3-642-14475-2_3

SPENCER, C.W. 1987. Hydrocarbon generation as a mechanism for overpressuring in Rocky Mountain region. *American Association of Petroleum Geologists Bulletin*, **71**, 368–388.

STROKER, T.M., HARRIS, N.B., ELLIOTT, W.C. & WAMPLER, J.M. 2013. Diagenesis of a tight gas sand reservoir: Upper Cretaceous Mesaverde Group, Piceance Basin, Colorado. *Marine and Petroleum Geology*, **40**, 48–68, https://doi.org/10.1016/j.marpetgeo.2012.08.003

STÜCK, H., SIEGESMUND, S. & RÜDRICH, J. 2011. Weathering behaviour and construction suitability of dimension stones from the Drei Gleichen area (Thuringia, Germany). *Environmental Earth Sciences*, **63**, 1763–1786, https://doi.org/10.1007/s12665-011-1043-7

SWARBRICK, R.E. & OSBORNE, M.J. 1998. Mechanisms that generate abnormal pressures: an overview. *In*: LAW, B.E., ULMISHEK, G.F. & SLAVIN, V.I. (eds) *Abnormal Pressures in Hydrocarbon Environments*. AAPG Memoirs, **70**, 13–34.

SWARBRICK, R.E., OSBORNE, M.J. & YARDLEY, G.S. 2002. Comparision of overpressure magnitude resulting from the main generating mechanisms. *In*: HUFFMAN, A.R. & BOWERS, G.L. (eds) *Pressure Regimes in Sedimentary Basins and Their Prediction*. AAPG Memoirs, **76**, 1–12.

Teufel, L.W. 1986. In situ stress and natural fracture distribution at depth in the Piceance Basin, Colorado: implications to stimulation and production of low permeability gas reservoirs. Paper presented at the 27th US Symposium on Rock Mechanics. American Rock Mechanics Association, 23–25 June, Tuscaloosa, Alabama, 702–708.

Thiercelin, M.J. & Plumb, R.A. 1994. A core-based prediction of lithologic stress contrasts in East Texas formations. *SPE Formation Evaluation*, **9**, 251–258, https://doi.org/10.2118/21847-PA

Townend, J. & Zoback, M.D. 2000. How faulting keeps the crust strong. *Geology*, **28**, 399–402, https://doi.org/10.1130/0091-7613(2000)28<399:HFKTCS>2.0.CO;2

Voight, B. & St Pierre, B.H.P. 1974. Stress history and rocks in stress. *Advances in Rock Mechanics*. Proceedings of the Third Congress of the International Society for Rock Mechanics, Denver, Colorado, 1–7 September 1974, 580–582.

Walls, J.D. & Dvorkin, J. 1994. Measured and calculated horizontal stresses in the Travis Peak Formation. *SPE Formation Evaluation*, **9**, 259–263, https://doi.org/10.2118/21843-PA

Warpinski, N.R. 1989. Elastic and viscoelastic calculations of stresses in sedimentary basins. *SPE Formation Evaluation*, **4**, 522–530, https://doi.org/10.2118/15243-PA

Warpinski, N.R. & Lorenz, J.C. 2008. Analysis of the Multiwell Experiment data and results: implications for the basin-centered gas model. *In*: Cumella, S.P., Shanley, K.W. & Camp, W.K. (eds) *Understanding, Exploring, and Developing Tight-Gas Sands*. AAPG, Hedberg Series, **3**, 157–176.

Warpinski, N.R. & Teufel, L.W. 1989. In-situ stresses in low-permeability, nonmarine rocks. *Journal of Petroleum Technology*, **41**, 405–414, https://doi.org/10.2118/16402-PA

Warpinski, N.R., Branagan, P.T. & Wilmer, R. 1985. In-situ stress measurements at U.S. DOE's Multiwell Experiment site, Mesaverde Group, Rifle, Colorado. *Journal of Petroleum Technology*, **37**, 527–536, https://doi.org/10.2118/12142-PA

Wong, T.F. & Brace, W.F. 1979. Thermal expansion of rocks: some measurements at high pressure. *Tectonophysics*, **57**, 95–117, https://doi.org/10.1016/0040-1951(79)90143-4

Yurewicz, D.A., Bohacs, K.M. *et al*. 2005. Controls on gas and water distribution, Mesaverde basin-centered gas play, Piceance Basin, Colorado. *In*: Cumella, S.P., Shanley, K.W. & Camp, W.K. (eds) *Understanding, Exploring, and Developing Tight-Gas Sands*. AAPG, Hedberg Series, **3**, 105–136.

Zoback, M.D. 2007. *Reservoir Geomechanics*. Cambridge University Press, Cambridge.

Zoback, M.D. & Healy, J.H. 1984. Friction, faulting and 'in situ' stress. *Annales Geophysicae*, **2**, 689–698.

Zoback, M.D., Townend, J. & Grollimund, B. 2002. Steady-state failure equilibrium and deformation of intraplate lithosphere. *International Geology Review*, **44**, 383–401, https://doi.org/10.2747/0020-6814.44.5.383

Geomechanical characterization of mud volcanoes using P-wave velocity datasets

RASHAD GULMAMMADOV*, STEPHEN COVEY-CRUMP & MADS HUUSE

School of Earth and Environmental Sciences, University of Manchester, Manchester M13 9PL, UK

**Correspondence: rashad.gulmammadov@alumni.manchester.ac.uk*

Abstract: Mud volcanoes occur in many petroliferous basins and are associated with significant drilling hazards. To illustrate the type of information that can be extracted from limited petrophysical datasets in such geomechanically complex settings, we use P-wave velocity data to calculate the mechanical properties and stresses on a two-dimensional vertical section across a mud volcano in the Azeri-Chirag-Guneshly field, South Caspian Basin. We find that: (1) the values of the properties and stresses calculated in this way have realistic magnitudes; (2) the calculated pore fluid pressures show spatial variations around the mud volcano that potentially highlight areas of fluid recharging after the most recent eruption; and (3) the information obtained is sufficient to provide helpful indications of the width of the drilling window. Although calculations of this kind may be readily improved with more sophisticated petrophysical datasets, the simplicity of this approach makes it attractive for reconnaissance surveys designed to identify targets worthy of further investigation in developing our understanding of mud volcano geomechanics, or which could be used to help formulate drilling strategies.

To date, around 6500 mud volcanoes have been identified worldwide, both onshore and offshore (Judd 2005). They are primarily developed where mudstone sequences are overlain by thick and rapidly deposited sands from modern and Tertiary deltas – for example, the Volga in the South Caspian Basin (SCB), the Baram in Borneo, the Niger in West Africa, the Mississippi in the USA and the Mackenzie in Arctic Canada (Allen & Allen 2013). Their occurrence is generally associated with an active tectonic setting, rapid sedimentation and high rates of gas generation (Milkov 2000).

As pathways for fluid release from deeply buried and overpressured sedimentary successions, mud volcanoes in petroliferous basins are important features to consider in reducing the risk and uncertainty within different parts of the exploration and production cycle. Their feeder pipes may rupture the seal and allow hydrocarbon fluids and entrained sediments to migrate up through the sealing sequences (Cartwright *et al.* 2007; Hong *et al.* 2013). This does not necessarily imply total failure of the seal because it is the timing and efficiency of mud volcano eruptions relative to the timing of petroleum charging that defines the failure level of the seal (Cartwright *et al.* 2007). In many cases petroleum accumulations are discovered because of seal breach and the subsequent leaking of hydrocarbon-rich fluids to the surface at the sites of mud volcanoes (Clarke & Cleverly 1991). Nevertheless, the presence of mud volcanoes and the scale, geometry and activity of the plumbing systems beneath them are clearly important factors to consider when formulating strategies for field development and the siting of the facilities.

For these reasons, among others, mud volcanoes have been systematically studied worldwide to develop an understanding of: (1) the controls on their internal structure and geomorphology (Hovland *et al.* 1997; Dimitrov 2002; Deville *et al.* 2003; Evans *et al.* 2007; Soto *et al.* 2011); (2) the structural controls on mud volcano locations (Roberts *et al.* 2011; Bonini 2013); (3) fluid/sediment flow under mud volcano complexes (Planke *et al.* 2003; Calvès *et al.* 2008); (4) the factors influencing the severity of mud volcano eruptions (Lerche & Bagirov 1999; Kopf *et al.* 2009; Contet & Unterseh 2015; Hill *et al.* 2015); and (5) controls on the geochemistry of the erupted fluids (Azzaro *et al.* 1993; Bristow *et al.* 2000; Guliyev *et al.* 2001; Mazzini *et al.* 2009; Feseker *et al.* 2010; Feyzullayev & Movsumova 2010; Carragher *et al.* 2013; Oppo *et al.* 2014). In offshore areas, numerous multi-scale near-surface geological studies have been performed to mitigate the risks to seabed facilities that are associated with mud volcano activity and its accompanying hazardous phenomena, such as the presence of shallow gas, slope failure and pockmarks (Contet & Unterseh 2015; Hill *et al.* 2015; Unterseh & Contet 2015). Yet the extent to which drilling in such zones has to be avoided because of risks related to mud volcanoes remains unclear.

From: Turner, J. P., Healy, D., Hillis, R. R. & Welch, M. J. (eds) 2017. *Geomechanics and Geology*. Geological Society, London, Special Publications, **458**, 273–292.
First published online May 24, 2017, updated June 9, 2017, https://doi.org/10.1144/SP458.2

Among the challenges posed by the complicated geology in and around mud volcanoes is the prediction of local pore fluid pressures, which has significant implications for drilling (e.g. borehole blowouts and instability). Understanding these manifestations of localized fluid flow from a geomechanical perspective requires an analysis of the fluid and pressure distribution, the deformation history, the distribution of fractures and the state of stress around the mud volcano. This, in turn, requires comprehensive petrophysical datasets and sophisticated data analysis. However, within these geomechanically complex areas there remains value in adopting a simpler reconnaissance-type approach to identify targets for more detailed investigation and key features that require better understanding.

We use P-wave velocity data available in the public domain to estimate the mechanical properties and stresses on a two-dimensional vertical section across a mud volcano structure located in the Azeri part of the Azeri-Chirag-Guneshly (ACG) field in the SCB. The aim of the study was to determine whether useful geomechanical information can be extracted from such a limited dataset.

Geological setting

Regional geology

The SCB, offshore Azerbaijan (Fig. 1a), is a deep Tertiary basin characterized by mobilized overpressured sediments that cause instability on the basin margins and in deeper strata. The initiation of the basin corresponds to closure of the Tethys Ocean as a result of Arabia–Eurasia convergence (Morton *et al.* 2003; Kopf *et al.* 2009). Subduction of the Arabian plate under Eurasia to the NNE generated an accretionary prism during the Mesozoic–Early Tertiary. Following the closure of Tethys (*c.* 20 Ma), continuing convergence and uplift to the north led to folding of a thick Oligocene to Holocene sequence deposited in front of the previously active accretionary prism (Jackson *et al.* 2002; Stewart & Davies 2006; Santos Betancor & Soto 2015) (Fig. 1b). Along the northern margin of the basin, anticlinal structures developed within the NW–SE-trending Absheron-Balkan deep-seated structural uplift, which is the offshore extension of the Caucasus fold belt (Fig. 1a).

The sedimentary succession in the basin (Fig. 2) mainly consists of Cenozoic clastic sediments deposited within three large delta systems: Kura from the west (sediments from the Lesser Caucasus), Amu Darya from the east (sediments from the Balkans) and Volga from the north (sediments from the Greater Caucasus and Urals) (Bredehoeft *et al.* 1988; Smith-Rouch 2006). These were deposited at remarkably high rates (up to 2.4 km Ma^{-1}) as the basin subsided, generating a sedimentary succession that is >25 km thick (Lerche & Bagirov 1999). A cover sequence, up to 10 km thick, consisting of sand–silt–shale intercalations was deposited during the Pliocene and Quaternary. The main source rock for the extensive hydrocarbon reserves within the basin is the Maykop, a kilometre-thick sequence of organic-rich mudstones deposited during the Oligocene and Early Miocene (Abrams & Narimanov 1997; Jones & Simmons 1997). The main producing unit, both onshore and offshore in the SCB, is the overlying Productive Series deposited during the Late Miocene to Early Pliocene. This succession is composed of alternating, regionally extensive, fluvio-deltaic sandstones, separated by laterally extensive lacustrine shales. The lacustrine shales act as major pressure seals within the basin (Javanshir *et al.* 2015).

Rapid sediment burial within the SCB has led to small geothermal gradients (13–18°C km^{-1}), setting the hydrocarbon generation depth at 5–10 km in the western shelf and continental slope and at 6–14 km in the deep-water region (Guliyev *et al.* 2010; Feyzullayev 2012). The presence of low-permeability seals coupled with the high rate of gas generation means that there are abnormally high pore fluid pressures within the mudstone units. Pore fluid pressures in the shales enclosing the regionally developed reservoirs are estimated to exceed hydrostatic pressures by a factor of *c.* 1.8, whereas in the sandstones within the basin the difference is a factor of *c.* 1.4 (Bredehoeft *et al.* 1988).The high rate of sedimentation and gas generation in the basin resulted in the slow removal of pore fluids from the compacting mudstones during burial and this has led to a high level of undercompaction (Buryakovsky *et al.* 2001). These geological conditions, coupled with the active tectonic regime, present a wide range of geological hazards for oil and gas operations (Lerche & Bagirov 1999). These include: mud flows and gas emissions, which can damage rigs and production equipment; hydrate dissociation, which is hazardous for drilling activities; and the presence of submarine banks, which are dangerous to marine traffic. In addition to the natural hazards present on the seabed and at shallow subsurface depths, significant challenges for drilling processes are presented at greater depths by deep earthquakes and areas of large fluid overpressure.

ACG field

The ACG field complex is located within anticlinal structures on the northern boundary of the SCB at water depths of 95–425 m (Fig. 1a). The cores of these anticlines contain mobile shales from the Maykop sequence; the depth to the top of this

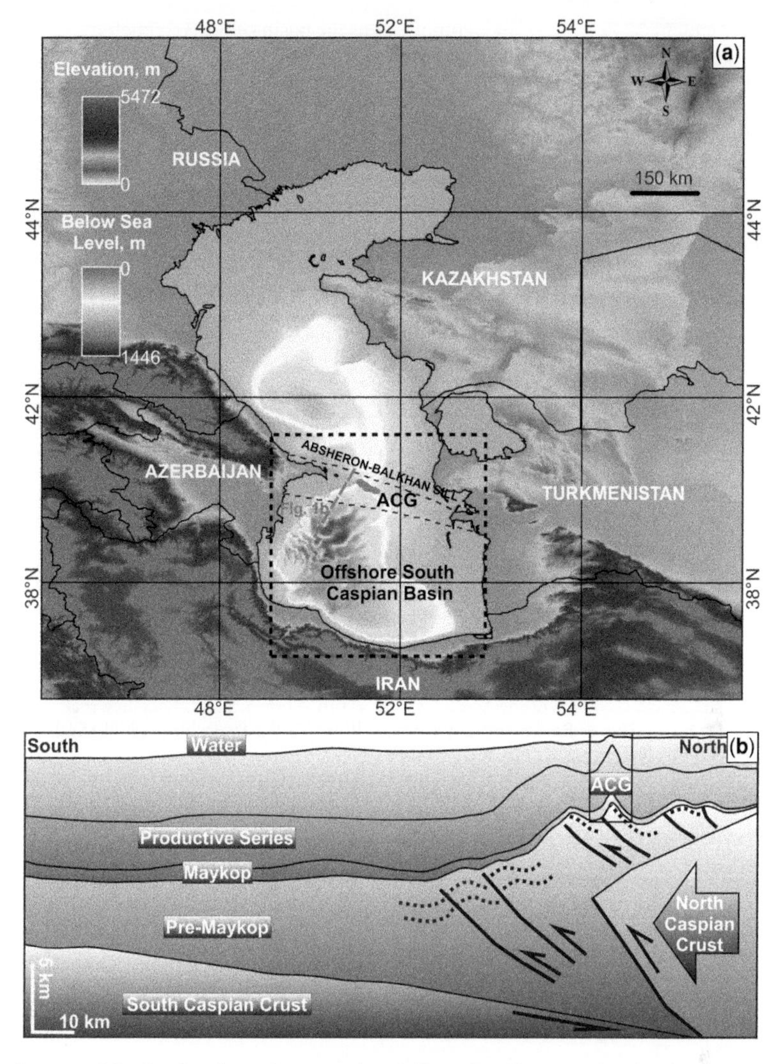

Fig. 1. (**a**) Bathymety of the Caspian Sea and topography of the surrounding countries showing the location of the offshore South Caspian Basin and the Azeri-Chirag-Guneshly (ACG) structure. Map extracted using GEBCO_2014 Grid – a global 30 arc second interval grid – and processed in ArcMap. (**b**) Simplified tectonic framework for the offshore South Caspian Basin (modified from Stewart & Davies 2006).

sequence is *c.* 5 km in the ACG. Mud volcanoes have formed where this mobile shale has exploited zones of weakness, resulting in the expulsion of mud and fluids, including hydrocarbons, at the seabed. These mud volcanoes are developed within three anticlinal culminations: Azeri, Chirag and Guneshly (Hill *et al.* 2015). Of these, the Chirag mud volcano is the most extensively studied (e.g. Lerche & Bagirov 1999; Stewart & Davies 2006).

The key geometric parameters and mechanical conditions of the Chirag mud volcano are illustrated in Figure 3. This mud volcano is located at a water depth of 120 m and contains several buried mud cones, which are stacked vertically, but share a common root system (Stewart & Davies 2006). The eruptive mud originates from the Maykop and is composed primarily of montmorillonite clay with some volcanic ash (Buryakovsky *et al.* 2001; Evans *et al.* 2006). Geochemical evidence suggests that the fluids within the mud volcano plumbing system also derive primarily from the Maykop (Mazzini *et al.* 2009; Kopf *et al.* 2009), but with a contribution from the Productive Series (Lerche & Bagirov 1999; Javanshir *et al.* 2015).

The pore fluid pressure gradients in the area are typically 0.012 MPa m^{-1} (Buryakovsky *et al.*

Era	Period	Epoch	Formation			Lithology	Average Thickness (m)	Petroleum Potential S	R	C
Cenozoic	Quaternary	Holocene	Recent			Mudstone and siltstone	260-820			
		Pleistocene								
			Absheron				650-1400		●	●
	Neogene	Pliocene	Akchagyl			Shale	30-50			
			Productive Series	Upper	Surakhany	Evaporite interbedded with shale	2600-3600	●	●	●
					Sabunchy	Shale				
					Balakhany	Fluvial sandstone with mudstone intercalations				
					Fasila					
		Miocene		Lower	NKG	Mudstone, siltstone & sandstone	800-1200			
					NKP					
					Kirmaki					
					Pod-Kirmaki					
			Kalin							
			Pontian			Marine shale	10-160			
			Diatom				75-310	●	●	●
			Chokrak				10-50			
			Tarkhan				30			
			Diatom				75-310			
			Maykop			Organic-rich shale	1000	●	●	●
	Palaeogene	Oligocene								
		Eocene				Marine shale	140-250			
		Paleocene					170-800			
Mesozoic	Cretaceous					Carbonates	>6000	●		
	Jurassic					Volcanics	>4500			

Fig. 2. Simplified stratigraphic column of the South Caspian Basin. Nomenclature: S, source rock; R, reservoir; C, cap rock (compiled from Yusifov & Rabinowitz 2004; Smith-Rouch 2006; Javanshir *et al.* 2015). NKG, Post-Kirmaky Shaly Suite; NKP, Post-Kirmaky Sandy Suite.

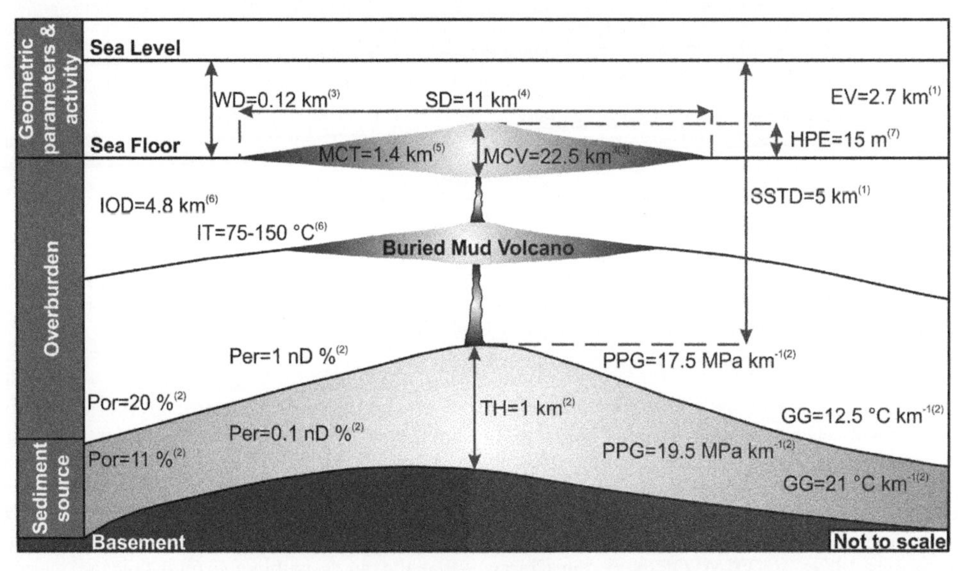

Fig. 3. Schematic diagram showing the geometry and key geomechanical properties of the Chirag mud volcano. EV, eruption volume; GG, geothermal gradient; HPE, highest point elevation; IOD, illitization onset depth; IT, illitization temperature; MCV, mud cone volume; MCT, mud cone thickness; Per, permeability; Por, porosity; PPG, pore pressure gradient; SD, surface diameter; SSTD, sediment source top depth; TH, thickness; WD, water depth. Superscripts in brackets refer to the references: (1) Evans *et al.* (2006), (2) Buryakovsky *et al.* (2001), (3) Stewart & Davies (2006), (4) Evans *et al.* (2006), (5) Davies & Stewart (2005), (6) Feyzullayev & Lerche (2009) and (7) Evans *et al.* (2008).

1995). Fluid overpressure in the area is generally associated with disequilibrium (gravitational) compaction. The smectite–illite transformation occurs at temperatures of 75–150°C, corresponding to depths >7 km (Feyzullayev & Lerche 2009).

Based on the eruption statistics for onshore mud volcanoes in Azerbaijan, it is estimated that the mean waiting time for weak eruptions of the Chirag mud volcano is 95 years and the mean waiting time for average and strong eruptions is 272 years (Lerche & Bagirov 1999). High-resolution geophysical imagery is currently being used to monitor hydrocarbon seepage, mud flows and the formation of slope failure scars to provide a better understanding of the activity of this mud volcano (Hill *et al.* 2015; Unterseh & Contet 2015).

Modelling background

Analytical and empirical correlations used in this study

Several analytical and empirical correlations between P-wave velocity and mechanical properties/*in situ* stresses have been developed that allow the latter to be estimated from the former (e.g. Zoback 2007, pp. 113–116; Mavko *et al.* 2009, pp. 386–388). The empirical correlations are intended to represent the average behaviour of a wide range of lithologies and so their usefulness is limited by how sensitive the correlated property is to the differences in lithology encountered in the region of interest, as well as to any other variable that has not been accommodated within the fitted equation. With this caveat, such correlations are being used to develop increasingly sophisticated geomechanical models, particularly when more comprehensive input data – such as pre-stack depth-migrated seismic inversion, S-wave velocities and borehole information – are also available to provide additional constraints (e.g. White *et al.* 2007; Sengupta *et al.* 2011; Gray *et al.* 2012).

In this study, we had only a very restricted dataset (primarily P-wave). The unavailability of more comprehensive datasets imposes limits on the extent to which we can validate our model results. However, the results can be viewed as representative for the context and methodology can be readily applied and tested for more sophisticated dataset in the Caspian and beyond. Therefore our comments in this respect are based on whether or not the model results seem realistic given the geomechanical context of the ACG.

The empirical correlations used to infer physical properties and stress states are listed in Table 1. Gradients of overburden, pore fluid pressure and fracture pressure have been evaluated as the change in magnitude of the given quantity over a given change in depth.

Given the limited dataset, the elastic rock properties were approximated as isotropic throughout the study. The matrix density ($\rho_{\text{matrix}} = 2600$ kg m^{-3}) in Table 1, equation (3) was approximated assuming that the rock is an aggregate of clay minerals consisting of 32.5% montmorillonite, 43.5% illite, 17.5% kaolinite and 6.5% chlorite, which is applicable for the NW SCB at a depth range of 1–2 km (Buryakovsky *et al.* 1995). The pore fluid density (ρ_{fluid}) was approximated as 1000 kg m^{-3} (Tozer & Borthwick 2010). The horizontal stress formula (Table 1, equation 11) makes the commonly used approximation of zero horizontal strain (no lateral expansion) which, together with the material isotropy, means that the local horizontal stresses are approximated as the same in all directions.

Fracture pressure (Table 1, equation 14) represents the pressure in the borehole that is needed to cause fracturing of the formation. Assuming zero tensile strength, fracture pressure is given by the minimum horizontal stress.

In an attempt to put bounds on the real variation in horizontal stress, the approach of using stress polygons, introduced by Zoback *et al.* (1986) and Moos & Zoback (1990), was implemented. Stress polygons show the permissible ranges of horizontal stresses at a given depth for a given pore fluid pressure for each of the three Andersonian fault regimes (Fig. 4). The 1upper and lower bounds of the maximum and minimum horizontal stresses on the stress polygons are constrained by the following relationships derived from the Coulomb failure criterion assuming that one of the principal stresses is vertical (Zoback 2007):

Normal fault:

$$\frac{\sigma_{\text{v}} - P_{\text{p}}}{\sigma_{\text{h}} - P_{\text{p}}} \leq [(\mu^2 + 1)^{1/2} + \mu]^2 \qquad (18)$$

Strike-slip fault:

$$\frac{\sigma_H - P_p}{\sigma_h - P_p} \leq [(\mu^2 + 1)^{1/2} + \mu]^2 \qquad (19)$$

Reverse fault:

$$\frac{\sigma_{\text{H}} - P_{\text{p}}}{\sigma_{\text{v}} - P_{\text{p}}} \leq [(\mu^2 + 1)^{1/2} + \mu]^2 \qquad (20)$$

where σ_{v} is the vertical principal stress, σ_{H} is the maximum horizontal principal stress, σ_{h} is the minimum horizontal principal stress, P_{p} is the pore fluid pressure and μ is the coefficient of friction. The diagonal line ($\sigma_{\text{H}} = \sigma_{\text{h}}$) in the diagram is intersected by vertical and horizontal lines that constrain

Table 1. *Analytical and empirical correlations used in this study*

Equation number	Property		Units	Equation	Note	Reference
Elastic parameters						
1	Shear wave velocity	V_s	km s^{-1}	$0.8621V_p - 1.1724$	Mudrock line for clastics	Castagna *et al.* (1985, equation 1)
2	Bulk density	ρ_b	kg m^{-3}	aV_p^m	Amended Gardner's equation (Gardner *et al.* 1974) for shales, where $a = 516.2$ and $m = 0.1869$	Quijada & Stewart (2007, table 2)
3*	Porosity	ϕ		$(\rho_{matrix} - \rho_b)/(\rho_{matrix} - \rho_{fluid})$	$\rho_{matrix} = 2600$ kg/m^3 and $\rho_{fluid} = 1000$ kg/m^3. Explanation follows Table 1	Avseth *et al.* (2010, p. 57, equation 2.10)
4	Theoretical porosity	ϕ_t		$\phi_0 e^{-\beta\sigma_v}$	ϕ_0 is the pre-compaction porosity; $\beta = 0.0421$, $\phi_0 = 0.4$	Rubey & Hubbert (1959, equation 16) Buryakovsky *et al.* (2001, p. 353, 374)
5*	Shear modulus	G	GPa	$\rho_b V_s^2$		Mavko *et al.* (2009, p. 81)
6*	Lamé's constant	λ	GPa	$\rho_b V_p^2 - 2G$		Mavko *et al.* (2009, p. 81)
7*	Poisson's ratio	ν		$(V_p^2 - 2V_s^2)/[2(V_p^2 - V_s^2)]$		Mavko *et al.* (2009, p. 81)
8*	Young's modulus	E	GPa	$G(3V_p^2 - 4V_s^2)/(V_p^2 - V_s^2)$		Mavko *et al.* (2009, p. 82)
9*	Bulk modulus	K	GPa	$\rho_b(3V_p^2 - 4V_s^2)/3$		Mavko *et al.* (2009, p. 82)
***In situ* stress and pressure**						
10*	Overburden stress	σ_v	MPa	$\rho_w g z_w + \rho_b g(z - z_w)$	ρ_w and z_w are the density and depth of the water, respectively	Zoback (2007, p. 8, equation 1.6)
11	Horizontal stress	σ_h	MPa	$\sigma_v \nu/(1 - \nu)$	In these calculations, we approximate σ_H to σ_h, where σ_H and σ_h are the maximum and minimum horizontal stresses, respectively. Explanation of their permissible magnitudes is given in the text	Iverson (1995, equation 4)
12*	Hydrostatic pressure	P_h	MPa	$\rho_w g z$		Zoback (2007, p. 28, equation 2.1)
13	Pore fluid pressure	P_p	MPa	$\sigma_v - (1/\beta)\ln(\phi_0/\phi)$	Pressure existing in the pores of the formation; derived from equation (4)	Rubey & Hubbert (1959, equation 16)
14	Fracture pressure	P_f	MPa	$P_p + (\sigma_v - P_p)(\sigma_h/\sigma_v)$	Explanation is given in the text	Matthews & Kelly (1967, p. 99)
Strength properties						
15	Friction angle	ϕ	°	$\sin^{-1}((V_p - 1)/(V_p + 1))$	For shales	Lal (1999, equation 17)
16	Cohesive strength	τ_0	MPa	$5(V_p - 1)/\sqrt{V_P}$	For shales	Lal (1999, equation 17)
17	Uniaxial compressive strength	C	MPa	$1.35V_p^{2.6}$	For shales, worldwide	Chang *et al.* (2006, equation 14)

V_p is in km s^{-1}.
* Relationship is analytical.

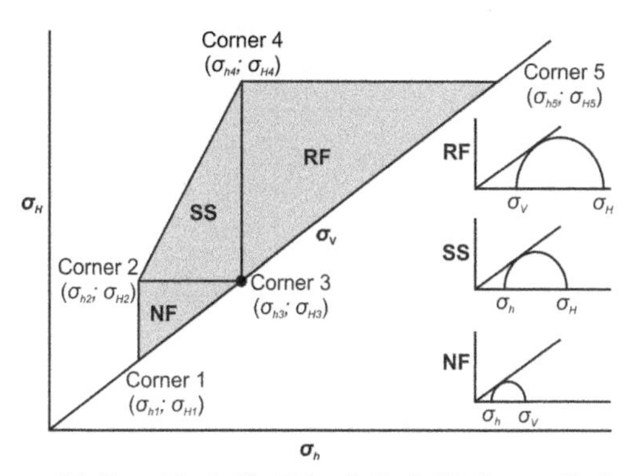

Fig. 4. Stress polygons defining the upper and lower bounds of the principal horizontal stresses in different fault regimes (modified from Zoback 2007).

the stress ranges for the different fault regimes. Stress polygons are always above the diagonal line because $\sigma_H \geq \sigma_h$.

In regions of excess pore pressure (overpressure), the differences between the magnitudes of the principal stresses are small and therefore small stress perturbations can lead to a change from one fault regime to another (Zoback 2007).

Feasibility calculations

To establish that the equations listed in Table 1 return realistic values of material properties and stresses within a SCB context, we evaluated these properties and stresses on an SCB mud volcano for which a structural model exists in the public domain. This is located within the Kurdashi-Araz-Deniz (KAD) anticlinal structure on the western margin of the SCB at a water depth of 30–770 m. The calculations were performed for depths of 500 and 1500 m below sea-level. These were selected from a seismic section across the mud volcano (Soto *et al.* 2011) to represent points on the structural crest and flank of the mud volcano, respectively (Fig. 5).

Hamilton (1979) established a generalized relationship between acoustic wave velocities and depth in marine sediments. This relationship was used to obtain the P- and S-wave values for the crest and flank locations (Fig. 6).

The SCB is characterized by abnormally high formation pressures and several previous studies have attempted to characterize shale compaction within the basin. The porosity–depth curve compiled by Bredehoeft *et al.* (1988) (Fig. 7) was used to obtain porosity values for the calculations.

The input parameters for the calculations are listed in Table 2. The material properties and stresses listed in Table 3 for the crest and flank of the mud volcano were obtained using these input parameters and the equations listed in Table 1. These are compared in Table 3 with typical ranges of these values for the material properties of clay minerals and poorly consolidated sandstones and mudstones, and with the stress states previously reported in the SCB. Our calculated values are consistent with those reported previously and so we have confidence that the empirical correlations detailed in Table 1 are not significantly affected by local factors specific to the SCB.

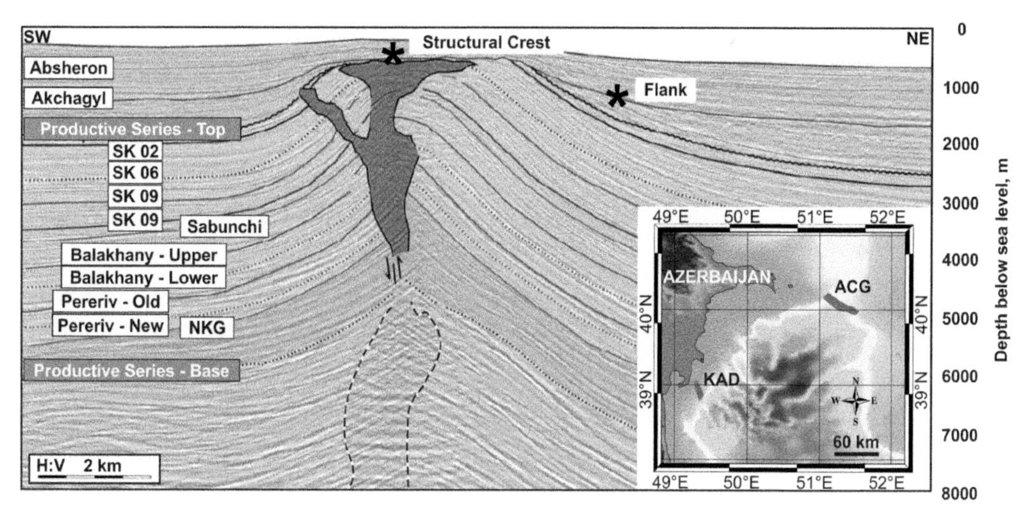

Fig. 5. Vertical seismic section of a mud volcano from Kurdashi-Araz-Deniz (KAD) structure in the offshore western SCB (modified from Soto *et al.* 2011). ACG, Azeri-Chirag-Guneshly structure.

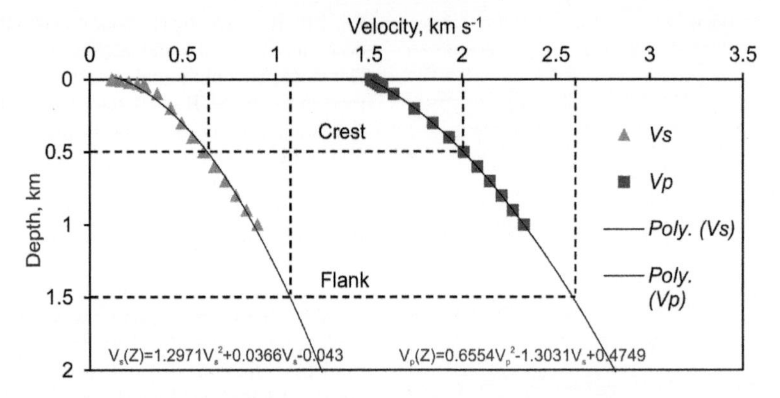

Fig. 6. Generic P- and S-wave velocity v. depth curves in marine sediments (modified from Hamilton 1979).

Two-dimensional model

Input parameters and procedure

The mechanical properties and stresses on a two-dimensional (2D) vertical section across an ACG mud volcano were modelled by digitizing the P-wave velocities presented on a full waveform inversion image published by Selwood *et al.* (2013). The seismic line was 10 km long by 5 km deep in an unknown orientation across one of the mud volcanoes in the Azeri part of the ACG field.

The digitization process involved:

(1) importing the image into MATLAB;
(2) reading red (R), green (G) and blue (B) values and replacing these RGB triplets with a single value per pixel;
(3) replacing each pixel value with the corresponding velocity obtained from the colour bar key to the image;
(4) generating the 2D synthetic seismic line and writing it as a SEG-Y file;
(5) importing the SEG-Y file into PETREL for calculations and visualization.

The resulting P-wave velocity section is shown in Figure 8. Values of density, porosity and mechanical properties (elastic properties and strength), together with the magnitudes of the principal stresses, pore fluid and fracture pressures were calculated from the P-wave velocities within the PETREL software package using the equations listed in Table 1; they are presented here as sections showing the 2D variation of these values. In addition, a vertical pseudo-well (RM-1) located on the structural crest was incorporated into the 2D model to assess the modelled parameters in one dimension along the well trajectory. The calculations were performed for an average water depth of 120 m, which is the average water depth in the Azeri field given by the bathymetry data of Hill *et al.* (2015).

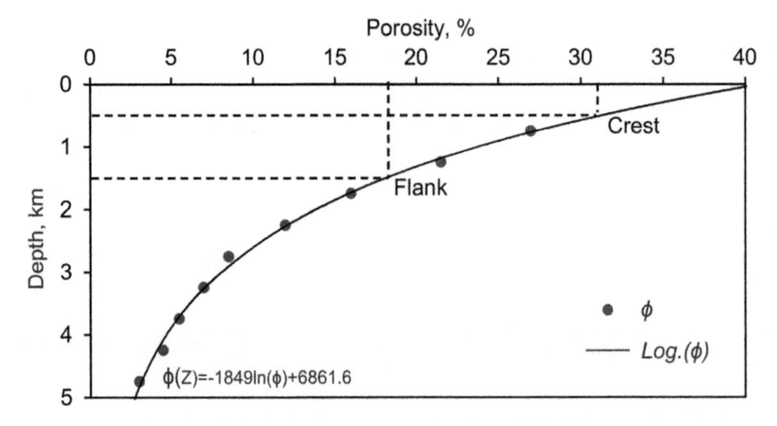

Fig. 7. Shale compaction curve in NW SCB (modified from Bredehoeft *et al.* 1988).

Table 2. *Feasibility model input parameters*

Structural position	Depth (m)	P-wave velocity (km s^{-1})	S-wave velocity (km s^{-1})	Porosity (%)
Crest	500	2.015	0.638	30
Flank	1500	2.591	1.076	17

Results

As the physical properties were calculated solely from P-wave velocity information, the spatial variation of these properties matches that of the P-wave velocity data (Fig. 8). The empirical correlations listed in Table 1 do, however, provide the magnitudes of the material properties and how these magnitudes vary across the mud volcano in the study area. The variation in elastic properties along the pseudo-well is illustrated in Figure 9. The values of these elastic properties at a depth of 500 m below sea-level are similar to those obtained at this depth on the structural crest of the KAD mud volcano.

Bulk density values were estimated using the method of Quijada & Stewart (2007) (Table 1, equation 2). They suggested that, in their equation, different values of the constants a and m are applicable for sands ($a = 224.9$ and $m = 0.2847$) and shales ($a = 516.2$ and $m = 0.1896$). In this study, the lithology was assumed to be an aggregate of clay minerals and hence the coefficients for shales were used. Figure 10 shows the variation in bulk density across the 2D section and along the pseudo-well. The values of bulk density are consistent with the bulk density values calculated at the corresponding positions on the KAD mud volcano. Relatively small bulk densities persist to greater depths in the vicinity of the mud volcano feeder system, presumably because the lithologies are in a brecciated and/or fluidized state.

Theoretical and inferred porosity values (ϕ_t and ϕ, respectively) were computed along the crestal pseudo-well RM-1 using Table 1, equations (3) and (4) (Fig. 11). The theoretical porosity curve assumes a normal compaction trend. The pressure transition zone, defined as the depth interval between when the inferred porosity curve starts to deviate from the theoretical porosity profile and when the rate of decrease of inferred porosity with depth significantly decreases (Swarbrick & Osborne 1996), lies between 620 m below sea-level (mbsl) and 2600 mbsl at the crestal pseudo-well RM-1.

Values of pore fluid pressure and fracture pressure were calculated along the pseudo-well RM-1 (Fig. 12). Estimated pore fluid pressures over the depth range 2–5 km are about 1.4–1.8 times hydrostatic pressure, in agreement with previous estimates of shale pore fluid pressure within the SCB (Bredehoeft *et al.* 1988; Javanshir *et al.* 2015). On the 2D section relatively small pore fluid pressures are seen to persist to a depth of 620 mbsl in areas close to the mud volcano (Fig. 13) and these perhaps

Table 3. *Modelled values of elastic properties, state of stress and rock strength on the crest and flank of a mud volcano from the KAD structure*

Estimated parameters		Structural position		Previously published values	
Name	Units	Crest	Flank	Range	Reference
Elastic properties					
Bulk density	kg m^{-3}	2140	2243	1580–2600	Mavko *et al.* (2009, p. 458, 459)
Shear modulus	GPa	0.87	2.60	0.2–5.5	Horsrud (2001, p. 71)
Lamé's constant	GPa	8.69	15.06	0.07–13.26	Islam & Skalle (2013, p. 1400)
Poisson's ratio		0.44	0.40	0.35–0.50	Schön (2011, p 162)
Young's modulus	GPa	2.52	7.25	3.2–9.5	Prasad (2002, p. 3)
Bulk modulus	GPa	7.53	11.59	6–12	Vanorio *et al.* (2003, p. 325)
Stress and pressures					
Vertical stress	MPa	13	35	10–35	Buryakovsky *et al.* (2001, p. 402)
Horizontal stress	MPa	8	22		
Hydrostatic pressure	MPa	5	15	4–14	Buryakovsky *et al.* (2001, p. 402)
Pore fluid pressure	MPa	6	16	12–20	Buryakovsky *et al.* (2001, p. 152)
Fracture pressure	MPa	11.57	28.41		
Rock strength					
Friction angle	°	19.67	26.30	21.75–38.09	Kohli & Zoback (2013, p. 5115)
Cohesive strength	MPa	3.58	4.94	0.3–38.4	Schön (2011, p. 256)
Uniaxial compressive strength	MPa	8.35	16.05	7.5–13.9	Schön (2011, p. 258)

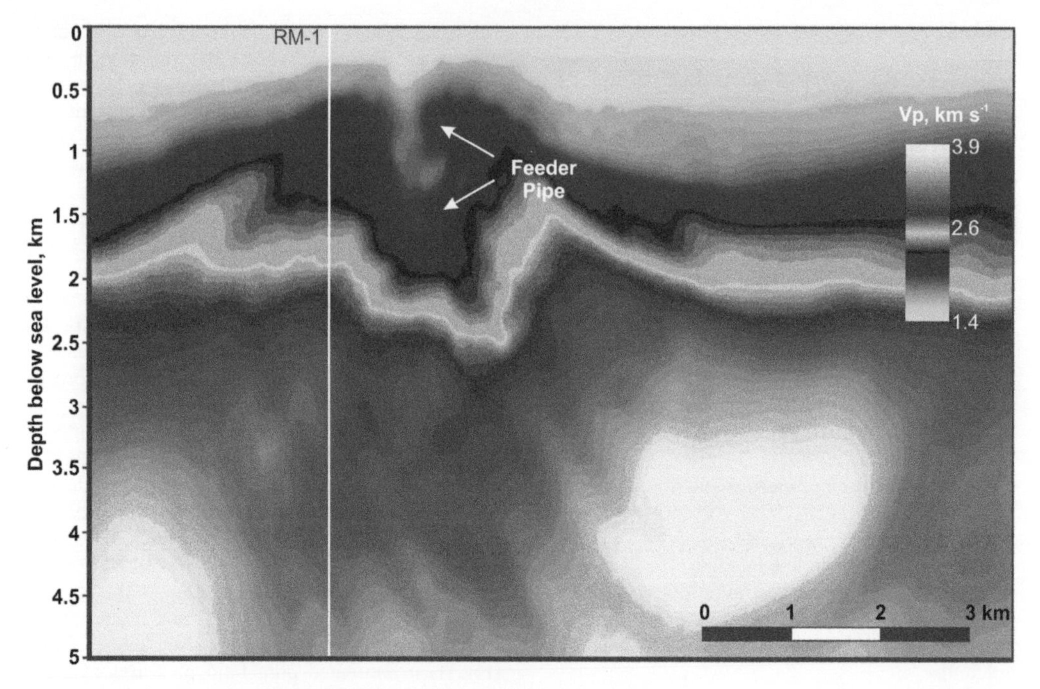

Fig. 8. P-wave velocity data generated from the full waveform inversion image published by Selwood *et al.* (2013). The feeder pipe of the investigated mud volcano is highlighted.

represent areas that have not yet fully recharged following recent eruptions.

The friction angle increases with depth, but with anomalously small values in the volcano vent area (Fig. 14), perhaps reflecting the relatively unconsolidated state of the sediments in this area. At depths greater than *c.* 2500 m, the friction angle values are in good agreement with the frictional properties given by Byerlee's law (Byerlee 1978; Schön 2011).

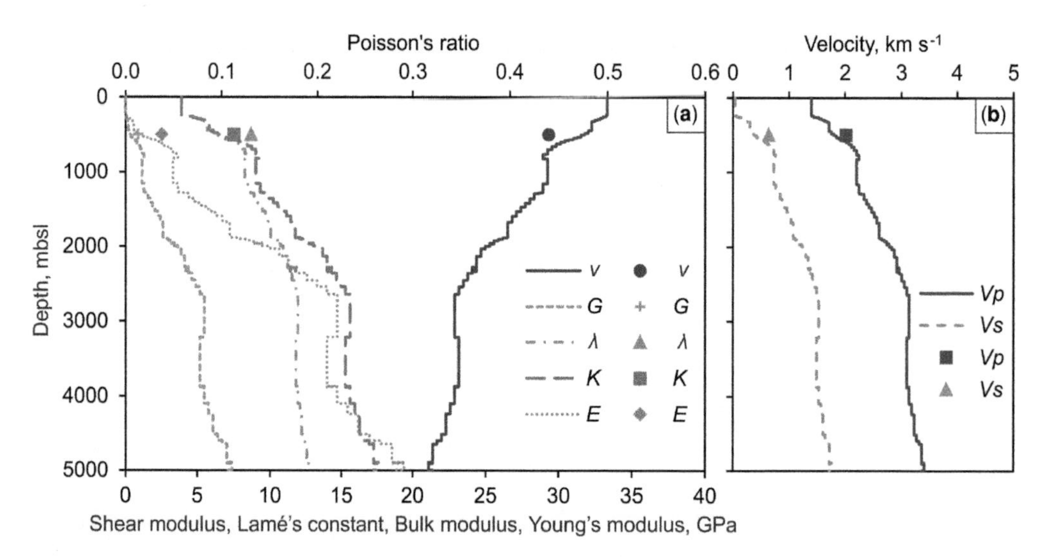

Fig. 9. (**a**) Profiles of elastic rock properties and (**b**) acoustic wave velocities along the RM-1 pseudo-well. Markers indicate the magnitudes of these properties obtained on the structural crest of the Kurdashi-Araz-Deniz mud volcano analysed in the feasibility modelling.

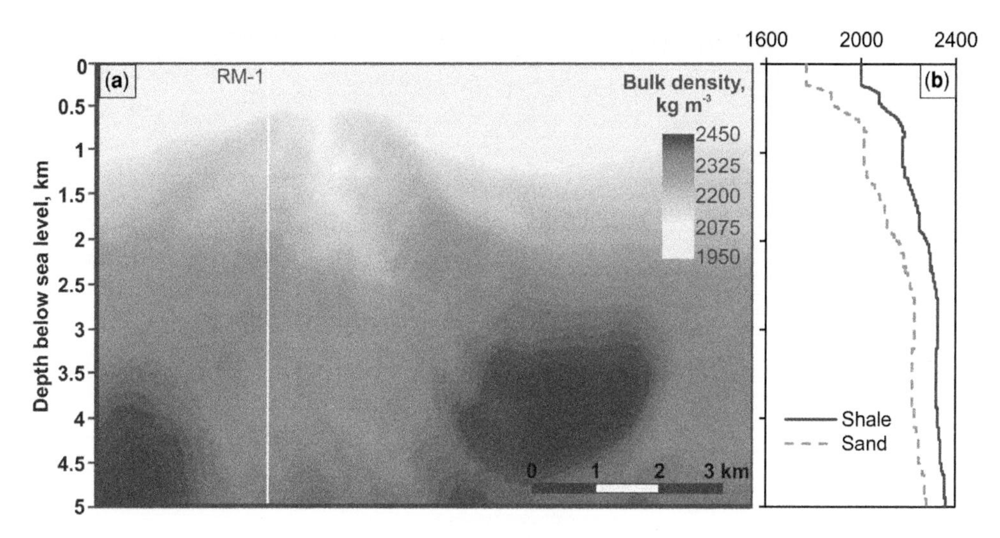

Fig. 10. (**a**) Vertical cross-section across the Azeri-Chirag-Guneshly structure showing the variation in bulk density obtained using the method of Quijada & Stewart (2007) with their parameters for shales. (**b**) The variation of bulk density with depth along the pseudo-well RM-1. The depth variation using the parameters of Quijada & Stewart (2007) for sands is also shown for comparison

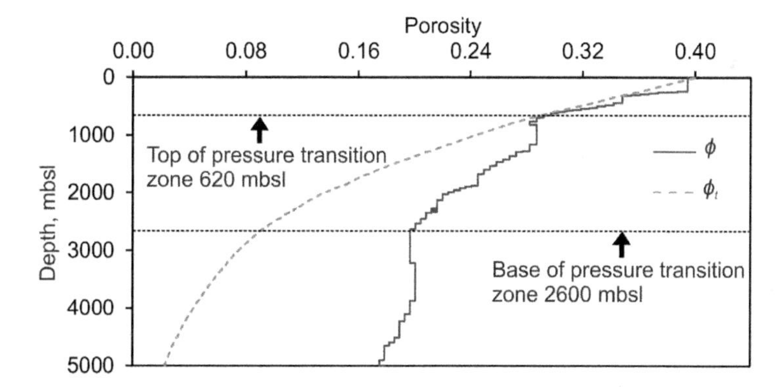

Fig. 11. Variation in inferred porosity and theoretical porosity along pseudo-well RM-1 showing the top and base of overpressure at 620 mbsl and 2600 mbsl, respectively.

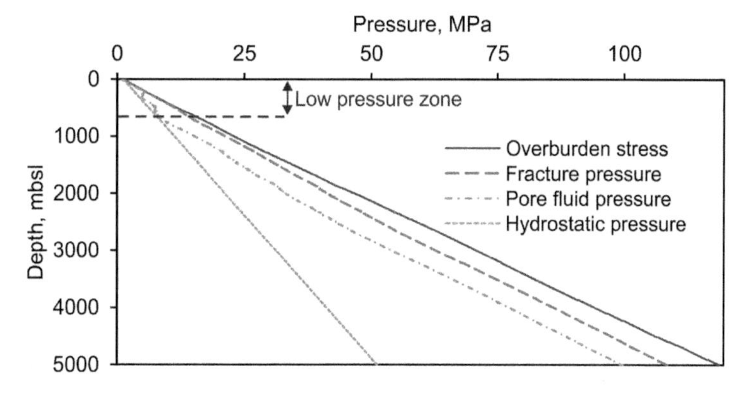

Fig. 12. Variation in overburden stress and pressure along pseudo-well RM-1.

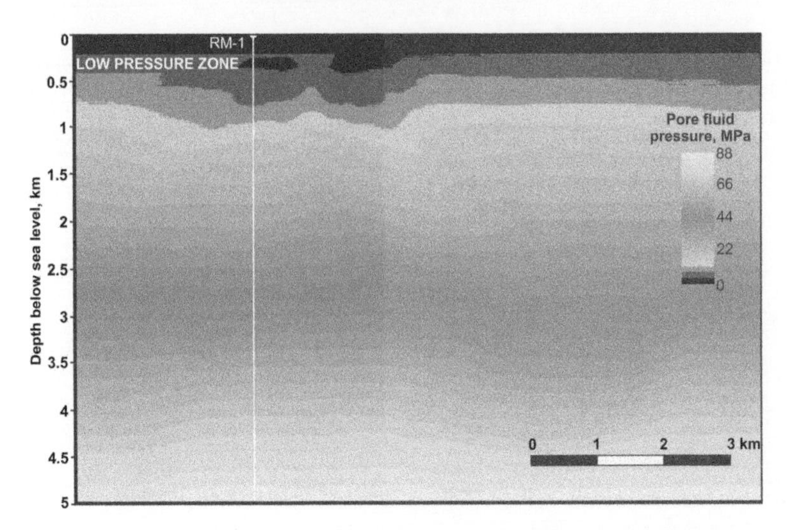

Fig. 13. Vertical cross-section across the Azeri-Chirag-Guneshly structure showing the variation in pore fluid pressure. This highlights the relatively small pore fluid pressures in the shallow unconsolidated sediments and in the vicinity of the mud volcano.

Discussion

Fluid flow

The spatial variation of fluid overpressure provides information about the direction of fluid flow near the mud volcano. Fluid overpressure is plotted in Figure 15 as the overpressure abnormality factor, which is defined as the ratio of pore fluid pressure to hydrostatic pressure. The study area has an abnormality factor of *c.* 1.2 in the first 620 mbsl, *c.* 1.5 in the depth range 620–2600 mbsl and *c.* 1.8 from 2600 to 5000 mbsl. As well as decreasing upwards, fluid overpressure decreases from the flanks towards the structural crest of the mud volcano, implying that a component of the regional fluid flow is

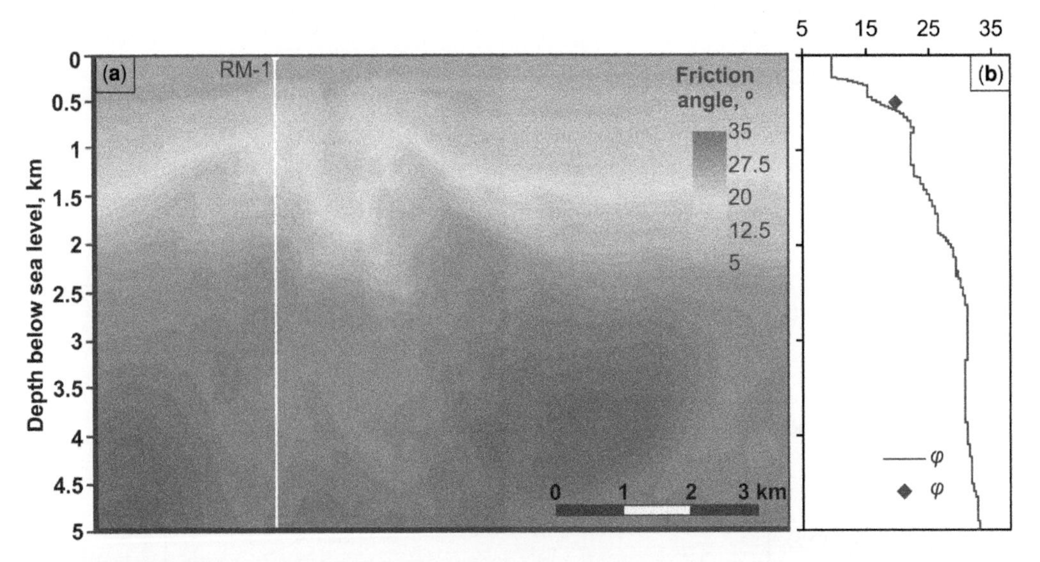

Fig. 14. (**a**) Vertical cross-section across the Azeri-Chirag-Guneshly structure showing the variation of friction angle and (**b**) the variation with depth of the friction angle along pseudo-well RM-1. The marker on the pseudo-well curve indicates the friction angle obtained on the structural crest of the Kurdashi-Araz-Deniz mud volcano analysed in the feasibility modelling.

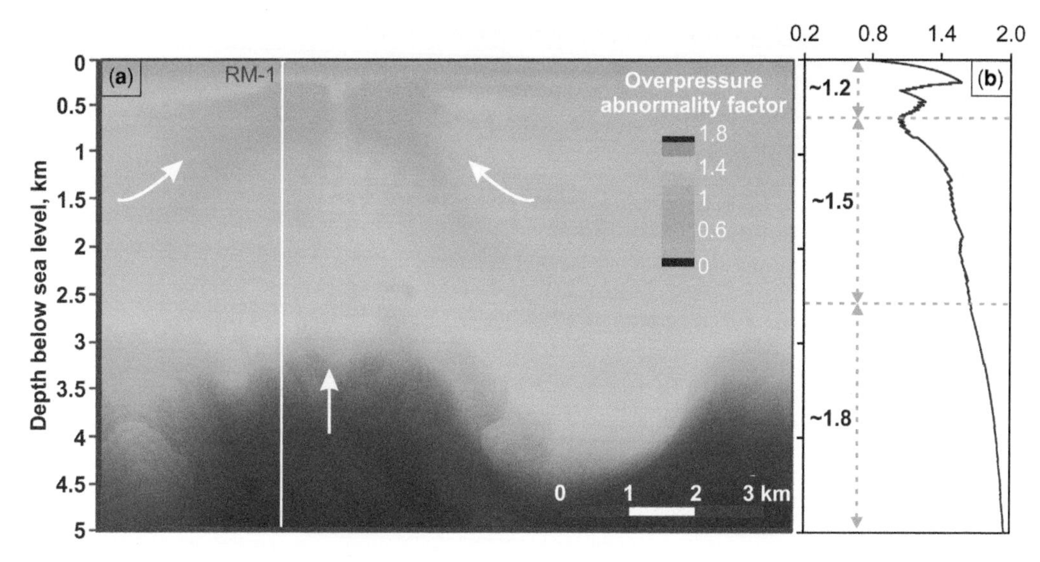

Fig. 15. (**a**) Vertical cross-section across the Azeri-Chirag-Guneshly structure and (**b**) along pseudo-well RM-1 showing the overpressure abnormality factor. Arrows indicate the inferred direction of fluid flow.

being directed laterally from the flanks to the crest. These observations support the suggestion that the perceived drive for the mud volcanoes in the offshore SCB involves lateral as well as upwards pressure transfer. They also point to the possibility of using P-wave data, particularly if supported with direct fluid pressure measurements, to assess fluid flow pathways within the stratigraphy.

Contemporary stress regime

The orientation of the present day stress field is commonly assessed using earthquake focal mechanism solutions and borehole stress orientation measurements. However, it has also been noted that when $\sigma_H \neq \sigma_h$, mud volcano calderas have a tendency to be elliptical, with the long axis oriented parallel to σ_h (Bonini 2012). While analysing the bathymetry image from the Azeri side of the ACG field (Hill *et al.* 2015), we observed that both the mud volcano calderas present in the region of interest are elliptical (Fig. 16a). In each case, the long axis of the caldera is oriented NW–SE, parallel to the orientation of the Absheron–Balkan uplift zone, while the short axis is oriented NE–SW. This implies that σ_h is oriented NW–SE and σ_H is oriented NE–SW. This is consistent with focal mechanism studies performed over the basin (Ritz *et al.* 2006; Jackson *et al.* 2002), with borehole breakout data in the World Stress Map database (Heidbach *et al.* 2008) and with the direction of maximum regional compressive stress inferred from the NE–SW-directed subduction of the South Caspian basement beneath the Absheron–Balkan uplift (Fig. 16b).

An analysis using stress polygons provides further constraints on the stress state. These have been calculated at three different depths along the pseudo-well RM-1 using equations (18–20). So that the three stress polygons can be compared on a single plot, following Zoback (2007) the stresses obtained using equations (18–20) have been normalized by the depth at which each was obtained and so are presented as MPa m^{-1}. The input values of vertical stress, pore fluid pressure and the coefficient of friction are those at the given depth in the pseudo-well RM-1, while the minimum and maximum horizontal stress values have been obtained by manipulating equations (18) and (20), respectively. The values obtained are listed in Table 4 and the resulting stress polygons are shown in Figure 17. We find that the stress polygons shrink with increasing depth as the overpressure increases. This finding is consistent with the notion that the principal stresses tend to become closer to the vertical stress in magnitude with increasing depth in overpressured areas and hence that relatively small changes in the stress field can lead to a shift from one Andersonian fault regime to another (Zoback 2007).

Implications for drillability

Pore fluid pressure and fracture pressure, together with their corresponding depth gradients, are central considerations when establishing safe drilling strategies. Although a knowledge of the actual magnitudes of these pressures is important for drilling activities, a knowledge of their gradients is more practical because the required drilling mud weight

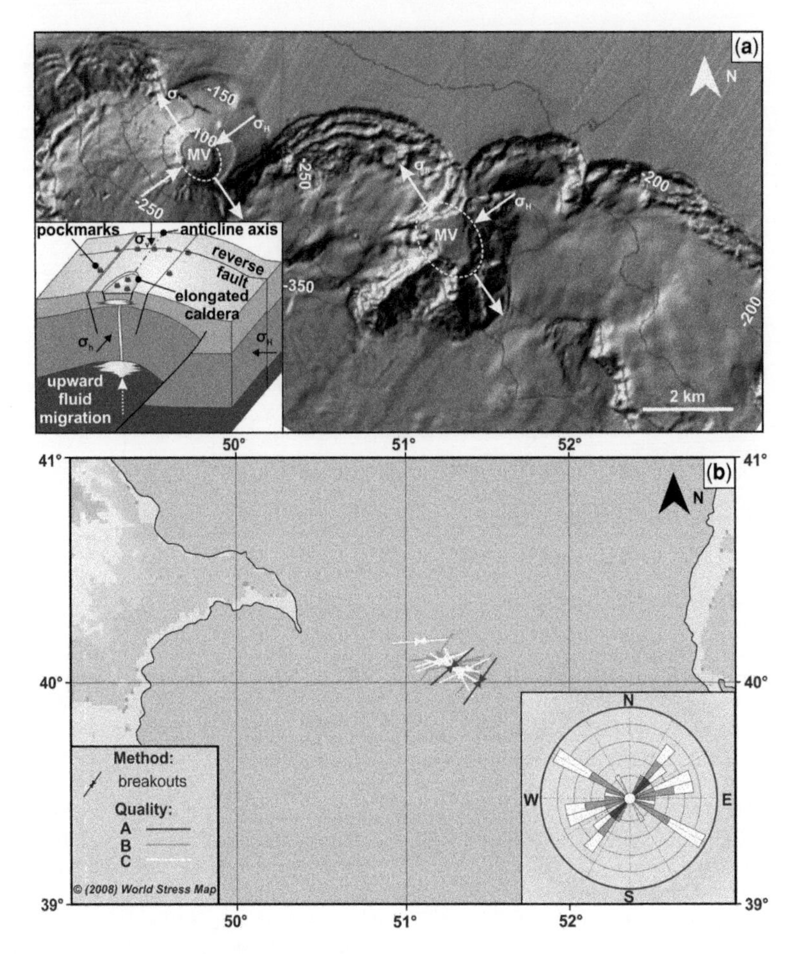

Fig. 16. Regional horizontal stress states in the Azeri-Chirag-Guneshly (ACG) field. (**a**) Elliptical mud volcano (MV) calderas drawn on the ACG bathymetry image of Hill *et al.* (2015). The inset figure is a conceptual diagram of stress states around an MV located in the structural crest of a larger scale antiform (modified from Bonini 2012). On the structural crest, outer arc extension means that the vertical stress is locally probably the greatest principal stress. (**b**) World Stress Map displaying borehole breakouts from the ACG overlain by a rose diagram of borehole breakout directions; the data are shaded according to confidence in their quality (with A being the highest quality).

is estimated in pressure gradients. The pore pressure gradient characterizes the minimum (or the lower bound) mud weight and the fracture gradient indicates the maximum (or the upper bound) mud weight (Eaton 1969). Identifying upper and lower bounds on the fracture gradient itself is generally good practice when using estimates of fracture gradient. The lower bound is defined as the fracture closure pressure, which is best measured by a leak-off test, whereas the upper bound indicates a

Table 4. *Input parameters and calculated minimum and maximum values of σ_h and σ_H used to construct the stress polygons shown in Figure 19*

Depth (m)	σ_v (MPa m^{-1})	P_p (MPa m^{-1})	μ	σ_H (MPa m^{-1})	σ_h (MPa m^{-1})
620	0.0235	0.0123	0.3718	0.0356	0.0177
2600	0.0236	0.0174	0.5957	0.0369	0.0194
5000	0.0238	0.0200	0.6445	0.0328	0.0211

Stress and pressure values are normalized by the corresponding depth.

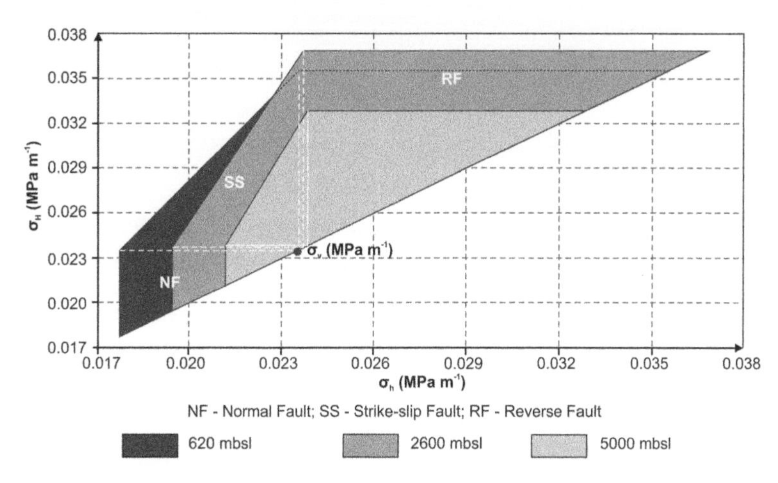

Fig. 17. Stress polygons at three depths in pseudo-well RM-1 showing the decreasing permissible ranges of horizontal stresses with increasing depth.

point at which there is mud loss from the borehole to induced fractures (Zhang 2011). Estimating these bounds requires a knowledge of the magnitudes of the horizontal stresses, the tensile strength and the thermal stress induced by the difference between the mud and formation temperatures. As we do not have these parameters, we have used a method reported by Matthews & Kelly (1967) (Table 1, equation 14) to determine the fracture pressure and its gradient. This method provides a value similar to the lower bound on the fracture gradient. Figure 18 shows the pressures and gradients estimated for the crestal pseudo-well RM-1. The large fluctuations in the pore fluid pressure gradient at shallow depths (>500 mbsl) are probably artefacts arising from the resolution of the P-wave velocity and how this affects the calculated porosity used to estimate the pore fluid pressure (Table 1, equation 13). However, the changes in the slope of the depth variation of pore fluid and fracture gradient that occurs at 620 and 2600 mbsl correlate with the top and base of the pressure transition zone identified in Figure 11.

We have attempted to define the safe drilling window (where the drilling window is defined as the difference between the fracture gradient and the pore fluid pressure gradient) using our results (Fig. 19). We observe that above a depth of *c.* 300 mbsl, the drilling window gradients are as small as *c.* 0.003 MPa m^{-1} on one flank of the mud volcano, whereas on the other flank the gradients are larger (up to 0.012 MPa m^{-1}). The model identifies some areas with large drilling window gradients that are close to the mud volcano feeder pipe. These may represent zones of fluid recharging and so may be transient features. Fluid venting pipes are known to extend down to around 2 km beneath the seabed in the ACG (Javanshir *et al.* 2015),

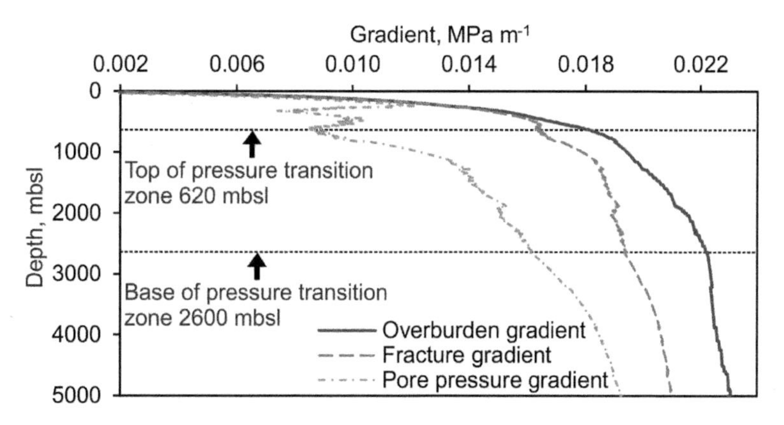

Fig. 18. Variation of pressure gradients along pseudo-well RM-1.

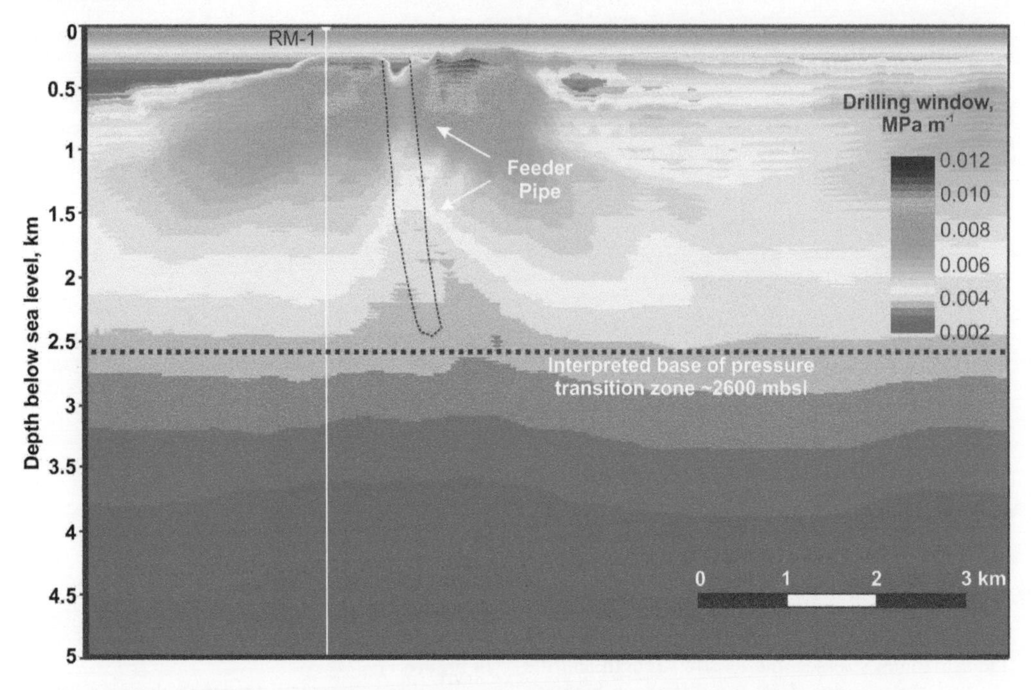

Fig. 19. Vertical cross-section across the Azeri-Chirag-Guneshly structure showing the width of the drilling window. The safest areas to drill are those with the widest drilling window.

which almost marks the base of the pressure transition zone (2600 mbsl) in this study. The areas below the pressure transition zone are characterized by drilling window gradients of 0.004 MPa m^{-1}, which decrease to 0.002 MPa m^{-1} with increasing depth. We interpret these values as estimates of the drilling window for deep overpressured sections where well-consolidated sediments reside.

Limitations of this study

Key sources of data for full geomechanical modelling include seismic and borehole data, while geological and drilling data are used for calibration purposes. Geological and seismic data provide regional-scale information for the entire section (overburden, underburden and zone of interest), whereas drilling and borehole data aid in focusing on a zone of interest with greater accuracy and higher resolution.

The analysis in this study is built almost entirely on P-wave velocity and is therefore sensitive to how tightly constrained the empirical correlations between P-wave velocity and the various mechanical properties and stresses are. A key limitation imposed by the nature of the data is the lack of opportunity to incorporate mechanical anisotropy. Given that the most significant causes of mechanical anisotropy are (1) oriented fractures, (2) the textural alignment of highly anisotropic minerals and (3) compositional banding, we can expect that the mechanical properties of a fractured, well-bedded sequence rich in clay minerals will be anisotropic. Hence considerable confidence could be added to the findings presented here if data that allowed mechanical anisotropy to be quantified (e.g. amplitude v. offset, vertical seismic profile, multi-, wide-, rich- and full-azimuth seismic data) were available. Nevertheless, even with the limited data available, these findings are consistent with the geodynamic context of this part of the SCB.

Conclusions

A 2D P-wave velocity dataset was used with empirical correlations between P-wave velocity and various mechanical properties to build a geomechanical model of the area around a mud volcano in the SCB. The key findings are:

- realistic values of elastic and brittle strength properties together with fluid pressures can be obtained using the empirical correlations;
- sections showing the spatial variation of pore fluid overpressure around the mud volcanoes calculated from P-wave velocity data have considerable potential for constraining models of fluid flow around these structures;

- preliminary estimates based on seismic velocities provide useful reconnaissance indications of regions that are safe to drill, regions that are risky and regions that should be avoided.

Taken together, these findings help to reinforce the observation that a considerable body of geomechanical information can be recovered from even very limited seismic datasets and that this can be useful both for defining targets for more comprehensive geomechanical studies and for providing guidance on drilling strategies.

This work was funded by the Ministry of Education of the Republic of Azerbaijan. The authors thank Peter Cook from BP Exploration (Caspian Sea) Limited for discussions that helped shape the preliminary concepts of the feasibility study. We gratefully acknowledge the comments of Ravan Gulmammadov on an earlier version of the manuscript and Andrew M.W. Newton for his assistance with extracting and processing the bathymetric map of the study area. We also acknowledge the use of SeisLab, which is a set of MATLAB codes for reading and writing standard SEG-Y files that has been placed in the public domain by Eike Rietsch. David Iacopini and an anonymous reviewer are thanked for their constructive feedback. We thank Schlumberger for the provision of PETREL.

Correction notice: The original version was incorrect. All instances of mbsf (metres below seafloor) should have read mbsl (metres below sea-level).

References

Abrams, M.A. & Narimanov, A.A. 1997. Geochemical evaluation of hydrocarbons and their potential sources in the western South Caspian depression, Republic of Azerbaijan. *Marine and Petroleum Geology Petroleum Geology*, **14**, 451–468, https://doi.org/10.1016/S0264-8172(97)00011-1

Allen, P.A. & Allen, J.R. 2013. *Basin Analysis: Principles and Application to Petroleum Play Assessment.* 3rd edn. Wiley-Blackwell, Chichester.

Avseth, P., Mukerji, T. & Mavko, G. 2010. *Quantitative Seismic Interpretation: Applying Rock Physics to Reduce Interpretation Risk.* 1st edn. Cambridge University Press, Cambridge.

Azzaro, E., Bellanca, A. & Neri, R. 1993. Mineralogy and geochemistry of Mesozoic black shales and interbedded carbonates, southeastern Sicily: evaluation of diagenetic processes. *Geological Magazine*, **130**, 191–202, https://doi.org/10.1017/S0016756800009857

Bonini, M. 2012. Mud volcanoes: indicators of stress orientation and tectonic controls. *Earth-Science Reviews*, **115**, 121–152, https://doi.org/10.1016/j.earscirev.2012.09.002

Bonini, M. 2013. Fluid seepage variability across the external Northern Apennines (Italy): structural controls with seismotectonic and geodynamic implications. *Tectonophysics*, **590**, 151–174, https://doi.org/10.1016/j.tecto.2013.01.020

Bredehoeft, J.D., Djevanshir, R.D. & Belitz, K.R. 1988. Lateral fluid flow in a compacting sand-shale sequence: South Caspian Basin. *American Association of Petroleum Geologists Bulletin*, **72**, 416–424, https://doi.org/10.1306/703C9A1E-1707-11D7-8645000102C1865D

Bristow, C.R., Gale, I.N., Fellman, E., Cox, B.M., Wilkinson, I.P. & Riding, J.B. 2000. The lithostratigraphy, biostratigraphy and hydrogeological significance of the mud springs at Templars Firs, Wootton Bassett, Wiltshire. *Proceedings of the Geologists' Association*, **111**, 231–245, https://doi.org/10.1016/S0016-7878(00)80016-4

Buryakovsky, L.A., Djevanshir, R.D. & Chilingar, G.V. 1995. Abnormally-high formation pressures in Azerbaijan & the South Caspian Basin (as related to smectite ↔ illite transformations during diagenesis & catagenesis). *Journal of Petroleum Science and Engineering*, **13**, 203–218, https://doi.org/10.1016/0920-4105(95)00008-6

Buryakovsky, L.A., Chilingar, G.V. & Aminzadeh, F. 2001. *Petroleum Geology of the South Caspian Basin.* 1st edn. Elsevier, USA, https://doi.org/10.1016/B978-088415342-9/50000-X

Byerlee, J. 1978. Friction of rocks. *Pure and Applied Geophysics*, **116**, 615–626, https://doi.org/10.1007/BF00876528

Calvès, G., Huuse, M., Schwab, A. & Clift, P. 2008. Three-dimensional seismic analysis of high-amplitude anomalies in the shallow subsurface of the northern Indus Fan: sedimentary and/or fluid origin. *Journal of Geophysical Research*, **113**, 1–16, https://doi.org/10.1029/2008JB005666

Carragher, P.D., Ross, A., Roach, E., Trefry, C., Talukder, A. & Stalvies, C. 2013. Natural seepage systems at Biloxi & Dauphin domes and Mars mud volcano, north east Mississippi Canyon protraction area, Gulf of Mexico. Paper OTC-24191-MS, presented at the Offshore Technology Conference, 6–9 May, Houston, TX, USA.

Cartwright, J., Huuse, M. & Aplin, A. 2007. Seal bypass systems. *American Association of Petroleum Geologists Bulletin*, **91**, 1141–1166, https://doi.org/10.1306/04090705181

Castagna, J.P., Batzle, M.L. & Eastwood, R.L. 1985. Relationships between compressional-wave in elastic silicate rocks and shear-wave velocities. *Geophysics*, **50**, 571–581, https://doi.org/10.1190/1.1441933

Chang, C., Zoback, M.D. & Khaksar, A. 2006. Empirical relations between rock strength and physical properties in sedimentary rocks. *Journal of Petroleum Science and Engineering*, **51**, 223–237, https://doi.org/10.1016/j.petrol.2006.01.003

Clarke, R.H. & Cleverly, R.W. 1991. Petroleum seepage and post-accumulation migration. *In*: England, W.A. & Fleet, A.J. (eds) *Petroleum Migration.* Geological Society, London, Special Publications, **59**, 265–271, https://doi.org/10.1144/GSL.SP.1991.059.01.17

Contet, J. & Unterseh, S. 2015. Multiscale site investigation of a giant mud-volcano offshore Azerbaijan – impact on subsea field development. Paper OTC-25864-MS, presented at the Offshore Technology Conference, 4–7 May, Houston, TX, USA, https://doi.org/10.4043/25864-MS

Davies, R.J. & Stewart, S.A. 2005. Emplacement of giant mud volcanoes in the South Caspian Basin: 3D seismic reflection imaging of their root zones. *Journal of the Geological Society, London*, **162**, 1–4, https://doi.org/10.1144/0016-764904-082

Deville, E., Battani, A. *et al.* 2003. The origin and processes of mud volcanism: new insights from Trinidad. *In*: Van Rensbergen, P., Hillis, R.R., Maltman, A.J. & Morley, C.K. (eds) *Subsurface Sediment Mobilization*. Geological Society, London, Special Publications, **216**, 475–490, https://doi.org/10.1144/GSL.SP.2003.216.01.31

Dimitrov, L.I. 2002. Mud volcanoes – the most important pathway for degassing deeply buried sediments. *Earth-Science Reviews*, **59**, 49–76, https://doi.org/10.1016/S0012-8252(02)00069-7

Eaton, B. 1969. Fracture gradient prediction and its application in oilfield operations. *Journal of Petroleum Technology*, **21**, 1353–1360, https://doi.org/10.2118/2163-PA

Evans, R., Davies, R. & Stewart, S. 2006. Mud volcano evolution from 3D seismic interpretation and field mapping in Azerbaijan. Presented at the AAPG/GSTT Hedberg Conference: Mobile Shale Basins – Genesis, Evolution & Hydrocarbon Systems, Abstracts, 4–7 June, Port-of-Spain, Trinidad and Tobago, http://www.searchanddiscovery.com/abstracts/html/2006/hedberg_intl/abstracts/evans.htm

Evans, R.J., Davies, R.J. & Stewart, S.A. 2007. Internal structure and eruptive history of a kilometre-scale mud volcano system, South Caspian Sea. *Basin Research*, **19**, 153–163, https://doi.org/10.1111/j.1365-2117.2007.00315.x

Evans, R.J., Stewart, S.A. & Davies, R.J. 2008. The structure and formation of mud volcano summit calderas. *Journal of the Geological Society, London*, **165**, 769–780, https://doi.org/10.1144/0016-76492007-118

Feseker, T., Brown, K.R. *et al.* 2010. Active mud volcanoes on the upper slope of the western Nile Deep-sea fan – first results from the P362/2 cruise of R/V Poseidon. *Geo-Marine Letters*, **30**, 169–186, https://doi.org/10.1007/s00367-010-0192-0

Feyzullayev, A.A. 2012. Mud volcanoes in the South Caspian Basin: nature and estimated depth of its products. *Natural Science*, **4**, 445–453, https://doi.org/10.4236/ns.2012.47060

Feyzullayev, A.A. & Lerche, I. 2009. Occurrence and nature of overpressure in the sedimentary section of the South Caspian Basin, Azerbaijan. *Energy Exploration & Exploitation*, **27**, 345–366.

Feyzullayev, A.A. & Movsumova, U.A. 2010. The nature of the isotopically heavy carbon of carbon dioxide and bicarbonates in the waters of mud volcanoes in Azerbaijan. *Geochemistry International*, **48**, 517–522, https://doi.org/10.1134/S0016702910050083

Gardner, G.H.F., Gardner, L.W. & Gregory, A.R. 1974. Formation velocity and density – the diagnostic basis of stratigraphic traps. *Geophysics*, **39**, 770–780, https://doi.org/10.1190/1.1440465

Gray, D., Anderson, P., Logel, J., Delbecq, F., Schmidt, D. & Schmid, R. 2012. Estimation of stress and geomechanical properties using 3D seismic data. *First Break*, **30**, 59–68.

Guliyev, I., Aliyeva, E., Huseynov, D., Feyzullayev, A. & Mamedov, P. 2010. Hydrocarbon potential of ultra deep deposits in the South Caspian Basin. Presented at the AAPG European Region Annual Conference, 1–66, http://www.searchanddiscovery.com/documents/2011/10312guliyev/ndx_guliyev.pdf

Guliyev, I.S., Feizulayev, A.A. & Huseynov, D.A. 2001. Isotope geochemistry of oils from fields and mud volcanoes in the South Caspian Basin, Azerbaijan. *Petroleum Geoscience*, **7**, 201–209, https://doi.org/10.1144/petgeo.7.2.201

Hamilton, E.L. 1979. Vp/Vs & Poisson's ratios in marine sediments and rocks. *Acoustical Society of America*, **66**, 1093–1101, https://doi.org/10.1121/1.383344

Heidbach, O., Tingay, M., Barth, A., Reinecker, J., Kurfess, D. & Müller, B. 2008. The World Stress Map database release 2008, https://doi.org/10.1594/GFZ.WSM.Rel2008

Hill, A.W., Hampson, K.M., Hill, A.J., Golightly, C., Wood, G.A., Sweeney, M. & Smith, M.M. 2015. ACG field geohazards management: unwinding the past, securing the future. Paper OTC-25870, presented at the Offshore Technology Conference, 4–7 May, Houston, TX, USA, https://doi.org/10.4043/25870-MS

Hong, W.L., Etiope, G., Yang, T.F. & Chang, P.Y. 2013. Methane flux from miniseepage in mud volcanoes of SW Taiwan: comparison with the data from Italy, Romania and Azerbaijan. *Journal of Asian Earth Sciences*, **65**, 3–12, https://doi.org/10.1016/j.jseaes.2012.02.005

Horsrud, P. 2001. Estimating mechanical properties of shale from empirical correlations. *SPE Drilling & Completion*, **16**, 68–73, https://doi.org/10.2118/56017-PA

Hovland, M., Hill, A. & Stokes, D. 1997. The structure and geomorphology of the Dashgil mud volcano, Azerbaijan. *Geomorphology*, **21**, 1–15, https://doi.org/10.1016/S0169-555X(97)00034-2

Islam, M.A. & Skalle, P. 2013. An experimental investigation of shale mechanical properties through drained and undrained test mechanisms. *Rock Mechanics and Rock Engineering*, **46**, 1391–1413, https://doi.org/10.1007/s00603-013-0377-8

Iverson, W.P. 1995. Closure stress calculations in anisotropic formations. Paper SPE-29598-MS, presented at the Low Permeability Reservoirs Symposium, 19–22 March, Denver, CO, USA, https://doi.org/10.2118/29598-MS

Jackson, J., Priestley, K., Allen, M. & Berberian, M. 2002. Active tectonics of the South Caspian Basin. *Geophysical Journal International*, **148**, 214–245, https://doi.org/10.1046/j.1365-246X.2002. 01005.x

Javanshir, R.J., Riley, G.W., Duppenbecker, S.J. & Abdullayev, N. 2015. Validation of lateral fluid flow in an overpressured sand-shale sequence during development of Azeri-Chirag-Gunashli oil field and Shah Deniz gas field: South Caspian Basin, Azerbaijan. *Marine and Petroleum Geology*, **59**, 593–610, https://doi.org/10.1016/j.marpetgeo.2014.07.019

Jones, R.W. & Simmons, M.D. 1997. A review of the stratigraphy of eastern Paratethys (Oligocene–Holocene), with particular emphasis on the Black Sea. *In*: Robinson, A.G. (ed.) *Regional and Petroleum Geology of the Black Sea and Surrounding Region*. AAPG Memoirs, **68**, 39–52.

JUDD, A. 2005. Gas emissions from mud volcanoes: significance to global climate change. *In*: MARTINELLI, G. & PANAHI, B. (eds) *Mud Volcanoes, Geodynamics & Seismicity*. Springer, Dordrecht, 147–157, https://doi.org/10.1007/1-4020-3204-8_13

KOHLI, A.H. & ZOBACK, M.D. 2013. Frictional properties of shale reservoir rocks. *Journal of Geophysical Research: Solid Earth*, **118**, 5109–5125, https://doi.org/10.1002/jgrb.50346

KOPF, A., STEGMANN, S., DELISLE, G., PANAHI, B., ALIYEV, C.S. & GULIYEV, I. 2009. In-situ cone penetration tests at the active Dashgil mud volcano, Azerbaijan: evidence for excess fluid pressure, updoming and possible future violent eruption. *Marine and Petroleum Geology*, **26**, 1716–1723, https://doi.org/10.1016/j.marpetgeo.2008.11.005

LAL, M. 1999. Shale stability: drilling fluid interaction and shale strength. Paper SPE-54356-MS, presented at the SPE Asia Pacific Oil and Gas Conference and Exhibition, 20–22 April, Jakarta, Indonesia, https://doi.org/10.2118/54356-MS

LERCHE, I. & BAGIROV, E. 1999. *Impact of Natural Hazards on Oil and Gas Extraction – the South Caspian Basin*. 1st edn. Springer, New York.

MATTHEWS, W.R. & KELLY, J. 1967. How to predict formation pressure and fracture gradient. *Oil & Gas Journal*, **20**, 92–106.

MAVKO, G., MUKERJI, T. & DVORKIN, J. 2009. *The Rock Physics Handbook – Tools for Seismic Analysis of Porous Media*. 2nd edn. Cambridge University Press, Cambridge.

MAZZINI, A., SVENSEN, H., PLANKE, S., GULIYEV, I., AKHMANOV, G.G., FALLIK, T. & BANKS, D. 2009. When mud volcanoes sleep: insight from seep geochemistry at the Dashgil mud volcano, Azerbaijan. *Marine and Petroleum Geology*, **26**, 1704–1715, https://doi.org/10.1016/j.marpetgeo.2008.11.003

MILKOV, A.V. 2000. Worldwide distribution of submarine mud volcanoes and associated gas hydrates. *Marine Geology*, **167**, 29–42, https://doi.org/10.1016/S0025-3227(00)00022-0

MOOS, D. & ZOBACK, M.D. 1990. Utilization of observations of well bore failure to constrain the orientation and magnitude of crustal stresses: application to continental, Deep Sea Drilling Project and Ocean Drilling Program boreholes. *Journal of Geophysical Research*, **95**, 9305–9325, https://doi.org/10.1029/JB095iB06p09305

MORTON, A., ALLEN, M. *ET AL*. 2003. Provenance patterns in a neotectonic basin: Pliocene and Quaternary sediment supply to the South Caspian. *Basin Research*, **15**, 321–337, https://doi.org/10.1046/j.1365-2117.2003.00208.x

OPPO, D., CAPOZZI, R., NIGAROV, A. & ESENOV, P. 2014. Mud volcanism and fluid geochemistry in the Cheleken peninsula, western Turkmenistan. *Marine and Petroleum Geology*, **57**, 122–134, https://doi.org/10.1016/j.marpetgeo.2014.05.009

PLANKE, S., SVENSEN, H., HOVLAND, M., BANKS, D.A. & JAMTVEIT, B. 2003. Mud and fluid migration in active mud volcanoes in Azerbaijan. *Geo-Marine Letters*, **23**, 258–267, https://doi.org/10.1007/s00367-003-0152-z

PRASAD, M. 2002. Measurement of Young's modulus of clay minerals using atomic force acoustic microscopy. *Geophysical Research Letters*, **29**, 2–5, https://doi.org/10.1029/2001GL014054

QUIJADA, M.F. & STEWART, R.R. 2007. Density estimation using density-velocity relations and seismic inversion. *CREWES Research Report*, **19**, 1–20.

RITZ, J.F., NAZARI, H., GHASSEMI, A., SALAMATI, R., SHAFEI, A., SOLAYMANI, S. & VERNANT, P. 2006. Active transtension inside central Alborz: a new insight into northern Iran–southern Caspian geodynamics. *Geology*, **34**, 477–480, https://doi.org/10.1130/G22319.1

ROBERTS, K.S., DAVIES, R.J., STEWART, S.A. & TINGAY, M. 2011. Structural controls on mud volcano vent distributions: examples from Azerbaijan and Lusi, east Java. *Journal of the Geological Society, London*, **168**, 1013–1030, https://doi.org/10.1144/0016-76492010-158

RUBEY, W.W. & HUBBERT, K.M. 1959. Role of fluid pressure in mechanics of overthrust faulting. Part II. Overthrust belt in geosynclinal area of western Wyoming in light of fluid-pressure hypothesis. *Geological Society of America Bulletin*, **70**, 167–206, https://doi.org/10.1130/0016-7606(1959)70[167:ROFPIM]2.0.CO;2

SANTOS BETANCOR, I. & SOTO, J.I. 2015. 3D geometry of a shale-cored anticline in the western South Caspian Basin (offshore Azerbaijan). *Marine and Petroleum Geology*, **67**, 829–851, https://doi.org/10.1016/j.marpetgeo.2015.06.012

SCHÖN, J. 2011. *Physical Properties of Rocks: A Workbook*. Elsevier, Amsterdam.

SELWOOD, C.S., SHAH, H.M. & BAPTISTE, D. 2013. The evolution of imaging over Azeri, from TTI tomography to anisotropic FWI. Extended Abstract, presented at the 75th EAGE Conference & Exhibition Incorporating SPE EUROPEC 2013, https://doi.org/10.3997/2214-4609.20130831

SENGUPTA, M., DAI, J., VOLTERRANI, S., DUTTA, N., RAO, N.S., AL-QADEERI, B. & KIDAMBI, V.K. 2011. Building a seismic-driven 3D geomechanical model in a deep carbonate reservoir. SEG Technical Program Expanded Abstracts 2011, 2069–2073, https://doi.org/10.1190/1.3627616

SMITH-ROUCH, L.S. 2006. Oligocene–Miocene Maykop/Diatom total petroleum system of the South Caspian Basin province, Azerbaijan, Iran and Turkmenistan. *US Geological Survey Bulletin*, **2201-I**, 1–27.

SOTO, J.I., SANTOS-BETANCOR, I., SANCHEZ BORREGO, I. & MACELLARI, C.E. 2011. Shale diapirism and associated folding history in the South Caspian Basin (Offshore Azerbaijan). AAPG Annual Convention and Exhibition, 10–13 April 2011, Houston, TX, USA, 30162, http://www.searchanddiscovery.com/abstracts/html/2011/annual/abstracts/Soto.html

STEWART, S.A. & DAVIES, R.J. 2006. Structure and emplacement of mud volcano systems in the South Caspian Basin. *American Association of Petroleum Geologists Bulletin*, **90**, 771–786, https://doi.org/10.1306/11220505045

SWARBRICK, R.E. & OSBORNE, M.J. 1996. The nature and diversity of pressure transition zones. *Petroleum Geoscience*, **2**, 111–116, https://doi.org/10.1144/petgeo.2.2.111

TOZER, R.S.J. & BORTHWICK, A.M. 2010. Variation in fluid contacts in the Azeri field, Azerbaijan: sealing faults or hydrodynamic aquifer? *In*: JOLLEY, S.J., FISHER, Q.J., AINSWORTH, R.B., VROLIJK, P.J. & DELISLE, S. (eds) *Reservoir Compartmentalization*. Geological Society, London, Special Publications, **347**, 103–112, https://doi.org/10.1144/SP347.8

UNTERSEH, S. & CONTET, J. 2015. Integrated geohazards assessments offshore Azerbaijan, Caspian Sea. Paper OTC-25911-MS, presented at the Offshore Technology Conference, Offshore Technology Conference, 4–7 May, Houston, TX, USA https://doi.org/10.4043/25911-MS

VANORIO, T., PRASAD, M. & NUR, A. 2003. Elastic properties of dry clay mineral aggregates, suspensions and sandstones. *Geophysical Journal International*, **155**, 319–326, https://doi.org/10.1046/j.1365-246X.2003.02046.x

WHITE, A., WARD, C., CASTILLO, D., MAGEE, M., TROTTA, J., MCINTYRE, B. & O'SHEA, P. 2007. Updating the geomechanical model and calibrating pore pressure from 3D seismic using data from the Gnu-1 Well, Dampier sub-Basin, Australia. Paper SPE-110926-PA, presented at the Asia Pacific Oil and Gas Conference and Exhibition, **12**, 408–418, https://doi.org/10.2118/110926-MS

YUSIFOV, M. & RABINOWITZ, P.D. 2004. Classification of mud volcanoes in the South Caspian Basin, offshore Azerbaijan. *Marine and Petroleum Geology*, **21**, 965–975, https://doi.org/10.1016/j.marpetgeo.2004.06.002

ZHANG, J. 2011. Pore pressure prediction from well logs: methods, modifications and new approaches. *Earth-Science Reviews*, **108**, 50–63, https://doi.org/10.1016/j.earscirev.2011.06.001

ZOBACK, M.D. 2007. *Reservoir Geomechanics*. 1st edn. Cambridge University Press, Cambridge.

ZOBACK, M.D., MASTIN, L. & BARTON, C. 1986. In-situ stress measurements in deep boreholes using hydraulic fracturing, wellbore breakouts and Stonely wave polarization. In: *Proceedings of the International Symposium on Rock Stress & Rock Stress Measurements*. Centek, Lulea, Stockholm, 289–299.

Index

Page numbers in *italics* refer to Figures. Page numbers in **bold** refer to Tables.